TREATISE ON
SOLID STATE CHEMISTRY

Volume 5
Changes of State

TREATISE ON SOLID STATE CHEMISTRY

Volume 1 • The Chemical Structure of Solids
Volume 2 • Defects in Solids
Volume 3 • Crystalline and Noncrystalline Solids
Volume 4 • Reactivity of Solids
Volume 5 • Changes of State
Volume 6 • Surfaces

TREATISE ON SOLID STATE CHEMISTRY

Volume 5
Changes of State

Edited by
N. B. Hannay
*Vice President
Research and Patents
Bell Laboratories
Murray Hill, New Jersey*

SPRINGER SCIENCE+BUSINESS MEDIA, LLC

Library of Congress Cataloging in Publication Data

Hannay, Norman Bruce, 1921-
  Treatise on solid state chemistry.

  Includes bibliographical references.
  CONTENTS: v. 1. The chemical structure of solids. — v. 2. Defects in solids. — v. 5. Changes of State.
  1. Solid state chemistry. I. Title.
QD478.H35   541'.042'1   73-13799
ISBN 978-1-4757-1122-6   ISBN 978-1-4757-1120-2 (eBook)
DOI 10.1007/978-1-4757-1120-2

Six-volume set: ISBN 978-1-4757-1122-6

© 1975 Springer Science+Business Media New York
Originally published by Plenum Press, New York in 1975
Softcover reprint of the hardcover 1st edition 1975

All rights reserved

No part of this book may be reproduced, stored in a retrieval system, or transmitted,
in any form or by any means, electronic, mechanical, photocopying, microfilming,
recording, or otherwise, without written permission from the Publisher

Foreword

The last quarter-century has been marked by the extremely rapid growth of the solid-state sciences. They include what is now the largest subfield of physics, and the materials engineering sciences have likewise flourished. And, playing an active role throughout this vast area of science and engineering have been very large numbers of chemists. Yet, even though the role of chemistry in the solid-state sciences has been a vital one and the solid-state sciences have, in turn, made enormous contributions to chemical thought, solid-state chemistry has not been recognized by the general body of chemists as a major subfield of chemistry. Solid-state chemistry is not even well defined as to content. Some, for example, would have it include only the quantum chemistry of solids and would reject thermodynamics and phase equilibria; this is nonsense. Solid-state chemistry has many facets, and one of the purposes of this *Treatise* is to help define the field.

Perhaps the most general characteristic of solid-state chemistry, and one which helps differentiate it from solid-state physics, is its focus on the chemical composition and atomic configuration of real solids and on the relationship of composition and structure to the chemical and physical properties of the solid. Real solids are usually extremely complex and exhibit almost infinite variety in their compositional and structural features.

Chemistry has never hesitated about the role of applied science, and solid-state chemistry is no exception. Hence, we have chosen to include in the field not only basic science but also the more fundamental aspects of the materials engineering sciences.

The central theme of the *Treatise* is the exposition of unifying principles in the chemistry, physical chemistry, and chemical physics of solids. Examples are provided only to illustrate these principles. It has, throughout, a chemical viewpoint; there is, perforce, substantial overlap with some areas of solid-

Foreword

state physics and metallurgy but a uniquely chemical perspective underlies the whole. Each chapter seeks to be as definitive as possible in its particular segment of the field.

The *Treatise* is intended for advanced workers in the field. The scope of the work is such that all solid-state chemists, as well as solid-state scientists and engineers in allied disciplines, should find in it much that is new to them in areas outside their own specializations; they should also find that the treatment of their own particular areas of interest offers enlightening perspectives.

Certain standard subjects, such as crystal structures, have been omitted because they are so well covered in many readily available standard references and are a part of the background of all solid-state scientists. Certain limited redundancies are intended, partly because they occur in different volumes of the series, but mainly because some subjects need to be examined from different viewpoints and in different contexts. The first three volumes deal with the structure of solids and its relation to properties. Volumes 4 and 5 cover broad areas of chemical dynamics in bulk solids. Volume 6 treats both structure and chemical dynamics of surfaces.

N.B.H.

Preface to Volume 5

A major area of solid-state chemistry deals with changes of state. This includes phase transformations, which are at the heart of much of the chemistry and metallurgy of complex inorganic solids. It also includes the growth of crystals, a subject of enormous importance to all of the solid-state sciences and their application. Finally, it includes many phenomena which have very different external manifestations but which have a common theoretical foundation, and the theory and experimental study of transitions in solids thus provides a linkage between apparently disparate areas of solid-state chemistry. This volume, together with Volume 4, which deals with the chemical reactivity of solids, thus covers the chemical dynamics of (bulk) solids.

Contents of Volume 5

Chapter 1
Critical Phenomena and Phase Transitions   1
    J. F. Nagle

1. Introduction 1
2. Phenomenology 3
   2.1. Order–Disorder Phenomena: Binary Alloys 3
   2.2. Ferromagnetism 5
   2.3. Fluids 9
   2.4. Antiferromagnetism 10
   2.5. Ferroelectrics and Antiferroelectrics 13
   2.6. Other Systems 15
3. Critical Exponents 17
   3.1. Definitions of Critical Exponents 17
   3.2. Difficulties in Obtaining Critical Exponents from Experiments 20
   3.3. Experimental Values of Critical Exponents 23
   3.4. Thermodynamic Inequalities for Critical Exponents .. 25
4. Theoretical Model Calculations 26
   4.1. Classical Model Calculations 27
   4.2. Two-Dimensional Ising Model Calculations 33
   4.3. Dimer Model Calculations 35
   4.4. Hydrogen-Bonded Model Calculations 37
   4.5. Spherical Model Calculations 43
   4.6. Series Expansion Calculations 45
   4.7. Other Calculations 51
5. General Theories 54
   5.1. Homogeneity-Scaling Theory 55
   5.2. Smoothness and Universality 57
   Acknowledgment 59
   References 59

Contents of Volume 5

Chapter 2
Solid-State Phase Transformations   67
   V. Raghavan and Morris Cohen

1. Introduction.. 67
2. Thermodynamic Considerations.................................... 68
   2.1. The Driving Force for a Phase Change........................ 68
   2.2. Thermodynamic Order of Phase Transformations............... 69
   2.3. First-Order Transformations without Compositional Change... 71
   2.4. First-Order Transformations with Compositional Change...... 72
   2.5. Second-Order Transformations............................... 74
3. Crystallography of Transformations.............................. 75
   3.1. Orientation Relationships.................................. 75
   3.2. The Interfacial Structure.................................. 77
   3.3. Phenomenological Description of Transformation Geometry... 78
4. Nucleation Kinetics... 81
   4.1. Random Nucleation.. 81
   4.2. Random Nucleation with Compositional Change............... 84
   4.3. Strain-Energy Effects...................................... 85
   4.4. Nonrandom Nucleation....................................... 89
5. Growth Kinetics... 92
   5.1. Interface-Controlled Growth................................ 92
   5.2. Diffusion-Controlled Growth................................ 94
6. Overall Transformation Kinetics................................. 96
   6.1. Empirical Relations.. 96
   6.2. The Johnson–Mehl and Avrami Equations...................... 97
   6.3. Transformation Kinetics for Diffusion-Controlled Processes. 98
   6.4. Transformation Kinetics for Special Nucleation Sites...... 100
   6.5. Variations of the Time Exponent for Different Situations.. 101
7. Particle Coarsening.. 101
   7.1. Driving Force for Coarsening.............................. 102
   7.2. Simple Kinetic Analysis................................... 102
8. Discontinuous or Cellular Precipitation........................ 105
   8.1. Mechanisms of Nucleation and Growth....................... 106
   8.2. Interlamellar Spacing..................................... 107
9. Martensitic Transformations.................................... 108
   9.1. Characteristics of Martensitic Transformations........... 108
   9.2. Kinetic Aspects... 109
   9.3. The Nucleation Problem.................................... 111
   9.4. Growth Path and Growth Kinetics........................... 113

10.	Massive Transformations	114
	10.1. Experimental Characteristics	114
	10.2. Nucleation and Growth Kinetics	115
11.	Polymorphic Transformations	116
	11.1. Introduction	116
	11.2. Buerger's Classification	116
12.	Recrystallization and Grain Growth	117
	12.1. Recrystallization	118
	12.2. Grain Growth	119
13.	Spinodal Decomposition	121
	13.1. Thermodynamics of Compositional Fluctuations	121
	13.2. Kinetics of Spinodal Decomposition	123
	13.3. Experimental Results	124
	Acknowledgments	125
	References	126

Chapter 3
Clustering Effects in Solid Solutions 129
D. de Fontaine

1.	Introduction	129
2.	Pairwise Internal Energy	131
	2.1. Average Concentrations	131
	2.2. Pair Interactions	132
3.	Configurational Free Energy	133
	3.1. Bragg–Williams Model	133
	3.2. Continuum Gradient Expansion	134
4.	Elastic Energy	135
5.	Stationary States	139
	5.1. Coherent Phase Equilibrium	139
	5.2. The Anharmonic Oscillator Analogy	143
	5.3. The Diffuse Interface	146
	5.4. The Critical Nucleus	147
	5.5. Higher Cluster Expansions	150
6.	Compositional Fluctuations	152
7.	Kinetics of Decomposition	153
	7.1. The Diffusion Equation	154
	7.2. Numerical Solutions	155
	7.3. The Linearized Equation	161
	7.4. The Intensity Equation	164
	7.5. Kinetics of Nucleation and Growth	165
8.	Morphology	168
9.	Conclusions	173
	References	174

Contents of Volume 5

Chapter 4
Relation of the Thermochemistry and Phase Diagrams of Condensed Systems   179
    Larry Kaufman and Harvey Nesor

1. Introduction 179
2. Current Efforts at Coupling Thermochemical and Phase Diagram Data 182
3. Calculation of the Iron–Chromium–Nickel System 190
4. Prediction of Partial Titanium–Aluminum–Gallium and Titanium–Aluminum–Tin Isothermal Sections 202
5. Calculation of the Chromium–Columbium–Nickel System 205
6. Prediction of Isothermal Sections in the Cb–Zr–Cr, Cb–Hf–Cr, Cb–Mo–Cr, and Cb–Ti–Zr Systems 211
7. Analysis of the Quasibinary Eutectic Iron–Titanium Carbide ... 218
8. Summary 228
    Acknowledgments 229
    References 229

Chapter 5
Theory of Crystal Growth   233
    Kenneth A. Jackson

1. Introduction 233
2. Mass and Heat Transfer 234
    2.1. Diffusion during Crystal Growth 234
    2.2. Convection 236
3. Interface Morphology 237
    3.1. Constitutional Supercooling 239
    3.2. Interface Stability 244
    3.3. Anisotropic Growth Kinetics 248
    3.4. Banding 250
4. Heat Flow 251
5. Diffusion-Limited Growth 251
    5.1. Dendritic Growth 252
6. Intrinsic Growth Kinetics 255
    6.1. Surface Roughness 258
    6.2. Multilevel Interface 262
    6.3. The Growth Rate Equation 267
    6.4. Rate Equations 271
    6.5. Clustering 272
    6.6. Small Crystals Using Periodic Boundary Conditions 273
    6.7. Monte Carlo Calculations 278
    References 281

Contents of Volume 5

Chapter 6
Growth from the Vapor   283
   Morton E. Jones and Don W. Shaw

1. Introduction 283
   1.1. Principles of Vapor Deposition 284
   1.2. Chemical Vapor Deposition 285
2. Thermodynamics of Gas–Solid Reactions 286
3. Kinetics of Gas–Solid Reactions 291
   3.1. Mass Transport Limiting Steps 292
   3.2. Surface Rate-Limiting Steps 295
4. Epitaxial Growth 302
   4.1. Silicon Deposition 303
   4.2. Gallium Arsenide Deposition 310
5. Summary 321
   References 322

Chapter 7
Crystal Growth from the Melt   325
   J. R. Carruthers

1. Introduction 325
2. Basic Principles and Techniques 326
   2.1. Principles of Solidification Processes 326
   2.2. Normal Freezing Techniques 340
   2.3. Zone Melting Techniques 348
   2.4. Nonconservative Growth Techniques 351
3. Materials Applications and Requirements 354
4. Studies of Crystal Growth from the Melt 360
   4.1. Morphological Stability of Crystal–Melt Interfaces .. 360
   4.2. Melt Statics and Dynamics 361
   4.3. Transient Segregation and Interface Demarcation 372
   4.4. Phase Equilibria and Nonstoichiometry 381
   4.5. Defect Studies 391
5. Control of Crystal Growth from the Melt 392
   5.1. Defect Control 392
   5.2. Compositional Control 393
   5.3. Diameter Control 395
   5.4. Shape Control 398
6. Future Directions 399
   References 400

Contents of Volume 5

Chapter 8
Solution Growth   **407**
    R. A. Laudise

1. Introduction 407
2. Underlying Principles 409
   - 2.1. Phase Equilibria and Distribution Constants in Solution Growth 409
   - 2.2. Kinetics of Solution Growth 413
3. Aqueous Solution Growth 417
   - 3.1. Holden Crystallizer 419
   - 3.2. Mason Jar Method 421
   - 3.3. Temperature Differential Method 423
   - 3.4. Other Methods of Aqueous Growth 423
4. Hydrothermal Growth 425
   - 4.1. General 425
   - 4.2. Equipment 427
   - 4.3. Phase Equilibria and Solubility 429
   - 4.4. Kinetics 432
   - 4.5. Perfection of Hydrothermal Crystals 435
   - 4.6. Other Hydrothermal Materials 437
5. Molten Salt Crystal Growth 437
   - 5.1. General 437
   - 5.2. Equipment 439
   - 5.3. Phase Equilibria and Solubility 439
   - 5.4. Crystal Growth by Slow Cooling 440
   - 5.5. Temperature Cycling and Accelerated Crucible Rotation . 444
   - 5.6. Flux Evaporation 447
   - 5.7. Seeding in Flux Growth 447
   - 5.8. Flux Growth by Electrolysis 449
6. Other Solvents 450
7. Liquid-Phase Epitaxy 452
8. Temperature-Gradient Zone Melting 454
9. Vapor–Liquid–Solid Growth 456
   - References 458

Chapter 9
New Phases at High Pressure   **463**
    J. C. Joubert and J. Chenavas

1. Introduction 463
2. High-Pressure Facilities Suitable for the Synthesis of New Materials 464

	2.1. The 1–10 kbar Pressure Range	464
	2.2. The High-Pressure Range	468
3.	Empirical Rules Which Determine the Crystal Structure of Solids under Pressure	472
4.	New Phases at High Pressure: Quenchable and Unquenchable Phases: Phase Diagrams under High Pressure	474
	4.1. Quenchable and Unquenchable Phases	474
	4.2. Phase Diagrams under High Pressure	476
	4.3. Examples of High-Pressure Synthesis	477
5.	Crystal Growth under High Pressure	497
	5.1. Crystal Growth in a Flux	497
	5.2. Crystal Growth by Hydrothermal Synthesis under Very High Pressure	501
6.	Conclusion	503
	References	503

Chapter 10
Transitions in Viscous Liquids and Glasses   513
David Turnbull and Brian G. Bagley

1.	Liquid ↔ Glass Transition	513
2.	Phase Separation of Viscous Melts	521
3.	Crystallization of Viscous Systems	523
4.	Electrical Threshold Switch	529
5.	Memories	533
	5.1. Electrical Memories	534
	5.2. Optical Memories	538
	References	546

Chapter 11
Transitions in Solid Polymers   555
Shogo Saito

1.	Introduction	555
2.	Multiple Transition Behavior in Solid Polymers	556
	2.1. Multiple Transitions	556
	2.2. Relaxation Phenomena	558
3.	Melting of Crystalline Polymers	561
	3.1. Melting Temperature for Homopolymers	561
	3.2. Effect of Molecular Weight on Melting Temperature	564
	3.3. Dependence of Melting Temperature on Chemical Structure	565
	3.4. Melting Temperature for Copolymers	568

Contents of Volume 5

4.	Glass Transition in Solid Polymers.	572
	4.1. Thermodynamic Phase Transition or Relaxation Phenomena.	572
	4.2. Glass Transition and Molecular Relaxation Time	574
	4.3. Glass Transition from the Point of View of Free Volume in Amorphous Phase	575
	4.4. Dependence of the Glass Transition Temperature on Cooling Rate.	577
	4.5. Dependence of the Glass Transition Temperature on Chemical Structure	578
	4.6. Glass Transition in Copolymers	582
5.	Subsidiary Transitions in Solid Polymers.	584
	5.1. Side-Chain Motion	584
	5.2. Torsional Oscillation of Frozen Backbone Chains	585
	5.3. Molecular Motion in the Crystalline Phase.	587
	5.4. Molecular Motion in the Intermediate Phase	589
	References.	591
Index.		**593**

Critical Phenomena and Phase Transitions

J. F. Nagle
Department of Physics
Carnegie-Mellon University
Pittsburgh, Pennsylvania

1. Introduction

The area of critical phenomena is a rather large one, both quantitatively, as measured by the number of researchers and the number of publications, and qualitatively, as judged by the percentage of "hard" results such as exact calculations and highly precise experiments. This makes it impossible to write an article or even a book which is completely self-contained and which includes all that has been accomplished.

The purpose of this chapter is to introduce the reader to the important results and ideas in the field, to help him to understand the significance of these results, to correlate these results with each other, and to point out the limitations so that he can appreciate what current researchers are trying to do. Stated somewhat differently, this chapter is meant to help the student find his way through the research literature rather than to replace or duplicate the literature. Accordingly, for details of experimental techniques and mathematical formalism the reader is referred to more specialized review articles and original papers which he must consult for an in-depth understanding of those topics which particularly interest him.

A description of the contents should be helpful in guiding the reader. Section 2 describes a number of systems with critical points and shows the similarity of these systems to each other. Also, theoretical

Chapter 1

models are introduced to describe these systems. Many readers with some knowledge of critical phenomena will find much of this section well known and will wish to skip it. Nevertheless, it is recommended that such initiated readers look at Sections 2.4 and 2.5, which contain ideas which may not be familiar.

Section 3 concentrates on the critical exponent description of critical points. The reader familiar with critical exponents may skip Section 3.1, but should find the example in Section 3.2 of some interest. This example involves the currently popular tricritical point and shows a complication which may occur in the measurement of critical exponents. Section 3.3 is essentially a review of experimentally measured values of the critical exponents for a number of the best-studied systems. This subsection should be useful to the reader as a source of experimental references, with an emphasis on the period 1968–1972. Finally, Section 3.4 is an elementary discussion of thermodynamic inequalities for the critical exponents.

In each of the subsections of Section 4 the reader is introduced to one of the most important theoretical calculations. Many of the subsections are further divided. The first division, such as 4.1.1, 4.3.1, 4.5.1, or 4.6.1, reviews the basic results on which the uninitiated reader should concentrate his attention. The already initiated reader may be interested in the further divisions of these sections, such as 4.1.2 on tricritical points, 4.2.2 on decorated Ising models, 4.3 on dimer models, 4.4 on hydrogen-bonded models, and especially 4.5.2 on the spherical model with long-range interactions. Section 4.6.2 gives a brief review of series expansion results for some of the best-studied models, which is analogous to the review given of the experimental values of the exponents in Section 3.3. Finally, in 4.7.2 and 4.7.3 the recent work using Wilson's methods is mentioned, although the reader is warned that the discussion given here may already be out of date because this work is so current.

In Section 5 two unifying theories, homogeneity–scaling and smoothness–universality, are discussed briefly. An interesting item for the initiated reader is the example discussed in Section 5.2, which at first sight appears to violate the smoothness conjecture.

It is my experience that the field of critical phenomena has been a very exciting and fruitful area of basic research. It has also been a truly interdisciplinary meeting ground for chemists, physicists, and applied mathematicians. I hope that this chapter will help the reader to enjoy some of this excitement and to appreciate the many and varied contributions to critical phenomena.

Critical Phenomena and Phase Transitions

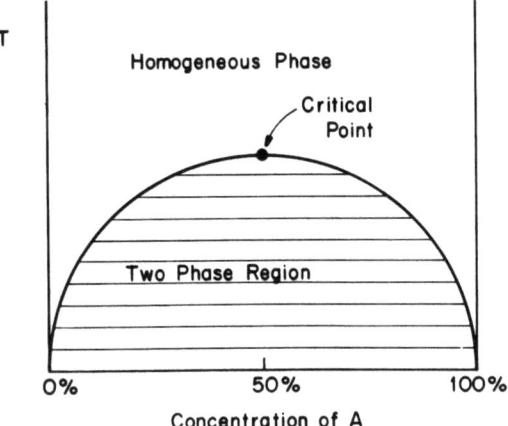

Fig. 1. The phase separation of an ideal AB alloy. In the two-phase region under the curve the alloy decomposes into the two phases whose concentrations are given at the ends of the horizontal tie lines.

2. Phenomenology

One of the most pleasing features of critical phenomena is the great similarity of critical point behavior in widely disparate systems. The first purpose of this section is to show this similarity for a variety of systems. In the course of doing this, some of the simple theoretical models and the rudiments of the critical exponent description will be introduced.

2.1. Order–Disorder Phenomena: Binary Alloys

It is conceptually simplest to start with the prototypical order–disorder phenomenon of substitutional binary alloys of metals A and B. Let us consider that class of binary alloys which, upon lowering the temperature T, separate into an A-rich phase and a B-rich phase, given time to come to thermodynamic equilibrium. Figure 1 shows the two-phase region of such an ideal binary alloy with a critical point at the critical temperature T_c and critical concentration of 50% A and 50% B.* This behavior is easy to understand qualitatively. To

---

* Examples of systems in this class are Al–Zn and Au–Pt.[1,2] Unfortunately, for such a solid alloy the time to come to equilibrium is very long, so Figure 1 must be considered as idealized. The experimental situation for binary fluid mixtures with a miscibility gap (two-phase region) is much cleaner experimentally but the idealized binary alloy case is theoretically simpler because the atoms are restricted to lattice sites.

Chapter 1

minimize the free energy $F = U - TS$ at high temperatures requires a thorough mixing of A and B to maximize the entropy S. However, as T is lowered, energy considerations become relatively more important and there will be a tendency to separate if atoms energetically prefer their own kind, in a sense which we now quantify.

Suppose for simplicity that the interactions extend only between nearest neighbors on a lattice and let the number of nearest neighbors of any atom be q, which is called the *coordination number* of the lattice. Let J_{AA}, J_{AB}, and J_{BB} be the interaction energies between AA, AB, and BB pairs of nearest-neighbor atoms, respectively, and let N_{AA}, N_{AB}, and N_{BB} be the number of such pairs. The energy will be

$$E = N_{AA}J_{AA} + N_{AB}J_{AB} + H_{BB}J_{BB} \quad (1)$$

Let us rewrite this energy in the form of an Ising Hamiltonian as follows: For lattice site i, $i = 1, \ldots, N$, where N is the total number of atoms in the crystal, let $\mu_i = +1$ if site i is occupied by an A atom and let $\mu_i = -1$ if site i is occupied by a B atom. Then the Ising Hamiltonian is

$$E = -J \sum \mu_i \mu_j - H \sum \mu_i + LN \quad (2)$$

where, most commonly, the first sum is restricted to nearest-neighbor pairs of sites. To make Eq. (2) equivalent to Eq. (1) requires

$$J = -\tfrac{1}{2}[\tfrac{1}{2}(J_{AA} + J_{BB}) - J_{AB}], \quad L = (q/4)[\tfrac{1}{2}(J_{AA} + J_{BB}) + J_{AB}] \quad (3)$$

and

$$H = (q/4)[J_{BB} - J_{AA}]$$

is the difference in chemical potentials of A and B. [This is easily shown using $N_{AA} + N_{AB} + N_{BB} = qN/2$ and $2N_{AA} + N_{AB} = qN_A$, $2N_{BB} + N_{AB} = qN_B$, giving $N_A - N_B = (2/q)(N_{AA} - N_{BB})$.]

If J in Eq. (3) is positive, the lowest energy state will exhibit phase separation. This follows because the total numbers N_A and N_B of A and B atoms are fixed, so only the first term on the right-hand side of Eq. (2) can be varied and it achieves a minimum when the number of AB pairs is minimized.

At finite temperatures below T_c the two phases as shown in Figure 1 are not completely ordered in the sense of being all A or all B. The degree of order is described by the order parameter Ψ. In this case of an AB alloy Ψ is proportional to the concentration $\rho_A^{(1)}$ of A atoms in the A-rich phase minus the concentration $\rho_A^{(2)}$ of A atoms in the B-rich phase, both measured at the same temperature and at opposite edges of the two-phase region,

$$\Psi(T) = \rho_A^{(1)}(T) - \rho_A^{(2)}(T) \tag{4}$$

As $T \to T_{c-}$ (from below T_c), $\Psi(T)$ vanishes and the way in which $\Psi(T)$ vanishes is characterized by a critical exponent β where $\Psi(T) \sim (T_c - T)^\beta$. Precisely, this means that

$$\beta = \lim_{T \to T_{c-}} \left\{ \frac{\ln|\Psi(T)|}{\ln|T_c - T|} \right\} \tag{5}$$

The existence of an order parameter is an almost universal feature of systems with critical points.*

2.2. Ferromagnetism

The basic microscopic picture of an insulating ferromagnet is that there are localized magnetic moments or spins arising from unpaired d or f electrons which interact via exchange interactions in such a way as to lower the energy when neighboring moments are parallel. (A good review series for magnetism is provided by Rado and Suhl.[7]) An oversimplified, non-quantum mechanical but still useful Hamiltonian is provided by the Ising model described in Eq. (2) with $J > 0$. Strictly speaking, the model in Eq. (2) is the spin-1/2 nearest-neighbor Ising model because the spin $S_i = \mu_i/2$ takes on only the two values $\pm 1/2$ and the interactions are restricted to nearest neighbors; both restrictions are easily generalized.

In any case a basic similarity with order–disorder binary alloys is apparent, where now the order parameter Ψ is the spontaneous

---

* A reasonable definition of a critical point is that some thermodynamic quantity shows nonanalyticity at the critical point. Experimentally, it seems that there is always an order parameter which disappears in a nonanalytic fashion at the critical point. However, it seems theoretically possible that there may be critical points for which there are no order parameters. One such case involving a fairly complicated Ising model was discussed by Wegner.[3] The dimer models[4,5] also seem to fall into this class. However, it is generally difficult to be certain that a given model does not have an order parameter. For example, the Ising model on the unionjack lattice[6] has, for one particular ratio of interaction strengths, a nonanalytic T_c, but for $T > T_c$ and also $T < T_c$ there is no order parameter, i.e., the system is paramagnetic. However, when one considers an extended phase diagram obtained by allowing the ratio of interaction strengths to vary, one finds that this peculiar critical point is, in fact, a boundary point of a region with antiferromagnetic order. The difficulty in the general case follows because it is not easy to catalogue systematically all possible extensions of the phase diagram.

Chapter 1

magnetization

$$M_s = \lim_{H \to 0} \sum_{i=1}^{N} \langle \mu_i \rangle \qquad (6)$$

Figure 1 is therefore also relevant to ferromagnets when the concentration axis is relabeled as magnetization M, and the boundary of the two-phase region is the value of the spontaneous magnetization M_s at temperature T. Unlike binary alloys, where the relative concentration is set by the experimenter, in ferromagnets it is the external magnetic field H which is set.* Thus one often describes the critical point and the first-order transition ending in the critical point by the H–T phase diagram shown in Figure 2, where the thick solid line indicates a first-order transition with discontinuous magnetization. (The entropy is continuous by symmetry or by the magnetic Clausius–Clapeyron equation $\Delta S/\Delta M = dH/dT$ along a phase line.)

The magnetic field H is said to couple directly to the order parameter Ψ because it is the product $H \sum \mu_i = HM$ which appears in the Ising Hamiltonian (2) or, more generally, HM appears in the magnetic thermodynamic free energies, e.g., $G = U - TS - HM$. The magnetic field will be called the ordering field for ferromagnets. In general the ordering field will be denoted by ζ. In the binary alloy system it is the difference in chemical potentials which couples to the order parameter and which will therefore, by analogy, be called the ordering field ζ.

One of the more striking characteristics of a critical point is the strong divergence to infinity as $T \to T_c$ of $(\partial \Psi/\partial \zeta)_T$, which for ferromagnets is the magnetic susceptibility in zero field $\chi_T = (\partial M/\partial H)_T$. To describe this divergence quantitatively the critical exponent γ is used, where

$$\chi_T \sim (T - T_c)^{-\gamma} \qquad (7)$$

---

*The experimental situation for magnetism is complicated, but not fundamentally altered, by demagnetization effects caused by the dipolar character of magnetism. For example, in zero applied field H below T_c a finite sample will form domains to reduce the energy stored in the induced external field set up by the spontaneous magnetization. Also, from basic magnetostatics, the internal field H is not the same as the external applied field H_0. Theoretically, we can ignore these complications by assuming an infinitely long, thin sample. Experimentally, one adjusts to this case by taking into account a demagnetization correction D to compute $H = H_0 - DM$, where D is shape dependent and equals $4\pi/3$ for a sphere. Explicit discussions of this phenomenon may be found in Refs. 8 and 9.

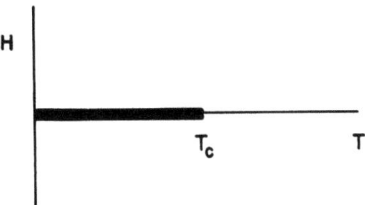

Fig. 2. The first-order phase line for a ferromagnet, ending in the critical point at T_c in zero field. The magnetization, but not the entropy, is discontinuous across the phase line.

as $T \to T_c$ from above T_c and

$$\chi_T \sim (T_c - T)^{-\gamma'} \qquad (8)$$

as $T \to T_c$ from below T_c.

The exponents γ and γ' are characteristically greater than or equal to one. It is easily shown for the Ising model by taking two derivatives of the free energy $= -kT \ln(\text{partition function})$ that in zero field

$$\chi_T = \left[\mu^2 + \sum_{i=2}^{\infty} \langle \mu_1 \mu_i \rangle \right] \Big/ kT \qquad (9)$$

where the correlation function is given by the thermal average

$$\langle \mu_1 \mu_n \rangle = Z^{-1} \sum_{\text{states}} \mu_1 \mu_n e^{-E/kT} \qquad (10)$$

where the partition function $Z = \sum_{\text{states}} e^{-E/kT}$. Since $|\langle \mu_1 \mu_n \rangle|$ is bounded by $+1$, in order for χ_T to become infinite either $\lim_{n \to \infty} \langle \mu_1 \mu_n \rangle \neq 0$ or right at the critical point the limit approaches zero slowly with n [see Eq. (26)]. This means that even those spins or magnetic moments that are separated by long distances become correlated at the critical point. This is therefore the onset of long-range magnetic order. However, in zero field, it is equally probable that the spins are pointed up as down, so increasingly larger regions of up spins as well as of down spins develop as $T \to T_c$ from above T_c. This gives rise to very large fluctuations in the average spin moment detectable by neutron scattering, which becomes anomalously large at T_c. Neutron scattering is a very effective tool for measuring critical properties of magnets. The reader is referred to the review by Marshall[10] for details of this technique.

Chapter 1

Although the Ising model provides a good basic description for ferromagnetism and its relation to other systems, it is necessary to consider other models as well. The most basic is the Heisenberg model, which is easily derived for a pair of overlapping spin-1/2 atoms (such as the hydrogen molecule) by noting that the "effective spin Hamiltonian"

$$\mathcal{H} = -J\mathbf{S}_1 \cdot \mathbf{S}_2 - H_z(S_1^{(z)} + S_2^{(z)}) \qquad (11)$$

simulates the singlet–triplet splitting of the interacting system.[11] The components of **S** are the Pauli spin matrices. The parameter J is related to the exchange integrals and is usually negative, so most systems are not ferromagnetic. However, even the sign of J is not easy to compute for real magnetic systems and superexchange via an intervening nonmetallic ion only makes it more difficult to compute.[12] The practical procedure is to evaluate J from experiments.[13] In many cases involving f electrons even the Heisenberg Hamiltonian is too simple and one must resort to models with many more parameters.[14] Fortunately, in some of these cases, such as dysprosium aluminum garnet[8] (often abbreviated DAG), the models actually reduce back to the Ising model!

A model which interpolates between the Ising and Heisenberg models is the anisotropic Ising–Heisenberg model

$$\mathcal{H} = -\sum_{i,j}[J_{xy}(S_i^{(x)}S_j^{(x)} + S_i^{(y)}S_j^{(y)}) + J_z S_i^{(z)}S_j^{(z)}] \qquad (12)$$

where $S_i^{(x)}$, $S_i^{(y)}$, and $S_i^{(z)}$ are the Pauli spin matrices. If $J_{xy} = J_z$, then Eq. (12) reduces to the Heisenberg model. If $J_{xy} = 0$, then Eq. (12) reduces to the Ising model. If $J_z = 0$, then Eq. (12) is called the *XY* model.

In the limit of infinite spin the quantum mechanical complications drop out of the Heisenberg model[15] and the spins become classical unit vectors with x, y, and z components. This model is known as the classical Heisenberg model. In the same limit the *XY* model becomes the classical planar model having classical unit vectors with only two components or spin dimensions. These are two of a sequence of "classical *n*-vector models" where the *n* refers to the spin dimensionality which is *independent* of the lattice dimensionality. The interaction remains the pair interaction

$$\mathcal{H} = -\sum_{i,j} J_{ij} \sum_{k=1}^{n} \mu_i^{(k)}\mu_j^{(k)} \qquad (13)$$

and the μ_i are now unit vectors in n dimensions. The case $n = 1$ is again the Ising model.

2.3. Fluids

Since the simple fluid system is the most familiar one, even though it is not a solid-state system, it is worth recalling this case and its similarity to the order–disorder and ferromagnetic cases. The ordinary gas–liquid first-order phase line in the pressure–temperature (P–T) phase diagram (Figure 3) is terminated by the triple point on one end and the critical point on the other end.

The Ising model can also be used to describe the fluid system, with a bit of conceptual stretching.[16] First, one divides the continuum into cells (or lattice points) and defines $\mu_i = -1$ if the cell has an atom and $\mu_i = +1$ if the cell has no atom. Multiple occupancy of a cell is not allowed. Interactions between atoms are reduced to interactions between occupied and unoccupied cells. The density of atoms in this lattice gas model of a fluid system is given by

$$\rho = (2Nv_0)^{-1} \sum_{i=1}^{N} \langle (1 - \mu_i) \rangle \qquad (14)$$

where v_0 is the volume of a cell. For this particular analog of the Ising model the proper order parameter is the difference in density of the liquid and the vapor, $\Psi = \rho_L - \rho_V$.

Also, the configurational chemical potential μ_c of the lattice gas is given by

$$\mu_c = -2H - 2E_0 \qquad (15)$$

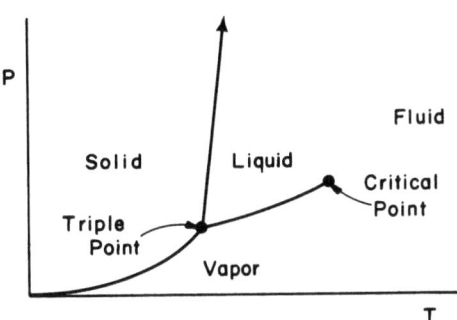

Fig. 3. The pressure–temperature phase diagram for a simple fluid system.

Chapter 1

where E_0 is the ground-state energy per spin in the Ising model, which is a constant. Thus in the lattice gas model the external ordering field ζ which couples with the order parameter is the chemical potential. However, in the lattice gas–Ising transformation one also has

$$Pv_0 = -(F/NkT) - H - (E_0/2) \qquad (16)$$

where the free energy F is not rapidly varying near the critical point, so one may also say roughly that the external ordering field is the pressure.[38]

This latter identification is the one suggested by thermodynamics, where the combination $PV = P/\rho$ always enters. Certainly for most purposes it is reasonable to say that H, M for a magnetic system are analogous to $-P, +V$ for a fluid system. For example, the isothermal compressibility $-(1/V)(\partial V/\partial P)_T$ diverges to infinity in the same characteristic way as $\chi_T = (\partial M/\partial H)_T$ and, accordingly, the same critical exponents γ and γ' are used to describe the divergence.

Thus the analogy between the phase diagrams in Figures 2 and 3 is complete for the critical point. The existence of a solid phase for systems like argon cannot be explained with a Hamiltonian involving just one interaction parameter, such as J in Eq. (2). However, phase diagrams such as the one in Figure 3 can be obtained for the case of competing interactions involving an attractive long-range part and a repulsive short-range part.[17,18]

2.4. Antiferromagnetism

Thus far we have restricted ourselves to the case when J is positive in Eq. (2). If $J < 0$, then in magnetic language the spins energetically prefer to align antiparallel and one has an antiferromagnet.* The H–T phase diagram shown in Figure 4 for an ideal antiferromagnet looks so much different from the phase diagram in Figure 2 for a ferromagnet that one is at first inclined to think that ferromagnets and antiferromagnets are conceptually not similar. But when it is noted that the field H is no longer the ordering field ζ because the order parameter is no longer the spontaneous magnetization, the dissimilarity of Figures 2 and 4 comes as a relief!

Many antiferromagnets have lattices of a "loose-packed" structure, which means simply that there are two sublattices L_A and

---

* In binary alloys or binary solutions if $J < 0$, no miscibility gap or macroscopically observable phase separation appears. Instead, long-range order of an ABAB $\cdots$ type sets in if one is at the critical concentration; the other domains of BABA $\cdots$ type are macroscopically indistinguishable from the former.

Critical Phenomena and Phase Transitions

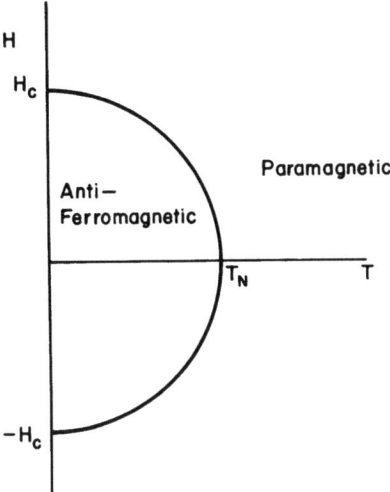

Fig. 4. The H–T (magnetic field–temperature) phase diagram for a simple antiferromagnet with a second-order phase line. The phase diagram for an actual antiferromagnet shows much more structure for low temperatures and high fields than is shown here. Typically, a "spin-flop" phase appears or the second-order phase line becomes first order at a point (which is currently called the tricritical point[19]) as shown in Figure 9. Here we are interested in the region near T_N.

L_B; any site on either sublattice, say L_A, has all its nearest neighbors on the other sublattice L_B. The order parameter for this kind of antiferromagnet is the difference in sublattice magnetizations,

$$\Psi = \sum_{i \in L_A} \langle \mu_i \rangle - \sum_{j \in L_B} \langle \mu_j \rangle \qquad (17)$$

The ordering field ζ which couples to Ψ directly is called the staggered field H' because a term $-\zeta\Psi = -H'\Psi$ added to the Hamiltonian can be thought of as a field directed along the positive z axis at the L_A sites and along the negative z axis at the L_B sites. Such an ordering field is not available experimentally, but the order parameter Ψ is detectable by virtue of extra peaks in neutron diffraction data, due to the larger unit cell of the crystal.[10]

Chapter 1

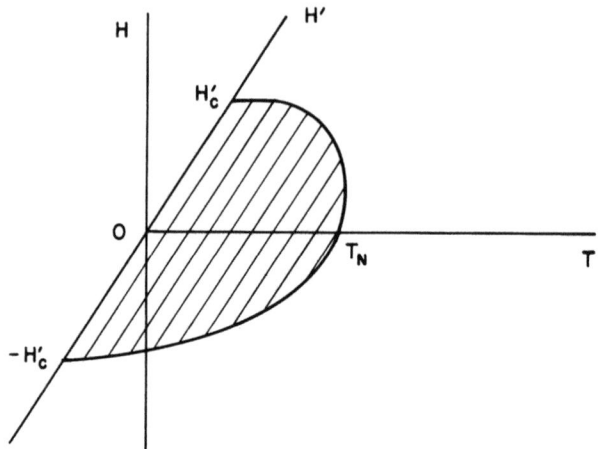

Fig. 5. The extended H–H'–T phase diagram for a ferromagnet with $J > 0$. The surface O, H_c', T_N, $-H_c'$ is a first-order surface since the spontaneous magnetization is discontinuous as H goes through zero. The line H_c', T_N, $-H_c'$ is a line of second-order critical points. This figure also applies to antiferromagnets upon transforming $H \to H'$, $H' \to H$.

In order to complete the analogy between ferromagnets and antiferromagnets, let us once again use the nearest neighbor Ising model for a two-sublattice structure. Starting with a ferromagnet with $J > 0$, simply redefine the signs of all spins on the L_A sublattice to have opposite sign and change the sign of J so that the spin–spin interaction energy in Eq. (2) remains the same for each spin state even though the model has been changed from ferromagnetic to antiferromagnetic. Further, the spin state energies, and therefore the partition function and free energy, remain exactly the same if the direct field H and the staggered field H' for the ferromagnet are replaced by the staggered field H' and the direct field H for the antiferromagnet (i.e., $H \to H'$, $H' \to H$). Therefore the loci of phase transitions in the H–T plane for an Ising ferromagnet are the same as the loci of phase transitions in the H'–T plane for an antiferromagnet and conversely. Figures 2 and 4 can now be used to synthesize an H–H'–T phase diagram for a ferromagnet which is an *extension* of the H–T phase diagram. This is shown in Figure 5. The surface in zero H field bounded by the H' axis and the line of critical points is a first-order surface because crossing this surface (by varying H through zero) results in a discontinuous change in spontaneous magnetization.

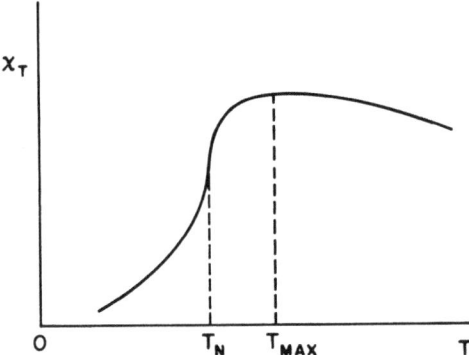

Fig. 6. χ_T versus T for a typical antiferromagnet.

In the presence of the extra staggered field there are now a number of additional susceptibilities which can be defined for an antiferromagnet. It is clear that the antiferromagnetic analog of the strongly divergent ferromagnetic χ_T is $(\partial \Psi/\partial H')_T$ and one associates the critical exponents γ and γ' with this quantity. It should be especially clear that the usual $\chi_T = (\partial M/\partial H)_T$ for an antiferromagnet is not likely to diverge at the Néel point. The behavior of χ_T for a typical antiferromagnet is shown in Figure 6. According to an exact Ising model calculation of Fisher,[20] there should be an infinite slope at the Néel temperature T_N, and T_N does not correspond to the maximum of χ_T in realistic models. Numerous experiments have confirmed this, including a relatively recent one[21] on $MnBr_2 \cdot 4H_2O$.

2.5. Ferroelectrics and Antiferroelectrics

Ferroelectric and ferromagnetic critical points are obviously conceptually similar, as revealed by replacing spontaneous magnetization M_s by spontaneous polarization P_s for the order parameter Ψ and replacing the magnetic field H by the electric field E for the ordering field ζ. An elementary microscopic picture for ferroelectrics is an assembly of orientable electric dipoles which interact. However, the interactions are certainly more complicated than simple dipole–dipole interactions because atomic rearrangements strain chemical bonds and run into steric restrictions.

Different ferroelectrics with different kinds of atomic rearrangements require different models to describe them. There are two main

Chapter 1

classes of ferroelectrics, the displacive class, including $BaTiO_3$, and the hydrogen-bonded class, including KH_2PO_4. Because they show greater similarity to other order–disorder types of critical points, this review will be restricted to the hydrogen-bonded class. A good discussion of ferroelectrics is given by Jona and Shirane.[22]

The prototypical hydrogen-bonded ferroelectric KH_2PO_4 has a three-dimensional, four-coordinated hydrogen bond network which is topologically isomorphic to the simple cubic diamond lattice although the geometric symmetry is lowered to tetragonal. Each hydrogen bond has a proton which sits in one of the two off-center positions on the bond as in ice. To first approximation the ice rule holds, i.e., each phosphate group has precisely two protons close to it on its four hydrogen bonds. Due to the tetragonal symmetry of the crystal, four of the six $H_2PO_4^-$ configurations should have equal energies, which differ from the other two configurations by ε, which is a phenomenological parameter in this model named after Slater. (See Figure 7.) At first glance this model looks equivalent to an Ising model where $\mu_i = \pm 1$ describes the position of the ith proton on its bond. However, infinitely strong spin interactions are required to maintain the ice rules. The Slater model does not allow the hydrogen-bonded network to reorient itself, on account of the rigidity imposed by the ice rules. The appropriate extension of the Slater model due to Takagi[23] allows three or one proton(s) to be close to a phosphate

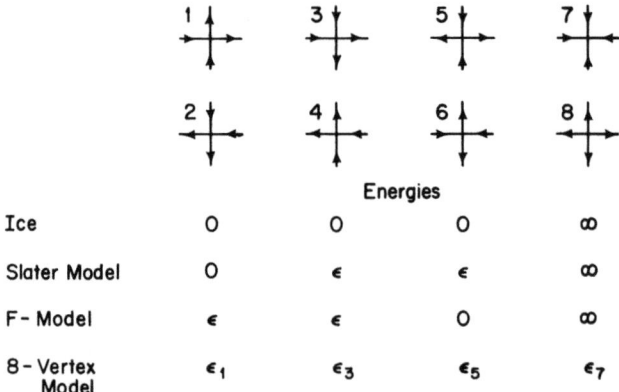

Fig. 7. The vertex configurations of ice-rule models and the eight-vertex model. If the arrow points toward (away from) the vertex, then the proton is close to that (the neighboring) vertex. The energies of the vertex configurations are given for various models. $\varepsilon > 0$.

group at the cost of an additional interaction energy $\varepsilon_1 \gg \varepsilon$. Further extensions to the basic Slater model have been proposed to allow for the dynamic tunneling of the protons along the bond.[24]

One experimental complication in KH_2PO_4 is that the transition in zero electric field is first order, although the latent heat and jump in polarization are small.[25] One of the usual theoretical assumptions in critical phenomena is that the lattice is rigid. The effect of a compressible lattice on the critical points of ferromagnets is still an open problem, but one of the favored possibilities is that a lattice instability prevents the rigid lattice critical point from being reached, because of the intervention of a first-order transition.[26,27] This effect would be even larger for ferroelectrics than for ferromagnets and may account for the small first-order transition in KDP. Even if this is so, it seems that for KH_2PO_4 the nonrigid lattice is still a perturbation to the basic rigid lattice models and not the dominant part of the theory as it may be for the displacive class of ferroelectrics.

The prototypical hydrogen-bonded antiferroelectric is $(NH_4)H_2PO_4$. The phenomenology of the critical point bears the same relation to ferroelectrics as antiferromagnets to ferromagnets. However, the experimental situation is even more complicated by the fact that the crystal shatters upon cooling below T_N![22]

2.6. Other Systems

Many systems such as superfluids, superconductors, and liquid crystals also exhibit phase transitions and critical points which show many analogies to the foregoing critical points when suitably formulated. This is not to say that all transitions are exactly alike; Section 4 emphasizes that they are not. But one should start with the basic similarities such as the order parameter and ordering fields, even if the latter cannot be realized in the laboratory.

2.6.1. Superfluid Helium

Understanding the superfluid helium transition is rather similar to understanding the antiferromagnet transition. In both cases there is a line of second-order transitions, in the H–T plane for antiferromagnets, and in the P–T plane for helium. In the helium case the order parameter Ψ is related to the density n_s of the superfluid component below the transition.[28] The existence of a superfluid component is related to a condensation of Bose particles into the

Chapter 1

ground state$^{(29)}$; this produces a coherent wave function with off-diagonal long-range order,$^{(30)}$ not spatial or configurational long-range order. Therefore the pressure is not the ordering field ζ for the superfluid transition in helium; rather the ordering field ζ is nonphysical,$^{(31)}$ analogous to the antiferromagnetic ordering field.

The analogy between the superfluid phase transition and the antiferromagnetic one has been made even more directly by Fisher$^{(32)}$ using the correspondence between quantum lattice gases and anisotropic Heisenberg antiferromagnets. Anisotropic antiferromagnets have spin-flop phases in which the ordering is in the x–y off-diagonal components of the spins rather than in the z diagonal components. This spin-flop phase is a reasonable analogy to the superfluid phase, and, indeed, a qualitative resemblance between the phase diagrams of the two systems can be inferred.$^{(32)}$

2.6.2. Superconductors

Just as is the case with superfluids, the ordering which occurs at the superconducting transition is basically a quantum mechanical ordering in momentum space rather than an ordering in real space. The order parameter becomes the gap parameter for Cooper pairs and the ordering field is nonphysical.$^{(31)}$ Nevertheless, there is an analogy to magnetic models in the use of the spin-analog method of solving the BCS equation, as reviewed by Kittel.$^{(33)}$ Because the spin interaction is long-ranged, the specific heat remains finite at the transition, as we will see in Section 4.1. This is in marked contrast to the superfluid helium case. Kadanoff *et al.*$^{(31)}$ relate this difference in thermal behavior to the difference behavior of the coherence lengths of the two problems.

2.6.3. Liquid Crystals

In liquid crystals the ordering is in real space and is due to the interactions between nonspherical molecules which tend to line up in the various (nematic, smectic, and cholesteric) phases. Thus the spatial directions of the various axes of the anisotropic molecules are correlated at large distances, although this is complicated by the effects of flows and walls, which may line up these molecules in different directions in different parts of the sample, somewhat like domains in magnets except that the domains can vary continuously in liquid crystals. Given a local alignment axis, then the order parameter can be defined by $\frac{1}{2}\langle 3\cos^2\theta - 1\rangle_{\text{thermal average}}$ where θ is the angle between the molecular axis and the local alignment axis. Any field (such as electric, magnetic, or flow fields) which tends to line the molecules up along the alignment axis is an ordering field ζ.

Like ferroelectrics, the liquid crystal transitions are usually first order, but, to quote Stinson and Litster,[34] "Over a wide temperature range the liquid crystal behaves as if it were going to undergo a second-order phase transition at a critical point." These authors and Chu et al.[35] use light scattering to effectively measure the divergence of the "susceptibility" $(\partial \Psi/\partial \zeta)_T$.

There are at least two possible reasons why liquid crystals exhibit more complicated transition behavior than some of the simple models already discussed. First, the excluded volume or steric interactions must be considered. Such interactions give rise to different phase behavior than the usual spin models, as we shall see in Sections 4.3 and 4.4. Second, McColl and Shih[36] estimate for their system, p-azoxyanisole, that the steric interactions and the attractive interactions are about equally important. Such competing interactions can give rise to a wide variety of phase behavior, as we shall see in Section 4.1.2.

3. Critical Exponents

One of the major modern advances in the study of critical phenomena is the critical exponent description.[32,37,38] This description reveals many essential features such as dependence of critical points on dimensionality, spin dimensionality, and range of interactions. The description was motivated largely by theoretical considerations as will be discussed in Section 4. However, it is a measure of the success of the description that the critical exponents for many systems have been measured reasonably successfully,[39] with some reservations to be noted in this section. This has allowed detailed comparison of theory and experiment with very gratifying results.

3.1. Definitions of Critical Exponents

If a function $f(x)$ is nonanalytic at the critical point $x = x_c$, then for $x > x_c$ and near x_c one says that $f(x) \sim (x - x_c)^\theta$, where θ is defined precisely by[32]

$$\theta = +\lim_{x \to x_c} [\ln f(x)/\ln(x - x_c)] \qquad (18)$$

Chapter 1

This definition is obvious when $f(x) = (x - x_c)^\theta$ for $0 < |\theta| < \infty$. Cases which are not so obvious include the following: (a) If $f(x) = \ln(x - x_c)$, then $\theta = 0$. (b) If $f(x) = f_1(x) + c$, where $c = f(x_c)$ is finite and nonzero, then $\theta = 0$ again. In case (a), $f(x)$ diverges to infinity, whereas in case (b), $f(x)$ remains finite. This rather dramatic difference is masked by the critical exponent description. Therefore it is usual to say that $\theta = 0$ (log) for case (a) where $f(x)$ diverges (even if the divergence goes as $\log|\log(x - x_c)|$), whereas it is usual to say that $\theta = 0$ (finite) in case (b). However, in case (b) consider the possibility that $f_1(x) \sim |x - x_c|^{1/2}$, which means that $f(x)$ has a cusp at x_c. This looks considerably different than the case where $f_1(x)$ is analytic at x_c and one may wish to include this information in the critical exponent as in the case of the specific heat exponent α to be discussed shortly. Finally, we have case (c): If $f(x) \sim \exp[-(x - x_c)^{-1}]$, then $\theta = \infty$ describes the essential singularity at $x = x_c$.*

Now let us review the specific critical exponents defined for specific physical quantities. We conform to the notation of Fisher,[30] who was instrumental in bringing about a uniform notation used by nearly all workers. The critical exponent for the order parameter Ψ is β, where

$$\Psi(T) \sim (T_c - T)^\beta \quad \text{as} \quad T \to T_{c-} \quad \text{(from below } T_c\text{)} \quad \text{at} \quad \zeta = \zeta_c \quad (19)$$

The critical exponents γ and γ' describe the divergence of the direct zero-field susceptibility $\chi_T = \lim_{\zeta \to 0} (\partial \Psi / \partial \zeta)_T$ above and below T_c:

$$\begin{aligned} \chi_T(T) &\sim (T - T_c)^{-\gamma} &\text{as} \quad T \to T_{c+} \\ \chi_T(T) &\sim (T_c - T)^{-\gamma'} &\text{as} \quad T \to T_{c-} \end{aligned} \quad (20)$$

where for $T < T_c$, $\chi_T(T)$ is measured on one side or the other of the coexistence curve (Figure 1). Strictly speaking, one may allow for different values of γ' on the different sides, although for simple magnetic models the up–down spin symmetry requires equality of the indices on both sides of the coexistence curve.

---

* There are contingencies which may complicate the simple critical exponent described above. A particular calculation on an Ising model with random impurities provides an example.[40,41] However, most systems are probably better behaved than random systems and we take the view that the critical exponent description should be used unless it is clearly inadequate to describe the experiments. References 40 and 41 also raise the possibility that different quantities such as Ψ and $(\partial \Psi / \partial \zeta)_T$ may have nonanalyticities at different temperatures. This is the case for the F model discussed in Section 4.4 and is possibly the case for the two-dimensional Heisenberg model discussed in Section 4.6.3.

Similarly the critical exponents for the specific heat are α and α′, where

$$C_\Psi \sim (T - T_c)^{-\alpha} \quad \text{as} \quad T \to T_{c+}$$
$$C_\sigma \sim (T_c - T)^{-\alpha'} \quad \text{as} \quad T \to T_{c-} \tag{21}$$

Here C_Ψ is defined as $T(\partial S/\partial T)_\Psi$ where the subscript Ψ means $M = 0$ for ferromagnetic systems, critical concentration for order-disorder systems, and critical volume for fluid systems. Also C_σ is defined as $T(\partial S/\partial T)_\sigma$ as the system moves along one or the other of the boundaries of the coexistence region σ. If the specific heat is finite but with a cusp at T_c, e.g.,

$$C_\Psi \sim \text{const} + (T - T_c)^{|\theta|}$$

then one may chose to say that the singular part of α, namely α_s, is given by $\alpha_s = -|\theta|$.

The shape of the critical isotherm in the Ψ–ζ plane has a critical exponent δ defined by

$$\zeta \sim \Psi^\delta \quad \text{for} \quad T = T_c \tag{22}$$

which for a ferromagnet is $H \sim M^\delta$ and for a fluid is $P \sim (V - V_c)^\delta \sim (\rho_c - \rho)^\delta$.

In addition to critical exponents describing macroscopic physical variables, there are critical exponents to describe microscopic correlation functions which may often be measured directly with suitable probes such as neutron diffraction.[10] For systems with short-range interactions it is expected that pair correlation functions decrease with distance in an exponential way. To be specific, consider the nearest-neighbor Ising model for which the pair correlation function of spin i and spin j is just the thermal average

$$\langle \mu_i \mu_j \rangle_{\text{thermal}} = Z^{-1} \sum_{\text{states}} \mu_i \mu_j e^{-\beta E} \tag{23}$$

Then we expect

$$\langle \mu_i \mu_j \rangle - \langle \mu_i \rangle^2 \sim e^{-r_{ij}\kappa(T)} \tag{24}$$

for $T > T_c$. However, as $T \to T_c$, the correlations become very long-ranged and this is described by the critical exponent $\nu > 0$, where

$$\kappa(T) \sim (T - T_c)^\nu \quad \text{above} \quad T_c \tag{25}$$

Chapter 1

and by $\kappa(T) \sim (T_c - T)^{v'}$ below T_c. Right at T_c the correlations fall off more slowly in a power law fashion

$$\langle \mu_i \mu_j \rangle - \langle \mu_i \rangle^2 \sim r_{ij}^{-(d-2+\eta)} \qquad (26)$$

which defines the critical exponent η.

Clearly, one can define many more critical exponents. One additional exponent of primary usefulness to the theoretician is the gap exponent Δ, which for magnets is related to additional derivatives of the susceptibility with respect to field; for example,

$$\partial^2 \chi_T / \partial H^2 \sim (T - T_c)^{-\gamma - 2\Delta} \qquad (27)$$

However, some quantities do not introduce new exponents. For example, for a fluid $C_P = (\partial U/\partial T)_P + PVK_T(\partial P/\partial T)_V$ and K_T always dominates the right-hand side so C_P also has the same γ divergence as K_T. For some systems, especially multicomponent systems, it is not always transparent which critical exponent goes with a given quantity such as, say, the specific heat at constant volume and constant composition. Griffiths and Wheeler[42] give an illuminating discussion of this question.

3.2. Difficulties in Obtaining Critical Exponents from Experiments

Experiments near critical points are among the most difficult to perform and a great deal of effort and sophistication has gone into them. For a more detailed account of the experiments and a good review of experimental determinations of critical exponents up to 1967 one should consult the review by Heller.[39] One of the problems is that measured quantities may become very large [susceptibilities $(\partial \Psi / \partial \zeta)_T$] or very small (order parameters Ψ) near the critical point. Another problem is that relaxation times become very long near the critical point, so care and time must be taken to ensure equilibrium. Also, physical quantities vary drastically with temperature and fields near the critical point, so inhomogeneities and lack of precise control of T and ordering field ζ result in errors which invalidate measurements made very close to the critical point. Another experimental difficulty is to determine T_c and other critical values precisely. One can easily imagine the problem of determining with precision that temperature at which Ψ disappears or $(\partial \Psi / \partial \zeta)_T$ diverges. Generally, it is best to measure as many critical properties as possible, such as Ψ, $(\partial \Psi / \partial \zeta)_T$, and C_Ψ for a single sample in the same laboratory and use all possible information, including even non-

Critical Phenomena and Phase Transitions

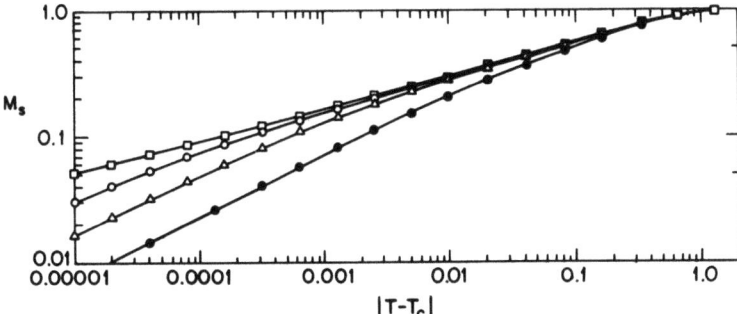

Fig. 8. Plot of $\log \Psi(T) = \log M_s$ versus $\log[T_N(H) - T] = \log|T - T_c|$ for four different fields H. The lowest curve is in zero field and the next two curves are in finite field; for these three curves the exact value of β is 1/2. However, the upper curve is at the tricritical field and $\beta = 1/4$ exactly. $T_N(H) \simeq 2°$ for all four curves.

equilibrium properties such as thermal conductivity, to determine T_c. Different samples often have slightly different critical points in solid systems and absolute temperature calibration of the precision needed for critical exponent measurement is difficult to achieve between different laboratories.

After the preceding, purely experimental difficulties are overcome one has the task of extracting the critical exponents from the data. For quantities such as Ψ [or $(\partial \Psi/\partial \zeta)_T$] a $\log \Psi$ versus $\log|T - T_c|$ plot is quite useful because the numerical value of β is simply the slope of the $\log \Psi$ versus $\log|T - T_c|$ line if $\Psi = (T_c - T)^\beta$. (For specific heat measurements which tend to diverge logarithmically a C_Ψ versus $\log|T - T_c|$ plot may be more useful.) However, matters are seldom as simple as this because $\Psi \sim (T_c - T)^\beta$ only asymptotically close to T_c.

To illustrate possible problems in obtaining critical exponents from experimental data, let us consider some exact calculations of Ψ versus T for an exactly solvable theoretical model of an antiferromagnet.[43] The advantage of this discussion is that we know the critical exponents exactly. Figure 8 shows a $\log \Psi(T)$ versus $\log[T_N(H) - T]$ plot for several fields H. For large values of $T_c - T$, i.e., at very low T, Ψ tends to saturate and the slope approaches zero. This clearly limits the critical region, defined as the region in T over which $\Psi \sim (T_c - T)^\beta$, on one end of the log–log plot. The other end is clearly limited by experimental difficulty in determining $T_c - T$, including imprecision in determining T_c, and in scatter of data very

21

Chapter 1

near the critical point. For the now classic case of superfluid helium, measurements have been made for $|t| = |T - T_c|/T_c$ down to 10^{-6} and a straight line in the specific heat versus $\log t$ plot is found over six decades in $|T - T_c|/T_c$.[44,45] Most systems are less favorable and one cannot obtain accurate data for t much less than 10^{-4} and often one only has t down to 10^{-3} or 10^{-2}.

Consider what this means in terms of the exact theoretical results shown in Figure 8. The exact results give $\beta = 1/2$ for the three lower curves at ordinary critical points and $\beta = 1/4$ for the upper curve. However, if one only had experimental data for $t > 10^{-3}$, even if T_c were known exactly, one would obtain values of $\beta = 0.25, 0.29, 0.33$, and 0.50 for the first through the fourth curves in Figure 8 by measuring the slopes at $t = 10^{-3}$. Presumably with these many decades of data one could see the curvature in the plots even through the experimental scatter and could perhaps extrapolate somewhat better values of β. Unfortunately, it is sometimes felt that these log–log plots should be perfectly straight and therefore T_c is adjusted to improve the linearity. For example, the third curve from the top could be straightened somewhat by increasing T_c but this only worsens the numerical value of β. If one has even fewer decades of data and if T_c is not known accurately, the problem of extracting β becomes very much harder.

This example may lead one to a rather dismal view of the possibility of determining critical exponents experimentally. However, this example is rather special because the critical exponents change discontinuously as a function of field. (Technically, a tricritical point is reached, as will be discussed in the next section.) In particular, notice that it is possible to obtain reliable values of β for the lowest curve, which is furthest from the tricritical point. Thus one hopes that the curvature in experimental log–log plots only becomes troublesome when one is near an unusual point, which in this theoretical case is a tricritical point at which the critical exponents take on different values than normal.

Nevertheless, it must be remembered that one can not always anticipate for a given substance whether such an unusual point is lurking somewhere in an extended phase diagram. For example, if the theoretical model in question were transformed to a ferromagnetic model, then the tricritical point could only be reached by application of a nonphysical staggered field. If the system were such that the tricritical point lay in the physically unattainable region close to the zero staggered field ($H' = 0$), plane then it could disturb the critical exponent determinations in the physical $H' = 0$ plane. Hopefully, a thorough experimental study to elucidate the interactions in a system and a careful theoretical study of the phenomenological model would allow one to predict such points and help in obtaining the values of the critical exponents from the critical point data. In fact, much current analysis of exponent data is done in conjunction with specific pertinent theoretical models.[46] However, it still seems highly advantageous for critical point data to be published in as complete a form as possible with no arbitrary and hidden assumptions in order that later workers with more complete information and new ideas can extract their own critical exponents from it.

Finally, the reader of the literature should not always take seriously the quoted errors for critical exponents. Often these quoted errors are much too small because of assumptions in treating the data such as linearizing the log–log plots by adjusting T_c or because the data do not extend far enough into the asymptotic critical region. It also follows that the best experiment is not necessarily the one with the lowest quoted errors, since extended data may show greater curvature and lead to a more conservative quoted error. Each researcher interested in a specific system should analyze the experimental data for himself.

3.3. Experimental Values of Critical Exponents

Despite the experimental difficulties in obtaining critical exponents, a number of important quantitative similarities between different systems have emerged. In particular, the critical exponent β clusters around 1/3 for many systems and γ clusters around 4/3. These results, which are now very firmly established, are decisive in eliminating the so-called classical theories which predict $\beta = 1/2$ and $\gamma = 1$. The deviations from $\beta = 1/3$ and $\gamma = 4/3$ are perhaps a bit less certain, but are often correlated very nicely with the results of theoretical calculations on models pertinent to the systems in question.

In this subsection some of the experimental results for numerical values of critical exponents are reviewed briefly with an emphasis on results since 1967. Earlier experiments are reviewed by Heller.[39]

For temperature control and resolution the best experimental system is the superfluid helium transition. In addition to the now classic specific heat measurements of Fairbanks et al.,[44] more recent work by Ahlers[45] gives $\alpha \approx 0$ (log) $\approx \alpha'$, in substantial agreement

Chapter 1

with the earlier work. Clow and Reppy[47] and others[48,49] measured the superfluid density $\rho_s \sim (T_\lambda - T)^{\beta^*}$ and find $\beta^* \sim 0.67 \pm 0.03 \sim 2/3$. (Note that $\rho_s \sim |\Psi|^2$, where Ψ is the superfluid wave function; so $\beta \sim 1/3$.)

The gas–liquid critical point of helium has also been carefully studied. For He^3 Meyer and colleagues[50-52] find $\alpha = \alpha' = 0.105 \pm 0.015$, $\beta = 0.361$, $\delta = 4.21 \pm 0.10$, $\gamma = 1.18$, and $\gamma' = 1.12$. For He^4 rather similar exponents are obtained. Moldover[53] finds $\alpha \sim \alpha' \sim 0.15$ and Roach[54] gives $\beta = 0.354 \pm 0.010$, $3.8 < \delta < 4.1$, and $\gamma = 1.1 \pm 0.1$.

Measurements on other simple fluid critical points give similar values. For CO_2 Chu[55] finds $\gamma = 1.32 \pm 0.1$ and $\nu = 0.67 \pm 0.05$, while the specific heat is found to diverge[56] in a manner consistent with $\alpha \sim 1/8 \sim \alpha'$ favored by theory, although Sengers and Chen[57] give $\alpha = 0.09$. Edwards et al.[58] find $\alpha = 0.08$ for Xe but $\alpha = 0$ and $\alpha = 1/8$ are also possible. For Xe Benedek and colleagues[59] find $\nu = 0.58 \pm 0.05$ and $\gamma = 1.21 \pm 0.03$. A composite set of exponents for simple fluids with theoretical input is given by Vicentini-Missoni et al.[60] to be $\beta = 0.35$–0.36, $\delta = 4.4$–4.6, $\gamma \sim 1.25$, and $\alpha \sim 0.05$.

Turning to the solid state, the best-studied order–disorder system is beta brass, for which Als-Nielsen[61] gives $\gamma = \gamma' = 1.25 \pm 0.02$, $\nu = 0.65 \pm 0.02$, and $\beta = 0.305 \pm 0.005$. More recent measurements of β on beta brass[62] also give $\beta = 0.2995 \pm 0.0035$, but FePd$_3$ and FePd seem to have significantly higher values of β, namely[63] $\beta \sim 0.371 \pm 0.01$ for FePd$_3$ and $\beta = 0.377 \pm 0.01$ for FePd. However, Fe$_3$Al behaves[64,65] much more like beta brass, with $\beta = 0.307$, $\nu = 0.649 \pm 0.005$, and $\eta = 0.080 \pm 0.005$. Other well-studied order–disorder transitions include NH$_4$Cl, but here the transition has a small first-order effect, presumably due to the lattice instability.[66]

There are many magnetic systems with ferromagnetic or antiferromagnetic critical points. Ferromagnetic Ni has been extensively studied and the earlier measurements are given in a recent paper by Benski et al.[67] It is somewhat amusing to see all the small quoted errors, many of which do not overlap with most of the others. Nevertheless, it seems that $\beta \sim 0.38$,[67] $\gamma \sim 1.31$–1.32,[68] $4.2 < \delta < 4.7$,[69] and $\alpha \approx -0.10 \approx \alpha'$,[70] which means that the specific heat has a finite cusp at T_c. Ferromagnetic Fe also has[71] $\gamma \approx 1.30 \pm 0.06$. The ferromagnet CrBr$_3$ is an insulator and is presumably a better Heisenberg ferromagnet than the metals because the spins are more localized on the lattice sites. Ho and Litster[72] find for CrBr$_3$ that $\beta = 0.368 \pm 0.005$, $\delta = 4.28 \pm 0.1$, and $\gamma = 1.215 \pm 0.02$.

Turning to the antiferromagnets, Heller and colleagues[73] find $\beta = 0.335 \pm 0.005$, $\gamma = 1.27 \pm 0.02$, $\gamma' = 1.32 \pm 0.06$, $\nu = 0.634 \pm 0.02$, and $\nu' = 0.56 \pm 0.05$ for the insulator MnF_2. Wolf and colleagues[8] find $\alpha = \alpha' \sim 1/8$, $\nu = 0.58$, and $\beta \sim 0.29$ for dysprosium aluminum garnet (DAG). The difference in the exponents for DAG and MnF_2 is likely due to the fact that DAG is better represented by an Ising model and MnF_2 by a Heisenberg model. Other insulating antiferromagnets are $RbMnF_3$, for which[74] $\beta = 0.316 \pm 0.008$, $\nu = 0.724 \pm 0.008$, $\gamma = 1.397 \pm 0.034$, and $\eta = 0.067 \pm 0.01$, $RbNiCl_3$, for which[75] $0.25 < \beta < 0.30$, and FeF_2, for which[76] $\gamma = 1.38 \pm 0.08$ and $\nu = 0.67 \pm 0.04$.

In contrast to magnetic systems, the state of the critical point art is somewhat retarded for ferroelectrics. This is partly because of first-order effects and lattice instabilities which prevent KH_2PO_4 from having a critical point in zero field. However, one must not expect ferroelectric critical points to be carbon copies of magnetic critical points. For example, triglycine sulfare seems to have a "classical" critical point with $\gamma = 1$, $\beta = 1/2$, and finite specific heat discontinuity,[77] although there is some disagreement on this.[78] On the other hand, it is reported[79] for $SrTiO_3$ that $\nu = 0.63 \pm 0.07$, $\beta = 0.333 \pm 0.010$, and $\gamma = 1.29 \pm 0.10$, which are quantitatively rather similar to the values of magnetic exponents.

3.4. Thermodynamic Inequalities for Critical Exponents

Whenever a new description of thermal behavior is introduced there is a reasonable possibility that quantitative relations can be established by the use of simple thermodynamics without having to turn to statistical theory. In the present case of critical exponents these quantitative relations appear in the form of inequalities among the exponents. The first such relation

$$\alpha' + 2\beta + \gamma' \geq 2 \qquad (28)$$

was discovered by Rushbrooke.[80] Thereafter numerous other such inequalities have been discovered which hold under varying conditions. Some of the more popular ones are

$$\alpha' + \beta(1 + \delta) \geq 2, \qquad \gamma' \geq \beta(\delta - 1), \qquad \gamma \leq (2 - \eta)\nu$$
$$2 - \eta \leq d(\delta - 1)/(\delta + 1) \qquad (29)$$

where d is the dimensionality of the system. For a review of these and other inequalities the reader should consult the review article by Griffiths.[81]

These inequalities may be very useful in evaluating the reliability of experimentally determined (and theoretically determined) critical exponents. For example, a few years ago it was found[54] that the experimentally determined critical exponents for ^{4}He did not obey $\beta(\delta + 1) + \alpha' \geq 2$. When this happens a close reexamination of the data and its interpretation is called for. If nothing is amiss, then it must simply be concluded that the true asymptotic range has not been reached experimentally.

Since the Rushbrooke inequality is easily derived for a ferromagnet, let us give the derivation. Starting with

$$dS = (\partial S/\partial T)_H \, dT + (\partial S/\partial H)_T \, dH \tag{30}$$

one has

$$C_M = C_H + T(\partial S/\partial H)_T (\partial H/\partial T)_M \tag{31}$$

Using the Maxwell relation $(\partial S/\partial H)_T = (\partial M/\partial T)_H$ and the standard relation $(\partial H/\partial T)_M = -(\partial H/\partial M)_T (\partial M/\partial T)_H = -(\partial M/\partial T)_H/\chi_T$, one has

$$C_H = C_M + T(\partial M/\partial T)_H^2/\chi_T \tag{32}$$

Since $C_M \geq 0$ by convexity or thermodynamic stability, one has

$$C_H \geq T(\partial M/\partial T)_H^2/\chi_T \tag{33}$$

Using this relation when $H = 0$ below T_c and the definitions of the critical exponents, one has

$$(T_c - T)^{-\alpha'} \geq T(T_c - T)^{2(\beta - 1) + \gamma'} \tag{34}$$

from which $\alpha' \geq -2\beta + 2 - \gamma'$. QED.

4. Theoretical Model Calculations

There are really two sources of critical phenomena data. The usual source of data is experiment. In addition, calculations on model systems may be regarded as an unusually good source of data. This view may be thought to be somewhat unorthodox because usually simplifying assumptions are made in a theoretical calculation.

But consider what assumptions are required for a calculation in equilibrium critical phenomena. The major assumption is that statistical mechanics is valid. The last serious doubt about this assumption was laid to rest with Onsager's two-dimensional Ising model calculation[82] which showed that critical points and phase transitions occur within the statistical mechanical framework.

The second theoretical input concerns the choice of model Hamiltonian. It is a truism that no real material is completely described by any model. Of course, the most interesting models are those which come close to describing some real system *and* which are amenable to mathematical analysis. Experiments and calculations can then be compared to help determine the model Hamiltonian of the real system and to characterize the features of an interacting system which give rise to certain kinds of behavior. Nevertheless, any model, no matter how simple or unrealistic, provides data points in the multidimensional space of all conceivable kinds of critical phenomena. Unrealistic models illustrate by contrast with real phenomena, rather than by agreement with real phenomena, features which do not appear in real Hamiltonians.

The third assumption in theoretical calculations is the reliability of the calculation. Fortunately, there are a number of really firm results in critical phenomena which are mathematically exact and rigorous, in which case no reliability assumption is necessary. Some of the results, such as the solution of the two-dimensional Ising model, are part of the bedrock upon which the structure of critical phenomenology is built. There are also many approximate results often based on series expansions, and here the assumption must be made that one can extract critical exponents from incomplete information, a difficulty not unlike that faced by the experimentalists.

This section, then, reviews the calculations made on model systems. It is to be emphasized that there is no (or should be no) *a priori* input concerning the values of the critical exponents or the nature of the phase transitions. The approach is analogous to the experimental one of doing what can be done and postponing theoretical conjecture until after the calculated results are obtained.

4.1. Classical Model Calculations
4.1.1. Basic Theory

Fisher[38] emphasized that a number of older calculations (which he labels "classical"; do not confuse with non-quantum mechanical), including the van der Waals gas–liquid theory, the mean field theory of magnetism, and the Bragg–Williams theory of order–disorder critical points, all give the same numerical values of the critical exponents, namely $\beta = 1/2$, $\gamma = 1 = \gamma'$, $\delta = 3$, and $\alpha = 0$ (finite discontinuity) $= \alpha'$. Since these values of the exponents are seldom seen in real systems, it might be supposed that one can immediately bypass such classical model calculations, especially since such calculations are usually advertised as approximations to real systems and such approximations are clearly wrong for the critical exponents.

However, the more fruitful perspective is to view these classical calculations, not as approximations to real systems, but as exact calculations for suitably constructed models, which we will call classical models (again, do not confuse with nonquantum mechanical). A clue to the nature of these models comes from the mean field theory of magnetism where it is assumed that every spin interacts equally strongly with every other spin. This model therefore has infinitely long-range interactions. Although there are some difficulties with this simple-minded view of classical models,* the picture has been put on a much more rigorous basis by Baker,[83] Kac et al.,[84] and Lebowitz and Penrose,[85] who show that an interaction of the form $\gamma e^{-\gamma r_{ij}}$ in the limit $\gamma \to 0$ reproduces the classical results.

The very long-range model, i.e., the classical model, is easy to solve because the energy depends only on a few macroscopic variables. For example, for a ferromagnet it is easy to show that the energy of a state is just

$$E = -JM^2 - HM \qquad (35)$$

where M is the magnetization of the state and H is the external field. The partition function is just

$$Z = \sum_M g(M) e^{-E\beta} \qquad (36)$$

where $g(M)$ is the number of states with magnetization M. The value of M for a given H and T is that value which maximizes $g(M)e^{-\beta E}$ or which equivalently minimizes

$$\beta E - \ln g(M) = \beta A \qquad (37)$$

---

* If every spin interacts equally strongly with coupling J with every other spin, then the ground state E decreases with the size N of the system as $-N^2 J$, which is nonthermodynamic since $\lim_{N \to \infty} E/N$ does not exist. Thus J must be taken to vary as $1/N$ and the interactions become infinitely weak, but the cumulative effect of all the interactions remains nonzero.

where A is the free energy as a function of M and T. Since

$$g(M) = \binom{N}{(N-M)/2}$$

then $\ln g(M)$ is a well-behaved function of M. Therefore at constant T the minimization of A gives H as a function of M and T, which may be expanded in a double power series

$$H = 2a_2 M + 4a_4 M^3 + 6a_6 M^5 + \cdots \quad \text{and} \quad a_{2n} = \sum a_{2n,m} T^m \quad (38)$$

Notice that $a_{\text{odd}} = 0$ by symmetry. Then $(\partial A/\partial M)_T = H$ and integrating this gives

$$A = a_0 + a_2 M^2 + a_4 M^4 + a_6 M^6 + \cdots \quad (39)$$

This last equation is the starting point for the Landau theory[86] of critical points and phase transitions. It seems like a perfectly general statement which makes only a very common, but not generally correct, assumption in scientific mathematics, namely that functions can be expanded in power series. However, the assumption is certainly true for classical models with very long-range interactions.

One can obtain the critical exponents for the classical model directly from the Landau expansion even without solving explicitly for the a_{2n}. Taking

$$1/\chi_T = (\partial H/\partial M)_T = 2a_2(T) + 12a_4(T)M^2 + 30a_6(T)M^4 + \cdots \quad (40)$$

gives, for $M = 0$, $\chi_T \sim 1/2a_2(T)$ where $a_2(T_c) = 0$ locates T_c and $a_2(T) \sim b_1(T - T_c) + b_2(T - T_c)^2 + \cdots$ is also assumed to be analytic at T_c. Thus $\gamma = 1$. Next, setting $H = 0$ in the equation for H gives

$$0 = 2(T - T_c)M + 4a_4(T)M^3 + \text{higher-order terms} \quad (41)$$

Near T_c this gives $M_s^2 \sim (T_c - T)$ or $\beta = 1/2$ provided $a_4 > 0$. (We will come back to the question of $a_4 \leq 0$ shortly.) Finally, if $T = T_c$, then $H \sim 4a_4 M^3$ near T_c and $\delta = 3$. The interested reader may calculate any other critical exponents he pleases using the expressions for H and A. (The specific heat jump comes because $M^2 \neq 0$ below T_c.)

The fact that experimentally observable critical exponents deviate considerably from the classical values is not too surprising for some systems such as magnets where the exchange interactions are of fairly short range. It is somewhat more surprising for the magnetic systems where the long-range dipole–dipole forces seem to

29

Chapter 1

dominate, as is also probable for some ferroelectric systems. Certainly for simple fluids with long-range van der Waals interactions one might hope *a priori* that the classical models would provide agreement with experiment, but this is not the case. Nevertheless, the classical models give us the exact results for one idealized extreme limit of simple systems, namely the limit of very long-range interactions.

4.1.2. Tricritical Points and Competing Interactions

Although classical models do not behave like real systems near critical points, they have the double value of being easy to solve and of giving realistic phase diagrams. Certainly, when a new kind of phenomenon is sought in a theoretical calculation, it makes sense to do a classical model calculation. Recently such calculations have been performed on (1) a model constructed to describe the ^{3}He–^{4}He mixture phase diagram with the λ line and also phase separation[87] and (2) on Ising models with competing interactions which may be used to describe the phase diagram of a metamagnet.[43] Both models have a higher-order critical phenomenon, variously called a hypercritical point[32] or a tricritical point.[88] For a metamagnetic system in a real field H and a staggered field H' Figure 9 shows tricritical points at $\pm t$. In Figure 9 the line $-tT_Nt$ is the usual line of critical points seen in antiferromagnets. The line tH_1 is a first-order transition line, thereby distinguishing the hypercritical point t from normal critical points such as the Néel point T_N. But in the classical model in a staggered field there are additional phase lines outside the H–T plane ($H' = 0$) such as the surface AH_1tB and its three images under $H \to -H$ and $H' \to -H'$. Thus t is a tricritical point because it is the meeting point of three critical lines, namely $-tT_Nt$, and tB, and its mirror image under $H' \to -H'$.

One naturally wishes to understand how tricritical points occur and what are the values of the tricritical exponents. These questions are not yet answered in general,[19] but they are easily answered for the classical models by examining the Landau expansion. The basic idea is that for the tricritical point T_t one has $a_4(T_t) = 0$ as well as $a_2(T_t) = 0$. The critical exponents are altered as follows. For $H' = 0$, $T < T_c$, one has

$$H' = 0 = 2(T - T_c)M' + 4(T - T_c)(M')^3 + 6a_6(M')^5 + \cdots \quad (42)$$

which is solved to lowest order (nearest the tricritical point) by $(M')^4 \sim T_c - T$, assuming $a_6 > 0$. This gives $\beta = 1/4$ instead of $1/2$.

Critical Phenomena and Phase Transitions

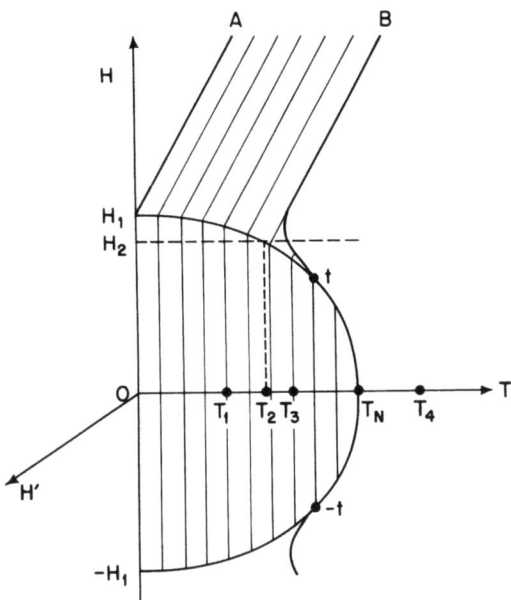

Fig. 9. The H–H'–T phase diagram for a metamagnet. The line $t, T_N, -t$ is a line of second-order critical points like the phase line shown in Figure 5. The line t, B and its three images under $H \to -H$ and/or $H' \to -H'$ are also lines of critical points. The points $\pm t$ are tricritical points because three critical lines meet at t and also at $-t$. The line $-H_1, -t$ is a first-order phase line and the shaded surfaces are first-order surfaces.

For $T = T_c$, $H' \sim 6a_6(M')^5$, giving $\delta = 5$. The γ exponent remains one and one can show that $\alpha = 0$ (finite) but $\alpha' = 1/2$.

At first sight these exponents seem to agree somewhat better with experiments than the ordinary classical exponents, but the agreement is still not very good and such tricritical points are the result of the special confluences $a_2(T) = 0 = a_4(T)$ which arise only occasionally when two parameters in the Hamiltonian are varied with respect to one another, such as direct field versus coupling constant J. When these ratios are adjusted so that $a_4(T) < 0$ when $a_2(T) = 0$, then the transition in zero staggered field $H' = 0$ becomes first order as shown in Figure 9. However, by using the double tangent construction for $H' \neq 0$ one sees from Figure 10 that there is a transition in a nonzero field $H = H_2$ for $T > T_2$. This first-order transition eventually ends in a critical point located on the line tB in Figure 9.

Chapter 1

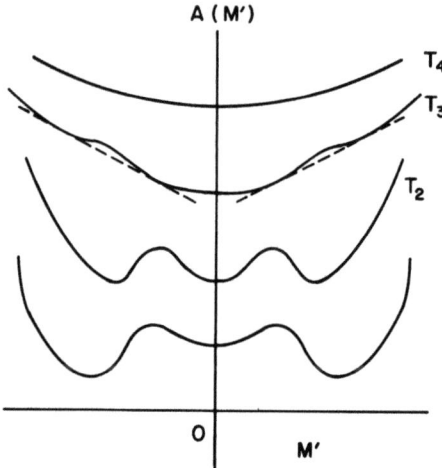

Fig. 10. The schematic free energy versus M' at constant T and $H = H_2$ for the same model as in Figure 9 where the temperatures T_1, T_2, T_3, and T_4 and the field H_2 are also shown in Figure 9. The concave portion of the curve for T_3 lies in a two-phase region with field H' given by the slope of the dashed line.

The particular model[43] used for the metamagnetic calculations shown in Figures 9 and 10 is an Ising model

$$\mathcal{H} = -J_{SR} \sum_{j=1}^{N} \mu_j \mu_{j+1} - (J_{LR}/N) \sum_{i=1}^{N} \sum_{j=1}^{N} \mu_i \mu_j - H' \sum_{i=1}^{N} \mu_i$$
$$- H \sum_{i=1}^{N} (-1)^i \mu_i \qquad (43)$$

The first term is a linear chain coupling, the second term is the very long-range classical term, the third term is the direct field interaction, and the fourth term is the staggered field term interaction. It is convenient to make $J_{LR} > 0$ so the model is ferromagnetic; the antiferromagnetic metamagnetic case is easily obtained by flipping all the spins on one sublattice, in which case H and H' interchange their meaning, with H' being the staggered field and H the direct field as in our discussion of Figure 9. It is the solution of this model which gives the log–log plot shown in Figure 8. The point was made in Section 3.2 that the estimation of critical exponents near special points such as tricritical points becomes difficult. Let us make the further point here that if J_{SR} is made increasingly negative, then the tricritical point can be made to approach the T axis for which

$H = H' = 0$. Suppose in a real material the tricritical point were just off the T axis for a ferromagnetic model so that it would not be reached experimentally since no one knows how to apply a staggered field. Then the log–log plots would show considerable curvature and the asymptotic region would be attained only for very small values of $|T - T_c|$. Such a case would be impossible to analyze for critical exponents unless one had considerable theoretical input.

It may also be remarked that Hamiltonians with more than one interaction term, such as the Hamiltonian in Eq. (43), show considerably richer phase behavior than is generally supposed,[43,89] especially when the interaction terms are in competition, e.g., $J_{SR} < 0$ and $J_{LR} > 0$ in Eq. (43). Stell and co-workers have extended this study of competing interactions to continuum fluids[90] and lattice gases with extended repulsive hard cores[91] plus attractive classical long-range competing interactions and have found some systems which have both a liquid–vapor phase line ending in a critical point and a solid–liquid phase line. Another model which has been solved in the classical limit which has very interesting phase behavior is a spin-1 model with biquadratic ($S_i^2 S_j^2$) interactions and fields (QS_i^2). This model can be used to describe ^{3}He–^{4}He solutions[87] and magnetic alloys.[92]

4.2. Two-Dimensional Ising Model Calculations
4.2.1. Basic Calculation

Onsager's solution[82] of the two-dimensional Ising model with nearest-neighbor interactions only and in zero magnetic field is easily the single most important and profound contribution in critical phenomena. This calculation showed clearly that the classical approximations were fundamentally incorrect for models with short-range forces, although in one form or another classical approximations, to be contrasted to classical models, continued to be taken seriously by many until the mid-1960's. However, many others realized the basic importance of exact calculations and the need for the critical exponent description which allows departures from classical behavior.

The conclusion that classical approximations are essentially incorrect for the two-dimensional Ising model is easily seen by examining the exact answer for the free energy as a function of T only, which may be written as

$$F = F_c + a_1(T - T_c) + a_2(T - T_c)^2 \ln|T - T_c| + \ldots \quad (44)$$

Chapter 1

This free energy cannot be expanded in a power series about T_c because of the $\ln|T - T_c|$ term and thus the Landau theory is inapplicable. The second derivative of F with respect to T gives the specific heat, which near T_c behaves as

$$C_{H=0} \sim A \ln|T - T_c| + B \tag{45}$$

which is symmetric about T_c with $\alpha = \alpha' = 0$ (log).

Unfortunately, the two-dimensional Ising model has still not been solved in a nonzero magnetic field H. Nevertheless, considerable progress has been made in calculating the pair correlation functions[93] and this enables one to find $\beta = 1/8$, $\nu = 1 = \nu'$, $\gamma = 7/4 = \gamma'$, and $\eta = 1/4$. The only major critical exponent which has not been evaluated exactly is δ. (But it is very likely on the basis of series expansions that $\delta = 15$.)

The values of critical exponents for the two-dimensional Ising model clearly disagree badly with the critical exponents found experimentally for the three-dimensional experimental systems reviewed in Section 3.3. However, there have recently been experiments[94] performed on antiferromagnetic K_2NiF_4, which has a layered structure in which the magnetic ions are strongly coupled with only weak coupling in the third dimension between layers. The results[94] for the critical exponents are $\beta = 0.138 \pm 0.004$, $\gamma = 1.0 \pm 0.1$, $\nu = 0.57 \pm 0.05$, and $\eta = 0.4 \pm 0.1$. The results for β and η are close to the two-dimensional Ising model results, although the results for γ and ν are not so favorable. Perhaps this latter disagreement is somehow related to the fact that the crystal really orders three dimensionally. Nevertheless, it seems that the critical exponents depend strongly on dimensionality of the interacting system. Much additional evidence supporting this conclusion comes from series expansions (Section 4.6).

Because of its intrinsic mathematical interest, the two-dimensional Ising model has been solved using several different techniques. There are two general methods, the transfer matrix method and the combinational method. The transfer matrix method was used by Onsager[82] and, with some simplification, by Kaufman and Onsager.[95] Schultz *et al.*[96] put the problem into fermion form and most recently Stephen and Mittag[97] have transferred in another direction following a suggestion of Onsager. The combinatorial approach was initiated by Kac and Ward[98] and finally resulted in the Pfaffian method, which is reviewed by Montroll.[99] The Pfaffian method seems to be the easiest method to learn and it has the advantage of being very useful for dimer problems, but it is clearly restricted to two-dimensional problems of a certain (nearest-neighbor) type.

4.2.2. Decorated Ising Models

Besides the basic Ising model calculation for the standard two-dimensional lattices (square, triangular, and honeycomb), there are a number of extensions of considerable interest. In a footnote to Section 2.1 we mentioned the calculation for the unionjack lattice[6] where there are two interactions which may be chosen to have opposite sign, i.e., competing interactions. We also mentioned in Section 2.4 the calculation of Fisher[20] for a superexchange antiferromagnet which throws considerable light on the behavior of χ_T near T_N. This calculation uses a "decorated" lattice, which technique is reviewed by Fisher.[38] Decorated lattices have also been used very ingeniously by Widom and co-workers [100–102] to describe binary and ternary lattice systems. The critical exponents at the plait points of such systems have altered values when compared to the original critical exponents of the undecorated models.

To elaborate on this latter result, let γ_x, β_x, and α_x be the plait point critical exponents. Then one has $\beta_x = \beta/(1 - \alpha)$, $\gamma_x = \gamma/(1 - \alpha)$, and $\alpha_x = -\alpha/(1 - \alpha)$ where α, β, and γ are the "true" critical exponents of the undecorated lattice. This is an example of what Fisher calls renormalization of the critical exponents.[103] The reason for the renormalization is that extra or "hidden" variables which correspond to densities (e.g., the density of one of the components) in a binary mixture) remain constant at the critical point (plait point) rather than the corresponding field (chemical potential). If the corresponding field were held constant, then one would obtain the "true" values of the critical exponents. This has possible implications for the interpretation of measured critical exponents in three-dimensional systems as conjectured by Fisher.[103]

4.3. Dimer Model Calculations

The Pfaffian method of solving the two-dimensional Ising model is even more convenient for solving dimer models defined as follows: Each state in a dimer model may be thought of as a covering of the lattice with diatomic molecules. Each lattice site is occupied by an atom and each atom is bonded to one and only one atom on a nearest-

Chapter 1

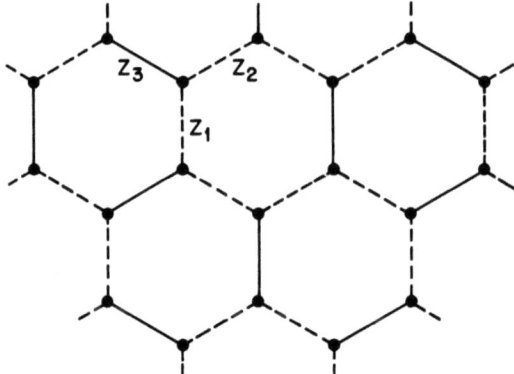

Fig. 11. A dimer state on a portion of the honeycomb lattice. The bonds covered by dimers are indicated by thick, solid lines and the vacant bonds are indicated by dashed lines. The three bonds in the three different directions may have different activities (Boltzmann factors) z_1, z_2, and z_3 as indicated.

neighbor site. (See Figure 11.) Not all orientations of bonds are required to be equivalent and this may be taken into account by assuming different energies (or different activities $z_i = e^{-\varepsilon_i/kT}$) for dimers which occupy different bonds on the lattice. The Pfaffian method can be used to obtain the partition function of any such problem provided that the lattice is planar with no crossing bonds.[99]

The dimer problem is the high-density limit of a more general problem, called the monomer–dimer problem, in which empty sites or monomers are allowed as well as dimers. The monomer–dimer problem is a relevant model for absorption of diatomic gas molecules on a solid surface. It is also a possible model for a nematic liquid crystal. In both cases the very important steric or hard-core repulsion is taken into account completely. Even in the absence of additional interactions, the monomer–dimer problem has been solved only for the one-dimensional lattice consisting of a linear chain for which there is no phase transition. However, by means of a rigorous qualitative argument Heilmann and Lieb[104] have shown that there is no phase transition as a function of density and temperature except at the limit of maximum density of dimers. Thus in order for this model to describe a nematic liquid crystal phase transition, it is essential that additional interactions between neighboring dimers be taken into account.

However, in the high-density limit of all dimers and no monomers there are indeed transitions as functions of temperature when different bonds have suitably different energies. Kasteleijn[4] pointed out that for the honeycomb lattice with activities z_1, z_2, and z_3 for the three different kinds of bonds (see Figure 11) there is a rather peculiar transition when $z_3 = z_1 + z_2$ (or $z_2 = z_1 + z_3$ by symmetry). When $z_3 > z_1 + z_2$, i.e., at low temperatures, the model is completely ordered with all dimers occupying the three bonds. (The reason for this is easily seen when one attempts to create a higher-energy perturbed state starting from the ground state.) Thus the specific heat is identically zero below T_c and the exponent $\alpha' = 0$ (finite). Above T_c the specific heat goes as $(T - T_c)^{-1/2}$, so $\alpha = 1/2$. This transition has no order parameter in the usual sense, so there is no γ, γ', β, or δ. At first glance this unusual critical behavior may be attributed to the "frozen-in" character of the low-temperature phase. However, another dimer model[5] has been constructed in connection with ferroelectric $NaH_3(SeO_3)_2$ and this dimer model is not "frozen in" at low temperature. Nevertheless, it also gives $\alpha' = 0$ and $\alpha = 1/2$ with no easily identifiable order parameter. In order to clearly distinguish these transitions from the usual λ and first-order transitions, this author[167] suggests that these transitions be called $\frac{3}{2}$-order transitions, since the lowest derivative of the free energy to be discontinuous is the third derivative with respect to $(T - T_c)^{1/2}$.

4.4. Hydrogen-Bonded Model Calculations

One of the major recent advances in exact statistical mechanical calculations has been in the area of one-dimensional anisotropic Heisenberg linear chains and two-dimensional ferroelectrics, both problems having resisted exact calculations since the 1930's. Major analytic advances were achieved by Yang and Yang[105] for the anisotropic Heisenberg linear chain at zero temperature. (Since it is a short-range model in one dimension, the anisotropic linear chain does not have a transition at nonzero temperature, but it does exhibit some strikingly singular behavior as a function of field H at $T = 0$.) Then Lieb[106] used the transfer matrix method to study the two-dimensional hydrogen-bonded models which obey the ice rules, and found that the eigenvectors of the transfer matrix could also be described by the Bethe ansatz method developed by Yang and Yang for the linear chain. Further, the eigenvectors are identical to the eigenvectors for the anisotropic Heisenberg linear chain. Although the eigenvalues are different, there is obviously a great similarity in the mathematical aspects of both problems. The most recent advance

Chapter 1

has been made by Baxter,[107] who has developed a new method to solve an even more general model known as the two-dimensional eight-vertex model. This model includes as special cases the nearest-neighbor Ising model, the dimer model, the Slater KDP model, and the F-model, all in two dimensions and zero field.

It is worth describing the eight-vertex model because of its great generality. The underlying lattice is the two-dimensional square lattice. The states are defined by putting an arrow on each nearest-neighbor bond. In hydrogen-bonded language the arrow indicates which side of the bond has the hydrogen. Not all arrow states are allowed. Consider an arbitrary lattice site and its four incident edges or bonds. Each possible incident choice of arrows on these four bonds is called a vertex (or site) configuration. If precisely one (or three) of the four incident arrows points toward the site, then in the eight-vertex model an infinite energy is assigned to these vertex configurations, which disallows such states. In hydrogen-bonded language this means assigning an infinite energy to single ions. Thus the Takagi KDP model[23] is *not* included in the eight-vertex model; the *most* general model requires all 16 vertex configurations to have finite energies. The eight remaining vertex configurations are shown in Figure 7 along with their energies for various special models. The first six vertex configurations in Figure 7 obey the ice rules in that precisely two protons are near each site, i.e., two arrows point toward the site. The last two vertex configurations represent doubly ionized groups in hydrogen-bonded language. Their inclusion and the exclusion of the singly ionized sites is clearly nonphysical for hydrogen-bonded models, but the inclusion of the doubly ionized vertices makes the eight-vertex model more general than the ice rule models.

To show that the eight-vertex model includes the dimer model on the square lattice as a special case, one establishes a several-to-one correspondence between the states of the two models. Consider Figure 12, which shows two superimposed square lattices rotated by 45° relative to each other. A dimer state is shown on the small diagonal lattice and an eight-vertex model state is shown on the larger (straight) lattice. The rule for obtaining an eight-vertex state from a dimer state is: Place an arrow on each edge of the large lattice with direction right or left (up or down) according to whether the dimer on the small lattice which overlaps that edge is right or left (above or below) the center of the edge on the large lattice. The resulting eight-vertex states consist only of vertex configurations of types 1–4, 7, and 8 in Figure 7, so the dimer model on the square lattice is less general than

Critical Phenomena and Phase Transitions

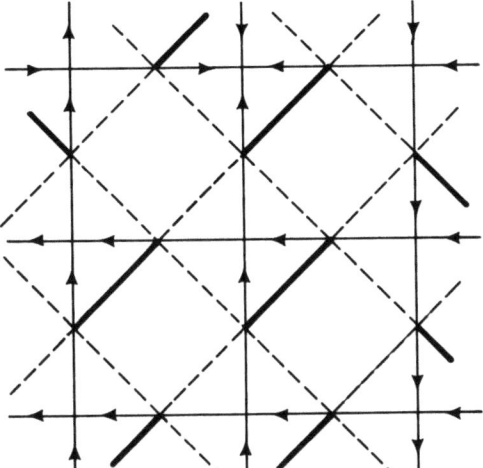

Fig. 12. Polymorphism of states of the dimer model on the square lattice and a special eight-vertex model state. A particular dimer configuration is shown on the small diagonal square lattice where the thick, solid lines represent dimers. The corresponding arrow configuration of the eight-vertex model is shown on the large, straight square lattice.

the eight-vertex model. The correspondence between the Ising model and the eight-vertex model is given by Wu[108] and Baxter.[107]

The mathematical details of calculations for the ice rule or six-vertex models are given in a review by Lieb and Wu,[109] and Baxter[107] has written an extensive paper on the eight-vertex calculation, so only the results will be discussed here. The phase diagram for the two-dimensional Slater KDP model is shown in Figure 13. In region I the model is rigidly ordered with all arrows up (complete polarization) and in region II all arrows are down. This frozen-in character of the Slater KDP model, which is similar to the frozen-in character of the Kasteleijn dimer model below T_c, has been known for many years. In fact, one can prove[110] that this behavior holds rigorously in any number of dimensions (including one dimension[111]) in zero field. The proof also gives $kT_c = \varepsilon/\ln 2$, where ε is the difference in vertex energies between vertex configurations 1 (and 2) and 3–6 (see Figure 7) and the proof gives a latent heat $\Delta H = \varepsilon/2$ (for dimensionality $d > 1$). Thus it seems unlikely that major changes will obtain for the three-dimensional Slater KDP

39

Chapter 1

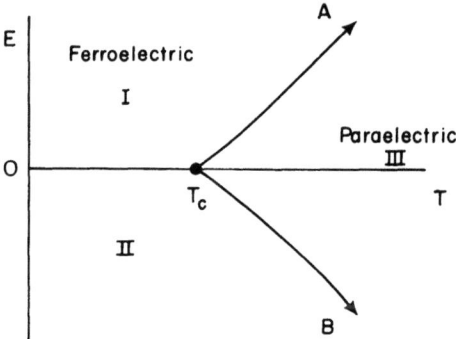

Fig. 13. The E–T (electric field–temperature) phase diagram for the two-dimensional Slater KDP model.[109] The lines A,T_c and B,T_c are second-order phase lines and the line O,T_c is a first-order phase line. The transition is first order at T_c.

model as compared to the two-dimensional calculations. In addition, the full calculation[109] for the two-dimensional model gives $\alpha = 1/2$ and $\alpha' = 0$, again reminiscent of the dimer model. Unlike the dimer model, there is a well-defined order parameter for the Slater KDP model. However, the transition is first order in zero field so β is undefined (or zero). Of course, from the phase diagram the transition point in zero field would hardly be expected to be an ordinary critical point. However, it is *not* a typical triple point either since the transition lines in the electric field are *not* first-order lines, i.e., there is no latent heat for transitions in a field, although the specific heat continues to diverge as $\alpha = 1/2$ and $\alpha' = 0$ in a field. The susceptibility diverges as $\gamma = 1$ and, of course, $\gamma' = 0$ at all transition points.[109]

The experimental situation for the phase behavior of KDP in an electric field is not as complete as one would like. However, Benepe and Reese[112] claim that a classical free energy expansion can be made in the vicinity of the transition in zero field as follows:

$$A = a_0 + a_2(T - \theta)P^2 + a_4P^4 + a_6P^6 + a_8P^8 \qquad (46)$$

It seems that a_4 is very small and negative, a_6 is nearly zero, and a_8 is large and positive. This implies that a tricritical or higher-order critical point is near, although it is probably not attainable by application of direct electric fields. However, if the classical theory is valid, there should be a first-order transition line in a positive field which should end in a critical point. Although the slopes of the first-order lines in a field should be similar to those in Figure 13, the classical theory gives $\alpha = 0$ and $\alpha' = 2/3$ at the higher-order critical point. Since the transition in zero field is very near this higher-order critical point, the specific heat should be dominated by apparent $\alpha = 0$ and $\alpha' = 2/3$ behavior until it gets cut off by the first-order instability. This latter behavior is much more consistent with experiment[25] than the Slater KDP behavior of $\alpha = 1/2$, $\alpha' = 0$.

Even though it is clear that the ice rule hydrogen-bonded models are too "rigid" to give good agreement with experiment, they still provide illuminating examples of theoretical critical behavior. The F model provides the two-dimensional antiferroelectric complement to the ferroelectric Slater KDP model. The vertex energies for the F model are shown in Figure 7 and it is easy to see that the ground state of this model is an "anti" type state. The phase behavior of this model is not so easy to predict, although the critical temperature in zero field, $kT_c = \varepsilon/\ln 2$, was conjectured[113] prior to Lieb's exact calculation. The phase behavior of the F model is shown in Figure 14, where the electric field is along the vertical axis of the model. Within the enclosed region the polarization is identically zero so this corresponds to a rather rigid (but not completely frozen-in) antiferroelectric phase analogous to the rigid phase of the Slater KDP model. In a field $\alpha' = 0$ and $\alpha = 1/2$, but in the limit of zero field the specific heat and all its derivatives are continuous although the free energy is nonanalytic at the critical temperature.

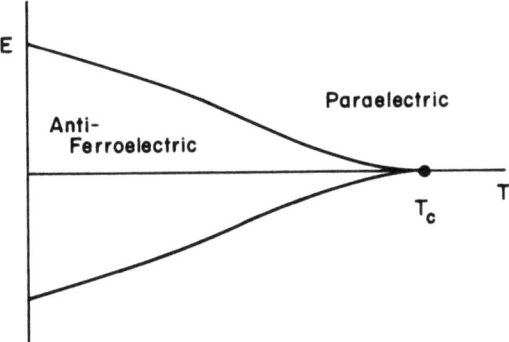

Fig. 14. The E–T phase diagram for the F model.[109] The phase lines are second order. There is also an infinite staggered susceptibility for field $E = 0$ and $T = 2T_c$.

Chapter 1

The ordering field ζ for the F model is the staggered field which energetically favors (disfavors) the vertices in Figure 7 of type 5 (6) on the A sublattice and vice versa on the B sublattice.[114] Although an exact solution has not yet been obtained with a staggered field, Baxter has produced two results which are especially illuminating. Very recently Baxter[115] has calculated the order parameter, i.e., the staggered polarization P_s, from the correlation functions. The result is that P_s vanishes at $kT_c = \varepsilon/\ln 2$ with an exponent $\beta = \infty$, i.e., P_s and all its derivatives also vanish at T_c. The second result is an exact calculation of the staggered susceptibility at one temperature, namely $T = 2\varepsilon/k \ln 2 = 2T_c$ and all staggered fields.[116] In the limit of zero staggered susceptibility $(\partial \Psi/\partial \zeta)_T$ diverges. Thus the critical temperature defined by the vanishing of the order parameter and the critical temperature defined by the divergence of $(\partial \Psi/\partial \zeta)_T$ are *not* the same for this model!

The solution of the eight-vertex model[107] gives some very interesting information about the specific heat critical exponents α and α'. As the parameters in Figure 7 are varied under the constraints $w_1 = w_2 = a$, $w_3 = w_4 = b$, $w_5 = w_6 = c$, and $w_7 = w_8 = d$, various phase boundaries are crossed. The KDP and F models constitute special degenerate cases in this multidimensional phase diagram. Disregarding these special cases, Baxter[107] finds that $\alpha = \alpha' = \pi/\cos^{-1}[(ab - cd)/(ab + cd)]$ when α is not an integer. This means that the specific heat critical exponents can vary over the entire allowed range from $+1$ (exclusive) to $-\infty$ (for the F model). This wide variation in α was not altogether expected before the exact solution was obtained and it certainly is different from experiment where α tends to be near zero. Also, the exponent ν has been found[117] to equal $1 - (\alpha/2)$ for the eight-vertex model.

One naturally wishes to understand why the simple hydrogen-bonded models, and also the dimer models, behave so differently from experiment and other "more normal" models. One clue comes from the "frozen-in" behavior in the ordered phases. Another clue comes from the correlation functions which, even for $T \neq T_c$, fall off only as power laws, as $1/r^2$ for the hydrogen-bonded model[109] and for the square lattice dimer models[118]; this is in distinct contrast to the Ising model, where the correlations fall off exponentially except at T_c, where the power law behavior pertains. Both these clues implicate the infinitely repulsive interactions, as characterized by the ice rule or the nonoverlapping of dimers, as the feature responsible for the different behavior of these models from the Ising model. A more complete explanation has been given by Sutherland,[119] who uses the analogy to the general Heisenberg chain for the hydrogen-bonded models. Sutherland concludes that the constrained ice rule model behaves for all temperatures like an unconstrained model at its critical point.

Although most real systems do not owe their critical points to infinitely repulsive hard cores, it is certainly valuable to know the behavior of this extreme sort of model just as it is valuable to know the behavior of the very long-range classical models.

4.5. Spherical Model Calculations
4.5.1. Basic Results

The spherical model was proposed and solved by Berlin and Kac in 1952.[120] Starting with the Ising Hamiltonian model, Kac relaxed the constraint that $\mu_i^2 = 1$ for each ith spin and instead required only the weaker "spherical" constraint $\sum_{i=1}^{N} \mu_i^2 = N$, which clearly allows each spin more freedom. This continuum spin model has certain obvious defects; in particular, the entropy at $T = 0$ is infinite. Nevertheless, the spherical model has reasonable critical behavior. It also provides a good contrast to the spin-1/2 Ising model with respect to the number of degrees of freedom of the individual spins, e.g., the Heisenberg models might be expected to fall between the Ising and spherical models in this regard. This notion has been made precise by Stanley,[121] who has shown that the spherical model solution is equivalent to the solution of the classical, n-dimensional spin model with spin dimensionality $n = \infty$. (See the end of Section 2.2 for the definition of the classical models with n-dimensional spins. Recall that $n = 1$ is the Ising model, $n = 3$ is the $S = \infty$ Heisenberg model, and that $n = 2$ is the $S = \infty$ xy model.)

Berlin and Kac[120] solved the spherical model with nearest-neighbor interactions and found no phase transition in one and two space dimensions, $d = 1$ and 2. However, in three dimensions the solution gives $\gamma = 2$, $\beta = 1/2$, $\nu = 1$, $\eta = 0$, $\delta = 5$, and $\alpha = \alpha' = 0$ (finite). (For some purposes it is better to say that α_s and $\alpha_s' = -1$ corresponding to $C_{H=0} \sim |T - T_c|$.) In four and higher dimensions the critical exponents take on their classical values.

4.5.2. Long-Range Interactions

Perhaps one of the most important uses of the spherical model is in the study of long-range power law interactions which fall off as

$r_{ij}^{-d-\sigma}$ where r_{ij} is the distance between spin i and spin j. Joyce[122] adapted the methods of Berlin and Kac[120] to this long-range interaction and computed critical properties for all values of σ and d. The values of the critical exponents fall into three categories or regions. When the range parameter σ is greater than two the critical exponents remain constant as a function of σ and they take on the same values as they do for the nearest-neighbor case, which corresponds to the $\sigma \to \infty$ limit. Let us call this the short-range or Onsager regime. When the range parameter σ is less than $d/2$ then the critical exponents again remain constant as a function of σ and they take on the same values they do for the very long-range classical model, which may be thought of as the $\sigma = -d$ limit. Let us call this the classical or van der Waals regime. For $d \geq 4$ for the spherical model the Onsager and van der Waals regimes overlap and have the same values of the exponents. But for $d < 4$ there is a third regime to be called the Joyce regime which occurs for $d/2 \leq \sigma \leq 2$ and which bridges the gap between classical and short-range critical behavior. In the Joyce regime the critical exponents depend only upon the ratio σ/d and not upon σ or d separately. For example,*

$$\gamma = (\sigma/d)/[1 - (\sigma/d)], \qquad \beta = 1/2, \qquad \delta = [1 + (\sigma/d)]/[1 - (\sigma/d)]$$
$$\alpha_s = [1 - 2(\sigma/d)]/[1 - (\sigma/d)], \qquad dv = [1 - (\sigma/d)]^{-1} \qquad (47)$$
$$(2 - \eta)/d = (\sigma/d)$$

A graph showing the behavior of the exponent γ as a function of σ/d is shown in Figure 15 for all d. The upper limit of the classical regime is always given by $\sigma/d = 1/2$, and this carries smoothly into the short-range regime for $d \geq 4$. For $d = 3$ the Joyce regime carries γ continuously from $\gamma = 1$ at $\sigma/d = 1/2$ to $\gamma = 2$ at $\sigma = 2$ or $\sigma/d = 2/3$, which marks the upper limit of the Joyce regime. For $\sigma > 2$ the exponent γ remains fixed at two. For $d = 1$ and 2 the Joyce regime continues to $\sigma/d = 1$, at which point the transition temperature is driven to $T_c = 0$. Thus only the $d = 3$ spherical model has all three distinct classical, Joyce, and Onsager regimes.

It has been conjectured by Nagle and Bonner[123] using approximate methods that the Ising model with $r^{-d-\sigma}$ interactions may also

---

* The use of dv and $(2 - \eta)/d$ instead of v and η is a bit unusual. The product dv is the critical exponent for a correlation volume rather than just the correlation length. Similarly $(2 - \eta)/d$ is the natural critical exponent to consider if volumes rather than lengths are considered basic.

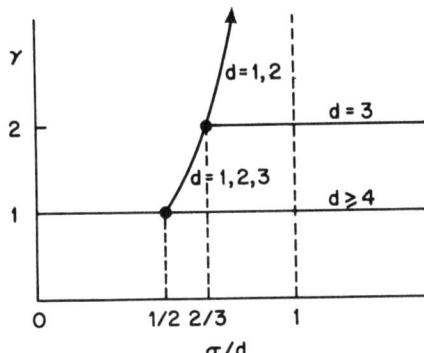

Fig. 15. The exponent γ versus σ/d for the spherical model with long-range interactions.

have a three-regime kind of behavior similar but not identical to that of the spherical model. Current work being done by Fisher[124] and Suzuki[125] on the Wilson model (to be discussed in Section 4.7) supports the conjecture, but some of the details are not yet clear. However, there is one conjecture which is generally agreed upon, namely that long-range r^{-6} van der Waals interactions in three dimensions should give rise to Onsager or short-range critical behavior rather than classical critical behavior. The best support for this conclusion remains the exact spherical model calculation, which puts $\sigma = 3$, $d = 3$ firmly into the short-range Onsager regime.

The spherical model can also be solved easily with a variety of competing interactions. Because the spherical model solution is so intimately related to ground-state properties, the phase behavior is simpler than the phase behavior of the Ising, XY, or Heisenberg models with competing interactions.[126] However, in three dimensions for antiferromagnetic nearest-neighbor interactions ($J_{NN} < 0$) and ferromagnetic long-range σ interactions ($J_\sigma > 0$) as a function of $J_{NN}/J_\sigma = x$ the transition changes character at the value of x for which the ground-state changes character. For small values of x the transition is ferromagnetic and for large values of x the order parameter Ψ and the ordering field ζ have an oscillatory character.

4.6. Series Expansion Calculations
4.6.1. Basic Method

Thus far in this section only exact calculations have been reviewed because those are very firm results mathematically. However, with the exception of the spherical model, the exact results exist only for two-dimensional models or classical models with very long-range interactions and none of the exact results gives quantitative agreement with experiment as evidenced, for example, by the values of the critical exponents. Clearly, calculations are needed on three-dimensional models with short-range interactions. After nearly thirty years of effort an exact solution of the three-dimensional Ising model seems as remote as ever, so it is necessary to turn to approximate calculations.

But it should be emphasized that exact calculations have not failed in the study of critical phenomena. First, they have certainly enlightened us concerning the diversity of possible behavior. Second, and of primary importance for this subsection, the exact calculations provide a stringent testing ground for approximate methods. In particular, the exact results for the two-dimensional Ising model have sunk many an intricate approximation method. The only approximation method which has emerged, not only unscathed, but enhanced by testing against the exact results is the series expansion method.* This test is especially convincing because there is no indication that the series expansion method should be more accurate in two dimensions than in three dimensions; in fact, the few differences indicate the opposite.

Suppose that the thermodynamic or correlation function $f(T)$ to be studied is expanded as a power series in J/kT:

$$f(T) = \sum_n a_n (J/kT)^n \qquad (48)$$

This expansion is called a high-temperature series expansion because of its region of convergence. Other expansion variables are $\tanh(J/kT)$ (also high temperature), $e^{-H/kT}$ (high field), and $e^{-J/kT}$ (low temperature). The first goal of a series expansion calculation is to calculate exactly as many of the terms a_n as possible in Eq. (48). To the uninitiated it may not appear altogether obvious how one sets about performing this task for a given model. In fact, straightforward

---

* We include, as a subcase of the series expansion method, another method best described as the sequence method. In this method exact calculations are performed for a sequence of systems which increase regularly in size. One example of the use of this method is provided by linear chain calculations[123] where the nth system in the sequence is a chain of n spins. Another example is the calculation of Runnels[127] on two-dimensional hard-core lattice gases using the transfer matrix for n by ∞ strips. The conceptual similarity with series expansions is extrapolation from the *regular* sequence of exact calculations to the infinite system in question.

calculations are usually very laborious and therefore many devious series expansion methods have been devised to simplify the calculations. These series expansion methods are described by Domb and Green[128] and many of the earlier methods are well described in the review by Domb.[129] The result of these methods is to reformulate $f(T)$ as

$$f(T) = \sum_{i \in L} g_i w_i \qquad (49)$$

where the sum is over linear subgraphs i (such as a square with four sites and four bonds). The quantity g_i is the number of ways that the subgraph i can be "embedded" or found in the lattice L (e.g., g_i for a square subgraph equals N for a square lattice). The quantity w_i is the "weight" of each graph, which is determined by the expansion method. The laborious part of the problem is to count the g_i. For this reason it is advantageous to devise an expansion for which many w_i are zero. To push these calculations as far as possible and to minimize error, high-speed computers are now used.

The second goal of a series expansion calculation is to attempt to estimate the critical behavior of $f(T)$ from the first m terms of the expansion in Eq. (48). Since a finite polynomial of m terms never has a nonanalyticity and therefore no critical point, the hope is to effectively extrapolate the a_n for $n > m$ from the first m of the a_n. The easiest case to consider is when all the a_n are positive, which means that the singularity in $f(T)$ nearest to the origin of the (J/kT) plane (letting J/kT be a general complex variable) occurs on the positive real axis. Consider for simplicity the example of

$$f(T) = [1 - (T_c/T)]^{-\gamma} \qquad (50)$$

Then the coefficients a_n are exactly

$$a_n = \left(-\frac{kT_c}{J}\right)^n \binom{-\gamma}{n} \qquad (51)$$

The ratio of two successive coefficients $r_n = a_n/a_{n-1}$ is

$$r_n = \frac{kT_c}{J}\left(1 + \frac{\gamma - 1}{n}\right) \qquad (52)$$

If r_n is plotted versus $1/n$, then the intercept at $1/n = 0$ gives the critical temperature kT_c/J and the slope of the straight line curve is $(\gamma - 1)(kT_c/J)$ as shown in Figure 16. Thus in this simple example T_c and γ are simply determined from the ratio plot of r_n versus $1/n$. In

Chapter 1

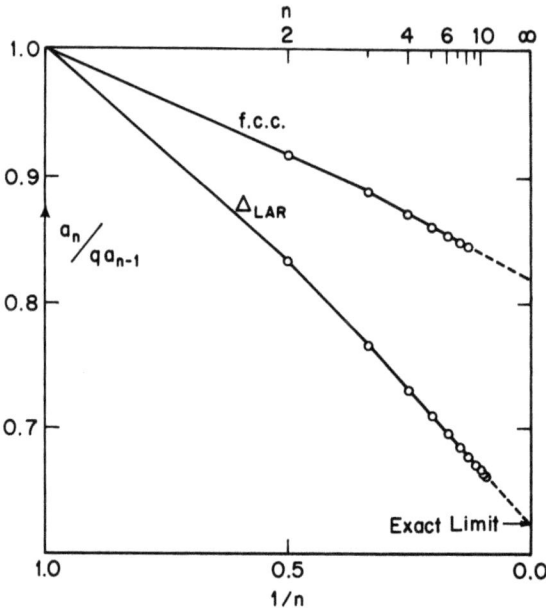

Fig. 16. Ratios a_n/qa_{n-1} versus $1/n$ for the susceptibility series of the triangular and fcc lattices. (From Domb and Sykes.[130])

the case when the $a_n > 0$ for all n the ratio plot continues to be useful although the presence of more complicated terms in $f(T)$ will mean that the ratio curve only becomes straight in the limit as $n \to \infty$. Thus when $f(T)$ is unknown the theoretician performing series expansions has some problems analogous to those of an experimentalist. Analogous to the problem of measuring closer to T_c is the problem of calculating more terms in the series expansions. Analogous to the problem of curvature in the experimental log–log plots is the problem of curvature in the $1/n$ ratio plots.

In the most general case the series do not have terms of a uniform sign or simply alternating sign. This means that the singularity nearest the origin is not the physical singularity on the positive real axis but a complex singularity. Although this complicates the extrapolation, more sophisticated techniques, particularly the Padé approximant, have been developed to analytically continue the series to the physical singularity. A very lucid, somewhat fuller but still elementary account of the ratio and Padé extrapolation methods is given by Fisher.[32,38] Extensive studies of the Padé approximant method are given by Baker and Hunter[131] and a general review of extrapolation methods is given by Gaunt and Guttmann (see Ref. 128).

4.6.2. Numerical Results

The series expansions for the Ising model are easier to calculate and the series are better behaved than for other models, so one expects that the Ising model results will be the most accurate approximate results. In two dimensions the series extrapolations reproduce the known exact results very well.[38,130] Since the two-dimensional Ising model has not been solved in a field, δ has been obtained only by series expansions which give $\delta = 15 \pm 0.08$.[132] It should be pointed out here that it is rather firmly established that critical exponents are the same for a model on all lattice structures of the same dimensionality,[37] such as the simple cubic and the body-centered cubic lattices. Therefore in reviewing the calculations for the critical exponents no mention will be made of the particular lattice. It may be noted, however, that the series extrapolations may be more accurate for one lattice than for another.

For the nearest-neighbor Ising model in three dimensions the exponent values $\gamma = 5/4 \pm 0.003$, $\alpha = 1/8 \pm 0.015$, and $\beta = 5/16 \pm 0.006$ are well established.[32] The originally suggested value[132] of $\delta = 5.20 \pm 0.15$ by Gaunt et al. has been lowered by Gaunt and Sykes[133] to $\delta = 5.00 \pm 0.05$. One incentive for changing the value of δ comes from homogeneity–scaling theory to be discussed in Section 5.1. However, the point to be made is that often authors are not willing to defend to the end their estimate of an exponent including uncertainty; this change in earlier estimates may reflect advances in extrapolation techniques or longer series or both. Therefore, the wary reader of the literature may allow a somewhat larger margin of error than the quoted one, although there are some researchers who are known for being overly cautious in their quoted uncertainties, so no hard and fast rule can be given. The values of the exponents α' and γ' also present difficulties because the low-temperature series is not as well behaved as the high-temperature series. Although it was originally suggested that $\alpha' = 1/16$, Gaunt and Domb[134] have argued that the true value is $\alpha' = 1/8$ but that the asymptotic behavior only becomes apparent very close to T_c, e.g., $t = (T_c - T)/T_c < 10^{-4}$. Further from T_c the specific heat appears to increase roughly logarithmically; this should serve as a good example to experimentalists who may feel that their experiments should show $\alpha = \alpha'$ for $t \gg 10^{-4}$. The history of the values of the critical exponent γ' is amusingly described by Guttmann et al.,[135] who opt for the value $\gamma' = 21/16$, in contrast to the currently favored value $\gamma' = 5/4 = \gamma$.

Chapter 1

The values of the critical exponents for the correlations were given by Fisher and Burford[136] as $v = 0.6430 \pm 0.0025 \simeq 9/14$ and $\eta = 0.056 \pm 0.008 \simeq 1/18$ and these have suffered only small displacements when more terms were added to the series and a different extrapolation was made by Moore et al.,[137] who give $v = 0.638^{+0.002}_{-0.001}$ and $\eta = 0.041^{+0.006}_{-0.003}$.

It is clear that the $d = 3$ Ising model agrees better with experiment than the previous theoretical results. This is especially true for β-brass (Section 3.3), which is expected to be a good Ising-like system. However, even allowing liberally for the numerical uncertainty, it seems that many of the magnetic systems have slightly larger values of β and γ than the $d = 3$ Ising model. This is not entirely unexpected since the Heisenberg model is more appropriate for many of the magnetic systems.

Although there is no exact solution for the Heisenberg model to use as a test of the series expansions, the series expansions should still be of value. Because of simplifications in obtaining the series, the $S = 1/2$ and $S = \infty$ (classical $n = 3$ vector model) cases are the most studied. Starting with spin-1/2 in three dimensions, Baker et al.[138] find $\gamma = 1.43 \pm 0.04 \sim 10/7$, although Bowers and Woolf[139] find the lower value of $\gamma = 1.3744 \pm 0.0008$ and Lee and Stanley[140] find $\gamma = 1.36 \pm 0.04$. There is as yet no way to obtain low-temperature series for the Heisenberg model but Baker et al.[141] manage to use the high-temperature series to obtain $\beta = 0.35 \pm 0.05$ and $\delta = 5.0 \pm 0.2$.

Turning to the classical $S = \infty$ Heisenberg model, Ferer et al.[142] find $\gamma = 1.405 \pm 0.020$, $v = 0.717 \pm 0.007$, $\alpha = -0.14 \pm 0.06$, and $\eta = 0.040 \pm 0.008$, which may be compared to the estimates of Ritchie and Fisher,[143] who give $\gamma = 1.375^{+0.02}_{-0.01}$, $v = 0.703 \pm 0.01$, and $\eta = 0.043 \pm 0.014$. Earlier estimates by Bowers and Woolf[139] using shorter series are in rough agreement. It may be noted that there is a general desire that the values of the critical exponents be independent of spin and the quoted errors seem to allow for this, although the earlier estimates are probably least influenced by this desire and these earlier estimates leave the uncomfortable feeling that γ for spin 1/2 is greater than γ for spin ∞.

4.6.3. Qualitatively Irregular Cases

The two-dimensional Heisenberg model was studied using series expansions by Rushbrooke and Wood,[144] who declined to commit themselves to the existence of a phase transition. However, Stanley and Kaplan[145] concluded from the short series of Rushbrooke and Wood that there was a transition, although they pointed out the possibility that the order parameter is zero below the transition. Shortly thereafter Mermin and Wagner[146] proved that there is no spontaneous magnetization for $T > 0$ and thus that there is no phase transition in the usual sense. This illustrates that care must be taken in drawing qualitative conclusions from series expansions, especially when the series are short. It is generally advisable to draw qualitative conclusions from intuition or rigorous proofs (see the review by Griffiths in Ref. 81) and to use the series expansions to obtain quantitative approximations for cases which are qualitatively understood. In the case of the two-dimensional Heisenberg model there may be a critical or nonanalytic temperature $T_c > 0$ at which the susceptibility diverges while the spontaneous magnetization remains zero, but this is still very conjectural. However, the interested reader should note the similarity between this conjecture and the rigorously established behavior of the F model (Section 4.4).

The series method has been applied to many other models too numerous to mention. However, the reviewer will perhaps be forgiven for mentioning his own series expansion work in connection with hydrogen-bonded models.[114] Although the series expansions were useful for calculating properties away from the critical point, it was clear from the behavior of the series near the critical point of the F model that accurate values of the critical exponents could not be obtained from the available number of terms, although recent exact results[115] show that the series were giving valid clues such as $\beta > 1$. The moral of the story is that series methods do not automatically give an answer, right or wrong; this encourages us to believe that when the method does give an answer, it may be reliable, subject to small fluctuations arising from choice of extrapolation procedure and length of series.

4.7. Other Calculations
4.7.1. One-Dimensional Cluster-Interaction Fluids

Before concluding this section on theoretical calculations two additional calculations should be noted. The first is an exact calculation for a model proposed by Fisher.[147] This model is a one-dimensional continuum model of particles with hard cores and an additional attractive interaction which is short range in character.

Chapter 1

Normally, such models have no transition in one dimension. The additional feature which causes the transition in these models is that the attractive interaction is a many-body interaction rather than just a two-body interaction. The phase diagrams of these models have liquid–gas first-order phase lines which end in critical points. However, these critical points are also the ends of lines of higher-order singularities and this feature is "nonphysical," although interesting. It is not clear whether these higher-order phase lines are due to the many-body interactions and/or to the one-dimensional character of the model. For reasonable potentials the models have critical exponents and display "scaling" (Section 5.1) although with certain sorts of breakdown.

4.7.2. Wilson Method

The second calculation which should be mentioned is due to Wilson.[148] This work is a relatively recent development and this reviewer is not yet certain of its ultimate status in the field of critical phenomena. But because of its current popularity and the belief in some circles that it may become the "ultimate" critical point calculation, a few remarks are in order.

The model treated by Wilson[148] is a continuum spin model, $-\infty < S_r < +\infty$, where a phase space weighting factor $\exp(-\frac{1}{2}|S_r|^2 - \frac{1}{4}f|S_r|^4)$ is introduced so that the average value of each spin squared is unity. If $f = 0$, then this reduces to the Gaussian model[120] which is known to give poor results. Adding higher moments like $|S_r|^4$ to the weighting factor exponential is supposed to make the model like the classical n-vector model discussed at the end of Section 2.2. This and other approximations induced Baker[149] to seek and find an explicit model for which one of the calculations of Wilson is exact.

The calculational innovation of Wilson is to reassemble the mathematics so as to focus the calculation onto the critical point rather than solving the entire model. At least two calculational techniques have been used, one based on the renormalization group[148] and one based on Feynman graph techniques.[150] The result of the calculations is to obtain expansions for the critical exponents directly, such as

$$\gamma = 1 + [(n + 2)\varepsilon/2(n + 8)] + O(\varepsilon^2) + \cdots \qquad (53)$$

where the expansions are in powers of $\varepsilon = 4 - d$, where d is the dimensionality of the lattice which is treated as a continuous variable and n is the spin dimensionality, e.g., $n = 1$ for the Ising-like model. For $d \geq 4$ the values of the exponents stick at the classical values while the series expansion applies for $d < 4$. It is rather worrying that the case $d = 4$ is clearly an essential singularity for γ as a function of ε and thus the expansions are likely to be asymptotic. This is clearly revealed in the earlier approximate renormalization group approach, which to eighth order for the $d = 3$ Ising-like model gives successive corrections of 0.167, 0.041, -0.016, 0.077, -0.2, 0.67, -2.5, and 10.3. Perhaps the exact Feynman graph expansion will yield a better behaved series, but it is still expected to be asymptotic.[150]

4.7.3. Long-Range Interactions

A valuable feature of the Wilson approach is that the corresponding models remain amenable to calculation when long-range interactions falling off as $r^{-d-\sigma}$ are included.[124,151] This reviewer's hope is that changes in the values of the critical exponents as the potential range σ varies may be obtained qualitatively, even if it turns out that neither the short-range nor the long-range cases may be calculated exactly by the Wilson method. At present there are three different types of expansion for the critical exponents in the presence of long-range interactions. First, there is an expansion[151] in powers of the inverse spin dimensionality $1/n$, where the zeroth-order term is the spherical model result ($n \to \infty$). Second, there is an expansion[124] in powers of $\varepsilon/\sigma = 2 - (d/\sigma)$ analogous to the short-range ε expansion and valid for $2 > d/\sigma$ with the zeroth-order term giving the classical values of the critical exponents. Third, there is an expansion[124] in a variable called $\Delta\sigma/d = (\sigma/d) - 1/2$. All three expansions are consistent to order ε/n. The expansions indicate that there is a "classical regime" (see Section 4.5) for $\varepsilon < 0$ and $\Delta\sigma < 0$ where the values of the exponents for all n and all d are given by $\eta = 2 - \sigma$, $\nu = 1/\sigma$, $\gamma = 1$, $\beta = 1/2$, $\delta = 3$, and $\alpha = 0$. There is also an intermediate "Joyce regime" in which the critical exponents vary monotonically with σ for a given n and d. Finally, when σ becomes larger than two the exponents stick at their short-range values. In these general respects these calculations are in agreement with numerical work for $d = 1$, $n = 1$ and with the conjectures[123] based upon that work for $d \geq 2$. In this connection it might also be noted that the expansion variables are functions only of the ratio σ/d and this is true of the coefficients to zeroth order in the $1/n$ expansion and the first-

Chapter 1

order coefficients in the ε and $\Delta\sigma$ expansions. However, the next higher coefficients contain factors of σ and d separately so "σ/d homogeneity," as we might call it, breaks down for $n < \infty$. However, the $d = 2$ and $d = 3$ second-order corrections are, respectively, about 90% and 80% as large as the $d = 1$ second-order correction. Therefore the breakdown of "σ/d homogeneity" is small in the region $1/2 < \sigma/d < 2/3$, which is the extent of the region in which the expansions agree with each other.

5. General Theories

The reader who has gone through the last two sections probably feels overwhelmed with facts and values of critical exponents and is yearning for some deeper understanding of it all. However, the reader has seen the great diversity and mathematical complexity of critical phenomena. It seems improbable that this can be compressed into a simple and complete prescription valid in all cases. More modest goals are to develop relations among the critical exponents of a system and to find families of systems with equal values of the exponents. A rigorous step toward the first goal lies in the thermodynamic inequalities of Section 3.4. The second step is the homogeneity-scaling theory to be discussed in the next subsection. A step toward the second goal is the conjecture that a model has identical critical exponents on different lattices of the same dimensionality. A second step involves the smoothness and universality conjectures to be discussed in Section 5.2.

Pedagogically, this section should have occurred before the values of the critical exponents were reviewed, to aid the reader in correlating and remembering the data. The reason for placing this section last is to emphasize that most of the results reviewed in Sections 3 and 4 are "the facts" which must take precedence over "conjectural theories." In particular, there is the danger that the conjectural theories may unduly influence theoreticians doing series expansion calculations and experimentalists who do not have many decades of data. To illustrate this kind of danger historically, one need only recall the long period of time during which the classical theory was generally accepted and seemed to provide agreement with experiment. However, it is certainly true that the classical theory was considerably better than no theory; likewise homogeneity-scaling theory and the universality conjecture are considerably better than no ideas at all, provided that it is recognized that they are conjectural theory and not laws.

5.1. Homogeneity-Scaling Theory

Since the classical equations of state do not allow a sufficiently wide range of values of the critical exponents, Widom[152] proposed a generalization equation of state which has one nonobvious restriction. This restriction is that the equation of state be homogeneous in both its variables. Other conjectures leading to similar results have been developed by many authors; the equivalences of the theories are reviewed by Griffiths.[153] The particular form of the homogeneous equation of state used by Griffiths[153] is, in magnetic language,

$$H(M, T) = M|M|^{\delta - 1} h(t|M|^{-1/\beta}) \tag{54}$$

where $t = (T - T_c)/T_c$ and h is an arbitrary function within certain thermodynamic restrictions. This equation of state has the critical exponents δ and β built into it. For example, along the critical isotherm, $t = 0$, so $H \sim M^\delta$. Along the coexistence curve $H = 0$, $M \neq 0$ so $h(t|M|^{-1/\beta}) = 0$ and this implies that $t|M|^{-1/\beta}$ equals a constant, which gives $M \sim t^\beta$. In addition, the assumption of homogeneity, which requires that h is a function only of the product $t|M|^{-1/\beta}$ and not of t and M separately, yields relations among the critical exponents. The easiest relation to obtain involves γ', β, and δ. Taking a derivative with respect to M of Eq. (54) and evaluating on the coexistence curve gives

$$\chi_T^{-1} = (\partial H/\partial M)_t = M|M|^{\delta - 1} t|M|^{-1 - 1/\beta} h'(-x_0) \tag{55}$$

which gives

$$t^{\gamma'} \sim t^\beta t^{\beta(\delta - 1)} t t^{-1 - \beta} = t^{\beta(\delta - 1)} \tag{56}$$

which gives $\gamma' = \beta(\delta - 1)$. Notice that the rigorous thermodynamic inequality is $\gamma' \geq \beta(\delta - 1)$.

Other exponent equalities are

$$\begin{gathered} \alpha = \alpha', \quad \gamma = \gamma', \quad \alpha + 2\beta + \gamma = 2, \\ \gamma = \beta(\delta - 1), \quad \Delta = \beta\delta \end{gathered} \tag{57}$$

The interested reader who wishes to see how these are derived may consult the original papers of Widom[152] and Griffiths[153] or the book by Stanley.[154]

Chapter 1

The theory of scaling was introduced by Kadanoff[155] for d-dimensional Ising models but is also applied to general spin models with short-range interactions. The principal fact used is that near the critical point the correlation length $\xi \sim (T - T_c)^{-\nu}$ becomes large. Thus cells of volume $(\xi/a)^d$ may be defined, where a is the nearest-neighbor distance, within which the spins are well correlated. The only important length is assumed to be the correlation length and the fine structure of the lattice is washed out of the theory, in accord with the conclusions of the calculations in Section 4. Kadanoff further assumes that near the critical point the system may be described as one of interacting *cells* rather than spins and that this new model may be scaled into the original spin model. This scaling assumption leads to a homogeneous equation of state which Griffiths[153] shows is equivalent to the Widom equation of state. In addition, Kadanoff's theory predicts the following equalities involving the correlation functions:

$$d\nu = 2 - \alpha, \qquad \gamma = \nu(2 - \eta), \qquad 2 - \eta = d(\delta - 1)/(\delta + 1) \qquad (58)$$

The predictions of homogeneity-scaling theory are satisfied by the exact results for the classical model, the two-dimensional Ising model, and the spherical model, in short, for all exact model calculations which have ordinary critical points (excluding the dimer and hydrogen-bonded models). In addition, numerous experiments and approximate calculations conform to or are not in blatant disagreement with homogeneity-scaling relations, so that, as conjectures go, this is a very good one.

Nevertheless, it is important that such a general theory be carefully tested on experimental systems for which one can measure the greatest number of decades of data, such as the superfluid transition in helium. Although only a few of the critical exponents of the superfluid transition can be measured, due to the unavailability of the ordering field ζ in the laboratory, Ahlers and co-workers[45,49] tested another scaling relation involving the coefficients A and A' of the specific heat singularities, $C \sim A \ln t$ above T_c or $C \sim A' \ln(-t)$ below T_c when $\alpha = \alpha' = 0$ (log). The theory requires that $A = A'$, whereas Ahlers[45] finds $A/A' > 1.04$. [Widom[166] states that this discrepancy can be removed by using slightly negative values of α' and α.]

It is also important to carefully test the predictions of homogeneity-scaling theory using the longest possible series expansions for three-dimensional models, and this requirement selects the Ising model for the most stringest test. As we pointed out in Section 4.7, there is some disagreement whether $\gamma = \gamma' = 5/4$ or $\gamma' > \gamma$. More important, there seems to be definite disagreement with the predictions $d\nu = 2 - \alpha$ and $2 - \eta = d[(\delta - 1)/(\delta + 1)]$, taking the accepted values $\alpha = 1/8$, $\delta = 5$, $\nu \sim 0.640$, and $\eta \sim 0.04$. Fisher points out in his review[32] that the exponent relations involving the dimensionality d get into trouble in the limit as d goes to infinity; in this limit one should obtain the classical results $\alpha = 0$, $\delta = 3$, $\eta = 0$, and $\nu = 1/2$. A series study by Moore et al.[137] indicates that a slightly weaker form of scaling of the correlations holds for the three-dimensional Ising model. This has led to the development of a modified theory called weak scaling theory which is characterized by a second important length in addition to the correlation length ξ. The interested reader is invited to consult the papers of Stell[156,157] and Theumann.[158]

5.2. Smoothness and Universality

The smoothness and universality conjectures state that the values of the critical exponents for two systems will be identical under appropriate perturbations in the model. It is frequently supposed that a perturbation which enters linearly into the Hamiltonian and which does not change the symmetry group of the order parameter in the ground state will not change the critical exponents. For example, consider the anisotropic Heisenberg model introduced in Section 2.2. As J_{xy}/J_z increases from zero (Ising limit) the ground state remains Ising-like with a symmetry group of two elements. According to the smoothness and universality conjectures, the values of the critical exponents will not change until $J_{xy}/J_z = 1$. At $J_{xy}/J_z = 1$ (Heisenberg limit) the spins in the ground state can point in any direction in three dimensions and the values of the exponents may change discontinuously from the values for $J_{xy}/J_z < 1$. This particular case was investigated for the classical Heisenberg model by Jasnow and Wortis,[159] who found consistency with the smoothness and universality conjectures. For $J_{xy}/J_z > 1$ the ground state becomes xy-like and the exponents again are conjectured to remain constant.

The smoothness conjecture was so named by Griffiths,[160] who considered a number of examples, including the important one that the values of the exponents do not depend upon the magnitude of the spin. Although there are indications from the series expansions that this may not be true for the Heisenberg model (see Section 4.7), for the Ising model Ferer et al.[161] and Fox and Gaunt[162] find, using series expansions, that smoothness appears to hold. However, for Ising models with long-range $r^{-d-\sigma}$ interactions Griffiths[160] shows that the smoothness hypothesis predicts that for $\sigma > 1$ the values of the exponents will be the same as those for the nearest-neighbor model. This conjecture is in direct contradiction to the calculations of Fisher et al.[124] and Suzuki[125] and a conjecture of Nagle and Bonner.[123]

The idea that the behavior of a model should depend on the symmetry properties of the Hamiltonian seems unarguable. But it is a much stronger statement to say that the critical behavior should depend *only* on the ground-state symmetry of the Hamiltonian, especially when the ground state has a negligible influence on the partition function at the critical temperature. A good counterexample is provided by a linear chain classical model

$$\mathcal{H} = -J_{SR} \sum_{i=1}^{N} \mu_i \mu_{i+1} - (J_{LR}/N) \sum_{i,j} \mu_i \mu_j \qquad (59)$$

This model has a classical critical point for both J_{SR} and J_{LR} positive and this continues to be the case for $J_{SR} < 0$ as long as $-J_{SR}/J_{LR} < R = 0.3171$. Notice that when $-J_{SR}/J_{LR} = R$ there is no change in the character of the ground state. Nevertheless, the exact solution[163] shows that β becomes one-fourth instead of one-half, δ becomes five instead of three, and α' becomes one-half instead of zero (finite). For $-J_{SR}/J_{LR} > R$ two classical critical points appear in a field and the point R is therefore a tricritical point at which three critical lines meet.[19] This is a rather drastic change in the critical point and the nature of the phase transition at the point R where the character of the ground state remains unchanged and ferromagnetic. A weaker statement of the smoothness conjecture which does not violate this result is to predict that the values of the critical exponents should not change as a linear parameter of the Hamiltonian is varied unless the character or nature of the underlying first-order phase transition changes *and/or* the symmetry of the ground state changes. This weaker statement is the main thrust of Griffith's smoothness conjecture; the use of the ground-state properties is an independent conjecture used by Griffiths as a diagnostic tool for predicting where *some* of the changes in critical exponents may occur. The main thrust of this critique is that the predictive powers of the smoothness conjecture are not as great as advertised.

One of the exciting aspects of Baxter's[107] solution of the eight-vertex model is that the values of the critical exponents vary continuously in apparent contradiction to the smoothness and universality conjectures. However, Kadanoff and Wegner[164] have argued that the eight-vertex model has the special property of having a term in the Hamiltonian which scales as r^{-d} and this allows the values of the critical exponents to vary, at least in Kadanoff's version of the universality conjecture,[165] which uses operator algebras and scaling to provide a physical derivation.

Acknowledgment

I wish to thank Drs. J. C. Bonner, S. R. Salinas and many other colleagues for useful discussions and for reading the manuscript.

References

1. J. W. Cahn, On spinoidal decomposition, *Acta Met.* **9**, 795–801 (1961).
2. T. J. Tiedema, J. Bouman, and W. G. Burgers, Precipitation in Au–Pt alloys, *Acta Met.* **5**, 310–321 (1957).
3. F. J. Wegner, Duality in generalized Ising models and phase transitions without local order parameters, *J. Math. Phys.* **12**, 2259–2272 (1971).
4. P. W. Kasteleijn, Dimer statistics and phase transitions, *J. Math. Phys.* **4**, 287–293 (1963).
5. J. F. Nagle and G. R. Allen, Models of the order–disorder transition in $NaH_3(SeO_3)_2$, *J. Chem. Phys.* **55**, 2708–2714 (1971).
6. V. G. Vaks, A. I. Larkin, and Yu. N. Ovchinnikov, Ising model with interaction between non-nearest neighbors, *Soviet Phys.—JETP* **22**, 820–826 (1966).
7. G. T. Rado and H. Suhl (eds.), *Magnetism*, Academic, New York (1963).
8. D. P. Landau, B. E. Keen, B. Schneider, and W. P. Wolf, Magnetic and thermal properties of DAG I, *Phys. Rev.* **B3**, 2310–2343 (1971).
9. W. P. Wolf, M. J. M. Leask, B. Mangum, and A. F. G. Wyatt, Ferromagnetism in gadolinium trichloride, *J. Phys. Soc. (Japan)* **17** (Suppl. B-1), 487–492 (1961).
10. W. Marshall and R. D. Lowde, Magnetic correlations and neutron scattering, *Rep. Prog. Phys.* **31**, 705–775 (1968).
11. D. C. Mattis, *The Theory of Magnetism*, Harper and Row, New York (1965).
12. P. W. Anderson, in *Magnetism* (G. T. Rado and H. Suhl, eds.), Vol. I, pp. 25–81, Academic, New York (1963).
13. K. W. H. Stevens, Spin Hamiltonians, in *Magnetism* (G. T. Rado and H. Suhl, eds.), Vol. I, pp. 1–23, Academic, New York (1963).
14. W. P. Wolf, Anisotropic interactions between magnetic ions, *J. de Physique* **32**, Cl-26–Cl-33 (1971).

Chapter 1

15. H. A. Brown and J. M. Luttinger, Ferromagnetic and antiferromagnetic Curie temperatures, *Phys. Rev.* **100**, 685–692 (1955).
16. T. D. Lee and C. N. Yang, Statistical theory of equations of state and phase transition II, Lattice gas and Ising model, *Phys. Rev.* **87**, 410–419 (1952).
17. G. Stell, H. Narang, and C. K. Hall, Simple lattice gases with realistic phase changes, *Phys. Rev. Letters* **28**, 292–294 (1972).
18. C. K. Hall and G. Stell, Phase transitions in two-dimensional lattice gases of hard core molecules with weak long-range attractions, *Phys. Rev.* **A7**, 1679–1689 (1973).
19. R. B. Griffiths, A proposal for notation at tricritical points, *Phys. Rev.* **B7**, 545–551 (1973).
20. M. E. Fisher, Lattice statistics in a magnetic field I. A two-dimensional superexchange antiferromagnet, *Proc. Roy. Soc.* **A254**, 66–85 (1960).
21. J. H. Schelleng and S. A. Friedberg, Thermal behavior of the antiferromagnet $MnBr_2 4H_2O$ in applied magnetic field, *Phys. Rev.* **185**, 728–734 (1969).
22. F. Jona and G. Shirane, *Ferroelectric Crystals*, Pergamon Press (1962).
23. Y. Takagi, Theory of the transition in KH_2PO_4 II, *J. Phys. Soc. Japan* **3**, 271–272 (1948).
24. R. Blinc and S. Svetina, Cluster approximation for order–disorder-type hydrogen bonded ferroelectrics II. Application to KDP, *Phys. Rev.* **147**, 430–438 (1966).
25. W. Reese, Studies of phase transitions in order–disorder ferroelectrics. III. The phase transition in KH_2PO_4 and a comparison with KD_2PO_4, *Phys. Rev.* **181**, 905–919 (1969).
26. L. W. Garland and R. Renard, First order phase transitions in Ising models due to lattice interactions, *J. Chem. Phys.* **44**, 1120–1129 (1965).
27. G. A. Baker and J. W. Essam, Lattice compressibility and critical behavior, *J. Chem. Phys.* **55**, 861–879 (1971).
28. B. D. Josephson, Relation between the superfluid density and order parameter for superfluid He near T_c, *Phys. Letters* **21**, 608–609 (1966).
29. O. Penrose and L. Onsager, Bose–Einstein condensation and liquid helium, *Phys. Rev.* **104**, 576–584 (1956).
30. M. D. Girandeau, Off-diagonal long range order and generalized Bose condensation, *J. Math. Phys.* **6**, 1083–1098 (1965).
31. L. P. Kadanoff *et al.*, Static phenomena near critical points, *Rev. Mod. Phys.* **39**, 395–431 (1971).
32. M. E. Fisher, The theory of equilibrium critical phenomena, *Rep. Progr. Phys.* **XXX** (Part II), 615–730 (1967).
33. C. Kittel, *Quantum Theory of Solids*, p. 157, Wiley, New York (1963).
34. T. W. Stinson and J. D. Litster, Pretransitional phenomena in the isotropic phase of a nematic liquid crystal, *Phys. Rev. Letters* **25**, 503–506 (1970).
35. B. Chu, C. S. Bak, and F. L. Lin, Coherence length in the isotropic phase of a room-room-temperature nematic liquid crystal, *Phys. Rev. Letters* **28**, 1111–1114 (1972).
36. J. R. McColl and C. S. Shih, Temperature dependence of orientational order in a nematic liquid crystal at constant molar volume, *Phys. Rev. Letters* **29**, 85–87 (1972).
37. M. E. Fisher, Correlation functions and the critical region of simple fluids, *J. Math. Phys.* **5**, 944–962 (1964).

38. M. E. Fisher, *The Nature of Critical Points*, Lectures in Theoretical Physics, Vol. VIIC, pp. 1–159 University of Colorado Press (1965).
39. P. Heller, Experimental investigations of critical phenomena, *Rep. Prog. Phys.* **XXX** (Part II), 731–826 (1967).
40. B. M. McCoy, Theory of a two-dimensional Ising model with random impurities III, *Phys. Rev.* **188**, 1014–1031 (1969).
41. B. M. McCoy, Incompleteness of critical exponent descriptions for ferromagnetic systems containing random impurities, *Phys. Rev. Letters* **23**, 383–386 (1969).
42. R. B. Griffiths and J. C. Wheeler, Critical points in multicomponent systems, *Phys. Rev.* **A2**, 1047–1064 (1970).
43. J. F. Nagle and J. C. Bonner, Ising chain with interactions in a staggered field, *J. Chem. Phys.* **54**, 729–734 (1971).
44. W. M. Fairbank and C. F. Kellers, the lambda transition in liquid helium, in *Critical Phenomena* (Nat. Bur. St. Misc. Publ. No 273) (M. S. Green and J. V. Sengers, (eds.), pp. 71–78, National Bureau of Standards, Washington, D.C. (1966).
45. G. Ahlers, Heat capacity near the superfluid transition in He^4 at saturated vapor pressure, *Phys. Rev.* **A3**, 696–716 (1971).
46. L. Kreps and S. A. Friedberg, Specific heat measurements of $MnBr_2 4H_2O$, to be published.
47. J. R. Clow and J. D. Reppy, Persistent-current measurements of the superfluid density and critical velocity, *Phys. Rev.* **A5**, 424–438 (1972).
48. J. A. Tyson, Superfluid density of He II in the critical region, *Phys. Rev.* **166**, 166–176 (1968).
49. D. S. Greywell and G. Ahlers, Second-sound velocity, scaling and universality in He II under pressure near the superfluid transition, *Phys. Rev. Letters* **28**, 1251–1254 (1972).
50. G. R. Brown and H. Meyer, Study of the specific heat singularity of He^3 near its critical point, *Phys. Rev.* **A6**, 364–377 (1972).
51. B. Wallace and H. Meyer, Critical isotherm of He^3, *Phys. Rev.* **A2**, 1610–1612 (1970).
52. B. Wallace and H. Meyer, Equation of state of He^3 close to the critical point, *Phys. Rev.* **A2**, 1563–1575 (1970).
53. M. R. Moldover, Scaling of the specific-heat singularity of He^4 near its critical point, *Phys. Rev.* **182**, 342–351 (1969).
54. P. R. Roach, Pressure–density–temperature surface of He^4 near the critical point, *Phys. Rev.* **170**, 213–223 (1968).
55. B. Chu and J. S. Lin, Small angle scattering from CO_2 in the vicinity of its critical point, *J. Chem. Phys.* **53**, 4454 (1970).
56. J. Lipa, C. Edwards, and M. Buckingham, Precision measurement of the specific heat of CO_2 near the critical point, *Phys. Rev. Letters* **25**, 1086–1090 (1970).
57. J. M. H. Levelt Sengers and W. T. Chen, The vapor pressure, critical isochore and some metastable states of CO_2, *J. Chem. Phys.* **56**, 595–608 (1972).
58. C. Edwards, J. A. Lipa, and M. J. Buckingham, Specific heat of xenon near the critical point, *Phys. Rev. Letters* **20**, 496–499 (1968).
59. I. W. Smith, M. Giglio, and G. B. Benedek, Correlation range and compressability of xenon near the critical point, *Phys. Rev. Letters* **27**, 1556–1560 (1971).

Chapter 1

60. M. Vicentini-Missoni, J. M. H. Levelt Sengers, and M. S. Green, Thermodynamic anomalies of CO_2, Xe, and He^4 in the Critical Region, *Phys. Rev. Letters* **22**, 389–393 (1969).
61. J. Als-Nielsen, Investigation of scaling laws by critical neutron scattering from beta-brass, *Phys. Rev.* **185**, 664–666 (1969).
62. D. R. Chipman and C. B. Walker, Long range order in β-brass, *Phys. Rev.* **B5**, 3823–3831 (1972).
63. G. Longworth, Temperature dependence of the ^{57}Fe hfs in the ordered alloys $FePd_3$ and FePd near the Curie temperature, *Phys. Rev.* **172**, 572–576 (1968).
64. L. Guttmans, H. C. Schnyders, and G. J. Arai, Variation of long-range order in Fe_3Al near T_c, *Phys. Rev. Letters* **22**, 517–519 (1969).
65. L. Guttmans and H. C. Snyders, Critical scattering of X-rays from Fe_3Al, *Phys. Rev. Letters* **22**, 520–522 (1969).
66. C. W. Garland and R. A. Young, Order–disorder phenomena. VI Anomalous changes in the volume of ammonium chloride, *J. Chem. Phys.* **48**, 146–148 (1968).
67. H. C. Benski, R. C. Reno, C. Hohenemser, and C. Abeledo, New Mössbauer effect on ^{57}Fe in a Ni host: the critical exponent for Ni, *Phys. Rev.* **B6**, 4266–4275 (1972).
68. J. E. Noakes and A. Arrott, Magnetization of nickel near its critical temperature, *J. Appl. Phys.* **39**, 1235–1236 (1968).
69. J. S. Kouvel and J. B. Comly, Magnetic equation of state for nickel near its Curie point, *Phys. Rev. Letters* **20**, 1237–1239 (1968).
70. D. L. Connelly, J. S. Loomis, and D. E. Mapother, Specific heat of nickel near T_c, *Phys. Rev.* **B3**, 924–934 (1971).
71. M. F. Collins, V. J. Minkiewicz, R. Nathans, L. Passell, and G. Shirane, Critical and spin-wave scattering of neutrons from iron, *Phys. Rev.* **179**, 417–430 (1969).
72. J. T. Ho and J. D. Litster, Faraday rotation near the ferromagnetic critical temperature of $CrBr_3$, *Phys. Rev.* **B2**, 4523–4532 (1970).
73. M. P. Schulhof, R. Nathans, P. Heller, and A. Linz, Inelastic neutron scattering from MnF_2 in the critical region, *Phys. Rev.* **B4**, 2254–2276 (1971).
74. L. M. Corliss, A. Delapalme, J. M. Hastings, H. Y. Lau, and R. Nathans, Critical magnetic scattering from $RbMnF_3$, *J. Appl. Phys.* **40**, 1278 (1969).
75. W. B. Yelon and D. E. Cox, Magnetic ordering in $RbNiCl_3$, *Phys. Rev.* **B6**, 204–208 (1972).
76. M. T. Hutchings, M. P. Schulhof, and H. J. Guggenheim, Critical magnetic neutron scattering from FeF_2, *Phys. Rev.* **B5**, 154–168 (1972).
77. K. Deguchi and E. Nakamura, Critical region in ferroelectric TGS, *Phys. Rev.* **B5**, 1072–1073 (1972).
78. R. Blinc, M. Burger, and A. Levstik, Critical behavior of ferroelectric TGS and deuterated TGS, *Solid State Commun.* **8**, 317–321 (1970).
79. Th. O. Waldkirch, K. A. Müller, W. Berlinger, and H. Thomas, Fluctuations and correlations in $SrTiO_3$ for $T \geq T_c$, *Phys. Rev. Letters* **28**, 503–506 (1972).
80. G. S. Rushbrooke, On the thermodynamics of the critical region for the Ising problem, *J. Chem. Phys.* **39**, 842–843 (1963).
81. R. B. Griffiths, Rigorous results and theorems, in *Phase Transitions and Critical Phenomena* (C. Domb and M. S. Green, eds.), Vol. I, pp. 7–109, Academic, London (1972).

82. L. Onsager, Crystal statistics I. A two-dimensional model with an order–disorder transition, *Phys. Rev.* **65**, 117–149 (1944).
83. G. A. Baker, Certain general order–disorder models in the limit of long range interactions, *Phys. Rev.* **126**, 2071–2078 (1962).
84. M. Kac, G. E. Uhlenbeck, and P. C. Hemmer, On the van der Waals theory of the vapor–liquid equilibrium, *J. Math. Phys.* **4**, 216–228 (1963).
85. J. Lebowitz and O. Penrose, Rigorous treatment of the van der Waals–Maxwell theory of the liquid–vapor transition, *J. Math. Phys.* **7**, 98–113 (1966).
86. L. D. Landau and E. M. Lifschitz, *Statistical Physics*, p. 135, Pergamon, Oxford (1958).
87. M. Blume, V. J. Emery, and R. B. Griffiths, Ising model for the transition and phase separation in He^3–He^4 mixtures, *Phys. Rev.* **A4**, 1071–1077 (1971).
88. R. B. Griffiths, Thermodynamics near the two fluid critical mixing point in He^3–He^4, *Phys. Rev. Letters* **24**, 715–717 (1970).
89. W. K. Theumann and J. S. Høye, Ising chain with several phase transitions, *J. Chem. Phys.* **55**, 4159–4166 (1971).
90. P. C. Hemmer and G. Stell, Phase transitions due to softness of the potential core, *Phys. Rev. Letters* **24**, 1284–1287 (1970).
91. G. Stell, H. Narang, and C. K. Hall, Simple lattice gases with realistic phase changes, *Phys. Rev. Letters* **28**, 292–294 (1972).
92. J. Bernasconi and F. Rys, Critical behavior of a magnetic alloy, *Phys. Rev.* **B4**, 3045–3048 (1971).
93. J. Stephenson, Ising model spin correlations on the triangular lattice IV, *J. Math. Phys.* **11**, 420–431 (1970).
94. R. J. Birgeneau, J. Skalyo, and G. Shirane, Critical magnetic scattering in K_2NiF_4, *Phys. Rev.* **B3**, 1736–1749 (1971).
95. B. Kaufman and L. Onsager, Crystal statistics III, *Phys. Rev.* **76**, 1244–1252 (1949).
96. T. Schultz, D. C. Mattis, and E. H. Lieb, Two-dimensional Ising model as a soluble problem of many fermions, *Rev. Mod. Phys.* **36**, 856–871 (1964).
97. M. J. Stephen and L. Mittag, A new representation of the solution of the Ising model, *J. Math. Phys.* **13**, 1944–1951 (1973).
98. M. Kac and J. C. Ward, A combinational solution of the two-dimensional Ising model, *Phys. Rev.* **88**, 1332–1337 (1952).
99. E. Montroll, in *Applied Combinatorial Mathematics* (E. F. Beckenbach, ed.), pp. 96–143, Wiley, New York (1964).
100. B. Widom, Plait points in two- and three-component liquid mixtures, *J. Chem. Phys.* **46**, 3324–3333 (1967).
101. R. K. Clark and G. A. Neece, Decorated lattice model for ternary systems, *J. Chem. Phys.* **48**, 2575–2582 (1968).
102. J. C. Wheeler, Behavior of a solute near the critical point of an almost pure solvent, *Ber. Bunsenges. Physikal. Chem.* **76**, 308–318 (1972).
103. M. E. Fisher, Renormalization of critical exponents by hidden variables, *Phys. Rev.* **176**, 257–272 (1968).
104. O. J. Heilmann and E. H. Lieb, Theory of monomer–dimer systems, *Commun. Math. Phys.* **25**, 190–232 (1972).
105. C. N. Yang and C. P. Yang, One-dimensional chain of anisotropic spin–spin interactions II, *Phys. Rev.* **150**, 327–339 (1966).

Chapter 1

106. E. H. Lieb, Exact solution of the problem of the entropy of two-dimensional ice, *Phys. Rev. Letters* **18**, 692–694 (1967).
107. R. J. Bacter, Partition function of the eight-vertex lattice model, *Ann. Phys.* **70**, 193–228 (1972).
108. F. Y. Wu, Exact solution of a model of an antiferroelectric transition, *Phys. Rev.* **183**, 604–607 (1969).
109. E. H. Lieb and F. Y. Wu, Two-dimensional ferroelectric models, in *Phase Transitions and Critical Phenomena* (C. Domb and M. S. Green, eds.), Vol. I, 332–487, Academic, London (1972).
110. J. F. Nagle, Proof of the first order phase transition in the Slater KDP model, *Commun. Math. Phys.* **13**, 62–67 (1967).
111. J. F. Nagle, The one dimensional KDP model in statistical mechanics, *Am. J. Phys.* **36**, 1114–1117 (1968).
112. J. W. Benepe and W. Reese, Electronic studies of KH_2PO_4, *Phys. Rev.* **B3**, 3032–3039 (1971).
113. J. F. Nagle, Lattice statistics of hydrogen bonded crystals II, *J. Math. Phys.* **7**, 1492–1496 (1966).
114. J. F. Nagle, Study of the F-model using low-temperature series, *J. Chem. Phys.* **50**, 2813–2818 (1969).
115. R. J. Baxter, Spontaneous staggered polarization of the F-model, *J. Phys.* **C6**, L94 (1973).
116. R. J. Baxter, Exact isotherm for the F model in direct and staggered electric fields. *Phys. Rev.* **B1**, 2199–2202 (1970).
117. J. D. Johnson, S. Krinsky, and B. M. McCoy, Critical index of the vertical arrow correlation length in the eight-vertex model, *Phys. Rev. Letters* **29**, 492–494 (1972).
118. M. E. Fisher and J. Stephenson, Statistical mechanics of dimers on a plane lattice II, *Phys. Rev.* **132**, 1411–1431 (1963).
119. B. Sutherland, Two dimensional hydrogen bonded crystals without the ice rule, *J. Math. Phys.* **11**, 3183–3186 (1970).
120. T. H. Berlin and M. Kac, The spherical model of a ferromagnet, *Phys. Rev.* **86**, 821–835 (1952).
121. H. E. Stanley, Spherical model as the limit of infinite spin dimensionality, *Phys. Rev.* **176**, 718–722 (1968).
122. G. S. Joyce, Spherical model with long-range ferromagnetic interactions, *Phys. Rev.* **146**, 349–358 (1966).
123. J. F. Nagle and J. C. Bonner, Numerical studies of the Ising chain with long-range ferromagnetic interactions, *J. Phys.* **C3**, 352–366 (1970).
124. M. E. Fisher, S. Ma, and B. G. Nickel, Critical exponents for long-range interactions, *Phys. Rev. Letters* **29**, 917–920 (1972).
125. M. Suzuki, Critical exponents for long-range interactions I—Dimensionality, symmetry, and potential range, *Progr. Theor. Phys.* **49**, 424–441 (1973).
126. J. C. Bonner and J. F. Nagle, Phase behavior of models with competing interactions, *J. Appl. Phys.* **42**, 1280–1282 (1971).
127. L. K. Runnels, J. P. Salvant, and H. R. Streiffer, Exact finite method of lattice statistics IV, *J. Chem. Phys.* **52**, 2352–2358 (1970).
128. C. Domb and M. S. Green (eds.), *Phase Transitions and Critical Phenomena*, Vol. 3, Academic, London (1973).

129. C. Domb, On the theory of cooperative phenomena in crystals, *Advan. Phys.* **9**, 149–361 (1960).
130. C. Domb and M. F. Sykes, Use of series expansions for the Ising model susceptibility and excluded volume problem, *J. Math. Phys.* **2**, 63–67 (1961).
131. G. A. Baker and D. L. Hunter, Methods of series analysis II, *Phys. Rev.* **7**, 3377–3392 (1973).
132. D. S. Gaunt, M. E. Fisher, M. F. Sykes, and J. W. Essam, Critical isotherms of a ferromagnet and of a fluid, *Phys. Rev. Letters* **13**, 713–715 (1964).
133. D. S. Gaunt and M. F. Sykes, Reanalysis of the critical isotherm of the Ising ferromagnet, *J. Phys. C: Solid State Phys.* **5**, 1429–1444 (1972).
134. D. S. Gaunt and C. Domb, The specific heat of the 3-d Ising model below T_c, *J. Phys. C* **1**, 1038–1045 (1968).
135. A. J. Guttman, C. J. Thompson, and B. W. Ninham, Determination of critical behavior from series expansions in lattice statistics IV, *J. Phys. C* **3**, 1641–1651 (1970).
136. M. E. Fisher and R. J. Burford, Theory of critical-point scattering and correlations I, *Phys. Rev.* **156**, 583–622 (1967).
137. M. A. Moore, D. Jasnow, and M. Wortis, Spin–spin correlation function of the 3-d Ising ferromagnet above the Curie temperature, *Phys. Rev. Letters* **22**, 940–943 (1967).
138. G. A. Baker, H. E. Gilbert, J. Eve, and G. S. Rushbrooke, High temperature expansions for the spin-1/2 Heisenberg model, *Phys. Rev.* **164**, 800–817 (1967).
139. R. G. Bowers and M. E. Woolf, Some critical properties of the Heisenberg model, *Phys. Rev.* **177**, 917–932 (1969).
140. M. H. Lee and H. E. Stanley, Spin-1/2 Heisenberg ferromagnet on cubic lattices, *Phys. Rev.* **B4**, 1613–1630 (1971).
141. G. A. Baker, J. Eve, and G. S. Rushbrooke, Magnetic phase boundary of the spin-1/2 Heisenberg ferromagnetic model, *Phys. Rev.* **B2**, 706–721 (1970).
142. M. Ferer, M. A. Moore, and M. Wortis, Some critical properties of the nearest neighbor, classical Heisenberg model for the fcc lattice in finite field for temperatures greater than T_c, *Phys. Rev.* **B4**, 3954–3963 (1971).
143. D. S. Ritchie and M. E. Fisher, Theory of critical point scattering and correlations II, *Phys. Rev.* **B5**, 2668–2692 (1972).
144. G. S. Rushbrooke and P. J. Wood, On Curie points and high temperature susceptibilities of Heisenberg model ferromagnets, *Mol. Phys.* **1**, 257–283 (1958).
145. H. E. Stanley and T. A. Kaplan, Possibility of a phase transition for the 2-d Heisenberg model, *Phys. Rev. Letters* **17**, 913–915 (1966).
146. N. D. Mermin and H. Wagner, Absence of ferromagnetism or antiferromagnetism in one- or two-dimensional isotropic Heisenberg models, *Phys. Rev. Letters* **17**, 1133–1136 (1966).
147. M. E. Fisher and B. U. Felderhof, Phase transitions in one-dimensional cluster-interaction fluids, *Ann. Phys.* **58**, 176–300 (1970).
148. K. G. Wilson, Renormalization group and critical phenomena II, *Phys. Rev.* **B4**, 3184–3205 (1971).
149. G. A. Baker, Ising model with a scaling interaction, *Phys. Rev.* **B5**, 2622–2633 (1972).
150. K. G. Wilson, Feynman-graph expansion for critical exponents, *Phys. Rev. Letters* **28**, 548–551 (1972).

Chapter 1

151. M. Suzuki, Critical exponents for long-range interactions, *Progr. Theor. Phys.* **49**, 1440–1450 (1973).
152. B. Widom, Equation of state in the neighborhood of the critical point, *J. Chem. Phys.* **43**, 3898–3905 (1965).
153. R. B. Griffiths, Thermodynamic functions for fluids and ferromagnets near the critical point, *Phys. Rev.* **158**, 176–187 (1967).
154. H. E. Stanley, *Introduction to Phase Transitions and Critical Phenomena*, Oxford Univ. Press, New York (1971).
155. L. P. Kadanoff, Scaling laws for Ising models near T_c, *Physics* **2**, 263–272 (1966).
156. G. Stell, Some implications of weak-scaling theory, *Phys. Rev.* **B2**, 2811–2813 (1970).
157. G. Stell, Scaling theory of the critical region for system with long-range forces, *Phys. Rev.* **B5**, 981–985 (1972).
158. W. Theumann, Phenomenological weak scaling theory, *Phys. Rev.* **B6**, 281–286 (1972).
159. D. Jasnow and M. Wortis, High-temperature critical indices for the classical anisotropic Heisenberg model, *Phys. Rev.* **176**, 739–750 (1968).
160. R. B. Griffiths, Dependence of critical indices on a parameter, *Phys. Rev. Letters* **24**, 1479–1482 (1970); Critical points dependent on parameters, in *Critical Phenomena in Alloys, Magnets and Superconductors* (R. E. Mills, E. Ascher, and R. I. Jaffee, eds.), pp. 377–391. McGraw-Hill, New York (1971).
161. M. Ferer, M. A. Moore, and M. Wortis, Universality of critical correlations in the three-dimensional Ising ferromagnet, *Phys. Rev.* **B3**, 3911–3914 (1971).
162. P. F. Fox and D. S. Gaunt, Critical isotherm of the Ising ferromagnet with spin 1/2, *J. Phys. C; Solid State Phys.* **5**, 3085–3096 (1972).
163. J. F. Nagle, Ising chain with competing interactions, *Phys. Rev.* **A2**, 2124–2128 (1970).
164. L. P. Kadanoff and F. J. Wegner, Some critical properties of the 8-vertex model, *Phys. Rev.* **B4**, 3989–3993 (1971).
165. L. P. Kadanoff, in *Proc. Enrico Fermi Summer School of Physics, Varenna, 1970* (M. S. Green, ed.), Academic, New York (1972).
166. B. Widom, Private communication.
167. J. F. Nagle, Lipid bilayer phase transitions, *Proc. Natl. Acad. Sci.*, to be published.

2

Solid-State Phase Transformations

V. Raghavan
Indian Institute of Technology Delhi
New Delhi, India

and

Morris Cohen
Massachusetts Institute of Technology
Cambridge, Massachusetts

1. Introduction

Recent reviews, conference reports, and symposia[1-11] concerning phase transformations provide a good measure of the research activity in this area of solid-state science. In the present chapter, we have chosen to emphasize basic concepts in the field, but wherever appropriate, experimental data are included. It is hoped that this chapter will provide a much needed link to the literature on solid-state chemistry and to more detailed references on specific phase transformations.[1-11]

Phase transformations can be broadly classified into two categories. A transformation that takes place more or less simultaneously in all parts of an assembly is regarded as a *homogeneous* transformation. Reactions in the gaseous phase are generally homogeneous. Some liquid and solid-state transformations are also homogeneous. Transitions comprising atomic rotations, electronic rearrangements, etc. can be homogeneous, taking place uniformly throughout the system. An interesting case of a homogeneous transformation in the solid-state is the spinodal decomposition to be discussed in Section 13 of this chapter. In that case, the transformation starts as a small compositional fluctuation spread over a large volume

Chapter 2

of the material. Initially, there is no sharp boundary between the parent and product phases. The compositional fluctuation grows in intensity with time to yield, finally, the equilibrium phases.

The second category of phase transformations is *heterogeneous* and of the nucleation-and-growth type. Tiny volumes of the product phase or nuclei, often assumed to be the same in structure and composition as the transformation product, form first. A sharp boundary delineates the nuclei from the surrounding matrix. These small regions subsequently grow by the outward movement of the boundary, with corresponding changes in composition and crystal structure behind the advancing front. The present coverage is largely devoted to transformations of the latter type. For more details on the first category of transformations, the reader is referred to Chapters 1 and 2 of this volume.

2. Thermodynamic Considerations

2.1. The Driving Force for a Phase Change

A phase transformation that occurs spontaneously is accompanied by a decrease in the total free energy of the system. If g_β is the Gibbs free energy per unit volume of the product phase β and g_α is that of the parent phase α, then during a spontaneous $\alpha \to \beta$ transformation at constant temperature and pressure

$$g_\beta - g_\alpha = \Delta g \leq 0 \qquad (1)$$

The free energies of phases such as α and β are functions of their composition and the external parameters, e.g., temperature and pressure. Data on the free energies of solid phases are often not complete or reliable, but in recent years, such thermodynamic properties have been computed for a number of phases and systems, mainly through a judicious combination of experimental information and theoretical analysis. This aspect of solid-state thermodynamics is covered in Chapter 4 by Kaufman and Nesor.

At the equilibrium temperature T_0 and pressure P_0, $\Delta g = 0$, i.e., the free energies of the parent and product phases are equal. For a transformation that occurs on cooling, an appreciable degree of supercooling below the equilibrium temperature may be necessary to initiate the transformation. The negative of the chemical free-energy change per unit volume at the transformation temperature ($-\Delta g$) is regarded as the *driving force* for the phase change.

The factors that retard the transformation are usually nonchemical in nature, such as the creation of an interface between the parent and product phases and the elastic strains resulting from the transformation. We can then talk of a *net* free-energy change ΔG during the transformation, which takes into account the nonchemical factors. The negative of the derivative of ΔG with respect to some size parameter of the product phase can be called the *transformational force*, and is to be distinguished from the *driving force* as defined above.

2.2. Thermodynamic Order of Phase Transformations

Although the Gibbs free energy per g-atom $\tilde{G}$ of a system changes *continuously* through a transition at the equilibrium temperature and pressure, the first derivatives of $\tilde{G}$ with respect to temperature and

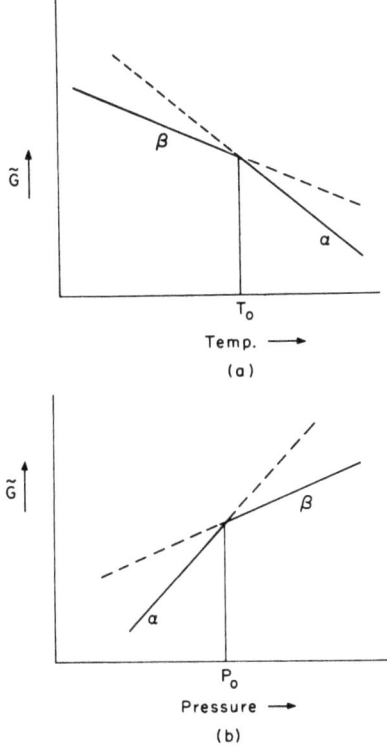

Fig. 1. Schematic variation of the Gibbs free energy $\tilde{G}$ through a first-order transformation (a) at temperature T_0 and (b) at pressure P_0.

Chapter 2

pressure often have discontinuities at the equilibrium temperature and pressure, as shown in Figure 1. From the basic thermodynamic laws, we can write

$$(\partial \tilde{G}/\partial T)_P = -\tilde{S} \qquad (2)$$

and

$$(\partial \tilde{G}/\partial P)_T = \tilde{V} \qquad (3)$$

where $\tilde{S}$ and $\tilde{V}$ are the entropy and volume per g-atom. Transformations that have a volume change and an entropy change associated with them are called first-order transformations.

A second-order transformation is one in which the free energy and its first derivatives change continuously through the equilibrium transition temperature T_c and pressure P_c but the second derivatives still have discontinuities. Through the relationships

$$(\partial^2 \tilde{G}/\partial T^2) = (-\partial \tilde{S}/\partial T)_P \qquad (4)$$

$$(\partial^2 \tilde{G}/\partial P^2) = (\partial \tilde{V}/\partial P)_T \qquad (5)$$

and

$$(\partial^2 \tilde{G}/\partial T\, \partial P) = (\partial \tilde{V}/\partial T)_P \qquad (6)$$

discontinuities in the second derivatives represent abrupt changes in specific heat, compressibility, and the volume coefficient of thermal

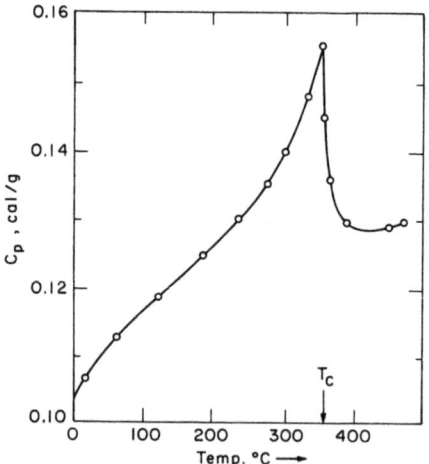

Fig. 2. Variation in the specific heat C_p of nickel with temperature through the second-order magnetic transition at the critical temperature T_c.

expansion. Figure 2 illustrates the corresponding changes in specific heat as nickel undergoes a second-order magnetic transition at the critical temperature T_c (358°C).

2.3. First-Order Transformations without Compositional Change

In one-component systems, there is obviously no change in composition during a phase transformation. In multicomponent systems, some transformations may also occur without compositional change.

Discussing solid-state transformations only, we note that close-packed crystal structures tend to be stable at lower temperatures, and a transformation to a more open structure often takes place on raising the temperature. Typically, the close-packed structure has stronger bonds and a smaller residual enthalpy $\tilde{H}_0$ or Gibbs free energy $\tilde{G}_0$ at 0°K than the open structure. On raising the temperature, however, the entropy term becomes increasingly important. Open crystal structures usually provide opportunity for larger amplitudes in the thermal vibrations of atoms and therefore have larger specific heats $\tilde{C}_p$, the basic free-energy equations being:

$$\tilde{G} = \tilde{H} - T\tilde{S} \tag{7a}$$

and

$$\tilde{G} = \tilde{H}_0 - \int_0^T \left(\int_0^T \tilde{C}_p \, dT/T \right) dT \tag{7b}$$

Equation (7b) shows that the free energy decreases more rapidly with increasing temperature, the larger the specific heat. In Figure 3 the

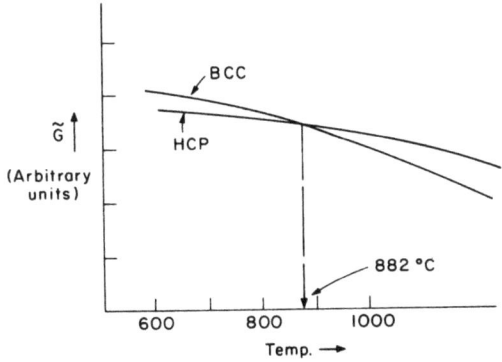

Fig. 3. Free energy vs. temperature relationship for titanium showing the hcp ⇌ bcc transition at 882°C.

Chapter 2

free energies of close-packed hcp titanium and more open bcc titanium are plotted schematically as a function of temperature. The hcp ⇌ bcc phase equilibrium occurs at 882°C. Similarly, CsCl, which has the 8:8 coordinated structure at low temperatures, transforms to the 6:6 coordinated NaCl structure on heating through 445°C. Quartz, the densest form of crystalline silica, is the stable phase at low temperatures.

The application of hydrostatic pressure at constant temperature increases the enthalpy of a phase since $\tilde{H} = \tilde{E} + P\tilde{V}$. This increase, if large enough, can raise the free energy of the existing phase sufficiently to give rise to a transformation from an open to a closer-packed structure. For example, the bcc alpha iron at room temperature transforms to close-packed hcp epsilon iron under hydrostatic pressure of about 125 kbar at room temperature.

The above discussion represents a rather simplified view of a complex situation. The assumption that bond energy is minimized by close-packing is not valid for bonds of a directional nature. Also, the free-energy equations may have to be modified by magnetic contributions in cases where the magnetic energy differences between phases are significant.

2.4. First-Order Transformations with Compositional Change

Let us consider a binary system of A and B. The free energies of two phases α and β in this system are shown schematically in Figure 4, plotted against the concentration of B component at constant temperature and pressure.* In Figure 4 the segment $ab = \Delta \tilde{G}$ represents the chemical free-energy change per g-atom for the $\alpha \to \beta$ transformation *without* a change in composition.

Now consider the formation of a small particle of β out of the α phase of composition $\bar{c}$. Let the particle have some general composition c_2, causing a small change in the composition of the α matrix from $\bar{c}$ to c_1. Then it can be shown that the change in chemical free energy per g-atom of the β particle formed is

$$\Delta \tilde{G}_2 = (G^\beta_{c_2} - \tilde{G}^\alpha_{\bar{c}}) - (c_2 - \bar{c})\frac{d\tilde{G}^\alpha}{dc} \qquad (8)$$

= segment *mn* in Figure 4

---

* Here the concentration of B is expressed as the atomic fraction of B. Later in the chapter, the concentration is expressed as g-atoms of B per cm³. The symbol c is used for both, but the difference is made clear in the text.

Solid-State Phase Transformations

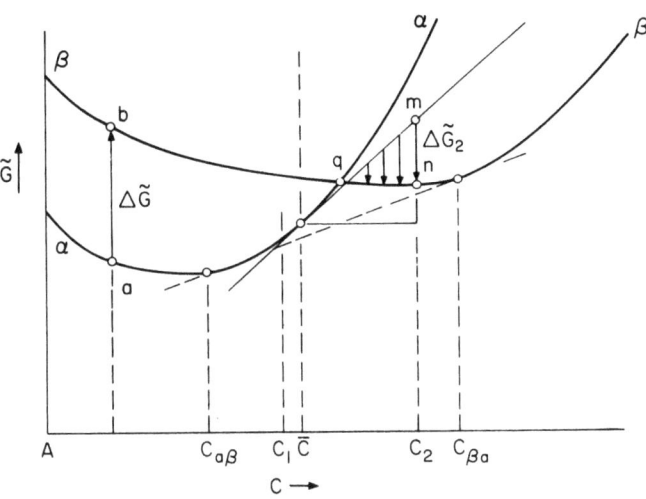

Fig. 4. Free energies of the α and β phases as a function of composition for a binary system of A and B.

(drawn at c_2 between the tangent to the α free-energy curve at $\bar{c}$ and the β free-energy curve). If the β particle has the equilibrium composition $c_{\beta\alpha}$, the segment mn will be shifted at $c_{\beta\alpha}$. It may be noted that the composition of the β particle c_2 must be significantly different from the matrix composition $\bar{c}$ in order for a finite driving force to be available; i.e., the segment mn must be beyond the point of intersection q of the two free-energy curves in order to be negative.

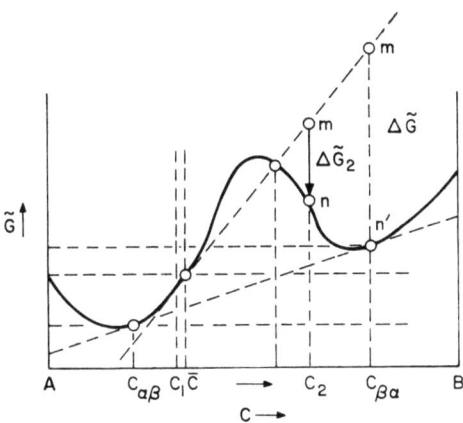

Fig. 5. Free energy vs. composition curve for a binary system of A and B with a miscibility gap.

Chapter 2

When we have a miscibility gap in a two-component system, the free-energy/composition curve is continuous between the two coexisting phases, as shown in Figure 5. Equation (8) is easily modified to be applicable here.

2.5. Second-Order Transformations

Except for spinodal decomposition and a few examples of order–disorder reactions, the transformations discussed in this chapter are of first order. However, for the sake of comparison some brief mention of second-order transformations is included here. A more comprehensive treatment is given in Chapter 1 of this volume.

The characteristics of a typical second-order transformation can be discussed with reference to the magnetic transition in nickel. The saturation magnetization of ferromagnetic nickel has a maximum value at $0°K$ and decreases on raising the temperature, slightly at first and then more rapidly until the critical temperature T_c is reached (Figure 6). Beyond T_c, nickel is paramagnetic. The maximum saturation magnetization at $0°K$ corresponds to a complete alignment of electron spins and therefore to a minimum residual enthalpy H_0 due to the associated exchange interactions.

As the temperature is raised, an increasing number of electrons set their spins antiparallel to their neighbors. The entropy gain resulting from the progressive randomization of the electron spins offsets the increase in $\bar{H}$ so as to minimize the free energy at any given temperature.

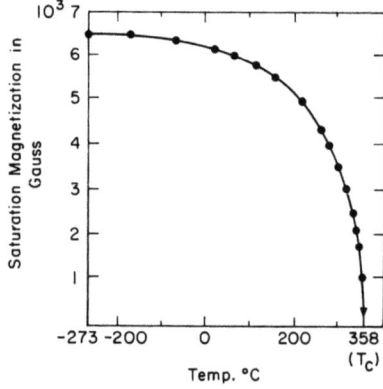

Fig. 6. Variation in saturation magnetization of nickel with temperature. The Curie point T_c is at $358°C$.

The disordering of the spins is an example of a second-order transformation and the cooperative phenomena involved. At low temperatures, when all spins are parallel, a large increase in $\tilde{H}$ is needed to reverse the spin of one electron against the combined restraints of all its neighbors. But as more and more electrons reverse their spins with increasing temperature, it becomes easier for others to do the same, since those already with reversed spins try to pull others into their orientation. When sufficient numbers of spins point in both directions, the change in H with further disordering becomes negligible and general disorder sets in; this occurs at the transition temperature or Curie point T_c. The cooperative action of the electron spins causes the saturation magnetization curve in Figure 6 to drop off precipitously as the temperature approaches T_c. Thus second-order transformations occur over a temperature range under equilibrium conditions, even when a composition change is not involved, and the associated excess specific heat is correspondingly spread over a temperature range, as in Figure 2. This behavior is in contrast with the case of a first-order transformation, as exemplified by Figure 1, where the excess specific heat becomes infinite at the transition temperature T_0.

3. Crystallography of Transformations

3.1. Orientation Relationships

Specific orientation relationships between parent and product phases occur quite frequently in solid-state transformations. There are at least two major factors which influence such crystallographic relationships. The first arises from the fact that a particular orientation may reduce the misfit at the interface between the two phases and thereby result in a lowering of the interfacial energy. In fcc $\rightleftharpoons$ hcp transformations, for example, an orientation having the close-packed planes and close-packed directions respectively parallel in the two phases will result in a one-to-one atom fit at the interface, thereby minimizing the interfacial energy. Such good fits are not feasible between two arbitrary crystal structures, and then the matching at the interface is accompanied by some misfits, even though it is often possible to reduce such misfit if certain orientations prevail. The orientation relationships observed in a number of transformations are listed in Table 1.[12] These transformations involve a change in

Chapter 2

composition, which is brought about by long-range diffusion. The observed orientation relationships are believed to minimize the interfacial energy.

TABLE 1
Orientation Relationships in Diffusion-Controlled Transformations[12]

Transformation*	Crystal structures of phases	Orientation relationships
$UC_{ssat} \rightarrow UC + UC_2$	UC: fcc (NaCl) UC_2: fct (CaF_2)	$(001)_{UC} \parallel (001)_{UC_2}$ $[100]_{UC} \parallel [100]_{UC_2}$
$\alpha_{ssat} \rightarrow \alpha + \theta'$ (Al–4% Cu)	α: fcc θ': fct	$(100)_{\theta'} \parallel (100)_{\alpha}$ $[001]_{\theta'} \parallel [001]_{\alpha}$
$\gamma_{ssat} \rightarrow \gamma + \alpha$ (Fe–C)	γ: fcc α: bcc	$(111)_{\gamma} \parallel (110)_{\alpha}$ $[1\bar{1}0]_{\gamma} \parallel [1\bar{1}1]_{\alpha}$
$US + O \rightarrow UOS$	US: fcc (NaCl) UOS: tetragonal	$(001)_{UOS} \parallel (001)_{US}$ $100_{UOS} \parallel 110_{US}$

* ssat = supersaturated.

The other major factor influencing orientation relationships between phases is intimately associated with the transformation mechanism and geometry, as discussed for shear transformations in Section 3.3. Examples of such relationships are shown in Table 2.[13]

TABLE 2
Orientation Relationships in Shear Transformations[13]

System	Shear transformation	Orientation relationships
52.5 Au–47.5 Cd	CsCl $\rightarrow$ twinned orthorhombic	$(110)_C \parallel (100)_O$ $[1\bar{1}0]_C \parallel [100]_O$
Fe–22 Ni–0.8 C (wt. %)	fcc $\rightarrow$ twinned bct	$(111)_F \parallel (101)_B$* $[1\bar{1}0]_F \parallel [11\bar{1}]_B$*
Zr, Ti, or Li	bcc $\rightarrow$ hcp	$(011)_B \parallel (0001)_H$ $[111]_B \parallel [11\bar{2}0]_H$

* Approximately parallel.

Occasionally, we encounter phase changes which have the properties of shear transformations, but which also involve diffusional processes. The orientation relationships in such cases may still be the result of the transformation mechanism and geometry.

3.2. The Interfacial Structure

The structure of the interface between parent and product phases is determined by the degree of matching between the two crystals on either side of the interface. Interfaces are classified into three groups, depending on the degree of matching.

When there is one-to-one lattice registry across the interface, as illustrated in Figure 7(a), the interface is said to be coherent even though elastic strains may be required to maintain such correspondence. Coherent interfaces usually have a low surface energy, in the range of 10–50 ergs/cm^2. If the crystal structures of the parent and product phases happen to match well across the common interface, nucleation of the new phase can be greatly facilitated as a result of the low interfacial energy.

Perfect matching, however, is rarely obtained, because the associated elastic strains may be excessively large. In such instances, regions

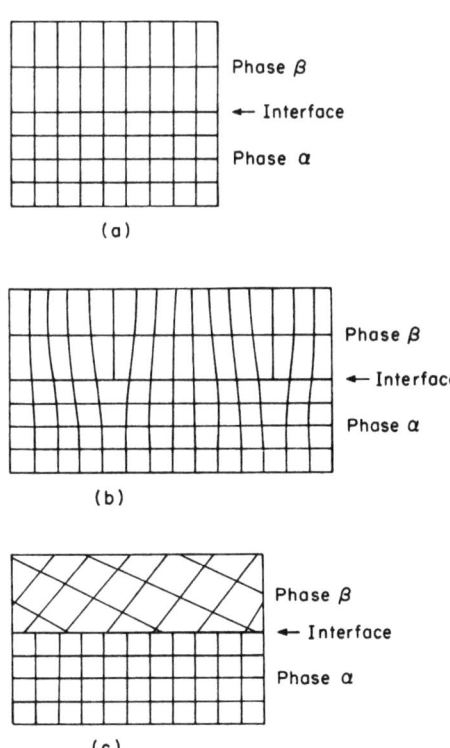

Fig. 7. Schematic sketches showing (a) coherent, (b) semicoherent, and (c) incoherent interfaces.

Chapter 2

of good coherency in the interface can be separated by localized regions of severe distortion comprised of interfacial dislocations, as illustrated in Figure 7(b). Here, along the interface, the spacing between seven atomic planes in the α phase is equal to the spacing between eight planes in the β phase. Accordingly, every eighth plane in the β phase forms an edge dislocation at the interface, allowing one-to-one registry of the intervening seven planes. Such an interface is referred to as semicoherent. Its energy is largely composed of the elastic and core energies of the interfacial dislocations, and lies in the range of 100–300 ergs/cm^2.

When the two crystal structures are very different from each other, relatively little matching is possible and a highly distorted, incoherent interface results, as shown schematically in Figure 7(c). The energy of this interface may be in the range of 500–1500 ergs/cm^2.

3.3. Phenomenological Description of Transformation Geometry

In transformations that are not accompanied by a compositional change, the parent crystal may transform to the product phase without any interchange of atomic neighbors. In other words, neither short-range nor long-range diffusion is involved. At least in simple systems, the net atomic displacements which bring about the transformation can be deduced in detail with reference to the crystal lattices of the parent and product phases.

For example, a 50 at.% Au–50 at.% Cd alloy[14] has a CsCl structure (alternatively referred to as an ordered bcc structure) at temperatures above 60°C. On cooling, the cubic structure transforms

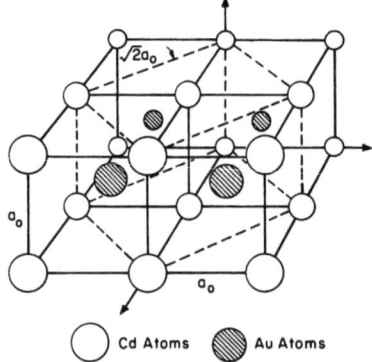

Fig. 8. An fct unit cell is outlined within four CsCl unit cells of an ordered Au–Cd alloy.

to an orthorhombic structure. In Figure 8, a face-centered tetragonal (fct) unit cell is outlined within four adjacent CsCl unit cells. (Note that the indicated cell does not have the same kind of atoms at all the face centers and should not strictly be called a fct unit cell.) A transformation from the cubic to the orthorhombic phase can now be brought about by simply altering the dimensions of the outlined fct unit, although there is no experimental method to verify that the transformation actually takes place in this manner. However, these displacements correspond to easily visualized atomic movements and we can deduce the orientation relationships between planes and directions in the two crystal lattices. For example, in the above Au–Cd case, the corresponding crystal directions are:

$$[1\bar{1}0]_C \rightarrow [100]_O$$
$$[110]_C \rightarrow [010]_O$$
$$[001]_C \rightarrow [001]_O$$

where the subscripts C and O stand for cubic and orthorhombic structures, respectively. A similar correspondence between crystal planes in the two lattices can also be inferred from Figure 8. This correspondence between planes and directions of the two structures is termed the lattice correspondence of the transformation.

Once the lattice correspondence is established, the lattice parameters enable us to specify the magnitude of the atomic displacements necessary to bring about the transformation in this way. These displacements are usually known as the Bain distortion, after E. C. Bain, who, in 1924, first suggested the nature of the atomic displacements operative during martensitic transformations in steels.[15] In the Au–Cd case, the Bain distortion expressed as strains along the principal axes of the orthorhombic cell are:

$$\varepsilon_1 = (b - \sqrt{2}a_0)/\sqrt{2}a_0 = 1.4\%$$
$$\varepsilon_2 = (c - \sqrt{2}a_0)/\sqrt{2}a_0 = 3.5\%$$
$$\varepsilon_3 = (a - a_0)/a_0 = -5.1\%$$

where a_0 and a, b, c are edge dimensions of the parent and the product unit cells, respectively. Note that the atomic displacements are only small fractions of the interatomic distances but the resulting strains are large. When more complicated structures are involved, atomic shuffles or adjustments are usually necessary in addition to the Bain distortion to bring about the final structure.

Chapter 2

However, an important condition still remains to be satisfied. This is the question of matching the crystal lattices at the common interface without building up unduly high elastic strain energy in the system. An effective compromise between interfacial energy and strain energy is attained in shearlike transformations, where the interface (or habit) plane remains undistorted and unrotated during the transformation. Phenomenologically, this is called an invariant-plane strain (IPS) transformation.

By combining the lattice correspondence and the Bain distortion with the IPS description, the transformation geometry can be worked out in considerable detail using matrix algebra or stereographic analysis.[16,17] Except in very simple transformations, such as those between close-packed structures, fcc → hcp, the Bain distortion alone cannot satisfy the IPS criterion. Also required are a lattice-invariant shear that will not change the crystal structure brought about by the Bain distortion, and a rigid-body rotation that keeps the interface plane unrotated as well as undistorted in a macroscopic sense.

The meaning of the lattice-invariant shear is illustrated in Figure 9.[3] The initial lattice shown schematically in Figure 9(a) undergoes a Bain distortion in Figure 9(b). In Figure 9(c) the original shape of the transformed region is restored by slip representing the lattice-

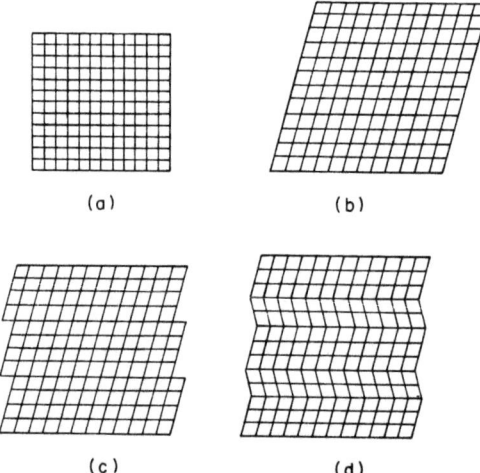

Fig. 9. Distinction between lattice deformation (Bain distortion) and lattice-invariant deformamation; see text.

invariant shear. Alternatively, the shape restoration can be brought about by twinning shears as in Figure 9(d). The rigid-body rotation is not shown in this two-dimensional representation.

Summing up the conceptual steps in the IPS transformation process, we have: (1) a Bain distortion, (2) a lattice-invariant shear, and (3) a rotation.

Given the lattice correspondence, the Bain distortion, and the right amount of lattice-invariant shear, the theory predicts an irrational habit plane in many instances, and this is found experimentally.

Another success of the phenomenological theory is the correlation between the predicted lattice-invariant shear and observations by electron microscopy. If the lattice-invariant shear is by slip, a high density of dislocations is noted in the transformation product. If it is by twinning, internal twins are found within the product phase, and the ratio in thickness of the twin-related regions agrees reasonably well with that theoretically calculated.

In recent years, the IPS description of transformational crystallography has been applied to shear transformations in a variety of systems,[18] such as V_3Si, $BaTiO_3$, and ZrO_2.

4. Nucleation Kinetics

Several excellent reviews of nucleation kinetics are available, particularly with regard to solid-state transformations.[3,19-23] The process of nucleation can take place randomly or nonrandomly. In the former case, the probability of nucleation at any given site is identical to that at any other site within the assembly. In the latter case, the probability of nucleation at certain preferred sites in the assembly is much greater than at other sites. Nucleation in the gaseous phase is usually random. Nucleation in liquids can be either random or nonrandom. Inclusions of solid foreign particles and the walls of the container often provide preferred nucleation sites during solidification. In solids, random nucleation is the exception rather than the rule; inclusions, grain boundaries, stacking faults, dislocations, and possibly point imperfections may form the preferred nucleation sites.

4.1. Random Nucleation

Volmer and Weber[24] and Becker and Doring[25] developed a theory of nucleation for the formation of liquid droplets in super-

Chapter 2

saturated vapors. A basic assumption in this theory is that, by random thermal fluctuations, very small droplets of the product phase, identical in structure and composition to the final bulk phase, form in the supersaturated parent (vapor) phase. The small droplets, which may contain of the order of 100 atoms, are assumed to have the same surface and thermodynamic properties as the bulk product phase. These ideas have been carried over to solid-state nucleation.

If ΔG is the free-energy change accompanying the formation of a new particle of the transformation product, we can then write

$$\Delta G = V \Delta g + S_A \sigma \qquad (9)$$

where V and S_A are the volume and surface area, respectively, of the particle, Δg is the chemical free-energy change per unit volume of the product formed, and σ is the specific surface energy of the interface separating the product and parent phases. When there is supersaturation in the parent phase, Δg is negative. The surface-energy term in Eq. (9) is, however, always positive. In the absence of strains, the particle tends to have a spherical shape to minimize surface energy. For a spherical particle of radius r

$$\Delta G = \tfrac{4}{3}\pi r^3 \Delta g + 4\pi r^2 \sigma \qquad (10)$$

This function passes through a maximum with increasing r whenever Δg is negative, as illustrated in Figure 10. The temperature dependence

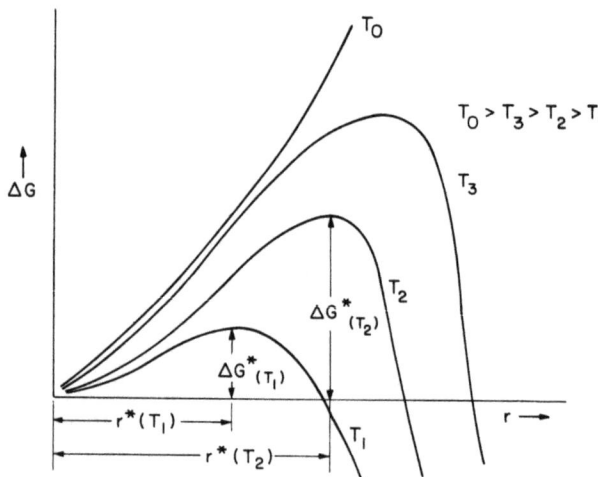

Fig. 10. Free-energy change ΔG accompanying the formation of a spherical particle of a new phase as a function of radius r at different temperatures.

of σ is usually small enough to be neglected. Setting $d(\Delta G)/dr = 0$, the following critical values (denoted by the asterisk) are obtained:

$$r^* = -2\sigma/\Delta g \tag{11}$$

$$\Delta G^* = 16\pi\sigma^3/3(\Delta g)^2 \tag{12}$$

The following simplification is often used:

$$\Delta g \simeq \Delta h(T_0 - T)/T_0 = \Delta h\, \Delta T/T_0 \tag{13}$$

where Δh is the enthalpy change per unit volume of the product formed, T_0 is the equilibrium temperature at which $\Delta g = 0$, and ΔT is the degree of supercooling. Combining Eqs. (12) and (13), we obtain

$$\Delta G^* = 16\pi\sigma^3 T_0^2/3(\Delta h)^2(\Delta T)^2 \tag{14}$$

Δh is usually taken to be insensitive to temperature in solid-state transformations.

In classical theory, the process of nucleation is defined by the addition of one atom to a critical-sized particle such that it becomes just supercritical. Let the total number of particles per unit volume of the matrix be N_T. Then, from Maxwell–Boltzmann statistics, the number of critical-sized particles N^* is approximated by

$$N^* = N_T \exp(-\Delta G^*/kT) \tag{15}$$

If the critical-sized particle is surrounded by s^* atoms in the matrix, the frequency v' with which these atoms can cross the interface to join the particle is given by

$$v' = s^* v \exp(-\Delta G_D/kT) \tag{16}$$

where v is the lattice vibration frequency and ΔG_D is the free energy of activation per atom for diffusion across the interface. Then the rate of random nucleation I (nucleation events per unit volume per unit time) is

$$I = N_T s^* v \exp[-(\Delta G^* + \Delta G_D)/kT] \tag{17}$$

The preexponential term $N_T s^* v$ is 10^{36} cm^{-3} sec^{-1} in order of magnitude. The exponent $(\Delta G^* + \Delta G_D)/kT$ must then be less than about 70 for a measurable rate of nucleation. The nucleation rate is very sensitive to the value of this exponent; if the latter is decreased to about 50, say by increasing the driving force $-\Delta g$ through a change in the reaction temperature, I can be increased by a factor of 10^6.

Chapter 2

As shown in Fig. 10, ΔG^* is a function of temperature. However, ΔG_D is independent of temperature. Thus the nucleation rate passes through a maximum as a function of temperature and $T_{I_{\max}}$ is determined by setting $dI/dT = 0$, from which the condition for $T_{I_{\max}}$ is

$$d(\Delta G^*)/dT = (\Delta G^* + \Delta G_D)/T_{I_{\max}} \qquad (18)$$

If ΔG^* is plotted against T, $T_{I_{\max}}$ can be determined graphically.

Transient effects may be observed during the early stages of nucleation before the time-independent value expressed by Eq. (17) is attained. On cooling from T_2 to a reaction temperature T_1, where $T_2 > T_0 > T_1$, it may take some time for N^* in Eq. (15) to build up to the steady-state value characteristic of T_1. During that period, I is expected to increase with time according to the relation

$$(I_t - I_i)/(I_s - I_i) = \exp(-\omega/t) \qquad (19)$$

where I_t is the nucleation rate at any time t, I_i is the initial nucleation rate, I_s is the nucleation rate attained at steady state, and ω is a time constant.

4.2. Random Nucleation with Compositional Change

Random nucleation with a compositional change in systems having two or more components has been considered by several authors. Becker[26] used an expression similar to Eq. (8) for the chemical free-energy change during the transformation when there is an attendant compositional change. He also assumed coherent nucleation of a particle of fixed composition. The surface energy here is solely due to the abrupt compositional change at the interface. It turns out to be proportional to the square of the difference in composition across the interface.

Borelius[27] considered the alternative possibility of a continuous variation in composition of a nucleus of a given size. Such a variation results in an energy barrier along the reaction path, occurring at that composition where the segment *mn* in Figure 5 has the maximum positive value between the free-energy curve and the straight line drawn tangent at $\bar{c}$.

Hobstetter[28] allowed both the nucleus composition and size to vary simultaneously, and obtained a saddle point corresponding to the critical nucleation barrier. The saddlepoint energy occurs at the critical radius r^* of the nucleus and the critical composition c_β^*.

However, it remained for Cahn[29] and his associates to generalize the treatment by dropping the assumption of a sharp interface for a transformation involving compositional changes. It was shown that a continuous variation in composition is possible across the interface for system compositions within the spinodal range. This treatment is described in Section 13.

4.3. Strain-Energy Effects

A change in volume accompanies most phase transformations. When the adjoining parent phase is a fluid, these volume changes can be accommodated easily by flow in the parent phase. If the reaction occurs entirely within the solid state, however, the volume change introduces strains in the phases and the nucleation process will be inhibited due to the associated strain energy. In martensitic transformations, besides any change in volume, the transforming region also undergoes a change of shape, resulting in additional strains. Estimates of the corresponding strain energies for different situations are available, as discussed below.

Nabarro[30] considered the uniform expansion or contraction which may occur during the formation of a new phase having an incoherent interface with the matrix. Assuming that all the strains are stored in the matrix, Nabarro showed that, for oblate and prolate spheroidal particles of semiaxes r and c, the strain energy per unit volume of particle formed is given by

$$\mathscr{E} = \tfrac{2}{3}\mu_m(\Delta V/V)^2 \Psi(c/r) \qquad (20)$$

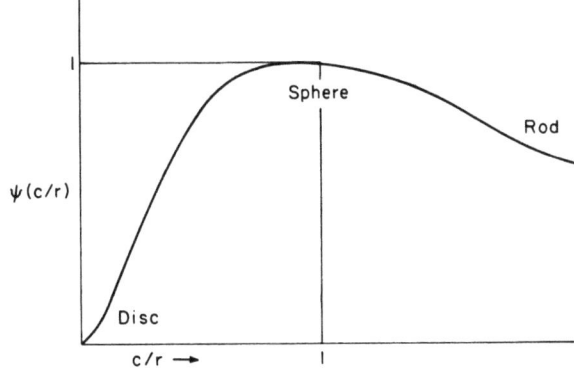

Fig. 11. Shape-dependent function $\Psi(c/r)$ vs. c/r for oblate and prolate spheroidal-shaped particles.

Chapter 2

where μ_m is the shear modulus of the matrix (taken to be elastically isotropic), and $\Delta V/V$ is the fractional volume change accompanying the transformation. $\Psi(c/r)$ is a function that depends on the particle shape, as illustrated in Figure 11:

$$\Psi(c/r) = \begin{cases} 1 & \text{for } c/r = 1 \quad \text{(sphere)} \\ 0.75 & \text{for } c/r \gg 1 \quad \text{(rod)} \\ \tfrac{3}{4}\pi c/r & \text{for } c/r \ll 1 \quad \text{(disk)} \end{cases}$$

Equation (20) can be rewritten as

$$\mathscr{E} = A_{\text{dil}}\Psi(c/r) \qquad (21)$$

where A_{dil} is the strain-energy factor due to dilational strains. For a disk with $c/r = 1/20$ and for a volume change of 5%, $\mathscr{E}$ has a value of $\sim 5 \times 10^7$ ergs/cm^3 for typical solid-state transformations in metallic systems.

Fisher et al.[31] have estimated the shear-strain energy resulting from a change in the shape of a transforming region. The strains are assumed to reside in the matrix within a sphere circumscribed about the disk-shaped particle, as shown in Figure 12. Assuming the strain to be uniformly distributed within the sphere, the shear strain there

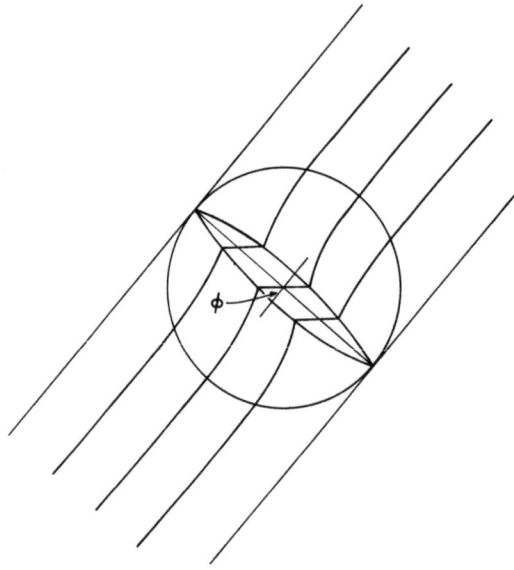

Fig. 12. Strains due to change of shape of a transforming region. The shear angle is ϕ.

Solid-State Phase Transformations

is $\sim c/r \tan \phi$, where ϕ is the shear angle of the transformation. Then the shear-strain energy $\mathscr{E}$ per unit volume of the particle turns out to be

$$\mathscr{E} = (\mu_m/2)[\tan^2 \phi](c/r) = (c/r)A_{\text{shear}} \qquad (22)$$

where $A_{\text{shear}} = (\mu_m/2) \tan^2 \phi$.

Strain energy may also arise due to coherency when the lattices of the product phase and the surrounding matrix are constrained to match. As already indicated in Section 3.2, moderate mismatches between the two structures can elastically strain the phases or can accommodate themselves in the form of interfacial dislocations. We shall consider the case where the mismatch results in both elastic coherency strains and misfit dislocations.

Let $a_m^\circ > a_p^\circ$, where a is the lattice constant or spacing, m and p stand for matrix and particle, respectively, and the superscript $^\circ$ designates the relaxed (unstrained) or natural (bulk) condition. We define the natural misfit as

$$\delta^\circ = (a_m^\circ - a_p^\circ)/a_m^\circ \qquad (23)$$

while the actual misfit existing during the nucleation event is

$$\delta = (a_m - a_p)/a_m \qquad (24)$$

Then the coherency strain in the particle is

$$\varepsilon_p = (a_p - a_p^\circ)/a_p^\circ \qquad (25)$$

and the coherency strain in the matrix is

$$\varepsilon_m = (a_m - a_m^\circ)/a_m^\circ \qquad (26)$$

The resulting coherency strain per unit volume of the particle depends on the relative stiffness of the matrix and the particle. Assuming isotropic elasticity, let E_m and E_p be the Young's moduli of the matrix and the particle, respectively, and v_m and v_p be the corresponding Poisson's ratios:

For equal stiffness ($E_m = E_p$)

$$\mathscr{E} = E\varepsilon^2/(1 - v) \qquad (27)$$

For a rigid matrix ($E_m \gg E_p$)

$$\mathscr{E} = 3E_p \varepsilon_p^2/2(1 - v_p) \qquad (28)$$

For rigid particles ($E_m \ll E_p$)

$$\mathscr{E} = 3E_m \varepsilon_m^2/(1 - v_m) \qquad (29)$$

These results are due to Eshelby.[32] Note that, under the assumptions adopted here, the coherency-strain energy is independent of particle shape, in contrast to the expressions derived by Nabarro[30] and Fisher et al.[31] in Eqs. (21) and (22).

For solid-state transformations, then, Eq. (9) for the free-energy change ΔG during nucleation must be modified to take the strain energy into account. In such cases, the balance between interfacial energy and strain energy results in a disk-shaped particle of semithickness c and radius r, having $V = \frac{4}{3}\pi r^2 c$ and $S_A \approx 2\pi r^2$ when $c \ll r$. If Eqs. (21) and (22) are applicable, then

$$\Delta G = \tfrac{4}{3}\pi r^2 c\, \Delta g + 2\pi r^2 \sigma + \tfrac{4}{3}\pi r c^2 A \qquad (30)$$

where $A = A_{\text{dil}} + A_{\text{shear}}$. By setting $\partial(\Delta G)/\partial c = 0$ and $\partial(\Delta G)/\partial r = 0$, we obtain the critical values:

$$c^* = -2\sigma/\Delta g \qquad (31)$$

$$r^* = 4A\sigma/(\Delta g)^2 \qquad (32)$$

$$\Delta G^* = (32\pi/3) A^2 \sigma^3/(\Delta g)^4 \qquad (33)$$

$$V^* = -(128\pi/3) A^2 \sigma^3/(\Delta g)^5 \qquad (34)$$

When misfit at the interface is accommodated both as elastic coherency strains and misfit dislocations, the expression for ΔG will include an additional variable δ, which is the optimum misfit that corresponds to the minimum total energy due to the elastic coherency strains and misfit dislocations. If $\delta°$ in Eq. (23) happens to be large, nucleation tends to occur without much coherency, because the associated coherency strain energy would then be excessive. Under such conditions, ΔG^* is minimized mainly with misfit dislocations. If $\delta°$ is small, nucleation tends to occur with a high degree of coherency, inasmuch the coherency strain energy is then relatively low. Here ΔG^* is minimized with a small value of the interfacial energy because of the high degree of coherency which can be tolerated.

For a given $\delta°$, the effect of particle size is also important, in that the strain energy arising from coherency is a function of the volume of the particle. On the other hand, if the misfit is accommodated entirely by interfacial dislocations, the energy due to this contribution is proportional to the interfacial area. As we have previously noted from Figure 10, the surface-area term dominates the volume term at small particle sizes, but the reverse is true at large sizes. This situation may result in coherent nucleation of a particle, but as it grows, the coherency strain energy becomes so large that the system can lower its free energy by forming interfacial dislocations to take up the misfit between the two phases. In other words, beyond a certain particle size, the decrease in coherency strain energy is greater than the increase in surface energy when the interface changes from coherent to semicoherent.

4.4 Nonrandom Nucleation

As already indicated, nucleation is usually quite heterogeneous in the solid state. Among the preferred nucleation sites are grain boundaries, grain edges and corners, dislocations, and surfaces of foreign particles embedded in the matrix.

Clem and Fisher[33] have derived the nucleation kinetics on grain boundaries, neglecting orientation, strain energy, and coherency effects. Equilibration of surface tension forces arising during nucleation is assumed. The new phase particle (β) nucleating in the matrix (α) is taken to be lens-shaped in the form of two spherical caps (Figure 13). The circular plane of intersection of the caps lies in the grain-boundary plane. The volume V and the surface area S_A of such a particle are given by

$$V = 2\pi R^3(2 - 3\cos\theta + \cos^3\theta)/3 \qquad (35)$$

and

$$S_A = 4\pi R^2(1 - \cos\theta) \qquad (36)$$

where R is the radius of curvature of the spherical surfaces and 2θ is the included lens angle as controlled by the operative surface tension $\sigma_{\alpha\alpha} = 2\sigma_{\alpha\beta}\cos\theta$. The radius of the circular intersection of the caps is $r = R\sin\theta$, and the semithickness of the disklike lens is $c = R(1 - \cos\theta)$.

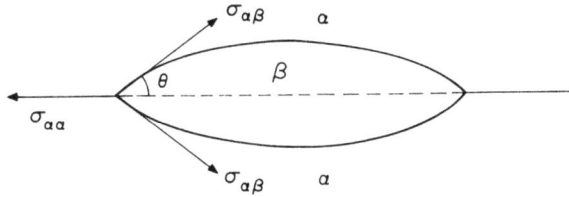

Fig. 13. Grain–boundary nucleation of a particle in the shape of two spherical caps forming a lens.

Chapter 2

The free-energy change during this type of nonrandom nucleation should now take into account the removal of grain-boundary energy equal to $\pi r^2 \sigma_{\alpha\alpha}$:

$$\Delta G_{non} = (\tfrac{2}{3}\pi R^3 \Delta g + 2\pi R^2 \sigma_{\alpha\beta})(2 - 3\cos\theta + \cos^3\theta) \quad (37)$$

At $d(\Delta G)/dR = 0$, the critical nucleus values are

$$R^* = -2\sigma_{\alpha\beta}/\Delta g \quad (38)$$

$$r^* = -2\sigma_{\alpha\beta}(\sin\theta)/\Delta g \quad (39)$$

$$\Delta G^*_{non} = 8\pi\sigma^3_{\alpha\beta}(2 - 3\cos\theta + \cos^3\theta)/3(\Delta g)^2 \quad (40)$$

$$= \Delta G^*_{ran}(2 - 3\cos\theta + \cos^3\theta)/2 \quad (41)$$

the last equality arising out of substitution from Eq. (12) for random nucleation. When $\cos\theta = 0$, $\sigma_{\alpha\alpha} = 0$, the lens becomes a sphere, and so the nucleation is then equivalent to the random case. For the special case of $\cos\theta = 1$, the lens degenerates to a vanishingly thin disk that spreads along the α–α interface without any nucleation barrier. Here $\sigma_{\alpha\alpha} \geq 2\sigma_{\alpha\beta}$ and $\Delta G^*_{non} = 0$.

It is easily seen from Eqs. (35), (37), and (38) that

$$\Delta G^*_{non} = -(\Delta g)V^*/2 \quad (42)$$

Nucleation at three-grain junctures (grain edges) and four-grain junctures (grain corners) leads to exactly the same result as in Eq. (42), if we use the volume of particles bounded by three and four spherical surfaces, respectively. Nucleation on foreign surfaces can be treated in a similar manner. The particle shape then corresponds to a spherical cap or half-lens and ΔG^*_{non} is one-half of that of the full-lens case given by Eq. (40).

When strain energy is taken into account, we drop the assumption of surface tension equilibration, and consider the particle at the grain boundary to be oblate-spheroidal instead of lens-shaped:

$$\Delta G_{non} = \tfrac{4}{3}\pi r^2 c \Delta g + 2\pi r^2 \sigma_{\alpha\beta} - \pi r^2 \sigma_{\alpha\alpha} + \tfrac{4}{3}\pi rc^2 A \quad (43)$$

From this we can derive

$$c^* = -2(\sigma_{\alpha\beta} - \tfrac{1}{2}\sigma_{\alpha\alpha})/\Delta g \quad (44)$$

$$r^* = 4A(\sigma_{\alpha\beta} - \tfrac{1}{2}\sigma_{\alpha\alpha})/(\Delta g)^2 \quad (45)$$

and

$$\Delta G^*_{non} = 32\pi A^2(\sigma_{\alpha\beta} - \tfrac{1}{2}\sigma_{\alpha\alpha})^3/3(\Delta g)^4 \quad (46)$$

If $\sigma_{\alpha\alpha} \approx 0$, Eq. (46) becomes Eq. (33) previously derived for the random case. If $\sigma_{\alpha\alpha} \approx \sigma_{\alpha\beta}$, $\Delta G^*_{non}/\Delta G^*_{ran} = \frac{1}{8}$. If $\sigma_{\alpha\alpha} \geq 2\sigma_{\alpha\beta}$, $\Delta G^*_{non} = 0$, that is, the nucleation barrier disappears.

Cahn[34] has dealt with the nucleation problem on dislocations. The β particle is pictured as cylindrical in shape, extending along a straight dislocation, and the strain energy of the dislocation is taken to be removed over a distance from the dislocation equal to the particle radius r. An incoherent interface is assumed and strain energy due to the particle formation itself is neglected. Then

$$\Delta G_{non}(\text{per unit length}) = \pi r^2 \Delta g + 2\pi r \sigma - A' \ln r \qquad (47)$$

where $A' \ln r$ approximates the strain energy of the dislocation per unit length. $A' = \mu b^2/4\pi(1 - v)$ for an edge dislocation and $\mu b^2/4\pi$ for a screw dislocation, where b is the Burgers vector of the dislocation and v is Poisson's ratio of the matrix.

Setting $d(\Delta G)/dr = 2\pi r \Delta g + 2\pi\sigma - A'/r = 0$, we find that ΔG may pass through a minimum and then a maximum as a function of r:

$$r(\text{for } \Delta G_{min}) = -[\sigma/(2\Delta g)][1 - (1 + 2A' \Delta g/\pi\sigma^2)^{1/2}] \qquad (48)$$

$$r(\text{for } \Delta G_{max}) = r^* = -[\sigma/(2\Delta g)][1 + (1 + 2A' \Delta g/\pi\sigma^2)^{1/2}] \qquad (49)$$

Since Δg is negative, $1 + 2A' \Delta g/\pi\sigma^2$ may be positive or negative. If it is positive, the above r values are real, and both a metastable minimum and then a barrier maximum appear in the ΔG vs. r curve. However, if $1 + 2A' \Delta g/\pi\sigma^2$ is negative, neither a minimum nor a maximum exists in the ΔG vs. r curve and there is no barrier to nucleation.

The rate of nonrandom nucleation can be expressed in a form similar to that of Eq. (17). In addition to the difference in the ΔG^* term as discussed above, the preexponential term will involve the number of preferred nucleation sites, being many orders of magnitude smaller than the number of atoms per unit volume used in the random case. In addition, we may have a distribution of sites of varying potencies (the greater is the potency, the lower is the corresponding ΔG^*). If N_i is the number of sites per unit volume having a nucleation barrier ΔG_i^*, then from Eq. (17)

$$I_{non} = \sum N_i v s_i^* \exp[-(\Delta G_i^* + \Delta G_D)/kT] \qquad (50)$$

Usually $I_{non} \gg I_{ran}$ because $\Delta G^*_{non} < \Delta G^*_{ran}$ dominates the result through the exponential factor, even though the preexponential

Chapter 2

factor may be much greater for the random case than for the preferred-nucleation case.

The factor N_i in Eq. (50) can be greatly modified during the course of a reaction by (a) the using up of available preferred sites as the nucleation proceeds, (b) the sweeping out of such sites through the volume occupied by the transformation product, and (c) the generation of new sites through autocatalysis.

5. Growth Kinetics

Growth is a continuation of the transformational process beyond the nucleation event. It is convenient to consider the growth process first *without* a compositional change and then *with* a compositional change. In growth without compositional change, we focus attention on the thermally activated atomic jumping across the interface separating the transformation product β from the parent phase α. The transformation here is said to be interface-controlled. When there is a compositional change, the growth rate is often controlled by diffusion in the parent phase. Though it is possible for growth to be controlled by combinations of interfacial and diffusional processes, we will deal only with the two limiting cases.

5.1. Interface-Controlled Growth

Let ΔG_D be the activation free energy for an atomic jump across the interface. The change in chemical free energy per atom transfer in the $\alpha \rightarrow \beta$ transformation is $v_a \Delta g$, where v_a is the volume per atom. The free-energy barrier for an atomic jump from α to β (in direction of the driving force) is ΔG_D, while the free-energy barrier for an atomic jump from β back to α (against the driving force) is $\Delta G_D - v_a \Delta g$, as illustrated in Figure 14. Note that $\Delta g < 0$ for a finite driving force. Thus the barrier in the reverse direction is larger than ΔG_D.

The net rate of atomic jumping from α to β per unit area of interface is

$$(dn^{\alpha \rightarrow \beta}/dt)_{\text{net}}$$
$$= sv \exp(-\Delta G_D/kT) - sv \exp[-(\Delta G_D - v_a \Delta g)/kT]$$
$$= sv \left[\exp(-\Delta G_D/kT)\right]\left[1 - \exp(v_a \Delta g/kT)\right] \quad (51)$$

Solid-State Phase Transformations

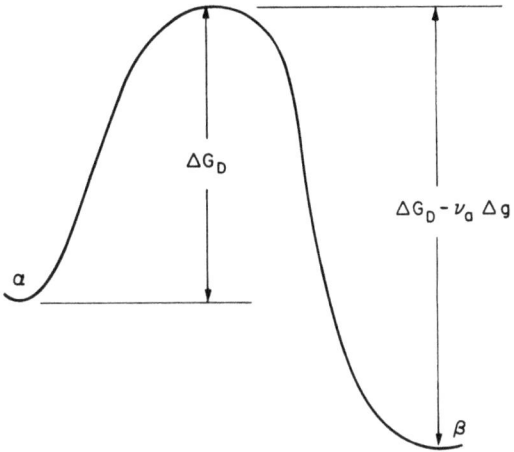

Fig. 14. Free-energy barriers for atomic jumps across the interface from α to β and from β to α during interface-controlled growth.

where s is the number of interfacial atoms in each phase per unit area of interface and v is the lattice vibration frequency.

The growth rate of the β phase U (cm/sec) can then be expressed as

$$U = dr_\beta/dt = dV_\beta/dt \text{ (per cm}^2 \text{ of interface)}$$
$$= (dn^{\alpha \to \beta}/dt)_{net} v_a$$
$$= \lambda_j v [\exp(-\Delta G_D/kT)][1 - \exp(v_a \Delta g/kT)] \quad (52)$$

where λ_j is the jump distance across the interface and is also the volume corresponding to 1 cm^2 of interface advancing through a single atomic growth step, i.e., $\lambda_j = sv_a$.

Inasmuch as the operative diffusion coefficient D_b is

$$D_b = \lambda_j^2 v \exp(-\Delta G_D/kT) \quad (53)$$

then

$$U = (D_b/\lambda_j)[1 - \exp(v_a \Delta g/kT)] \quad (54)$$

D_b is commonly taken as the diffusivity in the boundary between the two phases.

The above expression for U shows that the growth rate is zero at T_0, where $\Delta g = 0$. Just below T_0, Δg is negative but very small.

Thus

$$-v_a \Delta g \ll kT, \qquad \exp(v_a \Delta g/kT) \simeq 1 + v_a \Delta g/kT \qquad (55)$$

and

$$U \simeq \lambda_j v(-v_a \Delta g/kT) \exp(-\Delta G_D/kT) \simeq (D_b/\lambda_j)(-v_a \Delta g/kT) \qquad (56)$$

At temperatures where the driving force is very large

$$-v_a \Delta g \gg kT, \qquad \exp(v_a \Delta g/kT) \ll 1$$

and

$$U \simeq \lambda_j v \exp(-\Delta G_D/kT) = D_b/\lambda_j \qquad (57)$$

Hence, not only is the growth rate U small just below T_0, it is also small at low temperatures where D_b becomes small. Thus U passes through a maximum at some intermediate temperature.

5.2. Diffusion-Controlled Growth

Here compositions are expressed as molar concentrations: g-atoms of component B per cm³ of phase. The following notation will be adopted. c_α is the concentration in the α matrix phase as a function of time t and position y. $c_{\alpha\beta}$ is the concentration in the matrix phase at the interface in contact with a β particle; equilibrium is assumed to exist at this interface. $c_{\beta\alpha}$ is the concentration in a β particle, coexisting with the α phase. $\bar{c}$ is the initial concentration in the matrix phase and is usually the overall composition of the system. $c_{\alpha\beta}^{(\infty)}$ is the concentration in the matrix phase in equilibrium with β particles having a very large radius of curvature or planar interface.

Based on the conservation of B atoms, the g-atoms of B acquired by a growing β particle per unit area per unit time is

$$Uc_{\beta\alpha} = Uc_{\alpha\beta} + D_\alpha(\partial c_\alpha/\partial y)_{y=r} \qquad (58)$$

where U ($=dr/dt$) is the growth rate of the β particle, D_α is the diffusivity in the matrix phase, and $D_\alpha(\partial c_\alpha/\partial y)_{y=r}$ is the flux of B atoms at the interface due to the concentration gradient in the matrix phase. Hence

$$U = D_\alpha \frac{(\partial c_\alpha/\partial y)_{y=r}}{c_{\beta\alpha} - c_{\alpha\beta}} \qquad (59)$$

This is a rather general expression for the growth rate.

Solid-State Phase Transformations

For spherical particles of β, we can write

$$c_\alpha = \bar{c} - (r/y)(\bar{c} - c_{\alpha\beta}) \tag{60}$$

When $y = r$, $c_\alpha = c_{\alpha\beta}$, and when $y = \infty$, $c_\alpha = \bar{c}$. Also,

$$(\partial c_\alpha/\partial y)_{y=r} = (\bar{c} - c_{\alpha\beta})/r \tag{61}$$

Substituting Eq. (61) in Eq. (59), we obtain

$$U = D_\alpha(\bar{c} - c_{\alpha\beta})/r(c_{\beta\alpha} - c_{\alpha\beta}) \tag{62}$$

If $c_{\alpha\beta}$ is independent of r (when $r \gg r^*$), integration of Eq. (62) yields

$$r^2 = 2D_\alpha \frac{\bar{c} - c_{\alpha\beta}}{c_{\beta\alpha} - c_{\alpha\beta}} t \tag{63}$$

signifying parabolic growth; the growth rate decreases progressively as t and r increase.

For plane-front growth, the following relations apply:

$$\bar{c} - c_\alpha = (\bar{c} - c_{\alpha\beta})\left[1 - \text{erf}\frac{y - r}{2(D_\alpha t)^{1/2}}\right] \tag{64}$$

and

$$\left(\frac{\partial c}{\partial y}\right)_{y=r} = \frac{\bar{c} - c_{\alpha\beta}}{(\pi D_\alpha t)^{1/2}} \tag{65}$$

Substituting Eq. (65) in Eq. (59), we obtain

$$U = \left(\frac{D_\alpha}{\pi t}\right)^{1/2} \frac{\bar{c} - c_{\alpha\beta}}{c_{\beta\alpha} - c_{\alpha\beta}} \tag{66}$$

Noting that $c_{\alpha\beta}$ is independent of r for the plane-front case, integration of Eq. (66) yields

$$r^2 = 4\frac{D_\alpha}{\pi} \frac{(\bar{c} - c_{\alpha\beta})^2}{(c_{\beta\alpha} - c_{\alpha\beta})^2} t \tag{67}$$

again indicating parabolic growth.

However, returning to the case of spherical β particles, $c_{\alpha\beta}$ does depend on r when r is very small. Using the Thomson–Freundlich equation for solubility of very small particles [see Eq. (92) in Section 7], we can write Eq. (62) for spherical growth rate to include the effect

Chapter 2

of particle radius on solubility:

$$U = \frac{D_\alpha(\bar{c} - c_{\alpha\beta}^{(\infty)})}{c_{\beta\alpha} - c_{\alpha\beta}^{(\infty)}} \frac{1}{r}\left(1 - \frac{r^*}{r}\right) \tag{68}$$

According to this equation, U is negative for $r < r^*$; such particles are subcritical and tend to dissolve. For $r = r^*$, $U = 0$; such particles are in unstable equilibrium with the matrix of composition $\bar{c}$. At $r = 2r^*$, U reaches a maximum and then decreases with further growth of r. The maximum value of U is

$$U_{\max(r=2r^*)} = \frac{D_\alpha}{4r^*} \frac{\bar{c} - c_{\alpha\beta}^{(\infty)}}{c_{\beta\alpha} - c_{\alpha\beta}^{(\infty)}} \tag{69}$$

Recognizing that, approximately, $c - c_{\alpha\beta}^{(\infty)} \sim T_0 - T$, $r^* \sim 1/(T_0 - T)$, and $D_\alpha \propto \exp(-Q/RT)$, where Q is the activation energy of diffusion in the matrix phase, the temperature-dependent factors in Eq. (69) lead to

$$U_{\max} \propto (T_0 - T)^2 \exp(-Q/RT) \tag{70}$$

Then $U_{\max}$ is zero at $T = T_0$, and is also very small at low temperatures. The temperature that maximizes $U_{r=2r^*}$ is given by

$$2T^2 + QT/R - QT_0/R = 0 \tag{71}$$

6. Overall Transformation Kinetics

Overall transformation kinetics are often described by empirical equations. We will first discuss the empirical forms and then relate them to the nucleation and growth processes previously described.

6.1. Empirical Relations

In the exponential type of empirical kinetic equation, the volume fraction transformed X is expressed as a function of time as follows:

$$X = 1 - \exp[-(Kt)^n] \tag{72}$$

where K is the reaction-rate constant (time^{-1}) and n is the time exponent (dimensionless). When experimental data of X vs. t are available, a plot of $\log[1/(1 - X)]$ vs. t will be a straight line, if the above exponential equation holds. n and K are determined from the slope and intercept of the straight line. The transformation rate can be derived from the above equation:

$$dX/dt = \{\exp[-(Kt)^n]\}nK^n t^{n-1}$$
$$= nK^n t^{n-1}(1-X) \quad (73)$$

The empirical equation can also be written as

$$X = (Kt)^n - \frac{(Kt)^{2n}}{2!} + \frac{(Kt)^{3n}}{3!} - \cdots \quad (74)$$

If Kt is small, Eq. (74) can be approximated as

$$X = (Kt)^n \quad (75)$$

and the corresponding transformation rate becomes

$$dX/dt = nK^n t^{n-1} \quad (76)$$

Comparison of Eqs. (73) and (76) shows that the only difference between these two empirical rate laws is the factor $(1-X)$, the latter being the fraction of the untransformed phase. The factor $(1-X)$ is called the impingement factor because it takes into account the interference among the growing particles of the transformation product. Accordingly, the exponential form is considered to be corrected for impingement, while the simplified power form is not.

6.2. The Johnson–Mehl and Avrami Equations

The Johnson–Mehl model[35] of transformation kinetics for nucleation and growth processes assumes (a) that the nucleation rate $I [= (1-X)^{-1} dN_v/dt]$* and the growth rate $U (= dr/dt)$ are independent of X and t, (b) that nucleation occurs randomly in the untransformed phase, and (c) that the particles grow as spheres until impingement. N_v is the number of transformed particles per unit volume of the system.

At time t, the volume of a spherical particle that nucleated at time τ ($\tau < t$) is equal to $\frac{4}{3}\pi U^3(t-\tau)^3$. The volume fraction transformed is then given by

$$X_{\text{ext}} = \int_{\tau=0}^{\tau=t} \tfrac{4}{3}\pi U^3 (t-\tau)^3 I \, d\tau \quad (77)$$

---

*The $(1-X)^{-1}$ factor relates the nucleation rate to a unit volume of the untransformed phase, rather than to a unit volume of the total system.

Chapter 2

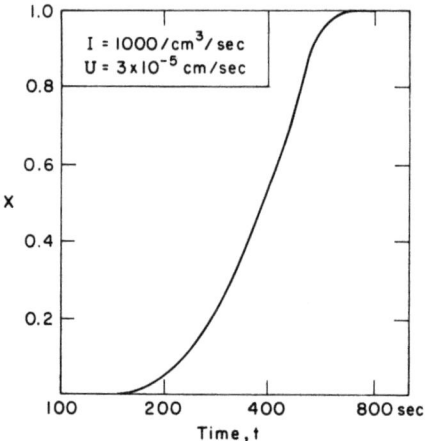

Fig. 15. Sigmoidal curve of fraction transformed vs. time, calculated for constant nucleation and growth rates of spherical particles.

The subscript "ext" stands for "extended," and denotes that the transformed volume is not yet corrected for impingement. Introducing the impingement factor $(1 - X)$, we obtain the Johnson–Mehl equation:

$$X = 1 - \exp(-\tfrac{1}{3}\pi I U^3 t^4) \tag{78}$$

Here the time exponent n is 4 and the reaction rate constant K is $(\tfrac{1}{3}\pi I U^3)^{1/4}$. A plot of X vs. t is sigmoidal in shape (Figure 15).

When the nucleation is so fast that all possible nucleation centers per unit volume (N_v) come into existence at $t \approx 0$, we obtain the Avrami equation[36]:

$$X = 1 - \exp(-\tfrac{4}{3}\pi N_v U^3 t^3) \tag{79}$$

6.3. Transformation Kinetics for Diffusion-Controlled Processes

We can derive expressions for the rate constant K and the time exponent n for the case of spherical growth with a compositional change, where diffusion in the matrix is rate controlling.[37] Let us assume that all the nucleation centers per unit volume (N_v) come into existence at $t \sim 0$, so that $I = 0$ during the transformation. Let X equal the number of g-atoms of B in the transformation product at any time, divided by the total g-atoms of B available for transformation:

$$X = V_\beta c_{\beta\alpha}/[V_0(\bar{c} - c_{\alpha\beta}^{(\infty)})] \tag{80}$$

where V_0 is the volume of the system. Now

$$X_{\text{ext}} = \tfrac{4}{3}\pi r^3 N_v c_{\beta\alpha}/(\bar{c} - c_{\alpha\beta}^{(\infty)}) \tag{81}$$

from which we obtain

$$r = \left[\frac{3X_{\text{ext}}(\bar{c} - c_{\alpha\beta}^{(\infty)})}{4\pi N_v c_{\beta\alpha}}\right]^{1/3} \tag{82}$$

Also,

$$\frac{dX_{\text{ext}}}{dt} = 4\pi r^2 \frac{dr}{dt} \frac{N_v c_{\beta\alpha}}{(\bar{c} - c_{\alpha\beta}^{(\infty)})} \tag{83}$$

where

$$\frac{dr}{dt} = U = \frac{D_\alpha(\bar{c} - c_{\alpha\beta}^{(\infty)})}{r(c_{\beta\alpha} - c_{\alpha\beta}^{(\infty)})} \quad \text{(for } r \gg r^*\text{)} \tag{84}$$

Then

$$\frac{dX_{\text{ext}}}{dt} = D_\alpha N_v^{2/3} \kappa_1 X_{\text{ext}}^{1/3} \tag{85}$$

where

$$\kappa_1 = \left[\frac{48\pi^2 c_{\beta\alpha}^2(\bar{c} - c_{\alpha\beta}^{(\infty)})}{(c_{\beta\alpha} - c_{\alpha\beta}^{(\infty)})^3}\right]^{1/3} \tag{86}$$

If we correct for impingement,

$$dX/dt = (1 - X)\,dX_{\text{ext}}/dt \tag{87}$$

where the $(1 - X)$ factor is used for both "soft" and "hard" impingement. In "soft" impingement, the diffusion fields of two β particles overlap, while "hard" impingement refers to the actual impingement of transformation products, as discussed in Sections 6.1 and 6.2. After substituting Eq. (85) into (87) and integrating, we obtain

$$X = 1 - \exp[-(\tfrac{2}{3}D_\alpha N_v^{2/3}\kappa_1)^{3/2} t^{3/2}] \tag{88}$$

Thus the time exponent is $\tfrac{3}{2}$ here and the reaction rate constant is $\tfrac{2}{3}D_\alpha N_v^{2/3}\kappa_1$.

6.4. Transformation Kinetics for Special Nucleation Sites

The Johnson–Mehl–Avrami equations are sometimes applicable even if the nucleation occurs at special sites such as grain boundaries, grain edges, or grain corners. Let us consider nucleation at such sites for the case of spherical growth at a constant growth rate. Grain corners, for example, can be considered to be randomly distributed within the material. If the nucleation rate is relatively small and constant, then Eq. (78) may be used. However, if the nucleation rate is large, so that all the preferred sites are exhausted during the early stages of the transformation, then Eq. (79) due to Avrami would be appropriate. Here N_v represents the number of grain corners per unit volume of the parent phase.

On the other hand, in the case of grain-boundary and grain-edge nucleation, the preferred sites cannot be regarded as randomly distributed.[38] For example, two particles forming along a given grain edge will impinge with a much higher probability than with particles growing out from another grain edge. If the sites are exhausted at an early stage, a rodlike product phase tends to grow along each grain edge. Similarly, in the case of grain-boundary nucleation, and early site exhaustion, the product grows in a slablike shape. These geometries lead to the following kinetic equations, which are significantly different from the Johnson–Mehl—Avrami equations. For grain-boundary nucleation at constant growth rate U

$$X = 1 - \exp(-2S_v U t) \tag{89}$$

and for grain-edge nucleation at constant growth rate

$$X = 1 - \exp(-\pi L_v U^2 t^2) \tag{90}$$

where S_v is the grain-boundary area per unit volume and L_v is the grain-edge length per unit volume.

In grain-boundary nucleation, when the boundaries become saturated at an early stage in the transformation, we obtain a simple kinetic law to describe the effect of grain size:

$$t_f \approx 0.5 \bar{D}/U \tag{91}$$

where t_f is the time for virtual completion of the transformation and $\bar{D}$ is the average grain diameter.

6.5. Variations of the Time Exponent for Different Situations

In diffusion-controlled growth, the ends of a rod-shaped particle or the rim of a platelike particle grow into new and untapped regions of the matrix. As long as soft impingement has not occurred, the growth rate of these ends or rims is constant, since they may be visualized to advance into a matrix with an unchanging concentration gradient directly ahead. However, the cylindrical faces of the rod or the flat faces of the plate grow parabolically, as in the case of a spherical particle under diffusion-controlled growth. In these instances, the concentration gradient in the matrix decreases progressively as the interface advances. Variations in the time exponent n for different situations are listed in Table 3.

TABLE 3
Time Exponents for Various Kinetic Conditions and Transformation-Product Shapes

Description	Time exponent n	
	$I = \text{const}$	$I = 0$
Growth of spheres		
Constant growth rate	4	3
Parabolic growth rate	5/2	3/2
Growth of plates		
Thickening and lengthening	7/2	5/2
Thickening only	3/2	1/2
Lengthening only	3	2
Growth of rods		
Thickening and lengthening	3	2
Thickening only	2	1
Lengthening only	2	1

7. Particle Coarsening

Coarsening is a structural change in which the average size of precipitate particles in a system increases with time, usually after the precipitation process is complete. Coarsening and Ostwald ripening are synonymous terms in the present context. Coarsening is of considerable interest in physical metallurgy; a typical case in point is a dispersion-hardened alloy where the optimum mechanical properties are obtained by producing a given particle-size range.

7.1. Driving Force for Coarsening

The reduction in the total interfacial energy of the system provides the driving force for coarsening. The equation basic to the kinetic analysis is the Gibbs–Thomson or Thomson–Freundlich solubility relationship:

$$c_{\alpha\beta}^{(r)} = c_{\alpha\beta}^{(\infty)} \exp(2\tilde{V}\sigma/RTr) \tag{92}$$

where $c_{\alpha\beta}^{(r)}$ is the composition of the α matrix in equilibrium with a spherical particle β of radius r (or with a local radius of curvature r), $c_{\alpha\beta}^{(\infty)}$ is the matrix composition in equilibrium with a particle of very large radius, $\tilde{V}$ is the molar volume of the precipitate, and σ is the specific interfacial energy. This equation is applicable to an ideal solution.

For the usual particle sizes and values of the other parameters, the term within the exponential is taken to be small compared to unity, and then Eq. (92) can be simplified to

$$c_{\alpha\beta}^{(r)} = c_{\alpha\beta}^{(\infty)}[1 + (2\tilde{V}\sigma)/RTr] \tag{93}$$

Also, for two particle sizes

$$c_{\alpha\beta}^{(r_1)} - c_{\alpha\beta}^{(r_2)} = c_{\alpha\beta}^{(\infty)} \frac{2\tilde{V}\sigma}{RT}\left(\frac{1}{r_1} - \frac{1}{r_2}\right) \tag{94}$$

Thus a concentration gradient will exist in the matrix between two particles of different sizes if local equilibrium at each particle prevails. Diffusional flow down this concentration gradient will cause the larger particle to grow at the expense of the smaller one.

7.2. Simple Kinetic Analysis

A general kinetic treatment of the particle-coarsening problem must consider particles of arbitrary shapes, variations in the number, size distribution, volume fraction, anisotropy of surface energy, and nonideality of the matrix solution. The rate-controlling step may be either the diffusion of solute atoms through the matrix or the atomic transfer step at the interfaces. Equations (92)–(94) can be applied only when the process is not interface controlled. If diffusion in the matrix is rate controlling, the variation in the diffusion coefficient as a function of concentration should also be considered. A rigorous treatment taking into account all these effects is extremely difficult. For our purposes, we will use a simple model due to Greenwood.[39]

In the Greenwood case, the following assumptions are made: (1) the particles are spherical, (2) the distribution is random, (3) the matrix solution is ideal, (4) diffusion in the matrix is rate controlling, (5) the diffusion coefficient is independent of concentration, and (6) the total volume of the precipitate remains substantially constant during the coarsening process.

From Eq. (93) we obtain

$$c_{\alpha\beta}^{(r_i)} = c_{\alpha\beta}^{(\infty)}[1 + (2\tilde{V}\sigma/RTr_i)] \tag{95}$$

where r_i is the radius of the ith particle. From the growth equation (59), we can write

$$U_i = \frac{dr_i}{dt} = D_\alpha \left(\frac{\partial c_\alpha}{\partial y}\right)_{y=r_i} \bigg/ (c_{\beta\alpha} - c_{\alpha\beta}^{(r_i)}) \tag{96}$$

$c_{\alpha\beta}^{(r_i)}$ can be replaced by $c_{\alpha\beta}^{(\infty)}$, in comparison with $c_{\beta\alpha}$. We now assume that each particle sees a composition c' at large distances, where $c' > c_{\alpha\beta}^{(\infty)}$ because of the interfacial energy contributed by all the other particles. Then

$$\left(\frac{\partial c}{\partial y}\right)_{y=r_i} = \frac{c' - c_{\alpha\beta}^{(r_i)}}{r_i} \tag{97}$$

and

$$U_i = \frac{D_\alpha}{r_i} \frac{c' - c_{\alpha\beta}^{(r_i)}}{c_{\beta\alpha} - c_{\alpha\beta}^{(\infty)}} \tag{98}$$

Equations (97) and (98) are analogous to Eqs. (61) and (62). Let there be N_p particles at time t. Since

$$\sum_i^{N_p} V_i = \text{const} \tag{99}$$

then

$$\sum_i^{N_p} dV_i/dt = 0 \tag{100}$$

and

$$4\pi \sum_i^{N_p} r_i^2\, dr_i/dt = 0 \tag{101}$$

Chapter 2

Substituting for $dr_i/dt = U_i$ from Eq. (98) into Eq. (101), we obtain

$$\sum_i^{N_p} r_i^2 \frac{D_\alpha}{r_i} \frac{c' - c_{\alpha\beta}^{(r_i)}}{c_{\beta\alpha} - c_{\alpha\beta}^{(\infty)}} = 0 \qquad (102)$$

and substituting for $c_{\alpha\beta}^{(r_i)}$ from Eq. (95) into Eq. (102), we get

$$\frac{1}{N_p}\sum_i^{N_p} r_i = \bar{r} = \frac{c_{\alpha\beta}^{(\infty)}}{c' - c_{\alpha\beta}^{(\infty)}} \frac{2\tilde{V}\sigma}{RT} \qquad (103)$$

or

$$c' = c_{\alpha\beta}^{(\infty)}[1 + (2\tilde{V}\sigma/RT\bar{r})] \qquad (104)$$

which equals $c_{\alpha\beta}^{(\bar{r})}$ by comparison with Eq. (93). Thus the assumed composition c' turns out to be that composition of the matrix that is in equilibrium with a particle of radius equal to the mean radius $\bar{r}$.

When this expression for c' is inserted in Eq. (98) for U_i, we obtain

$$\frac{dr_i}{dt} = \frac{D_\alpha}{r_i} \frac{c_{\alpha\beta}^{(\infty)}}{c_{\beta\alpha} - c_{\alpha\beta}^{(\infty)}} \frac{2\tilde{V}\sigma}{RT}\left(\frac{1}{\bar{r}} - \frac{1}{r_i}\right) \qquad (105)$$

According to Eq. (105),

$$U_i < 0 \quad \text{for} \quad r_i < \bar{r}$$
$$U_i > 0 \quad \text{for} \quad r_i > \bar{r}$$
$$U_i = 0 \quad \text{for} \quad r_i = \bar{r}$$

These relationships indicate that (1) particles smaller than $\bar{r}$ will dissolve, (2) particles larger than $\bar{r}$ will grow, and (3) particles having the mean radius $\bar{r}$ will tend to persist.

$U_{\max}$ can be obtained from

$$\frac{dU_i}{dr_i} = \frac{d}{dr_i}\left[\frac{1}{r_i}\left(\frac{1}{\bar{r}} - \frac{1}{r_i}\right)\right] = 0 \qquad (106)$$

Solving, we find that the particle size that grows fastest in a given array of particles is

$$r_m = 2\bar{r} \qquad (107)$$

Substituting Eq. (107) into Eq. (105), we obtain

$$U_{\max} = D_\alpha \frac{c_{\alpha\beta}^{(\infty)}}{c_{\beta\alpha} - c_{\alpha\beta}^{(\infty)}}\left(\frac{2\tilde{V}\sigma}{RT}\frac{1}{r_m^2}\right) \qquad (108)$$

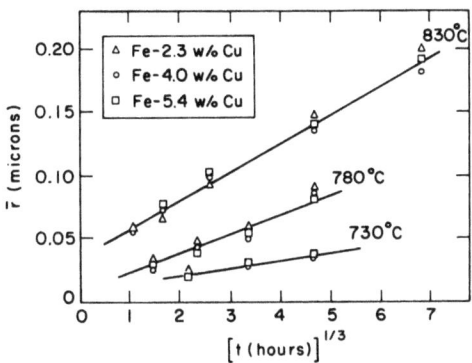

Fig. 16. Linear variation of mean radius with cube root of coarsening time for copper particles in an iron matrix of three Fe–Cu alloys.[41]

Thus the maximum growth rate is inversely proportional to r_m^2 or $\bar{r}^2$.

The rate of coarsening can be expressed as a functional relationship between $\bar{r}$ and t. Consider particles of radius $r_m = 2\bar{r}$. Integrating Eq. (108), we obtain

$$\int_{r_m^0}^{r_m} r_m^2 \, dr_m = D_\alpha \frac{c_{\alpha\beta}^{(\infty)}}{c_{\beta\alpha} - c_{\alpha\beta}^{(\infty)}} \frac{2\tilde{V}\sigma}{RT} \int_0^t dt \qquad (109)$$

where $r_m^0 = 2\bar{r}^0$ at the beginning of the coarsening process. Solving for $\bar{r}$, we find

$$\bar{r} = \left[(\bar{r}^0)^3 + \frac{3}{8} D_\alpha \frac{c_{\alpha\beta}^{(\infty)}}{c_{\beta\alpha} - c_{\alpha\beta}^{(\infty)}} \frac{2\tilde{V}\sigma}{RT} t \right]^{1/3} \qquad (110)$$

$\bar{r}$ can be determined by quantitative metallography from the Fullman relation[40]:

$$\bar{r} = \pi/4\overline{M}$$

where $\overline{M}$ is the mean of the reciprocals of the diameters of the circular intersections of the particles with a random test plane.

The cube-root dependence on time of coarsening suggested by Eq. (110) has been observed experimentally, as in Figure 16.[41]

8. Discontinuous or Cellular Precipitation

In continuous precipitation, particles of the product phase (rich in solute) nucleate and grow, with a continuous decrease (via concentration gradients) in solute concentration of the supersaturated matrix, until all the excess is depleted. The parent phase, except for decrease in solute concentration, retains its crystal structure and matrix orientation. The kinetic equations given in Section 6.3 are applicable to this type of precipitation.

In discontinuous precipitation, on the other hand, cells or colonies of lamellae or rods lying generally parallel to one another within a given cell precipitate from a supersaturated solution. The average composition of the two (or sometimes more) product phases within each cell is equal to the overall composition of the parent phase. As the transformation proceeds, the product phases advance edge-on or end-on and so the cellular configuration grows at the expense of the parent phase, with local changes in matrix concentration at the interfaces.

Typical examples of discontinuous precipitation are the eutectoidal decomposition of austenite (γ-iron) in steel to lamellar ferrite (α-iron) and cemetite (Fe_3C), and the cellular precipitation of tin-rich and lead-rich phases from lead–tin solid solutions.

8.1. Mechanisms of Nucleation and Growth

Smith[42] has proposed that the lamellar reaction is started by the formation of a nucleus at a grain boundary of the parent phase, with the nucleus having a low-energy (coherent or semicoherent) interface relative to one of the adjoining grains. This situation leads to an incoherent interface between the nucleus and the other grain, thereby facilitating more rapid growth into the second grain. The resulting morphology tends to be hemispherical.

Fisher[43] has provided a simple picture for the development of a lamellar configuration. In essence, the nucleation and growth of either phase in the transformation product alter the local composition of the surrounding matrix in such a way as to favor the nucleation and growth of the other phase in the product. The combined process then results in the simultaneous advance of the two phases in cellular fashion, as depicted in Figure 17.

During growth of the lamellar product, the rate-controlling step can be the diffusion of the solute (a) in the parent phase ahead of the advancing cellular region, (b) in one of the two product phases, or (c) in the interface between the parent and product phases. By making simplifying assumptions, the general growth equation for diffusion-controlled growth [Eq. (59)] can be applied to both of the precipitating

Solid-State Phase Transformations

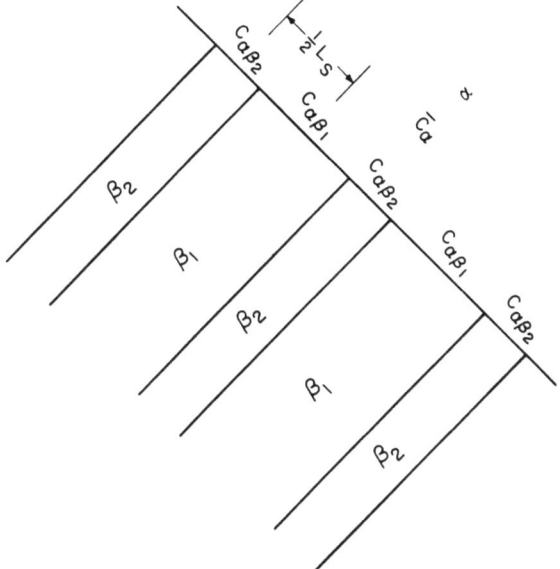

Fig. 17. Schematic representation of lamellar configuration in cellular precipitation. α is the supersaturated parent phase of composition $\bar{c}_\alpha$, while β_1 and β_2 are the product phases. The local composition of the parent phase is $c_{\alpha\beta_1}$ near phase β_1 and $c_{\alpha\beta_2}$ near phase β_2.

phases, giving

$$U = \frac{D}{c_{\beta_2} - c_{\beta_1}} \frac{dc}{dy} \qquad (111)$$

Here D is the diffusion coefficient that pertains to the operative diffusing species and path, as described above. The effective concentration gradient dc/dy can be approximated by taking $\Delta c = c_{\alpha\beta_2} - c_{\alpha\beta_1}$ (see Figure 17) and $\Delta y = L_s/2$, where L_s is the interlamellar spacing. Usually, L_s reaches a steady-state value for isothermal reactions, and then dc/dy and the growth rate become constant with time.

8.2. Interlamellar Spacing

Because of the lamellar morphology arising in cellular transformations, a considerable part of the decrease in free energy per unit volume (Δg) attending the reaction may be absorbed as interfacial energy between the two product phases (β_1 and β_2 in Figure 17).

Chapter 2

Thus

$$\Delta g_{net} = \Delta g + 2\sigma/L_s \qquad (112)$$

where the Δg's are negative.

If the spacing is small enough, the entire driving force can be consumed by the interfacial energy, and then

$$\Delta g_{net} = 0 = \Delta g + 2\sigma/L_s^{min} \qquad (113)$$

or

$$\Delta g = -2\sigma/L_s^{min} \qquad (114)$$

Substituting Eq. (114) into (112), we obtain

$$\Delta g_{net} = -2\sigma[(1/L_s^{min}) - (1/L_s)] \qquad (115)$$

It is usually assumed that, among all the possible interlamellar spacings, the one that will dominate in a given cellular reaction is the one that maximizes the growth rate in Eq. (111). Through arguments similar to those leading to the maximum growth rate in Eq. (69), it can be shown that interlamellar spacing of $L_s = 2L_s^{min}$ will result in the maximum growth rate of the cellular front in Figure 17.

Because of the way that Δg and the operative diffusivity vary with temperature, the growth rate is nil at the equilibrium temperature ($\Delta g = 0$ at T_0), passes through a maximum with decreasing temperature, and then decreases at still lower temperatures. This behavior is analogous to the temperature dependence of the growth rate of precipitated particles, as reflected by Eqs. (70) and (71).

9. Martensitic Transformations

9.1. Characteristics of Martensitic Transformations

Martensitic transformations are diffusionless. Each atom tends to retain its original neighbors during the reaction and there is no interchange among the atoms. The transformation does not involve individual atomic jumps, which are characteristic of a diffusion-controlled or an interface-controlled transformation. The diffusionless characteristic means that the martensitic product has the same composition as the parent phase, and if the parent phase is ordered, this will also be carried over into the transformation product.

Martensitic transformations are displacive in nature, and occur by a shearlike process. A general shear can be separated into a simple shear parallel to the interface (habit) plane (producing a shape change) and a dilatation normal to the interface plane (producing a volume change). The displacive aspect of the transformation results in surface upheavals; that is, fiducial lines or scratches on a prepolished surface undergo shifts in the transformed region but without losing their continuity across the interfaces (see Figure 12).

The martensitic product often has a platelike shape which tends to minimize the strain energy resulting from the displacive transformation. Martensitic growth rates can be extremely rapid ($\sim 10^5$ cm/sec) even at temperatures approaching 0°K. This signifies that the interface is highly mobile, and does not require thermally activated atomic jumping for its motion, in contrast to interface-controlled growth (Section 5.1).

The mechanism and crystal geometry of martensitic transformations have already been discussed in Section 3. Here we will devote some attention to the kinetic aspects of martensitic transformations.

9.2. Kinetic Aspects

Most martensitic transformations occur primarily on cooling, that is, with increasing driving force. Figure 18 shows the changes in electrical resistance as a function of temperature for 70 Fe–30 Ni and 52.5 Au–47.5 Cd alloys.[44] In both cases, the martensite has a lower

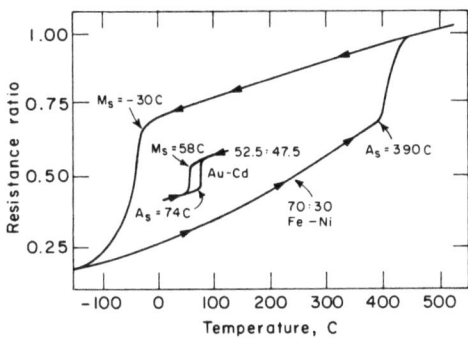

Fig. 18. Resistivity changes during cooling and heating of Fe–Ni and Au–Cd alloys. M_s and A_s represent the temperatures at which the martensitic and the reverse reactions start on cooling and heating, respectively.[44]

Chapter 2

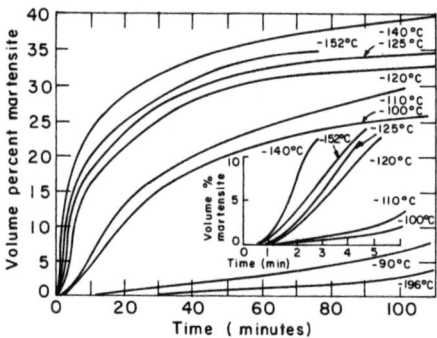

Fig. 19. Isothermal martensitic transformation at different reaction temperatures for an Fe–24 Ni–3 Mn alloy.[45]

resistivity than the parent phase. The start of the martensitic transformation at the temperature M_s on cooling and the start of the reverse transformation at the temperature A_s on heating are also indicated in Figure 18. The cooling and heating transformations are both insensitive to time. As is evident from Figure 18, the temperature hysteresis between the martensitic transformation on cooling and its reversal on heating may be large or small, depending on the system.

In some instances, however, martensitic transformations occur isothermally, and then time becomes an important variable in the kinetics along with temperature, as shown in Figure 19 for an Fe–24 Ni–3 Mn alloy.[45] Even then, the martensitic plates are found to grow rapidly to full size in a fraction of a second. Therefore the increase in transformation rate observed during the early stages can only be attributed to an increase in the nucleation rate. This feature is thought to be due to autocatalysis, in which the first plates formed stimulate new nucleation centers. The inclusion of this autocatalytic effect in the transformation kinetics is treated in the following discussion.[46,47]

The nucleation rate I is, as before, expressed by

$$I = \frac{1}{1-X}\frac{dN_v}{dt} \qquad (116)$$

where N_v is the number of nucleation events (or number of martensitic plates formed) per unit volume. If plates are in the form of thin, circular disks, N_v can be measured as a function of reaction time by using techniques of quantitative metallography.[40]

Letting ΔW_a be the activation energy per nucleation event, we have

$$dN_v/dt = n_t v \exp(-\Delta W_a/kT) \qquad (117)$$

where n_t is the number of nucleation sites per unit volume at time t. n_t can be expressed as

$$n_t = (n_i - N_v + pX)(1 - X) \qquad (118)$$

where n_i is the initial number of nucleation sites per unit volume. The factor N_v in Eq. (118) represents the nucleation sites already activated into martensitic plates, and the $(1 - X)$ factor corrects for the nucleation sites swept out by the transformation product. Here p is the autocatalytic factor, and is defined as the number of new nucleation sites generated per unit volume of martensite formed.

The nucleation rate given by Eq. (117) can be converted to a transformation rate, provided the volume v_t of a martensitic plate formed at any time t is known. This quantity normally decreases with the time of transformation due to the partitioning by the prior-formed martensite. Partitioning formulas are available to calculate v_t using various geometric assumptions,[31] but this procedure does not often yield a realistic description of the transformation process. Alternatively, it is possible to measure the mean volume per plate of all the existing plates, $\bar{v}$, at various fractions of martensite in the transformation by employing quantitative metallography. $\bar{v}$ is related to v_t in the following way[47]:

$$v_t = \bar{v} + N_v \, d\bar{v}/dN_v \qquad (119)$$

and since

$$dX/dt = v_t \, dN_v/dt \qquad (120)$$

we have

$$dX/dt = (n_i - N_v + pX)(1 - X)v[\exp(-\Delta W_a/kT)] \\ \times (\bar{v} + N_v \, d\bar{v}/dt) \qquad (121)$$

This equation can then be fitted to experimental X vs. t curves to determine the activation energy ΔW_a and the autocatalytic factor p.[47]

9.3. The Nucleation Problem

In iron-base alloys, the interface between martensite and the parent phase turns out to be semicoherent if the lattice-invariant

Chapter 2

shear (see Section 3.3) takes place by dislocation motion. A typical value for σ in such cases is ~ 200 ergs/cm^2. The strain-energy parameter A for the shape strain in iron-base alloys is $\sim 2 \times 10^{10}$ ergs/cm^3. With these values of σ and A, Eq. (33) yields an unacceptably high value for ΔG^*, namely several thousand electron volts at typical martensitic transformation temperatures.[44] Any appreciable relaxation of the strains and consequent reduction in strain energy during the transformation does not appear feasible at low temperatures approaching 0°K, where the transformation can proceed vigorously. At the same time a semicoherent interface is dictated by the transformation geometry,[16,17] and so coherent nucleation to reduce σ is not realistic for these diffusionless, displacive transformations. This dilemma has led to the concept of an operational nucleation process, where a preexisting nucleation site simply grows (without requiring activation over the classical nucleation barrier) as soon as the free-energy conditions become sufficiently favorable. The nature of these preexisting sites is a subject of considerable speculation, but to facilitate calculations, they are often assumed to be tiny regions of the product phase, stabilized by the coalescence of line imperfections at elevated temperatures.[44]

One of the recent studies[48] of this problem identifies the operational nucleation step with the nucleation of new dislocation loops at the tips of a preexisting embryo, thus generating the lattice-invariant shear as the particle grows. The total energy change ΔU_{total} accompanying the formation of the loop is the sum of the transformational free-energy change during loop nucleation ΔU_{trans} and

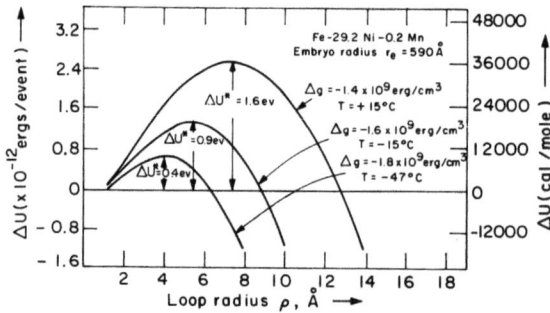

Fig. 20. Free-energy change for the nucleation of a dislocation loop at the tip of a martensitic particle at three different temperatures for an Fe–29 Ni–0.2 Mn alloy.[48]

the elastic and core energy (self-energy) of the nucleation loop ΔU_{disl}. The contribution ΔU_{trans} turns out to be proportional to $\partial(\Delta G)/\partial r$, the free-energy gradient in the radial direction of the plate-like martensitic particle. This partial derivative has to be negative in order for the loop nucleation to occur, and in this respect, is similar to Δg in Eq. (9). The activation barrier ΔU^* is then

$$\Delta U^* = (5/16)\mu_m b^2 \rho^*[\ln(\rho^*/b) - 0.6] \tag{122}$$

where μ_m is the shear modulus of the matrix, b is the Burgers vector of the dislocation loop, and ρ^* is the critical loop radius. The resulting temperature dependence of the activation barrier shown in Figure 20[48] is very much in line with the ΔW_a values found experimentally through the kinetic equation (121).

9.4. Growth Path and Growth Kinetics

The very high growth rate observed for martensitic plates even at subzero temperatures[49] rules out the possibility of a thermally activated mechanism such as outlined in Section 5. The martensitic growth process is more aptly compared with plastic deformation in which the dislocations move with high velocities under a suitably applied stress. The equivalent stress in martensitic growth arises from transformational forces, as introduced in Section 1.1. In the thickening direction of the plate, the transformational stress τ_{th} is shown to be[48]

$$\tau_{th} = \tfrac{2}{3}(\delta r/b)(\Delta g + 2Ac/r) \tag{123}$$

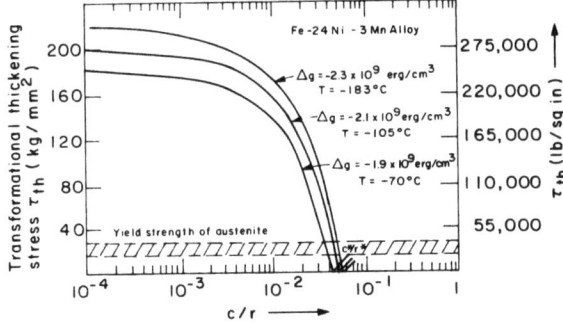

Fig. 21. Transformational stress in the thickening direction as a function of the semithickness-to-radius ratio of a martensitic particle at three driving forces Δg.

Chapter 2

where δr is the spacing between the interfacial dislocations, and c/r is the semithickness-to-radius ratio of the martensitic plate. This stress is plotted in Figure 21 for an Fe–24 Ni–3 Mn alloy at three transformation temperatures. A similar stress for the radial growth processes can be evaluated.[48] It is interesting to note that the transformational stress is commensurate with the yield stress of the parent phase (austenite) when c/r is about that experimentally observed for fully grown martensitic plates ($\sim 1/20$–$1/30$).

The transformational stresses can be related to growth velocities through the equation[50]

$$v_d = v_{\text{lim}} \exp(-d/\tau) \tag{124}$$

where v_d is the dislocation velocity under a shear stress τ, v_{lim} is the limiting dislocation velocity, and d is the drag coefficient. According to this model, the calculated time for the growth of a martensitic plate to full size is of the order of a few microseconds, which is consistent with actual measurements.[49]

10. Massive Transformations

10.1. Experimental Characteristics

Massive transformations form a class of solid-state reactions characterized by no compositional change (like martensitic transformations) and by no shear displacements (unlike martensitic transformations). In 1930, Philips[51] was the first to describe this type of transformation occurring in Cu–Zn alloys within a narrow composition range. Some years later, the term "massive structure" was adopted to describe the blocky appearance of the fcc α crystals obtained by quenching the high-temperature bcc β phase in the Cu–Al system. Knowledge concerning the massive transformations has increased substantially in recent years,[52] and the transformation has been identified in a number of systems, such as Ag–Cd, Cu–Ga, Fe–C, and Fe–Ni.

Segment ba in Figure 4 represents the free-energy change for a massive transformation (from β to α in this case). To obtain a massive transformation, other competing reactions must not intervene. Slow cooling rates can result in decomposition of the parent phase by transformations involving compositional changes; very fast cooling rates may lead to a martensitic reaction. Accordingly, intermediate cooling rates tend to favor the massive transformation.

Solid-State Phase Transformations

Nucleation of a massive reaction usually occurs at the grain boundaries of the parent phase. Except for special instances, there is usually no particular orientation relationship between the product and parent phases, and both of these phases have the same composition. The massive product grows roughly on a spherical front with irregular interfaces, and may sweep across the grain boundaries of the parent phase. The growth rates for massive reactions (~ 1 cm/sec) are typically much higher than those for polymorphic transformations ($\sim 10^{-3}$ cm/sec) or for recrystallization ($\sim 10^{-4}$ cm/sec), but are much smaller than for martensitic transformations (up to $\sim 10^5$ cm/sec).

10.2. Nucleation and Growth Kinetics

Measurements of the isothermal growth rate of the massive transformation in a Cu–38 at. % Zn alloy have been carried out in some detail.[53] A sample plot of growth distance vs. time at a given temperature is shown in Figure 22. The linear relationship indicates a constant growth rate of 0.8 cm/sec, after an initial delay time of 5 msec.

The growth equation for an interface-controlled transformation [Eq. (52) in Section 5.1] has been applied to these measurements. The value of ΔG_D thus determined is consistent with grain-boundary diffusion in this alloy, suggesting that the growth rate is controlled by thermally activated atomic jumping across the transformation interface.

The massive transformation in Cu–Zn alloys is assumed to proceed from preexisting submicroscopic particles that precipitate

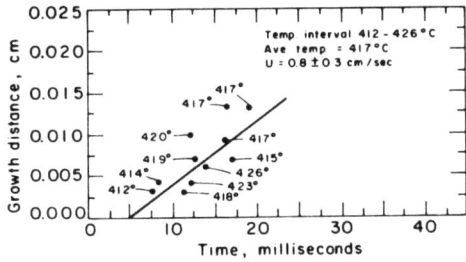

Fig. 22. Linear relationship between massive growth distance and holding time at $\sim 417°C$ in a Cu–38 at. % Zn alloy.[53]

Chapter 2

from the parent phase while passing through a two-phase region during the quenching of the parent phase to the reaction temperature. These precipitated particles form via a diffusion-controlled process (Section 5.2), by rejecting excess zinc into the adjacent matrix. The delay time in Figure 22 can be identified with the time required to absorb the excess zinc from the surrounding matrix; during this "incubation period" there is rapid acceleration of the growth rate to a steady-state value corresponding to the constant interface-controlled rate shown in Figure 22.

11. Polymorphic Transformations

11.1. Introduction

Polymorphic transformations are of particular interest to the chemist. The experimental characteristics of polymorphic changes in a number of systems have been covered in recent reviews,[4,7] and so no attempt will be made here to cover the ground that way. However, it may be worthwhile to describe the classification of such transformations proposed by Buerger.[54]

11.2. Buerger's Classification

Buerger[54] has given a structural classification of solid-state transformations based on changes in bond type and coordination. The categories are:

1. Transformations of secondary coordination
    Displacive   rapid
    Reconstructive   sluggish
2. Transformations of disorder
    Rotational   rapid
    Substitutional   sluggish
3. Transformations of first coordination
    Dilatational   rapid
    Reconstructive   sluggish
4. Transformations of bond type   (usually sluggish)

Transformations of secondary coordination (category 1) occur in network structures having atoms of low primary coordination.

For example, in silicate structures, a transformation of secondary coordination takes place by rearrangement of the silicate tetrahedra relative to one another but without change in the primary coordination within the tetrahedra. This transformation can be brought about in two ways: (a) by displacing the tetrahedra with respect to one another without breaking the bonds between them, i.e., a displacive transformation, and (b) by breaking the bonds between the tetrahedra completely and then rearranging them in a new array corresponding to the product crystal structure, i.e., a reconstructive transformation.

Category 2, transformations of disorder, also has two subsets. Groups of tightly bound atoms in an ordered structure can rotate relative to the rest of the structure and so induce disorder. Likewise, interchanging of positions among atoms in a random way can cause disordering. These two possibilities correspond respectively to the rotational and substitutional types in Buerger's classification.

As in category 1, transformations of first coordination can also occur with or without breaking of bonds. These two possibilities are designated as reconstructive and dilatational types in category 3.

It is instructive to compare reconstructive transformations with those of interface-controlled growth. Polymorphic transformations, massive transformations, and recrystallization (discussed in Section 12) have many similarities in this context. The displacive and dilatational types described by Buerger are comparable to the martensitic transformations in Section 9. No (long-range) diffusion-controlled transformations are included in Buerger's classification.

12. Recrystallization and Grain Growth

Recovery, recrystallization, and grain growth are phenomena intimately associated with the annealing of cold-worked materials. In crystalline materials, the density of dislocations and point imperfections (vacancies and interstitialcies) increases with the extent of plastic deformation at temperatures below about 0.3–0.4 of the melting point. All such structural defects have extra energy associated with them, and so the plastically deformed state has a higher free energy than the undeformed (or annealed) state. Between 1% and 10% of the deformation energy remains stored in the material, and this provides the driving force for various relaxation processes during subsequent annealing. After recovery and/or recrystallization, a further reduction of free energy in the system can occur by

Chapter 2

grain growth, which reduces the grain-boundary area (and hence the grain-boundary energy) per unit volume.

During recovery, the excess point imperfections produced by plastic deformation anneal out in various ways, such as by being absorbed at grain boundaries and dislocations, by clustering into vacancy and interstitial loops, or by the self-annihilation of vacancies and interstitialcies. Correspondingly, the dislocations become more mobile on heating and rearrange themselves into lower energy configurations or undergo annihilation by combinations of opposite signs. The study of recovery in single crystals has also brought to light the phenomenon of polygonization, in which dislocations of the same sign can adopt a lower energy configuration by arranging themselves in rows to produce small-angle tilt and twist boundaries.

The kinetics of recovery can be followed by measuring the change in some physical property, such as electrical resistivity, as a function of time and temperature. However, if the annealing temperature is raised to a point where the dislocations become quite mobile, large-scale rearrangements ensue and result in recrystallization.

12.1. Recrystallization

Recrystallization is the nucleation and growth of relatively strain-free crystals, progressively consuming the cold-worked matrix. The growth occurs by the migration of high-angle grain boundaries between the distorted and strain-free states, until impingement occurs between the strain-free crystals.

The kinetics of recrystallization can be quantitatively described by nucleation and growth formulations such as those due to Johnson and Mehl[35] and Avrami[36] [see Eqs. (78) and (79) in Section 6.2]. The nucleation and growth rates have been measured in a number of metallic systems undergoing recrystallization.

The simple nucleation kinetics outlined in Section 4 [Eq. (10)] can be applied to the type of nucleation that takes place during recrystallization. It should be noted, however, that there is no change in crystal structure during recrystallization; the driving force $(-\Delta g)$ arises from the difference in the free energy between the strain-free and distorted states. For heavily cold-worked materials, the driving force has a value of $\sim 10^9$ ergs/cm^3. The nucleation rate can then be described by Eq. (17), where ΔG_D represents the grain-boundary diffusivity. The calculated critical size from Eq. (11) is about $r^* = 100$ Å for slightly worked copper and 5 Å for highly worked copper.

Nevertheless, there are some uncertainties about applying standard nucleation theory to recrystallization. The formation of a strain-free nucleus by thermal fluctuations during recrystallization has long been in question. Two alternative mechanisms are grain-boundary migration and subgrain growth. In the first case, high-angle grain boundaries simply migrate by the motion of an irregular front, leaving behind strain-free material of the same orientation as the strained grain from which the boundary is moving away. Thus no nucleus of a new orientation is formed, and so there is very little interfacial energy involved that would create a nucleation barrier. Presumably, the free energy in the region into which the grain boundary advances is somewhat higher than that of the originating grain.

The second mechanism involves subgrain growth with the gradual annihilation of small-angle boundaries by the dislocations moving out of them into the advancing high-angle interfaces between the distorted and strain-free regions. Such a process produces a strain-free crystal where several subgrains existed initially, and so generates a "nucleus" for growth in recrystallization.

12.2. Grain Growth

By definition, grain growth refers to the increase in the average grain size upon further annealing after the recrystallization process is complete. The grain-size distribution often remains approximately the same during grain growth. Any radical change in the grain-size distribution is termed discontinuous grain growth or secondary recrystallization.

When the average grain size increases, the grain-boundary area per unit volume decreases, and the corresponding decrease in interfacial energy per unit volume supplies the driving force for grain growth. Let us consider the idealized case of a spherical grain of radius R. In that case, the grain-boundary energy per unit volume is $4\pi R^2 \sigma / \frac{4}{3}\pi R^3 = 3\sigma/R$. For a curved boundary, the radius of curvature is positive from one side and negative from the other. The free energy difference per unit volume is

$$\Delta g = C\sigma[(1/R_1) - (1/R_2)] \qquad (125)$$

where C is a constant. If $R_1 = -R_2 = R$,

$$\Delta g = 2C\sigma/R \qquad (126)$$

Chapter 2

The latter equation is analogous to that for the pressure difference due to interfacial curvature.

The grain-growth kinetics can be described by the following empirical equation (when the initial grain size is small enough to be neglected):

$$\bar{D} = C_1 t^n \quad (127)$$

where $\bar{D}$ is the average grain diameter. Simple theory indicates that the time exponent n should be $\frac{1}{2}$. If we assume that the rate of change of $\bar{D}$ is proportional to Δg, then

$$d\bar{D}/dt = C_2 \Delta g \quad (128)$$

From Eqs. (126) and (128), we can write

$$d\bar{D}/dt = C_3 \sigma/\bar{D} \quad (129)$$

$$(\bar{D}_t^2 - \bar{D}_0^2) = 2C_3 \sigma t \quad (130)$$

where the subscripts t and 0 refer to time $t = t$ and $t = 0$. This parabolic grain-growth law has been observed experimentally.

Grain growth is usually hampered by the presence of second-phase particles. During grain-boundary migration the combined interfacial energy of a grain boundary and a precipitate particle is the same before the boundary meets the particle and after it has left the particle behind. However, when the particle lies in the boundary, the grain-boundary area is reduced by the cross section of the particle, and has to be increased again if the boundary is to pull itself away from the particle in the grain-growth process. With N_v particles of average radius r_p in a unit volume, the driving force in Eq. (126) is reduced to

$$\Delta g = 2C\sigma[(1/R) - N_v \pi r_p^2] \quad (131)$$

assuming, for the sake of simplicity, that the specific interfacial energy between the particle and matrix is the same as the specific grain-boundary energy σ. The volume fraction of particles X is

$$X = \tfrac{4}{3}\pi r_p^3 N_v \quad (132)$$

Combining the above two equations, we obtain

$$\Delta g = 2C\sigma[(1/R) - (3X/4r_p)] \quad (133)$$

Thus, at $R = 4r_p/3X$, $\Delta g = 0$, and so grain growth will cease at a finite grain size, which is found experimentally.

13. Spinodal Decomposition

As pointed out in Section 1, spinodal decomposition starts with infinitesimal compositional fluctuations spread out through large volumes undergoing transformation. The compositional fluctuations grow in amplitude as a function of time, eventually yielding the final product phases. This mode of transformation is basically different from the nucleation and growth type of reactions discussed earlier, and has been well summarized by Cahn.[29] Spinodal decomposition proceeds with a progressive decrease in free energy from the very beginning, without any need for mounting a nucleation barrier.

In binary systems exhibiting a miscibility gap of the type represented by Figure 5, there are two inflection points in the free-energy vs. composition curves. For all compositions between such inflection points, the second derivative of free energy with respect to composition is negative. When the inflection-point compositions are plotted as a function of temperature (on a phase diagram), we obtain the chemical spinodal curve. It follows that the second derivative of free energy with respect to composition is zero along the spinodal curve, is negative within the spinodal, and is positive outside.

13.1. Thermodynamics of Compositional Fluctuations

The free-energy change during a transformation with compositional change was discussed in Section 2.4. We now address the case of compositional change in which nucleation and growth are not operative.

The free energy of a solid solution of nonuniform composition can be divided into two terms[55–57]: (1) the free energy of an infinitesimal volume of composition c integrated over the whole volume under consideration, and (2) the free energy associated with the compositional gradient across any infinitesimal volume, again integrated over the whole volume. The second part is analogous to the usual surface energy term. If, for example, there is a coherent interface between the two phases of different composition (α_1 and α_2), surface energy is associated with this interface. When compositional fluctuations occur, the interfacial bond distribution changes and the corresponding energy change in any infinitesimal volume is proportional to the square of the concentration gradient across the volume. If this change is integrated over the whole system, we have a gradient

Chapter 2

energy term which can be considered as the surface energy of the diffuse interfaces.

Consider for the sake of simplicity a one-dimensional compositional fluctuation along the length x of a bar of overall composition $\bar{c}$ and of constant cross-sectional area A'. The (Helmholtz) free energy of this bar is then given by

$$F = A' \int f(c)\, dx + A' \int \kappa (dc/dx)^2\, dx \qquad (134)$$

where $f(c)$ is the free energy per unit volume of composition c, and κ is the gradient energy coefficient.

If the fluctuation is sinusoidal, the composition c at position x can be written as

$$c - \bar{c} = Y \cos kx \qquad (135)$$

where Y is the amplitude of the fluctuation and k is the wave number, related to the wavelength λ through $k = 2\pi/\lambda$. Expanding $f(c)$ about the mean composition $\bar{c}$ and neglecting terms higher than second order, we have

$$f(c) = f(\bar{c}) + (c - \bar{c})\left(\frac{df}{dc}\right)_{c=\bar{c}} + \tfrac{1}{2}(c - \bar{c})^2 \left(\frac{d^2f}{dc^2}\right)_{c=\bar{c}} \qquad (136)$$

From the above three equations, we can derive the free energy per unit volume:

$$f = f(\bar{c}) + \frac{Y^2}{4}\left(\frac{d^2f}{dc^2}\right)_{c=\bar{c}} + \frac{Y^2}{2}\kappa k^2 \qquad (137)$$

Therefore the free-energy change due to a sinusoidal fluctuation is given by

$$\Delta f = \frac{Y^2}{4}\left[\left(\frac{d^2f}{dc^2}\right)_{c=\bar{c}} + 2\kappa k^2\right] \qquad (138)$$

It is easy to see from Eq. (138) that $\Delta f = 0$ for a composition on the spinodal curve only when $k = 0$ (i.e., $\lambda = \infty$). For a fluctuation of a finite wavelength, Δf can equal zero only within the spinodal. For a given composition and temperature within the spinodal, there is a critical k_c (and λ_c) at which $\Delta f = 0$. We have $\Delta f < 0$ when $\lambda > \lambda_c$; that is, only wavelengths longer than the critical can form spontaneously. By setting $\Delta f = 0$ in Eq. (138) and using $k = 2\pi/\lambda$, we obtain

$$\lambda_c = \left[-8\pi^2 \kappa / \left(\frac{d^2 f}{dc^2}\right)_{c=\bar{c}} \right]^{1/2} \quad (139)$$

When the lattice parameter of the parent phase varies with composition, a compositional fluctuation will introduce elastic strains, inasmuch as coherency is maintained during the fluctuation. Cahn[56] has considered the modification of Eq. (138) when strain energy is present, with the following result for isotropic solids:

$$\Delta f = \frac{Y^2}{4} \left[\left(\frac{d^2 f}{dc^2}\right)_{c=\bar{c}} + \frac{2\eta^2 E}{1-\nu} + \kappa k^2 \right] \quad (140)$$

where η is the strain per unit composition change, E is Young's modulus, and ν is Poisson's ratio. Equation (139) for λ_c will now include the strain-energy term. Also, the anisotropic elastic properties of crystals lead to the interesting consequence that fluctuations along certain crystallographic directions will be energetically favored.

13.2. Kinetics of Spinodal Decomposition

The compositional fluctuations are brought about by diffusion, but the diffusion is uphill here, in contrast to the usual downhill diffusion through the matrix during the growth of a particle (see Section 5.2). Therefore chemical-potential gradients are used in place of concentration gradients in the flux equation.

The flux J under a chemical-potential gradient is

$$J = -M[\partial(\mu_B - \mu_A)/\partial x] \quad (141)$$

where M is the mobility, and μ_A and μ_B are the chemical potentials of the A and B components in the binary solution. It can be shown that[29]

$$\mu_B - \mu_A = \partial f / \partial c \quad (142)$$

However, the gradient energy associated with a compositional fluctuation tends to inhibit the fluctuation and the net chemical-potential difference (driving force) available is smaller than that given by Eq. (142). When the gradient energy is taken into account, Eq. (142) becomes

$$\mu_B - \mu_A = \partial f / \partial c - 2\kappa \, \partial^2 c / \partial x^2 \quad (143)$$

Chapter 2

Substituting in Eq. (141), we obtain

$$J = -M\frac{\partial}{\partial x}\left(\frac{\partial f}{\partial c} - 2\kappa\frac{\partial^2 c}{\partial x^2}\right) \quad (144)$$

$$= -M\left(\frac{\partial^2 f}{\partial c^2}\frac{\partial c}{\partial x} - 2\kappa\frac{\partial^3 c}{\partial x^3}\right) \quad (145)$$

For nonsteady-state conditions

$$\frac{\partial c}{\partial t} = -\frac{dJ}{dx} = M\left(\frac{\partial^2 f}{\partial c^2}\frac{\partial^2 c}{\partial x^2} - 2\kappa\frac{\partial^4 c}{\partial x^4}\right) \quad (146)$$

provided that $\partial^2 f/\partial c^2$ can be taken as independent of position (and therefore of composition). This approximation is valid only during the early stages of decomposition, when the amplitude of the fluctuation is still small. Equation (146) can also be modified to include the coherency strain energy.

A general solution of Eq. (146) is

$$A(k, t) = A(k, 0)\exp[R(k)t] \quad (147)$$

where $A(k, 0)$ is the initial amplitude of the Fourier component of wave number k, and $R(k)$ is the so-called amplification factor defined by:

$$R(k) = -Mk^2\left[\left(\frac{\partial^2 f}{\partial c^2}\right)_{c=\bar{c}} + 2\kappa k^2\right] \quad (148)$$

By setting $dR(k)/dk = 0$, we find that the maximum in the amplification factor occurs when

$$\lambda_{\max} = \sqrt{2}\,\lambda_c \quad (149)$$

$\lambda_{\max}$ represents the wavelength that will grow the fastest in intensity among the compositional fluctuations.

13.3. Experimental Results

Several experimental checks of the diffusion equation (146) are now available for spinodal decomposition. The early observations of satellites about the Bragg peaks and small-angle x-ray scattering experiments confirm the compositional modulations during spinodal decomposition.[58]

Any arbitrary compositional fluctuation due to thermal agitation can be expressed as the sum of a series of sinusoidal fluctuations

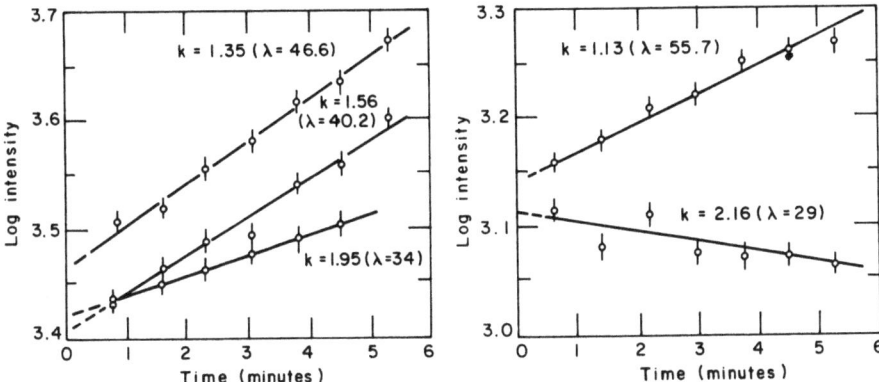

Fig. 23. Plot of log of intensity vs. time for five wave numbers; data are for small-angle x-ray spectra of Al–22 at.% Zn specimens, spinodally transformed at 65°C.[58] The wavelengths of the compositional waves are indicated in angstrom units. The slope of each line is the amplification factor $R(k)$.

of different Y and k values. Since $R(k)$ occurs within the exponential in Eq. (147), the wavelength λ_{max} corresponding to R_{max} will easily outgrow the other wavelengths. This accounts for the uniform spacing of precipitate particles that are believed to originate from spinodal decomposition. Moreover, this relatively uniform spacing is expected to resist coarsening tendencies (see Section 7.2). The precipitate morphology observed in isotropic materials such as glass undergoing spinodal decomposition has been simulated by computer.[29] In crystalline systems exhibiting elastic anisotropy, the eventual precipitation tends to occur parallel to certain crystallographic directions.

The kinetics of spinodal decomposition has been studied in detail by Rundman and Hilliard,[58] using a small-angle x-ray scattering technique. The log of intensity plotted against time in Figure 23 shows a linear relationship, as expected from Eq. (147), and the slope of each line is the amplification factor $R(k)$. For the smallest wavelength indicated ($\lambda = 29$ Å), $R(k)$ is negative, which means that this wavelength is shrinking in intensity. Alternatively, $R(k)$ is positive for the longer wavelengths shown, and so the latter are growing in intensity.

Acknowledgments

The authors express their appreciation to the office of Naval Research for their support over the years of research at the

Chapter 2

Massachusetts Institute of Technology related to the field of solid-state phase transformations, thereby providing a framework for this review and perspective.

One of the authors (V.R.) also wishes to convey his thanks to the Director, Indian Institute of Technology, Delhi, for support during the preparation of this chapter.

References

1. M. Cohen *et al.*, in *Perspectives in Materials Research* (L. Himmel, J. J. Harwood, and W. J. Harris, eds.), p. 309, U.S. Office of Naval Research, Washington, D.C. (1963).
2. *Conference on Physical Properties of Martensite and Bainite, Special Report No. 93*, Iron and Steel Institute, London (1965).
3. J. W. Christian, *The Theory of Transformations in Metals and Alloys*, Pergamon (1965).
4. C. N. R. Rao and K. J. Rao, in *Progress in Solid-State Chemistry*, p. 131, Pergamon (1967).
5. Proceedings of the First International Conference on Metal–Nonmetal Transitions, *Rev. Mod. Phys.* **40**, 673 (1968).
6. C. M. Wayman, in *Advances in Materials Research* (H. Herman, ed.), p. 147, Interscience, New York (1968).
7. T. J. Gray and V. D. Frechette, eds., *Kinetics of Reactions in Ionic Systems*, Plenum, New York (1969).
8. *Symposium on Mechanisms of Phase Transformations in Crystalline Solids*, Institute of Metals Monograph No. 33, London (1969).
9. *Phase Transformations*, ASM, Ohio (1970).
10. Symposium on Formation of Martensite in Iron Alloys, *Met. Trans.* **2**(9) (1971).
11. V. Raghavan and M. Cohen, *Recent Developments in Metallurgical Science and Technology* (*Physical Metallurgy*), p. 255, Indian Institute of Metals, New Delhi (1972).
12. H. J. Aaronson, C. Laird, and K. R. Kinsman, in *Phase Transformations*, p. 313, ASM, Ohio (1970).
13. D. S. Lieberman, in *Phase Transformations*, p. 1, ASM, Ohio (1970).
14. L. C. Chang and T. A. Read, *Trans. AIME* **197**, 1503 (1953).
15. E. C. Bain, *Trans. ASME* **70**, 25 (1924).
16. M. S. Weschler, D. S. Lieberman, and T. A. Read, *Trans. AIME* **197**, 1503 (1953).
17. J. S. Bowles and J. D. Mackenzie, *Acta Met.* **2**, 129, 138, 224 (1954).
18. M. Di Domenico, Jr. and S. H. Wemple, *Phys. Rev.* **155**, 539 (1967).
19. J. H. Hollomon and D. Turnbull, *Prog. Metal Phys.* **4**, 333 (1953).
20. D. Turnbull, *Solid State Physics* **3**, 225 (1956).
21. J. N. Hobstetter, *Decomposition of Austenite by Diffusional Processes*, p. 1, Interscience, New York (1962).
22. K. C. Russell, in *Phase Transformations*, p. 217, ASM, Ohio (1970).
23. R. B. Nicholson, in *Phase Transformations*, p. 269, ASM, Ohio (1970).
24. M. Volmer and A. Weber, *Z. Physik. Chem.* **119**, 277 (1926).

25. R. Becker and W. Doring, *Ann. Physik* **24**, 719 (1935).
26. R. Becker, *Ann Physik* **32**, 128 (1938).
27. G. Borelius, *Arkiv. Mat. Astron. Fys. A* **32** (1945).
28. J. N. Hobstetter, *Trans. AIME* **180**, 121 (1949).
29. J. W. Cahn, *Trans. TMS-AIME* **242**, 166 (1968).
30. F. R. N. Nabbarro, *Proc. Roy. Soc. A* **175**, 519 (1940).
31. J. C. Fisher, J. H. Hollomon, and D. Turnbull, *Trans. AIME* **185**, 691 (1949).
32. J. D. Eshelby, *Proc. Roy. Soc. A* **241**, 376 (1957).
33. P. J. Clem and J. C. Fisher, *Acta Met.* **3**, 347 (1958).
34. J. W. Cahn, *Acta Met.* **4**, 449 (1956).
35. W. A. Johnson and R. F. Mehl, *Trans. AIME* **135**, 416 (1939).
36. M. Avrami, *J. Chem. Phys.* **7**, 1103 (1939).
37. C. Wert, *J. Appl. Phys.* **20**, 943 (1949).
38. J. W. Cahn, *Trans. AIME* **209**, 140 (1957).
39. G. W. Greenwood, *Acta Met.* **4**, 243 (1956).
40. R. L. Fullman, *Trans. AIME* **197**, 447 (1953).
41. G. R. Speich and R. A. Oriani, *Trans. TMS-AIME* **233**, 623 (1965).
42. C. S. Smith, *Trans. ASM* **45**, 562 (1953).
43. J. C. Fisher, *Thermodynamics in Physical Metallurgy*, p. 201, ASM, Ohio (1950).
44. L. Kaufman and M. Cohen, *Prog. Met. Phys.* **7**, 165 (1958).
45. C. H. Shih, B. L. Averbach, and M. Cohen, *Trans. AIME* **203**, 183 (1955).
46. V. Raghavan and A. R. Entwisle in *Conference on Physical Properties of Martensite and Bainite*, Special Report No. 93, p. 30, Iron and Steel Institute, London (1965).
47. S. R. Pati and M. Cohen, *Acta Met.* **19**, 1327 (1971).
48. V. Raghavan and M. Cohen, *Acta Met.* **20**, 333 (1972).
49. R. F. Bunshah and R. F. Mehl, *Trans. AIME* **197**, 1951 (1953).
50. V. Raghavan and M. Cohen, *Acta Met.* **20**, 779 (1972).
51. A. J. Philips, *Trans. AIME* **89**, 194 (1930).
52. T. B. Massalski, in *Phase Transformations*, p. 433, ASM, Ohio (1970).
53. D. A. Karlyn, J. W. Cahn, and M. Cohen, *Trans. TMS-AIME* **245**, 197 (1969).
54. M. J. Buerger, *Phase Transformations in Solids*, p. 183, Wiley, New York (1951).
55. J. W. Cahn and J. E. Hilliard, *J. Chem. Phys.* **28**, 258 (1958).
56. J. W. Cahn, *Acta Met.* **9**, 795 (1961).
57. J. E. Hilliard, in *Phase Transformations*, p. 497, ASM, Ohio (1970).
58. K. B. Rundman and J. E. Hilliard, *Acta Met.* **15**, 1025 (1967).

3

Clustering Effects in Solid Solutions

D. de Fontaine

School of Engineering and Applied Science
University of California at Los Angeles

1. Introduction

Solid-state transformations can be classified into two main categories: (a) *homogeneous* reactions, which occur in the bulk crystalline (or amorphous) phases, and (b) *heterogeneous* ones, which occur at structural defects such as grain boundaries, dislocations, etc. This chapter covers homogeneous reactions only, and particularly those transformations that are caused by instabilities or metastabilities of solid solutions to local compositional changes. Such *replacive* reactions can be of two types: (a) *clustering* transformations, which favor like-atom bonds, and (b) *ordering* transformations, which favor unlike bonds. Only clustering reactions are treated here, although the formalism can easily be extended to cover order–disorder reactions as well.

The pioneering work on phase stability was of course that of Gibbs,[1] who introduced the idea of compositional changes that are large in degree but small in extent (nucleation type) and those that are small in degree but large in extent (spinodal type). Nucleation theory was developed by Volmer and Weber[2] and Becker and Döring.[3] Excellent reviews of nucleation in solids have been given by Turnbull,[4] Christian,[5] and Russell.[6]

It seemed for some time that further theoretical work in phase transformations would be confined to improving nucleation theories until, in 1956, Hillert[7,8] proposed a radically new idea which consisted in imposing a plane wave compositional modulation in the solid solution and calculating its energy by bond counting within and across the planes of the wave. The configurational entropy was of Bragg–Williams type. In this way Hillert was able to study stable or metastable one-dimensional concentration modulations as well as the approach to equilibrium through a suitable diffusion equation. Hillert's was a discrete, one-dimensional model and the equations to be solved were nonlinear. The model was in principle universal and included as special cases nucleation and growth, spinodal decomposition, and order–disorder reactions.

Recognizing the importance of Hillert's formalism, Cahn and Hilliard[9] then proposed a continuum, three-dimensional treatment leading to a very elegant theory of nucleation.[10] In 1961 Cahn[11] introduced the concept of spinodal decomposition characterized by periodic solutions of a linearized partial differential equation which governs the early-stage kinetics of unstable solid solutions. One of Cahn's most remarkable contributions was the inclusion in the diffusion equation of an anisotropic elastic energy term[12] which was found to account successfully for observed morphologies. Cahn also introduced the concept of a *coherent miscibility gap*,[13] thereby providing theoretical justification for the experimentally observed "G.P. zone solvus."

Something was lost in the discrete → continuum translation, however: the order–disorder reaction. It was reintroduced by Cook et al.[14,15] through a discrete-space version of Cahn's linear diffusion equation and further extended to multicomponent systems by de Fontaine[16–18] and Morral and Cahn.[19] These later developments are not covered here, but will be treated in a forthcoming review.[20]

Attempts at solving the general nonlinear diffusion equation governing all clustering processes have not met with a great deal of success, although the mathematical models of Cahn[21] and Langer[22] and the numerical calculations of Hillert,[8] de Fontaine,[16] Shendalman and O'Toole,[23] and Swanger et al.[24] have yielded some interesting partial results.

Previous reviews of the Hillert–Cahn–Hilliard theories were given by Cahn,[25,26] Hilliard,[27] de Fontaine,[28] and Bonfiglioli (in Spanish).[29] Of particular interest are the historical introduction in Cahn's 1968 review paper and the discussion of experimental results in Hilliard's review paper. Neither of these aspects will be covered extensively here. Topics included in the present article, but not previously reviewed, are the conceptually unifying anharmonic oscillator analogy of Khachaturyan and Suris[30] and Langer[22] (Section 5.2), and the troublesome case of the vanishing spinodal in Kikuchi's cluster expansions[31] (Section 5.5). The free energy functional is derived in Sections 3 and 4, stationary states (coherent equilibrium, diffuse interface, and critical nucleus) are discussed in Section 5, compositional fluctuations are briefly treated in Section 6, the kinetics of decomposition are given in Section 7, and decomposition morphologies are described in Section 8.

2. Pairwise Internal Energy

The internal energy of a solid solution is generally approximated by a sum of pair energies.[32] Since in this approximation many-body interactions are neglected, the internal energy is expressed mathematically as a quadratic form in the suitable concentration variables. For various applications (fluctuations, kinetics) it is desirable to express this quadratic form in terms of continuously variable concentrations. This can be accomplished by the averaging technique described below.

2.1. Average Concentrations

Consider a binary crystalline solid solution each site* (p) of which can be occupied by an A or a B atom. Actual site occupancy can be described by the following concentration (or occupancy) operator[18]:

$$\hat{c}(p) = \begin{cases} 1 & \text{if atom B is at } (p) \\ 0 & \text{if atom A is at } (p) \end{cases}$$

The overall average concentration of B is thus

$$c_0 = (1/N) \sum_p \hat{c}(p) \qquad (1)$$

where the sum extends over all N lattice sites in the crystal.

Imagine that a portion of sites is masked so that only a region R centered on point (p) remains. The average concentration in R is

---

* Parentheses are used around symbols such as (p) to indicate triplets of numbers. Thus, an arbitrary lattice point is $(p) \equiv (p_1, p_2, p_3)$. If a primitive coordinate system is used, the p_i are integers.

obtained as in Eq. (1):

$$c(p) = (1/N_R) \sum_{r \in R} \hat{c}(p + r) \tag{2}$$

where the sum now extends only to the N_R lattice sites in R centered on (p). In this abstract definition region R could be, for example, a small volume element $d\mathbf{x}$ centered about point $\mathbf{x}(p)$ in crystal space. The continuous variable $c(\mathbf{x})$ then represents the local concentration in a nonuniform binary solid solution.

2.2. Pair Interactions

In the pair interaction model the total internal energy E_t of a solid solution can be expressed in terms of concentration operators as follows[18,33]:

$$E_t = \tfrac{1}{2} {\sum_{p,p'}}' [v_{AA}(p' - p)\hat{c}_A(p)\hat{c}_A(p') + v_{BB}(p' - p)\hat{c}_B(p)\hat{c}_B(p')$$
$$+ 2v_{AB}(p' - p)\hat{c}_A(p)\hat{c}_B(p')] \tag{3}$$

where the subscripts indicate the atomic species explicitly and where the double sum in p and p' is over all lattice sites, with $(p = p')$ excluded by the accent on the summation sign. The pair bond energies are designated by $v_{AA}, \ldots,$ and depend in this approximation only on the lattice distance $\mathbf{x}(p' - p)$.

The following steps are now carried out: (a) The concentration operators $\hat{c}$ are replaced by local average concentrations c according to the averaging method outlined above, and (b) average concentrations are replaced by concentration variations $\xi(p)$, or deviations from the mean defined by

$$\xi(p) = c(p) - c_0 \tag{4}$$

The resulting energy expression is

$$E_t = E_A + E_B + E$$

where E_A and E_B are the energies of the pure components A and B and where E is the energy of mixing

$$E = N\omega_0 c_0(1 - c_0) - {\sum_{p,p'}}' \omega(p' - p)\xi(p)\xi(p') \tag{5}$$

with

$$\omega(r) = v_{AB}(r) - \tfrac{1}{2}[v_{AA}(r) + v_{BB}(r)] \tag{6}$$

and

$$\omega_0 = \sum_r{}' \omega(r)$$

The interaction parameters ω have the familiar form of a difference between unlike-pair and average like-pair energies, as shown by Eq. (6).

3. Configurational Free Energy

To obtain the Helmholtz free energy of the solution, it suffices, in principle, to write down the partition function of the system and to compute its natural logarithm multiplied by $-kT$, k being Boltzmann's constant and T the absolute temperature. If it is assumed for simplicity that the vibrational and configurational components of the partition function can be decoupled, the total free energy can be written as

$$F_t = F_A + F_B + F$$

in which F_A and F_B contain the internal energies (E_A, E_B) and the vibrational entropies of the pure components, the energy of mixing F containing only the energy of mixing E and a configurational entropy S.

The simplest possible case is that of a binary crystalline substitutional solution with nearest-neighbor interactions; but even for this case, which is isomorphous to the three-dimensional Ising model, obtaining a closed form of the free energy of mixing F is out of the question. Approximations must therefore be introduced at this stage, the simplest one being that of the Bragg–Williams model. Although this model has serious limitations, it is adequate for the purpose of presenting the basic principles of solid-solution clustering and ordering in a qualitative manner. More quantitative results require the use of various cluster expansion methods (see Section 5.5).

3.1. Bragg–Williams Model

A general Bragg–Williams model is obtained by using the averaging method outlined in Section 2.1 combined with the assumption of random distribution of atomic constituents in the averaging regions R. The free energy of mixing is then

$$F = E - TS \qquad (7)$$

with internal energy of mixing given by Eq. (5) and entropy of mixing given by[18,34]

$$S = -k \sum_p \{c(p) \ln c(p) + [1 - c(p)] \ln[1 - c(p)]\} \qquad (8)$$

The simple model proposed here is thus one of "average atoms" interacting through (nonlocal) pairwise forces out to arbitrary neighbors, combined with an ideal (local) configurational entropy acting at individual points (p).

The nonlocal two-body interactions of Eq. (5) can be converted to (local) self-interactions in k space by the simple expedient of a Fourier transformation. For that purpose the crystal is assumed to contain N primitive unit cells, with periodic boundary conditions imposed. A discrete space (or lattice) Fourier transform operator is then defined by[15]

$$\mathscr{F} \equiv (1/N) \sum_p \exp[-i\mathbf{k}(h) \cdot \mathbf{x}(p)]$$

with $\mathbf{x}$ and $\mathbf{k}$ being real-space and k-space vectors, respectively, the symbol (h) denoting a site in k space (within the first Brillouin zone). In general, upper case symbols will henceforth denote the Fourier transforms of the corresponding lower case symbols; thus

$$X(h) = \mathscr{F}\xi(p)$$

The nonlocal energy can then be diagonalized in Fourier space to yield the expression

$$\sum_p \sum_r \omega(r)\xi(p)\xi(p+r) = N \sum_h \Omega(h)|X(h)|^2 \qquad (9)$$

with k-space interaction coefficient given by

$$\Omega = N\mathscr{F}\omega$$

3.2. Continuum Gradient Expansion

The original model of nonuniform solutions formulated by Cahn and Hilliard[9,10] was a continuum one. Their free energy functional contained a local and a nonlocal contribution and was derived through the use of an expansion in concentration gradients which can be obtained as follows: The concentration derivative at ($p + r$) can be expressed in terms of that at (p) by the expansion[15]

$$\xi(p+r) = \xi(p) + (1/2)x_i(r)x_j(r)\xi_{,ij}(p)$$
$$+ (1/4!)x_i(r)x_j(r)x_k(r)x_l(r)\xi_{,ijkl} + \cdots \qquad (10)$$

with

$$\xi_{,ij} = \partial^2 \xi / \partial x_i \, \partial x_j, \quad \text{etc.}$$

In Eq. (10) $x_i(r)$ denotes the ith Cartesian component of lattice vector **x** from (p) to $(p + r)$, summation over repeating Cartesian subscripts ($i = 1, 2, 3$) being implied. By symmetry, odd derivatives do not appear in Eq. (10). The gradient expansion of Eq. (10) combined with Eq. (5) leads to second-rank, fourth-rank,..., completely symmetric tensors $K_{ij}, K_{ijkl}, \ldots$. In cubic crystals further symmetry arguments[35] indicate that K_{ij} must be a scalar coefficient K and that K_{ijkl} must have only two independent components.

The following series of steps is now carried out: (a) $\xi(p + r)$ from Eq. (10) is inserted in Eq. (5); (b) the local entropy is Taylor-expanded about $\xi(p)$ to second order; (c) the free energy at (p) is summed over the whole crystal; (d) the summation is replaced by an integral; (e) the divergence theorem is used to replace the Laplacian by a gradient squared; (f) higher-order gradients are discarded for simplicity. The resulting free energy is

$$F = \int_V [f(c) + K(\nabla c)^2] \, d\mathbf{x} \tag{11}$$

the integration being performed over the volume V of the solution. Equation (11) was first derived by Cahn and Hilliard[9] in their treatment of the free energy of nonuniform solutions in the continuum approximation. The first term in the integrand is a local free energy per unit volume $f(c)$. The second term is the gradient energy with K, the gradient energy coefficient, having units of energy per length. For cubic crystals we have the relation[15]

$$z \sum_r{}' \omega(r) x_i(r) x_j(r) = 2a^3 \delta_{ij} K$$

where z is the number of atoms per cubic unit cell of lattice parameter a. In this approximation the gradient energy term depends only on the pair interaction parameters $\omega(r)$. If a nonlocal configurational entropy had been used instead of Eq. (8), then $K(\nabla c)^2$ would have been a gradient *free* energy.

4. Elastic Energy

If atoms A and B have different "sizes," clustering (i.e., incipient phase separation) or ordering in the solution will produce lattice distortions. In principle, the resulting strain energy contribution can be included implicitly in the quadratic pairwise term of Eq. (5), as will be shown presently. However, the pairwise expansion in the case of large strains in highly anisotropic crystals converges very slowly with (r) so that the gradient expansion leading to Eq. (11) is no longer valid. Furthermore if in Eq. (11) the free energy f is obtained by macroscopic thermodynamic measurements in the gas phase of a well-equilibrated phase-separated crystal or by extrapolation from high-temperature thermodynamic measurements supplemented by phase diagram data, then f will *not* contain strain energy contributions since the equilibrium studied will correspond to practically stress-free phases separated by incoherent interfaces, for which published phase diagrams are generally constructed. Finally, it may be possible to obtain pairwise interaction energy coefficients from first principles by suitable quantum mechanical calculations; but then, even in the case of ordering systems, it will be necessary to add explicit elastic energy terms. For these reasons it is advantageous to treat elastic effects separately.

Elastic interactions are more difficult to treat mathematically than pairwise "chemical" interactions. This is because the important parameters in the elastic calculations are vectors (forces, displacements) or tensors (stresses, strains) rather than the simpler scalars denoting "composition." Because of these difficulties, misconceptions and errors abound in the published elastic energy calculations, although completely general (and correct) formulas have been known for some time and are applicable to wide classes of problems in phase transformations. General formulas will be given below, along with simplified ones which are particularly useful in treating clustering problems.

In clustering reactions the major contribution to free energy changes resides at the long-wavelength portion of the composition spectrum, so that continuum elasticity may generally be used with confidence. However, concentration modulations of wavelength less than about ten lattice parameters, such as encountered in ordering reactions, require the use of a discrete (or microscopic) elasticity model.[36-39] The simplest approach to both microscopic and macroscopic elasticity problems is through the Fourier transform method: In essence, a three-dimensional concentration profile (nucleus, coherent precipitate, spinodal modulation, ordered structure, etc.) is decomposed in harmonic concentration plane waves. A formula is obtained for the elastic energy of each Fourier component of the spectrum, then all significant contributions are added with proper amplitude-squared. This superposition procedure is valid for all problems treated by linear elasticity; its application to clustering problems was introduced by Cahn.[11,12]

The general elastic energy formula is found to be[36]

$$E_e = \tfrac{1}{2} N \sum_h G(h)|X(h)|^2 \qquad (12)$$

which parallels Eq. (9) derived for pairwise "chemical" interactions in the Fourier representation. The k-space elastic energy coefficient G is the elastic energy of a harmonic concentration wave of wave vector $\mathbf{k}(h)$, of amplitude unity, and of arbitrary phase.

Formulas for $G(h)$ for arbitrary crystal classes, for arbitrary elastic anisotropy, and for arbitrary atomic misfit can be written down quite generally in terms of the inverse of a matrix of rank three which contains the elastic constants of the material and the components of the $\mathbf{k}$ vector. This inverse matrix is known as the Green's tensor of the material; its components in the case of crystals, treated as elastic continua of cubic symmetry, were given by Zachariasen as early as 1944.[40] When the cubic Green's tensor is introduced into the general formula for $G(h)$ along with the distortion parameter (relative lattice parameter change per unit compositional change)[11]

$$\eta = (1/a)\, da/dc \qquad (13)$$

one obtains the following expression:

$$G(\mathbf{h}) = 2\eta^2 Y(\mathbf{n})/N_v \qquad (14)$$

where N_v is the number of atoms per unit volume and where the "effective modulus" $Y(\mathbf{n})$ is given by[30]

$$Y(\mathbf{n}) = \frac{1}{2}(C_{11} + 2C_{12})\left\{3 - \frac{C_{11} + 2C_{12}}{C_{12} + C_{44}} \right.$$

$$\left. \times \left[\frac{\beta^2 + 2\beta(\alpha - 1)I_4 + 3(\alpha - 1)^2 I_6}{\beta^2(\alpha + \beta) + \beta(\alpha^2 - 1)I_4 + (\alpha - 1)^2(\alpha + 2)I_6}\right]\right\} \qquad (15)$$

in which $\mathbf{n}$ is a unit vector along $\mathbf{k}(h)$ and where

$$\alpha = (C_{11} - C_{44})/(C_{12} + C_{44}), \qquad \beta = C_{44}/(C_{12} + C_{44})$$

$$I_4 = n_2^2 n_3^2 + n_3^2 n_1^2 + n_1^2 n_2^2, \qquad I_6 = n_1^2 n_2^2 n_3^2$$

the n_i being the direction cosines of $\mathbf{n}$. Equation (15) is exact. Formulas such as (14) can be written down for other crystal classes, but the matrix inversions are generally very difficult. An approximate formula for Y in cubic crystals was given by Cahn[12]

$$Y(\mathbf{n}) = \tfrac{1}{2}(C_{11} + 2C_{12})\{3 - [(C_{11} + 2C_{12})/(C_{11} + 2AI_4)]\} \quad (16)$$

where A is the anisotropy parameter,

$$A = 2C_{44} - C_{11} + C_{12}$$

Equation (16) is exact for the three directions $\langle 100 \rangle$, $\langle 110 \rangle$, and $\langle 111 \rangle$. The elastic energy of a concentration modulation along $\langle 100 \rangle$ will have minimum energy, and one along $\langle 111 \rangle$ will have maximum energy, for solids with $A > 0$, the usual case. For $A < 0$ (Nb, V, Mo) the opposite conclusion holds. In isotropic solids ($A = 0$) a further simplification of Eqs. (15) or (16) occurs:

$$Y_0 = C_{11} + C_{12} - 2(C_{12}^2/C_{11}) = E/(1 - v) \quad (17)$$

where E is Young's modulus and v is Poisson's ratio.

Note that in an elastic continuum the effective modulus depends on the direction but not on the wavelength of the concentration modulation; hence the notation $Y(\mathbf{n})$. This conclusion holds also for crystals of symmetry lower than cubic. In an isotropic continuum even the dependence of Y on the orientation of the wave disappears, as expected; hence the notation Y_0. The elastic energy of an isotropic solid is thus expressed simply by back-transforming Eq. (12):

$$E_e = \eta^2 Y_0 \int_V \xi^2(\mathbf{x}) \, d\mathbf{x} \quad (18)$$

so that the elastic energy of a small volume $d\mathbf{x}$ depends only on the local concentration at $\mathbf{x}$. For anisotropic crystals the back-transform of Eq. (12) leads to a convolution integral in real space so that the elastic energy in volume $d\mathbf{x}$ is nonlocal. The elastic energy (14) is local in k space, however, hence the advantage of the Fourier transform method.

Equation (18) will be used in Section 5.1 to derive the notion of a "coherent miscibility gap"; Eq. (16) will be used implicitly in conjunction with the spinodal theory of Section 7.1, and the consequences of the orientation dependence of the effective modulus $Y(\mathbf{n})$ will be explored in Section 8 dealing with the morphology of decomposition products. These and other formulas are useful in a variety of circumstances as well, since virtually all transformations in solids are accompanied by elastic strain energy effects.

5. Stationary States

Stationary states of a binary solid solution are stable or metastable concentration fields obtained by minimizing the free energy functional (11) under specified boundary conditions and for given average concentration c_0 and temperature T. The stationary states to be investigated here are: (a) the equilibrium between two coherent phases, (b) the critical nucleus, and (c) metastable periodic concentration modulations. From now on, since attention will be focused on clustering solutions, the generally adequate continuum approximation of the free energy F will be used.

5.1. Coherent Phase Equilibrium

According to the mean field approximation of Section 3, the local free energy density of a clustering system below the critical temperature should resemble that of Figure 1(a). The concentrations of c_α, c_β of bulk homogeneous phases in equilibrium are given by the common tangent rule, itself derived from the equilibrium condition of equality of chemical potentials

$$\mu_A(c_\alpha) = \mu_A(c_\beta), \qquad \mu_B(c_\alpha) = \mu_B(c_\beta)$$

Gradient terms need not be considered here since these have appreciable values only in narrow interfacial regions between the homogeneous phases, and therefore contribute little to the total free energy.

The locus of common tangency points of free energy curves for varying T defines the miscibility gap of Figure 1(b). The locus of inflection points of $f(c)$, i.e., the locus of points c_s satisfying the equation

$$f'' \equiv d^2f/dc^2 = 0$$

is called the *spinodal* and is shown as a dashed line. In actual binary systems free energy curves and miscibility gaps are not expected to exhibit the symmetry of those of Figure 1, but this is of no consequence to the qualitative discussion which follows.

The spinodal and miscibility gap define three regions in the phase diagrams: (I) region of stability, (II) region of metastability, and (III) region of instability. The meaning of these terms is clarified by considering a solution containing local compositional fluctuations that conserve average concentration. Let the extreme values of the

Chapter 3

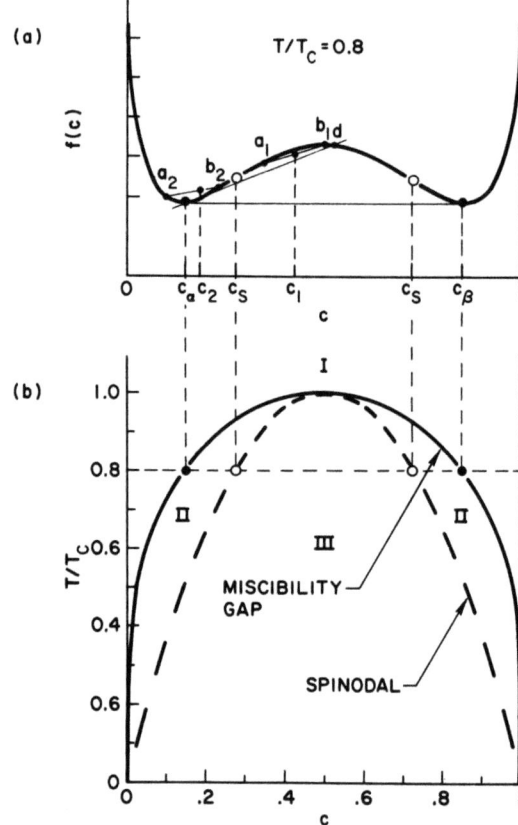

Fig. 1. Regular solution model: (a) free energy curve $f(c)$ at $T/T_c = 0.8$; (b) miscibility gap (locus of common tangent points c_α, c_β) and spinodal (locus of c_s) as a function of reduced temperature T/T_c. Here a_1–b_1 is unstable fluctuation about c_1, a_2–b_2 is stable fluctuation about c_2. Point d is on tangent at c_2. Regions I, II, and III are stable, metastable, and unstable, respectively.

fluctuation be a_1–b_1 about average concentration c_1 (Figure 1a). Since the free energy of the fluctuation is represented approximately by the intersection of the chord a_1–b_1 and the vertical through c_1, the fluctuation is seen to lower the free energy, and the solution is therefore unstable to perturbations of arbitrarily small amplitudes a_1–b_1. This general conclusion holds throughout region III of the phase diagram. The opposite conclusion holds in region II, where small concentration fluctuations such as a_2–b_2 raise the free energy. However, the free energy can decrease for local compositional

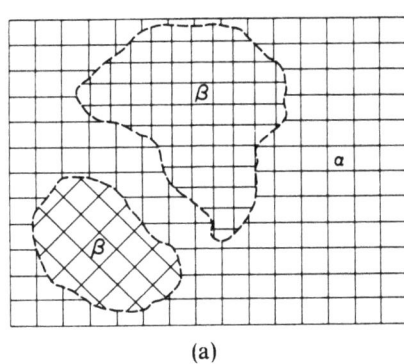 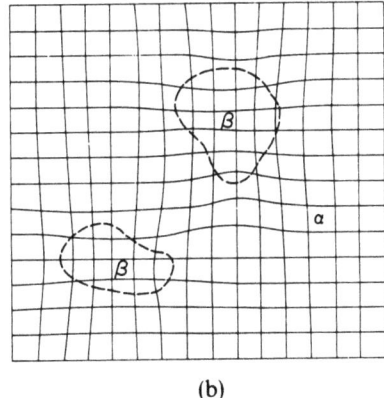

Fig. 2. (a) Incoherent two-phase equilibrium showing lattice discontinuities; (b) coherent two-phase equilibrium showing lattice distortions.[26]

fluctuations beyond point d (obtained by constructing the tangent at c_2). The homogeneous solution in region II is therefore said to be in a metastable state. In region I no lowering of the free energy can occur and the homogeneous solution is therefore unconditionally stable.

For phase-separating solid solutions, free energies and phase diagrams are generally determined for incoherent phase equilibria, i.e., for bulk phases separated by incoherent interfaces across which the lattice is discontinuous, as shown schematically in Figure 2(a).[26] If the phases in coherent equilibrium have different lattice parameters, lattice distortions (rather than discontinuities) will occur, as depicted in Figure 2(b). Solute-rich coherent clusters, or G.P. zones, had been postulated by Guiner[41] and by Preston[42] on the basis of experimental findings and it was known that clusters which formed homogeneously at low temperatures tended to dissolve (revert) above a certain temperature which was below the equilibrium phase boundary given by the phase diagram. The locus of reversion temperatures, known then as the G.P. zone solvus, was an ill-defined concept until Cahn[13] gave it its *raison d'être*, rebaptized it "coherent solvus," and showed how it could be calculated and plotted on a phase diagram.

Since coherent phases must generally be under stress, Cahn argued, an elastic energy contribution must be added to the "incoherent" free energy of the solid. For isotropic solids the total elastic energy of a nonuniform solid solution is given by Eq. (18), so that the local bulk free energy $f(c)$ must be altered to[13]

$$f(c) \to f(c) + \eta^2 Y_0 (c - c_0)^2 \qquad (19)$$

Chapter 3

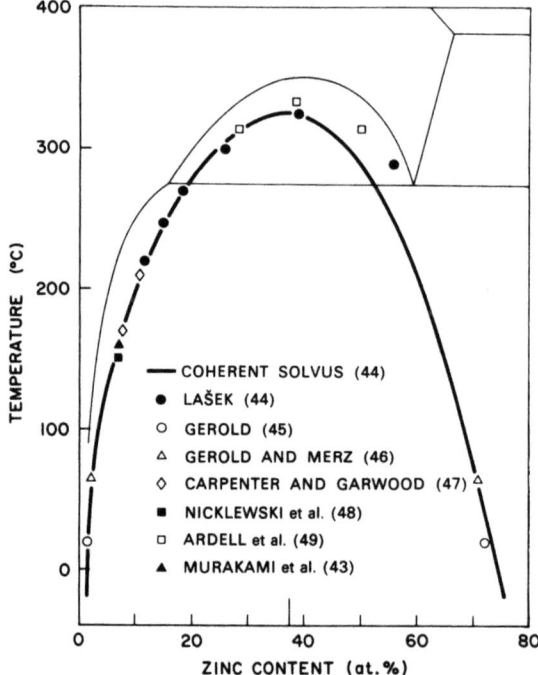

Fig. 3. Coherent solvus (heavy line) calculated by Lašek[44] for the Al–Zn system (light lines) and experimental points (adapted from Ref. 43).

Since the "coherent" free energy now contains a positive term quadratic in concentration, the points of common tangency move inward with respect to those of the unmodified $f(c)$, resulting in coherent states located totally within the incoherent equilibrium states.

The best-documented evidence supporting these ideas is found on the Al-rich side of the Al–Zn phase diagram. Figure 3, taken from the work of Murakami et al.,[43] shows the coherent miscibility gap calculated by Lašek[44] according to the method of Cahn, along with experimental points obtained by various investigators using a variety of techniques: hardness and electrical resistivity measurements, X-ray small-angle scattering, and electron microscopy.[44–49] The elastic misfit parameter η of Eq. (13) is small in Al–Zn so that the depression of the coherent solvus is likewise small. For large misfit the depression can be considerable, notably in the Au–Ni system.[50]

In anisotropic crystals the elasticity correction E_e to the free energy is essentially nonlocal, being given either by a general discrete-space formula or, in the continuum approximation, by a slowly converging gradient expansion. It follows that a local correction such as that of Eq. (19) will not suffice and that a unique coherent miscibility gap can no longer be defined. Coherent equilibrium is still possible, of course, but depends on the shape and relative orientations of the coherent particles. To simplify the discussion of stationary states, isotropic elasticity will be assumed in the remainder of Section 5, and the function $f(c)$ will represent the "coherent" free energy of Eq. (19).

Coherent equilibrium is always a metastable one, although very small coherent particles can be stable relative to incoherent ones of the same size.[26] However, a finely dispersed structure must eventually coarsen to reduce gradient (or interfacial) energy, and a critical particle size must be reached at which the increase in energy due to the creation of high-energy incoherent interfaces is more than offset by the reduction in bulk coherent elastic free energy.[51] If suitable dislocations can then be nucleated at or captured by the interfaces, coherence will be lost and the true equilibrium structure will prevail.

5.2. The Anharmonic Oscillator Analogy

The general features of coherent phase equilibrium have just been described. The details of concentration profiles of (coherent) stationary states are obtained by minimizing the appropriate free energy functional, holding the average concentration c_0 constant. It is particularly illuminating to study the mathematically simple case of one-dimensional concentration profiles $c(x)$. For unit cross-sectional area of an infinite solid the free energy functional to be minimized is

$$F = \int_{-\infty}^{\infty} [f(c) + Kc_x^2] \, dx$$

with $c_x \equiv dc/dx$, with auxiliary condition

$$\int_{-\infty}^{\infty} (c - c_0) \, dx = 0$$

and with appropriate boundary conditions.

Chapter 3

The correct stationary-state profile is found by solving the Euler–Lagrange equation

$$\frac{d}{dx}\frac{\partial L}{\partial c_x} - \frac{\partial L}{\partial c} = 0 \qquad (20)$$

where L is

$$L(c, c_x) = f(c) - (c - c_0)\mu + Kc_x^2 \qquad (21)$$

and where the Lagrangian multiplier μ is found to equal a difference of chemical potentials, $\mu_B - \mu_A$. The mathematical formalism has suggested to Khachaturyan and Suris[30] and Langer[22] the following elegant mechanical analogy: let x play the role of time, c the role of distance, and $-f + \mu c$ the role of potential energy. Then the

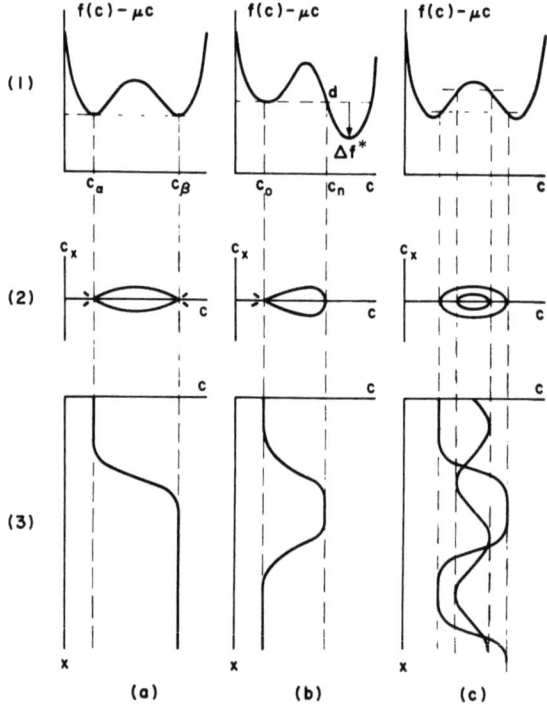

Fig. 4. Anharmonic oscillator analogy for stationary states (adapted from Refs. 22 and 30). Column (a): stable two-phase equilibrium; column (b): one-dimensional nucleation; column (c): metastable periodic solutions. Row 1: anharmonic potential $f - \mu c$ (horizontal dashed lines indicate energy levels H); row 2: trajectories in phase space; row 3: one-dimensional concentration profiles.

gradient term becomes a kinetic energy and L the Lagrangian for the anharmonic oscillator. The "first integral of motion" yields the Hamiltonian of the system[52]

$$c_x \partial L/\partial c_x - L \equiv K c_x^2 - f(c) + \mu c = H \tag{22}$$

where the arbitrary constant H has units of energy. A value of H having been selected, the constant μ is determined by conservation of average concentration: $\mu = \mu(c_0, H)$.[30] The trajectories in phase space (c_x, c) can then be obtained from Eq. (22) and from these the stationary-state concentration profiles can be found:

$$x - x_0 = \int \frac{dc}{\{[H - f(c) + \mu c]/K\}^{1/2}} \tag{23}$$

where x_0 is a constant of integration. A good approximation to $f(c)$ is often the quartic curve[16]; the profiles of Eq. (22) are then given by normal elliptic integrals of the first kind.[53]

Figure 4, adapted from the work of Khachaturyan and Suris,[30] depicts the following:

1. Free energy curves $f(c) - \mu c$, which, in the oscillator analogy, are the negative of the anharmonic potentials seen by the particle (horizontal dashed lines represent energy levels H).
2. Trajectories in phase space, i.e., plots of concentration gradient versus concentration.
3. Stationary-state profiles $c(x)$.

Three types of physical situations (for which the multiplier μ can be determined by inspection) are illustrated:

(a) Common-tangent thermodynamic equilibrium between bulk phases of concentration c_α and c_β separated by a diffuse interface.
(b) A one-dimensional nucleus of central concentration c_n in (unstable) equilibrium with an infinite matrix phase of concentration c_0.
(c) Two metastable periodic profiles of amplitude less than the equilibrium one.

Cases (a) and (b) will be examined in more detail in Sections 5.3 and 5.4 respectively. Case (c) was discussed by Langer[22] for c_0 at the center of a symmetric free energy curve for which $\mu = 0$ by symmetry. For low values of the constant $-H$ the system sees only the harmonic part of the potential and sinusoidal composition modulations result. For larger values of $-H$ the system "spends more time" near the extreme concentration values so that the profile tends toward the square wave. In all cases the period of the modulation is calculated from the complete elliptic integral corresponding to Eq. (23).[30] Thus the amplitude of a metastable periodic state cannot increase unless the wavelength is adjusted accordingly, as was originally demonstrated by Hillert.[8] It is therefore possible for short-wavelength periodic modulations to develop for kinetic reasons and to remain "hung up" in low-amplitude metastable states due to the resistance to coarsening of the periodic structure. Such effects have been exhibited in computer simulation of one-dimensional spinodal decomposition,[16] but clear-cut experimental evidence of the existence of these states is presently lacking.

5.3. The Diffuse Interface

The equilibrium profile and free energy of a flat interface between two coexisting isotropic phases were determined by Cahn and Hilliard.[9] Since this is a one-dimensional problem, the results of the previous section are immediately applicable. It can be shown that the value of H corresponding to the common tangent construction must be $-\mu_A$, the negative of the chemical potential of species A in the bulk equilibrium phases. Equation (22) then yields

$$Kc_x^2 = \Delta f(c) \tag{24}$$

where $\Delta f(c)$ is the free energy referred to a standard state consisting of phases α and β in equilibrium. The interfacial face energy σ per unit area is the total free energy of the system minus that of the equilibrium phases of constant composition throughout:

$$\sigma = \int_{-\infty}^{\infty} [\Delta f(c) + Kc_x^2] \, dx \tag{25}$$

which, by Eq. (24), can also be written[9]

$$\sigma = 2K \int_{-\infty}^{\infty} c_x^2 \, dx = 2 \int_{-\infty}^{\infty} (K\Delta f)^{1/2} \, dx$$

As above, the equilibrium interface profile is given by Eq. (23). If $f(c)$ is a symmetric quartic in c, the profile $c(x)$ is given by the inverse of a particularly simple elliptic integral, i.e., by a hyperbolic tangent.[9,22] A typical shape for diffuse interface profiles is given in

Figure 4(a3). The mathematical results of the variational problem posed by the diffuse interface may be interpreted physically as follows: By Eq. (25) the interfacial free energy is the sum of a local and of a gradient energy. The correct profile is that which combines the smallest possible integrated "steepness" of $c(x)$ combined with the introduction in the solution of the least amount of nonequilibrium material of intermediate concentrations between c_α and c_β.

Cahn and Hilliard[9] have shown that the thickness of the diffuse interface increases with temperature and becomes infinite at the critical temperature T_c (top of the coherent miscibility gap). In the critical region σ was found to vary as $(T_c - T)^{3/2}$, in accordance with experimental measurements.

5.4. The Critical Nucleus

It was mentioned that a solution in the metastable region II can decompose by forming embryos of a new phase of concentration beyond that of point d in Figure 1. The critical nucleus is that embryo requiring least work of formation W which is just in equilibrium with an infinite matrix phase of concentration c_0. This equilibrium is an unstable one since expansion or contraction of the nucleus decreases the free energy of the system.

Classical nucleation theory, or droplet model, assumes the existence in the disappearing phase of homogeneous embryos of the new phase separated from the old by an infinitely sharp interface of interfacial energy σ which is independent of concentration in hydrostatic systems. This interfacial tension causes the embryo to be under a pressure greater than that P_0 of the surrounding phase. The equality of chemical potentials in the critical nucleus and in the surrounding phase determines the pressure P^* inside the critical nucleus. This pressure is exactly equilibrated by the surface tension of the droplet when the following equality holds:

$$\Delta P \equiv P^* - P_0 = 2\sigma/r^*$$

where r^* is the radius of the spherical critical nucleus. Embryos with radius less than the critical one must shrink, those with radius greater than r^* must grow.

For given undercooling or supersaturation the critical pressure P^*, and hence r^*, can be evaluated from the thermodynamics of the system. For liquid droplets condensing from the pure vapor ΔP can be calculated directly from vapor pressure data and from a knowledge of the molar volume of the liquid phase as a function of pressure. For incompressible solutions ΔP is given by the bulk Helmholtz free energy per unit volume Δf^* obtained by the construction of Figure 4(b1), while for nucleation in isotropic solids Δf^* must be taken from the coherent free energy curve given by Eq. (19). Other cases require additional thermodynamic data, but the problem is always determinate since the condition of thermodynamic equilibrium between critical nucleus and surrounding phase fixes all necessary parameters.

The classical theory has serious limitations: First, the classical theory makes the ad hoc assumption of a "square-well" profile for the nucleus; second, at large undercoolings the calculated r^* is so small that the notion of a distinct interface of energy σ breaks down completely; third, the classical theory predicts a finite nucleation barrier inside the spinodal where the system should be completely unstable to long-wavelength concentration fluctuations. The model developed by Cahn and Hilliard[10] does not require the assumption of a homogeneous nucleus and does predict the expected properties at the spinodal. This model will now be briefly described.

The case of a one-dimensional nucleus was illustrated in Figure 4(b). For the critical nucleus in equilibrium with the surrounding phase $-H$ must again equal μ_A, the chemical potential of A in the homogeneous matrix phase of concentration c_0. The Lagrangian multiplier is found to be $\mu_B - \mu_A$. The stationary-state profile is given by Eq. (23) with the quantity in brackets given by $\Delta f(c)$, the free energy referred to that of the metastable homogeneous state c_0 as standard state. A one-dimensional nucleus is not physically meaningful, however, so that Cahn and Hilliard[10] solved instead the stationary-state variational problem in three dimensions under the assumption of spherical symmetry. In spherical coordinates this is still a one-dimensional problem but the anharmonic oscillator analogy breaks down and a "first integral of motion," such as Eq. (22), can no longer be obtained simply.

Calculated concentration profiles for a spherical nucleus are shown in Figure 5[10] for average concentrations progressively approaching the spinodal. The variables X and t are here reduced concentrations and radius. Near the coherent phase boundary (upper curve) the profile resembles that of the classical nucleus, as expected. As the spinodal composition is approached, however, the profile becomes progressively more diffuse and the concentration at the center of the nucleus decreases. Finally, for average concentra-

Clustering Effects in Solid Solutions

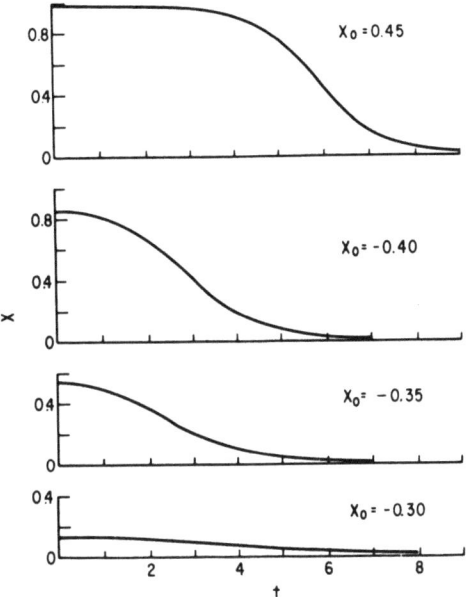

Fig. 5. Theoretical concentration profiles for spherically symmetric critical nucleus for concentrations X_0 progressively approaching the spinodal. Variables X and t are reduced concentration and radius, respectively.[10]

tion at the spinodal ($c_0 = c_s$) the homogeneous solution of concentration c_0 itself is the critical nucleus: infinitely shallow and infinitely diffuse. Cahn and Hilliard further showed that the critical radius goes to infinity at the phase boundary, as in the classical treatment. For increasing supersaturations the critical radius reaches a minimum and then goes to infinity at the spinodal, in contradistinction to the finite value of r^* predicted by the classical theory.

The work of formation W of the critical nucleus can be determined by inserting the calculated stationary-state profile back into the energy functional (11):

$$W = \int_V [\Delta f(c) + K(\nabla c)^2] \, dx \qquad (26)$$

It is found that W increases without limit as the coherent equilibrium phase boundary is reached, as in the classical theory, but correctly goes to zero at the spinodal, whereas $W = \tfrac{4}{3}\pi r^2 \sigma$ still has finite value according to the classical theory. The Cahn and Hilliard model is

Chapter 3

thus internally consistent and represents a considerable improvement over the classical description. Unfortunately, the mathematical formalism is correspondingly more complicated.

To close this section, let us note a significant difference between a one-dimensional and a three-dimensional nucleus. For the three-dimensional critical nucleus $\Delta f(c_n)$ of the material at the center of the nucleus must be negative,[10] since this free energy drop is required to offset the pressure difference ΔP. For the one-dimensional critical nucleus ΔP vanishes, resulting in $\Delta f = 0$, which means that the concentration at the center of the nucleus must be that of point d in Figure 4(b2) (or Figure 1). This is borne out by the oscillator analogy: The turning point of the classical oscillator must be at d since the system cannot tunnel through the potential energy barrier.

5.5. Higher Cluster Expansions

The foregoing results are based on free energy curves which have the characteristic shape shown, for example, in Figure 1. Such curves result from the mean field approximation, leading to the well-known regular solution and Bragg-Williams models. Since these models are known to constitute very poor approximations, Kikuchi[31] sought to recalculate free energy curves for binary clustering systems on the basis of his "cluster variation method."[54] Because of the complexity of the calculations, Kikuchi was only able to pursue the computation on a two-dimensional triangular lattice using three-point and W-shaped five-point clusters.

The resulting free energy curves, compared to that derived from the pair approximation, are shown in Figure 6.[31] Surprisingly, even these small-cluster expansions tend to drop the free energy curve close to the common tangent construction, which is the exact solution when all clusters are included in the partition function expansion. Thus, improved approximations to $f(c)$ would tend to invalidate the whole treatment of stationary states along with many of the kinetic results of Section 7, since the metastable portion of the curve and the spinodal compositions eventually disappear into the common tangent. It must be recalled, however, that except for the diffuse interface calculation, the stationary states considered above are unstable or metastable ones. The starting point for these calculations is the homogeneous, i.e., quasirandom solid solution for which the pair approximation is expected to hold quite well. In fact, Kikuchi has shown[31] that a disordered solution rapidly quenched to low temperatures retains

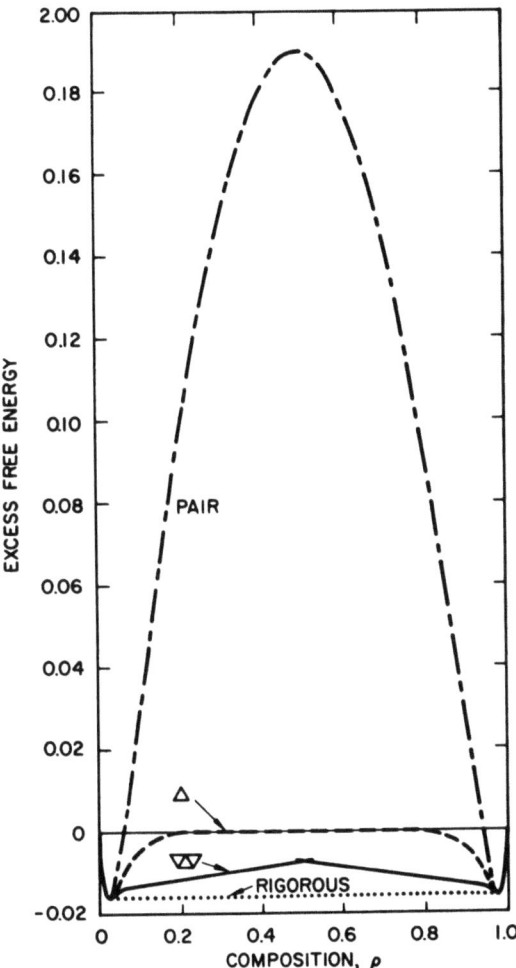

Fig. 6. Configurational free energy (in kT units per lattice point) versus concentration ρ calculated by Kikuchi[31] for various cluster approximations.

the general features of the mean-field-type curves, even in the more rigorous models, although the central "hump" is not as large as expected on the basis of the pair model, while the spinodal points move toward the equilibrium concentrations. Thus the results of the previous sections are still qualitatively valid.

There remains the troublesome question of the diffuse interface profile, which is a truly equilibrium state. If the central hump of the free energy curve disappears, how then can the minimization of the free energy functional (11) be justified since it would now lead to an infinitely wide interface with vanishing interfacial energy? The answer was recently given by Kikuchi[55]: Although no central hump exists for the free energy curve in an exact calculation for bulk phases, this hump reappears in a rigorous interface calculation because of the constraints imposed by the existence of coexisting neighboring homogeneous phases. In approximate calculations the relative magnitude of the hump is greater than in the rigorous calculation because in small-cluster expansions clusters that contribute significantly to the partition function are discarded, and more clusters are missed around the central concentrations than close to the equilibrium concentrations. Thus calculations based on hump-backed free energy curves are actually self-consistent though subject to quantitative improvement when larger cluster models are used.

6. Compositional Fluctuations

The much maligned mean-field free energy model yields useful formulas for the expected intensity $\langle |X|^2 \rangle$ of concentration fluctuations in stable solid solutions: The local entropy is first expanded in a Taylor series to second order in the concentration variations ξ, then added to the nonlocal internal energy, Eq. (9). The resulting free energy quadratic is diagonalized in the Fourier representation to yield

$$\Psi(h) = [kT/c_0(1-c_0)] - 2\Omega(h)$$

in which the Fourier transform Ω of the pair interaction parameters is assumed to contain the elastic terms implicitly. This expression inserted in the Landau fluctuation formula[56]

$$\langle |X(h)|^2 \rangle = kT/N\Psi(h) \qquad (27)$$

gives the required expression:

$$N\langle |X(h)|^2 \rangle / c_0(1-c_0) = 1/[1 - 2c_0(1-c_0)\Omega(h)/kT]$$

This equation, except for a normalization constant in the numerator, is just the expression derived by Clapp and Moss[57] for the Fourier transform of the pair correlation function above the critical temperature. Similar formulas have been given by Krivoglaz[58] and Wilkins.[59] Yet another variant has been proposed,[60] based on concentration intensities $|X|^2$ considered as condensing bosons.

Regardless of the model used, these formulas predict Lorentzian intensity peaks above the points in k space where $\Omega(h)$ has minimum value. For clustering systems these points are the nodes of the reciprocal lattice including the origin of k space. Surprisingly, though several clustering binary systems with accessible critical points in the solid state are known, critical scattering has never been observed unambiguously. The reason for this apparent anomaly was given by Cahn[13]: In crystalline solids the top of the (incoherent) equilibrium miscibility gap is not associated with true cooperative phenomena in the bulk solid. The true critical point is associated with coherent fluctuations, whose free energy includes a strain energy contribution. Since the function $G(h)$ is positive throughout k space for mechanically stable crystals, the coherent critical point is depressed to lower temperatures. Intense critical fluctuations should therefore only be observed just above the top of the coherent miscibility gap, i.e., within the incoherent two-phase region of the phase diagram. The failure to observe critical fluctuations in crystals is thus due to the experimental difficulty of maintaining a fluctuating solid solution in the (incoherent) two-phase region at high temperature for a time sufficient to perform the required intensity measurements. In anisotropic crystals the critical intensity should be highly anisotropic, further complicating its detection.

7. Kinetics of Decomposition

With Cahn's pioneering papers,[11,12] spinodal decomposition was introduced as another fundamental kinetic process besides the more familiar one of nucleation and growth. All too often, however, the spinodal mechanism subsequently came to be regarded as a competing mechanism to coherent nucleation, or as a distinct mechanism whose domain of existence in the phase diagram was sharply restricted by the spinodal line (dashed curve in Figure 1). These simplistic views are untenable for the following reasons:

1. Spinodal decomposition cannot be defined thermodynamically, but only kinetically. The spinodal itself is not an equilibrium thermodynamic concept and in fact disappears from the free energy curve in a rigorous calculation (see Figure 6). Thus the position of the spinodal curve depends on the free energy model used and on the state of quenched-in disorder in the solution.

2. A nonequilibrium solid solution will approach the equilibrium state along irreversible paths which depend on past history.

Chapter 3

These paths are diverse and complex and cannot be neatly categorized.

In some instances, however, it may be possible to recognize "typical" decomposition modes for which ad hoc models may serve as useful first approximations. Nucleation and growth, on the one hand, and spinodal decomposition, on the other, are two such modes located at the extreme ends of the decomposition spectrum. The former mode is characteristic of solutions evolving just inside the phase boundary, the latter is characteristic of solutions near the center of the coherent miscibility gap, i.e., at large supersaturations. Typical nucleation morphologies suggest random events, whereas typical spinodal morphologies display a regular periodicity. These observations are in keeping with a general principle enunciated by Glansdorff and Prigogine[61]: For small departures from equilibrium a system will evolve through random fluctuations; for large departures from equilibrium a system will prefer to dissipate its excess free energy by creating (metastable) periodic structures.

Ideally, the kinetics of clustering solid solutions should be studied by deriving an appropriate diffusion equation without reference to specific models. A suitable diffusion equation is presented in Section 7.1. This nonlinear partial differential equation can only be solved by numerical methods (Section 7.2), so that tractable *ad hoc* models must be introduced if analytical solutions are desired. One such model, the linear spinodal theory, is discussed in Sections 7.3 and 7.4. The other, that of nucleation and growth, is covered briefly in Section 7.5. Both models are also discussed in this volume by Raghavan and Cohen (Chapter 2).

7.1. The Diffusion Equation

The general continuity equation with vanishing source term reads

$$N_V \, \partial c/\partial t = -\text{div } \mathbf{J} \tag{28}$$

where the left hand side of Eq. (28) represents a rate of accumulation of B atoms per unit volume and where $\mathbf{J}$ is a flux of B atoms with respect to a moving reference plane for which the total flux is zero.[27] The flux is given by

$$\mathbf{J} = -N_V M \,\text{grad}\, \mu \tag{29}$$

where μ is the difference of chemical potentials $\mu_B - \mu_A$. An expression for the atomic mobility M was given by Huston *et al.*[62] Thus far

Eqs. (28) and (29) fit any classical diffusion treatment; the physics of the problem are introduced with the nature of the potential μ, the general form of which was already determined for the stationary-state problem of Section 5. By analogy, the required potential is thus found by combining the three-dimensional forms of Eqs. (20) and (21):

$$\mu(c) = f'(c) - 2K\nabla^2 c \qquad (30)$$

where the prime denotes differentiation with respect to concentration. In kinetic problems μ is a function of local concentration $c(\mathbf{x}, t)$ and therefore varies, through c, in space and time t.

The required diffusion equation is obtained by combining Eqs. (28)–(30):

$$\partial c/\partial t = \nabla M \nabla \mu(c) \qquad (31)$$

In order to reproduce approximately the shapes of clustering free energy curves (Figure 1), $f(c)$ must be at least a quartic in c,[16] which means that f' must be a cubic in c. Equation (31) is therefore highly nonlinear. A general solution in closed form of the diffusion equation is thus out of the question, although certain basic properties of the solution have been examined by Cahn[21] by a perturbation method. Computer-generated numerical solutions of Eq. (31) have provided some insight into the clustering kinetics in binary solutions, and some results of these computations are presented in the next section.

7.2. Numerical Solutions

When information about the approximate free energy curve is incorporated via $\mu(c)$ into Eq. (31) the solution to the diffusion equation correctly obeys the phase diagram without requiring additional constraints. Examples of such well-disciplined solutions are given in Figures 7–9, which represent computer simulations of the evolution of one-dimensional concentration profiles in Al–Zn solid solutions during isothermal aging at 100°C.[16] The initial concentration variation $c(x, 0)$ was a small-amplitude fluctuation which is not shown in these figures. The thermodynamic parameters used were those calculated by Rundman and Hilliard[63] for the Al–Zn system. The numerical solution $c(x, t)$ of Eq. (31) was obtained by a Fourier space forward difference iteration scheme.[16]

Each figure shows concentration profiles as a function of distance for successive aging times. The horizontal full lines indicate the average composition c_0, the dashed lines indicate the (coherent) equilibrium compositions. The results for a 37.5 at.% Zn alloy (i.e.

Chapter 3

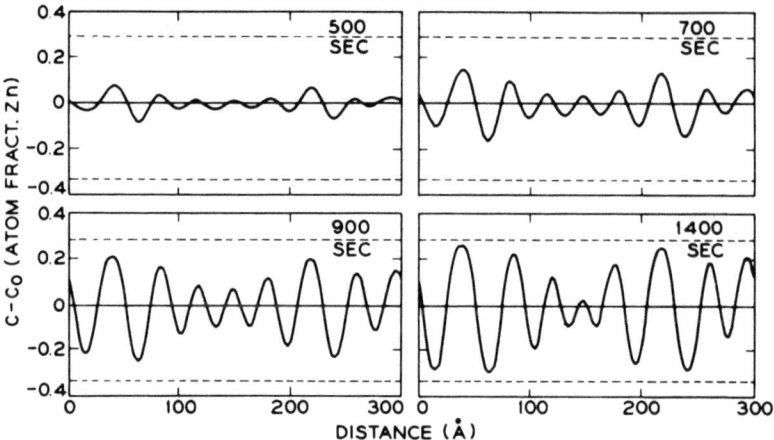

Fig. 7. Computed one-dimensional concentration profiles for an Al–37.5 at.% Zn alloy aged at 100°C (center of coherent miscibility gap) for the indicated durations.[16]

for c_0 close to the center of the miscibility gap) are shown in Figure 7: The profiles are quasisinusoidal with bounded amplitude. Note that the extreme concentrations of this periodic structure do not reach the coherent equilibrium boundaries, even after repeated iterations. This behavior is undoubtedly an illustration of the effect shown in Figure 4(c): the development of a metastable stationary-state short-wavelength periodic structure of low amplitude. Further growth of the composition modulation requires a lengthening of the psuedoperiod, which does not take place readily due to the extreme resistance to coarsening of such regular one-dimensional structures.

Figures 8a and 8b are relative to a 22.5 at.% Zn solution, which is closer to the spinodal composition at 100°C. A striking feature of the profiles is the appearance at an early stage of "Guinier zones"[64] consisting of Zn-rich peaks surrounded by denuded Al-rich regions. This result, which confirms the qualitative arguments put forward by Bonfiglioli and Guinier,[65] is a direct consequence of the asymmetric location of c_0 with respect to the miscibility gap boundaries. At this composition the Guiner zone structure is eventually followed by a well-defined periodic profile which correctly obeys the phase diagram. After prolonged aging the solution displays coarsening effects (Figure 8b), which are apparently less sluggish than those of the symmetric composition (center of the miscibility gap). To quote

Clustering Effects in Solid Solutions

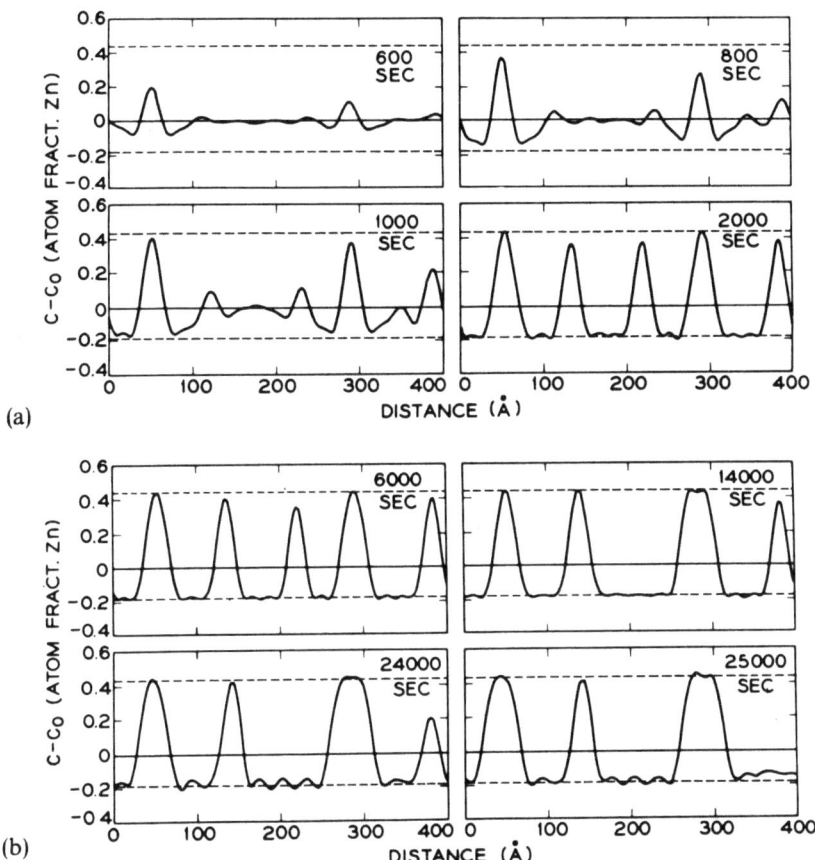

Fig. 8. (a) Computed one-dimensional concentration profiles for an Al–22.5 at.% Zn alloy aged at 100°C. Note the appearance of "Guinier zones." (b) Continuation of Fig. 8(a), longer aging times. Note the coarsening reaction.[16]

Hilliard,[27] "It is rather remarkable that a single equation can depict the complete life cycle, from birth to death, of a particle."

For compositions still closer to the spinodal (Fig. 9; 20 at.% Zn) an early-stage coarsening reaction sets in and the secondary maxima of the Guinier zones dissolve before they have a chance to develop into full-fledged precipitates as they did in the previous case. Thus a periodic structure fails to develop for this asymmetric composition although the average composition is still within the spinodal. Actually, the final concentration profile peaks at the spots where the first maxima make their appearance; in other words, the location of the precipitates is dictated almost entirely by the vagaries of the initial

Chapter 3

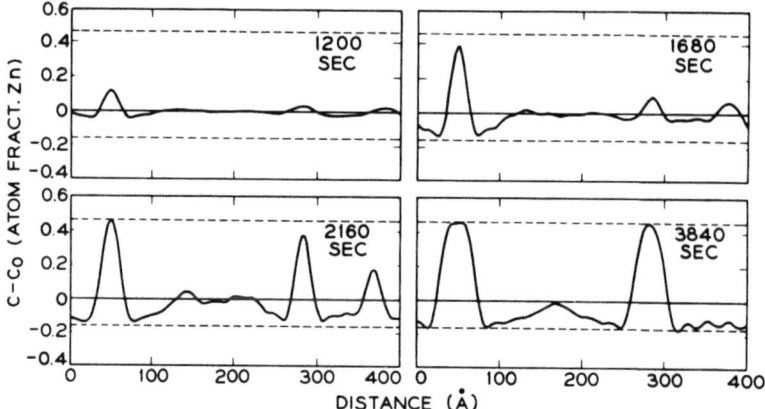

Fig. 9. Computed one-dimensional concentration profiles for an Al–20.0 at.% Zn alloy aged at 100°C (close to the coherent miscibility gap boundary). The final morphology is almost completely dictated by the initial conditions.[16]

fluctuation, a characteristic more reminiscent of nucleation than of "typical" spinodal decomposition.

Some of these numerical results have received experimental confirmation[16,27]: Figure 10(a) shows a sequence of experimental small-angle X-ray scattering spectra from an Al–22 at.% Zn alloy aged at 150°C in a silicone oil bath, and Fig. 10(b) shows a sequence of Fourier intensity spectra obtained by solving Eq. (31) numerically for the same alloy aged "inside the computer." Considering that—except for an arbitrary factor in the intensity scale—there are no adjustable parameters, the agreement between experiment and theory is very encouraging.

Two-dimensional computer calculations have been performed by Shendalman and O'Toole,[23] and two of their computer printout sequences are shown in Figures 11 and 12. The first sequence is for an isotropic solution located in the metastable region (between the miscibility gap and the spinodal). As explained by the authors, the initial "random noise" input develops into fairly small patches as time proceeds, with little connectivity between composition peaks. Larger patches form at later times. Classical nuclei are nowhere evident. The second sequence (Figure 12) is for the system within the spinodal region: Here the initial random noise evolves into a highly connected morphology, followed by its breaking up into distinct

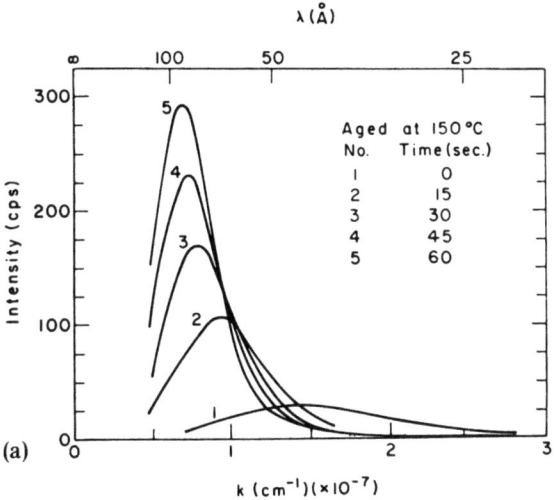

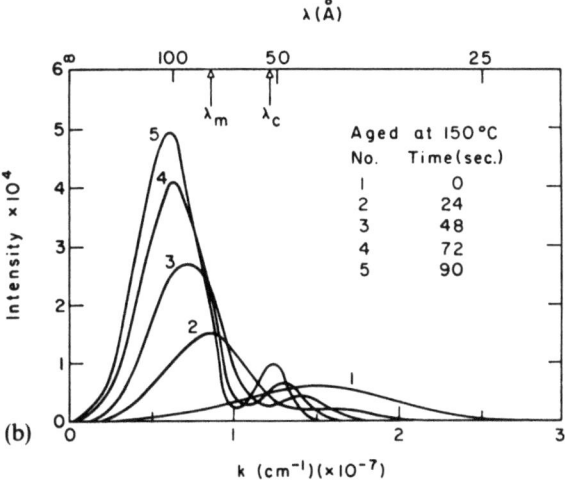

Fig. 10. (a) Small-angle scattered intensity spectra for an Al–22.5 at.% Zn alloy aged at 150°C; (b) one-dimensional computer simulation of the same process.[27]

circular droplets, which have a concentrated core surrounded by a ring of intermediate composition. Early-stage coarsening is seen to occur.

From the foregoing it is apparent that nucleation versus spinodal decomposition is not an either/or situation. Nevertheless, there is such a thing as "typical" spinodal decomposition: It occurs for

Chapter 3

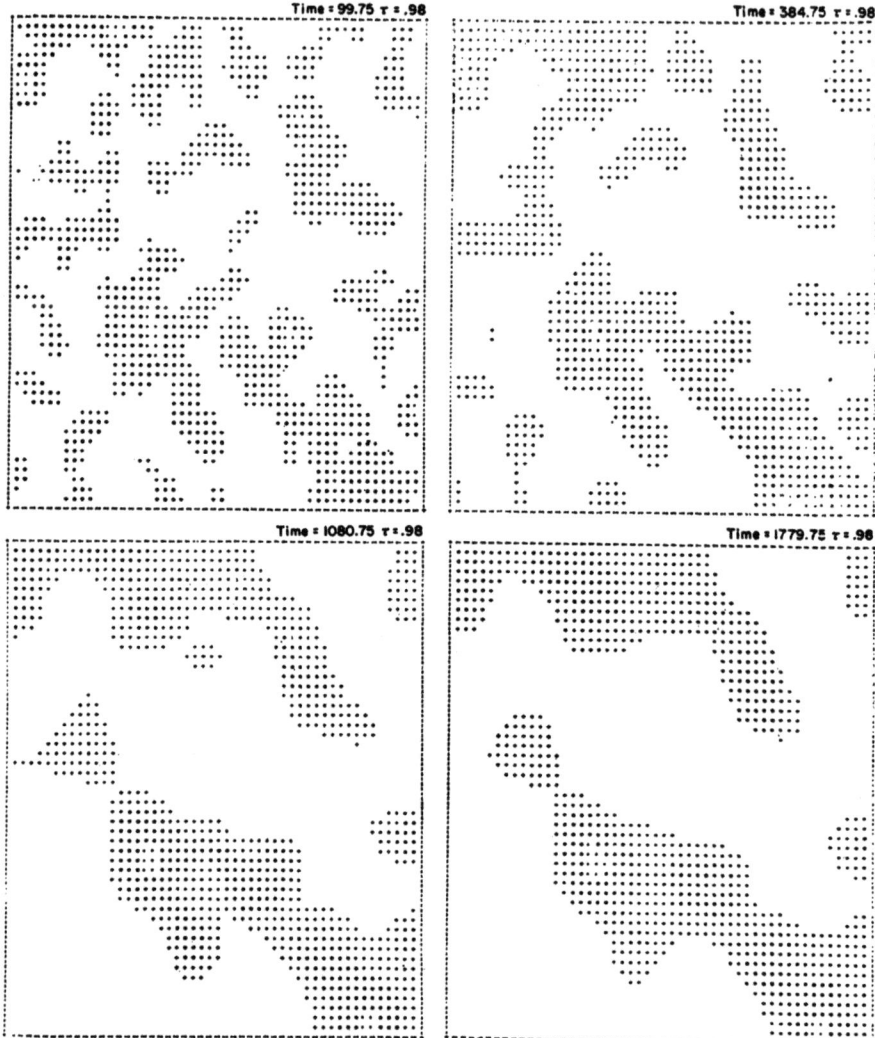

Fig. 11. Computed two-dimensional concentration contours for an isotropic solution with $T/T_c = 0.98$ (in metastable region II).[23]

compositions well inside the spinodal, i.e., at large supersaturations or undercoolings, and owes its popularity to the fact that for these cases the nonlinear terms of Eq. (31) can be neglected for the initial stages of the process, and the trimmed diffusion equation can then readily be solved analytically, as will be shown in the next section.

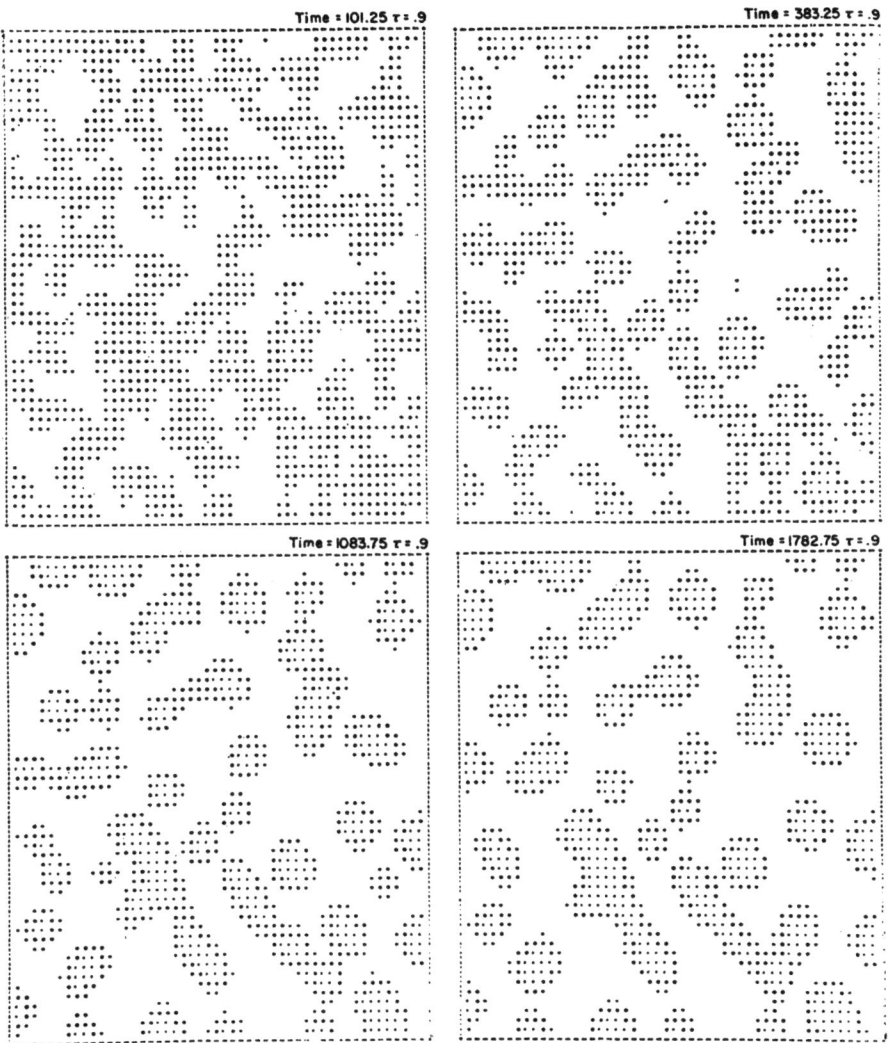

Fig. 12. Computed two-dimensional concentration contours for an isotropic solution with $T/T_c = 0.90$ (in unstable region III).[23]

7.3. The Linearized Equation

For average compositions and aging temperatures well inside the spinodal and, in particular, for c_0 in the vicinity of the center of the miscibility gap the absolute value of the ratio of the third to the second derivative of $f(c)$ is small compared to unity and the second-degree term in c can be initially neglected in the potential $\mu(c)$. The third-degree term in the potential can also be neglected initially,

Chapter 3

since its magnitude is proportional to the third power of the composition variation, a small quantity. It follows that under the above conditions the potential function $\mu(c)$ can be linearized and this leads to the following diffusion equation[11,12]:

$$\partial c/\partial t = M(f_0'' + 2\eta^2 Y)\nabla^2 c - 2MK\nabla^4 c \tag{32}$$

where M and K are assumed to be concentration and time independent, and where f_0'' represents the second derivative of the "incoherent" free energy evaluated at $c = c_0$. Equation (32) can be Fourier-transformed to an ordinary differential equation whose solution is found to be

$$X(\mathbf{k}, t) = X(\mathbf{k}, 0) \exp[\alpha(\mathbf{k})t] \tag{33}$$

with *amplification rate*

$$\alpha(\mathbf{k}) = -Mk^2[f_0'' + 2\eta^2 Y(\mathbf{k}) + 2Kk^2] \tag{34}$$

In these equations k^2 is the square of the wave vector $\mathbf{k}$ and X is, as in Section 3, the Fourier transform of the concentration deviation $\xi = c - c_0$, $X(\mathbf{k}, 0)$ being the Fourier transform of the initial concentration fluctuation. The k-space elastic modulus $Y(\mathbf{k})$ is given by an appropriate formula of Section 4.

Equation (33) shows that the Fourier amplitudes of concentration waves initially grow ($\alpha > 0$) or decay ($\alpha < 0$) in time according to an exponential law. A typical plot of the amplification rate $\alpha(k)$ [or $R(\beta)$ in the original notation] as a function of wave number is given in Figure 13[27] along with the amplification rate one would have obtained from the "classical" equation, i.e., the one obtained from Eq. (32) by setting the gradient energy coefficient K equal to zero. Two important quantities are introduced in Figure 13: the critical wave number k_c, solution of

$$\alpha(k) = 0$$

and the optimum wave number k_m, solution of

$$d\alpha/dk = 0$$

for a particular direction $\mathbf{k}$.

The physical meaning of these concepts is best discussed in terms of wavelengths λ, inversely related to wave number by

$$\lambda = 2\pi/k$$

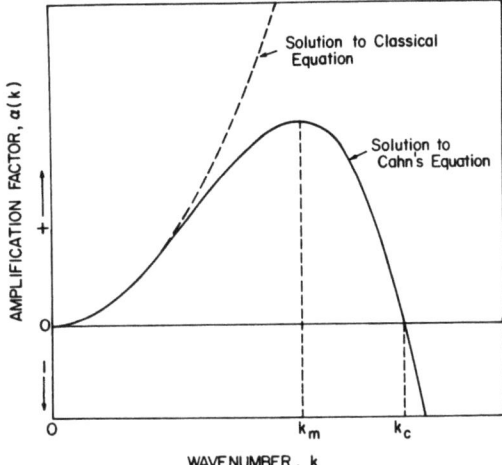

Fig. 13. Dependence of the amplification factor $\alpha(k)$ on wave number k (full curve). k_m and k_c are the optimum and critical wave numbers, respectively.[27,28]

Within the framework of the linear theory all harmonic composition waves of wavelength λ larger than the critical wavelength

$$\lambda_c = 2\pi[-2K/(f_0'' + 2\eta^2 Y)]^{1/2} \qquad (35)$$

will grow in amplitude as isothermal aging progresses, while those with $\lambda < \lambda_c$ will decay. Maximum growth in a given direction **k** will occur for the optimum wavelength

$$\lambda_m = \lambda_c\sqrt{2} \qquad (36)$$

These considerations are valid only for the initial stages of spinodal decomposition, well inside the coherent spinodal. Outside the spinodal the linear approximation breaks down completely since, by Eq. (35), λ_c and therefore also λ_m are imaginary quantities. This is because, since K is positive for a clustering system, the quantity in brackets in Eq. (35) can only be positive for

$$f_0'' + 2\eta^2 Y < 0$$

which defines the region inside the coherent spinodal.

Equation (35) predicts that the quantity $\alpha(\mathbf{k})/k^2$ should plot as a straight line versus k^2. This property of the dependence of amplification rate on reciprocal wavelength was verified experimentally by

Chapter 3

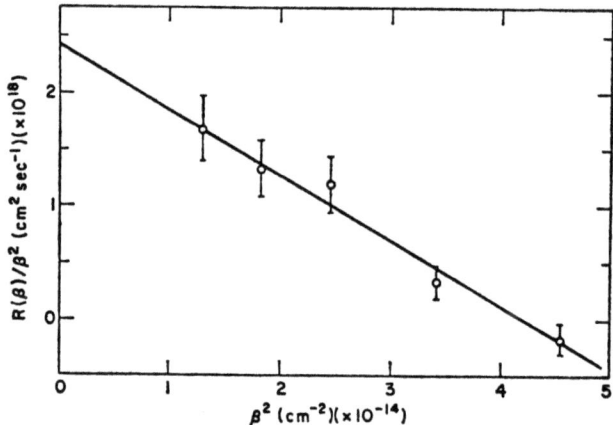

Fig. 14. Plot of $\alpha(k)/k^2$ versus k^2 [$R(\beta)/\beta^2$ versus β^2 in original notation] for an Al-22 at.% Zn alloy aged at 65°C.[63]

Rundman and Hilliard[63] on an Al-22 at.% Zn alloy. Figure 14, taken from their work, shows the expected linear behavior. Similar experiments have since been reported on metallic and glass systems with varying degrees of success (see Ref. 27). Observed anomalies generally appear to be due to the very limited applicability of the linear spinodal theory.

In further extensions[14,15] of Cahn's original theory, a differential-difference diffusion equation was derived directly from the discrete Fourier-space free energy functional of Section 3.1 without recourse to the usual continuum gradient expansion. The mathematical formalism turned out to be remarkably simple and had the significant advantage of providing early-stage solutions to order–disorder processes as well as to clustering ones. This feature is particularly important in multicomponent systems where ordering and clustering can occur simultaneously.[16-20]

7.4. The Intensity Equation

Compositional fluctuations can be introduced formally into the framework of the linear spinodal theory by means of an intensity (rather than amplitude) equation as shown by Cook.[66,67] More elaborate treatments were subsequently given by Langer[22] and Yamauchi.[68] The required intensity equation has solution[66]

$$|X(\mathbf{k}, t)|^2 = [|X(\mathbf{k}, 0)|^2 - \langle |X(\mathbf{k})|^2 \rangle] \exp[2\alpha(\mathbf{k})t] + \langle |X(\mathbf{k})|^2 \rangle \qquad (37)$$

with the fluctuation term given, for example, by Eq. (27). It follows from the above equation that the critical wavelength in a particular direction is no longer given by Eq. (35) but by

$$\lambda_c' = 2\pi/k_c'$$

with k_c' a solution of

$$|X(\mathbf{k}, 0)|^2 = \langle |X(\mathbf{k})|^2 \rangle$$

The critical wavelength λ_c' thus depends on the initial condition, i.e., on the quenched-in concentration intensity spectrum. As shown by Cook,[67] the present correction is particularly important for very small initial intensity spectra so that, all things being equal, the more perfect the quench, the more the value of the actual critical wavelength will deviate from that predicted by Cahn's theory. The application of Eq. (37) requires that very careful absolute intensity measurement be made. Such experiments have recently been conducted by Acuña,[69] who emphasized that proper verification of the theory requires an *a priori* knowledge of f_0'', a quantity not generally available.

7.5. Kinetics of Nucleation and Growth

In the metastable region of the phase diagram (region II, Figure 1) the homogeneous solution is stable with respect to small-amplitude concentration modulations but unstable with respect to a class of localized departures from the average concentration called supercritical fluctuations. Nonlinear concentration and gradient terms in the master diffusion equation (31) can then no longer be neglected for even a qualitative approach to solid-solution decomposition in the metastable region. Equation (31) as it stands also lacks an appropriate fluctuation source term. In contrast to the case of typical spinodal decomposition of the previous two sections, the nonlinear diffusion equation can therefore no longer be simplified, so that ad hoc models must be invoked in order to obtain a tractable kinetic theory. A suitable model is obtained by extending to solid solutions the classical theory of nucleation first developed by Volmer and Weber[2] and Becker and Döring[3] to treat the problem of the nucleation of a pure liquid from its vapor.

To obtain an approximate theoretical model for the kinetics of a solid solution decomposing in the metastable region, one first derives an expression for the rate of production of supercritical fluctuations in the solution. This rate is assumed to be governed by the rate of creation of very special type of fluctuation: the critical nucleus, treated as a stationary state in Section 5.4. The critical nucleus, being in unstable equilibrium with the surrounding matrix phase, has equal probability of decaying or of growing. Overall kinetic laws are therefore obtained by combining the derived nucleation rate with an appropriate expression for the rate of growth of nuclei that have just become supercritical.

The nucleation rate I must be equal to the expected number of attempts of forming the critical nucleus per unit volume per unit time, times the probability that an attempt is successful:

$$I = N_v v e^{-W_R/kT} \quad (38)$$

In this equation N_v is the number of available nucleation sites per unit volume, v is an appropriate "frequency factor," and W_R is the reversible (minimum) work of formation of the critical nucleus in thermodynamic equilibrium with a very large volume V of the parent phase. The work W_R is given by Eq. (26) in the Cahn and Hilliard nucleation theory and by

$$W_R = \tfrac{4}{3}\pi(r^*)^2 \sigma$$

in the classical theory. In this equation r^* and σ are the critical radius and interfacial energy, respectively, as defined in Section 5.4, where the relation connecting σ, r^*, and ΔP was given along with methods of evaluating the pressure difference ΔP. The work of the overall volume change against the outside pressure is neglected.

The frequency factor in Eq. (38) is difficult to evaluate. However, uncertainties of even two or three orders of magnitude in v are generally dwarfed by the exponential dependence of uncertainties in W_R in the Boltzmann factor. In early work on the nucleation of liquid droplets from a supersaturated vapor[2,3] the frequency factor was obtained by considering a distribution of embryos of the condensing phase evolving past the critical size. For the present problem of nucleation in solid solutions the frequency factor must contain in addition the rate at which solute atoms can transfer across the matrix–embryo interface. Let Δg be the activation energy required for this transfer; then the frequency factor is given approximately by

$$v = (kT/h)e^{-\Delta g/kT}$$

where h is Planck's constant. This equation was derived by Turnbull and Fisher[70] from absolute reaction rate theory but, as shown by Christian,[5] can be considered as an extension of the Becker–Döring theory. More recent derivations include the finite probability of dissolution of supercritical nuclei, thereby introducing the so-called Zeldovich factor,[6] which is not really numerically significant, however.

Growth of the new phase from evolving nuclei is generally handled by yet another *ad hoc* model: that of Avrami[71] and Johnson and Mehl.[72] These authors introduce the concept of "extended volume" at time t which is the total volume that the growing phase would have occupied at time t had it been allowed to grow unimpeded by other transforming regions, and assuming that nucleation would take place anywhere within the sample, including the previously transformed regions. Let Z_x denote the extended volume fraction. It is related to the true volume fraction Z by the equation

$$dZ = (1 - Z)\,dZ_x$$

where $(1 - Z)$ is the probability of overlap of a newly transformed region onto a previously transformed one. The integrated form of this equation is

$$Z = 1 - e^{-Z_x} \qquad (39)$$

The reason for the introduction of this fictitious volume fraction Z_x is that its growth law can be determined in a straightforward manner, whereas that of Z cannot. If $G(t)$ is the growth rate of any linear dimension of the new phase, then for isotropic (spherical) growth one obtains the following equation for the extended volume at time t:

$$Z_x(t) = \tfrac{4}{3}\pi \int_0^t I(\tau) G^3(\tau)(t - \tau)^3 \, d\tau$$

This expression, when introduced into Eq. (39), yields the required growth law. Under very general assumptions the resulting growth law will be of the form

$$Z = 1 - \exp(-Bt^n) \qquad (40)$$

where B is an appropriate constant, and where $3 \leq n \leq 4$ for three-dimensional nucleation and growth. An interesting table of values of n for a wide variety of other situations is given on p. 489 of Christian's excellent treatise.[5]

Equation (40) plots as a sigmoid curve exhibiting incubation, growth, and saturation stages. In clustering reactions saturation sets in with the exhaustion of the supply of available solute atoms and

Chapter 3

occurs when the volume fraction Z has reached its maximum value dictated by the lever rule. Further evolution of the phase-separated system can then only take place through a coarsening reaction in which the larger precipitated clusters grow at the expense of smaller ones. Classical coarsening theories governing this stage of the decomposition process were given independently by Wagner[73] and Lifshitz and Slyozov.[74]

8. Morphology

Computer studies of the nonlinear diffusion equation (31) indicate that both the kinetics and the morphology of solid-solution decomposition should be continuous throughout the phase diagram. A "typical" spinodal morphology develops in the central unstable region of the phase diagram and the main features of the resulting structure can be predicted approximately from the linear theory alone. The argument runs as follows: Because of the exponential dependence of the kinetics of growth (or decay) of the harmonic concentration waves, the morphology will quickly be dominated by concentration waves **k** corresponding to the largest amplification rate $\alpha(k_m)$. If the anisotropy factor A is positive, $\langle 100 \rangle$ waves will dominate and the resulting structure will be described approximately by three sinusoidal waves of wavelength λ_m in the directions [100], [010], and [001], λ_m being the optimum wavelength defined above (typically of the order of 40–100 Å in metallic systems). These $\langle 100 \rangle$ composition modulations produce a characteristic "basket weave" or "tweed" structure. Figure 15 shows a sequence of electron micrographs obtained by Butler and Thomas[75] for the case of a Cu–Ni–Fe alloy of central (or symmetric) composition with respect to the pseudo-binary miscibility gap. Figure 15(a) illustrates the early-stage tweed structure, in good agreement with the predictions of the linear theory. Prolonged aging causes coarsening, as shown in micrographs 15(b) and 15(c), eventually followed by loss of coherence, as is apparent from the interface dislocations seen in micrographs 15(d–f).

For alloy compositions located closer to, but still inside, the spinodal the expected decomposition morphologies should be less regular than those of symmetric compositions. This is borne out by a comparison between Figures 15 and 16(a), the latter electron micrograph, also from the Cu–Ni–Fe work of Butler and Thomas, being representative of a more asymmetric composition. The later-stage alignment of particles (Figure 16b) is probably a coarsening effect.

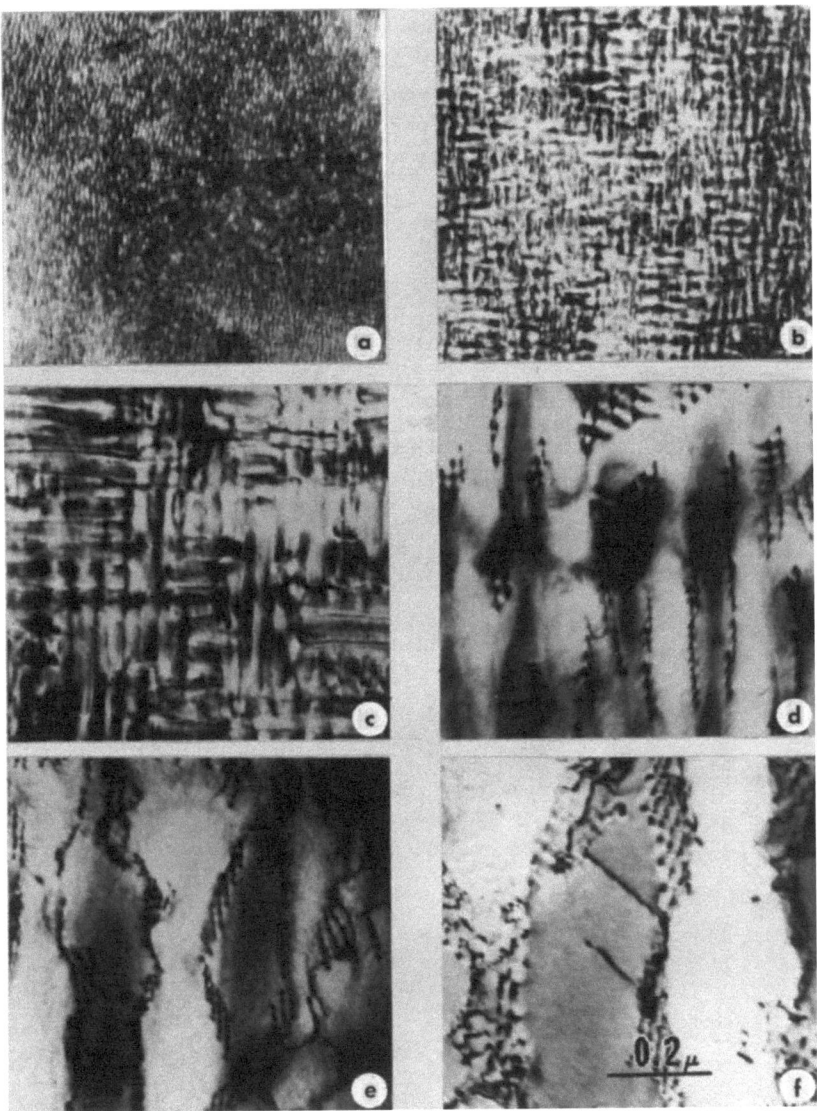

Fig. 15. Electron micrographs of CuNiFe foils aged near the center of the miscibility gap at 775°C for (a) 2 min, $\lambda = 150$ Å; (b) 15 min, $\lambda \approx 254$ Å; (c) 5 hr, $\lambda \approx 635$ Å; (d) 17 hr, $\lambda \approx 970$ Å; (e) 40 hr, $\lambda \approx 1330$ Å; (f) 200 hr, $\lambda > 5000$ Å. Direction [200] is horizontal, with foil orientation $\langle hk0 \rangle$.[75]

Cadoret and Delavignette reached similar conclusions in their independent study of the Cu–Ni–Fe system.[76]

From the foregoing considerations it is evident that spinodal structures can coarsen when subjected to prolonged aging treatments.

Chapter 3

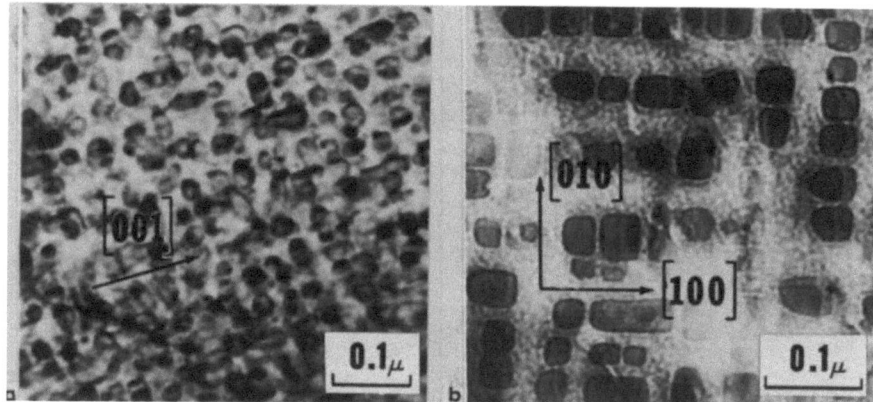

Fig. 16. Electron micrograph of CuNiFe alloy of asymmetric composition aged at 625°C: (a) early stage; (b) aged 300 hr.[76]

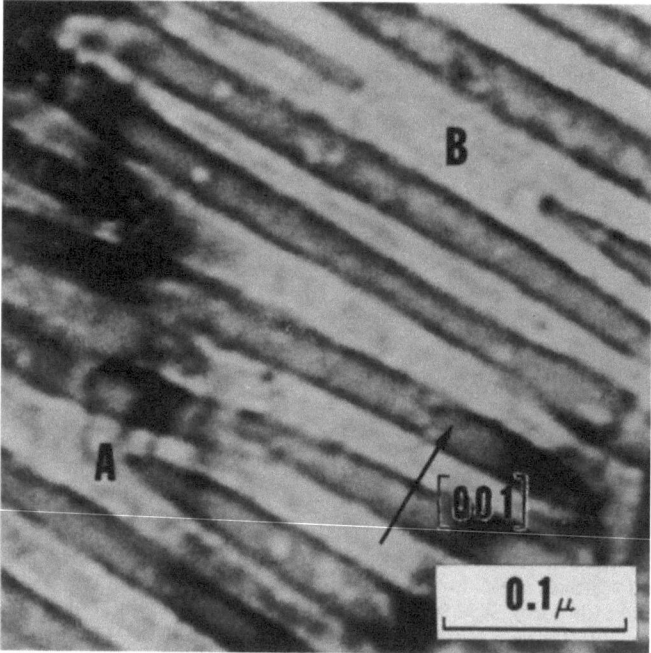

Fig. 17. Imperfections in periodicity responsible for the coarsening of the modulated structure (at A and B, for example) in a CuNiFe alloy of symmetrical composition aged 5 hr at 775°C. The orientation of the foil is [110] with the (001) planes perpendicular to the foil surface.[75]

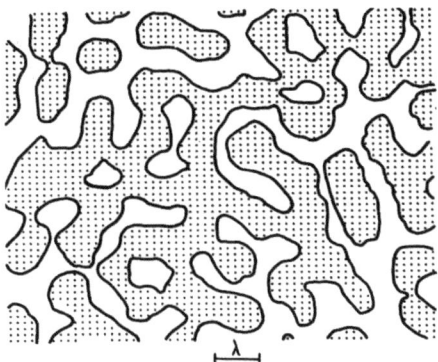

Fig. 18. Spinodal morphology in isotropic solutions obtained by superposition of optimum (λ_m) concentration plane mass with random phases and directions.[79]

This coarsening does not occur through the homogeneous expansion of the characteristic wavelength in accordion fashion, but rather by a local doubling of the wavelength which on a one-dimensional plot manifests itself by the dissolution of the weaker precipitates as illustrated by the computer-generated sequence of composition profiles shown in Figure 8 and which in three dimensions can be regarded as the climb of precipitate-edge-dislocations visible at points A and B of Figure 17. The coarsening kinetics apparently obey the $t^{1/3}$ law until coherence is lost.[75]

For negative anisotropy ($A < 0$) a $\langle 111 \rangle$ morphology is expected, but its precise nature is difficult to predict because of the uncertainty concerning the phases of the four dominant $\langle 111 \rangle$ waves.[77] The condition $A < 0$ is apparently verified for spinodal decomposition in symmetric Al–Zn alloys, electron micrographs of which show peculiar $\langle 111 \rangle$ modulations.[78]

Preferred orientation of decomposition products should not occur in isotropic solids, such as glasses. For symmetric compositions the expected morphology can be conveniently simulated by summing concentration waves of optimum wavelength but of random orientations and phases. The resulting structure is shown in Figure 18,[79] which is remarkably similar to Figure 12 obtained, as explained in Section 7.2, by solving the nonlinear diffusion equation numerically. The linear theory, on which Figure 18 is based, is powerless to produce coarsening effects, however. Note also the similarities in Figures 12, 18, and 19(a), the latter micrograph having been obtained

Chapter 3

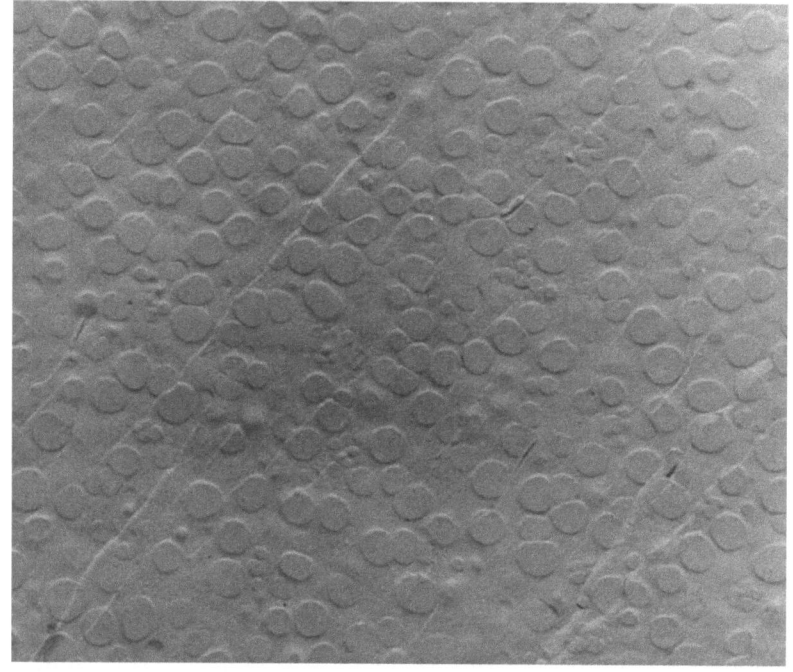

(b)

(a)

Fig. 19. Electron micrographs of Vycor glass formed (a) in the spinodal region (650°C, 100 hr); (b) in the metastable region (715°C, 24 hr).[80]

from a Vycor glass transformed in the spinodal region.[80] For comparison Figure 19(b) shows the morphology of the same Vycor glass transformed in the metastable region by nucleation and growth. In the latter case the correspondence between experimental (Figure 19b) and theoretical (Figure 11) morphologies is not as good. The computed structure is somewhat suspect, however, as nucleation is far more difficult to computer-simulate than decomposition inside the spinodal because of the difficulty of simulating realistic time-dependent composition fluctuations. Spinodal morphologies in B_2O_3–PbO–Al_2O_3 glasses have been studied in detail by Zarzycki and Naudin[81] using the interesting technique of optical transforms and spatial filtering of electron micrographs.

9. Conclusions

The kinetics of clustering in solid solutions are governed by the nonlinear partial differential equation (31) with fluctuating source term. Numerical solutions of the equation show no discontinuity in the kinetics or in the morphology of the decomposition process across the spinodal. The morphology varies from that of quasi-randomly dispersed solute clusters (G.P. zones) for solid solutions close to the coherent miscibility gap boundaries, all the way to connected periodic solute-rich and solute-poor regions for solutions in the center of the coherent miscibility gap. Solutions of intermediate concentrations exhibit, in the early stages of decomposition, quasi-periodic, partially connected structures recalling the Guinier zone model[64]: clusters with solute-rich cores surrounded by solute-poor denuded shells (Figure 8). Later-stage morphologies are strongly dependent on initial fluctuations in the case of systems located in or close to the metastable region but only weakly dependent on initial fluctuations in the unstable region. Approximate solutions to the kinetic equation can be constructed with the help of models. One such model, valid for the early stages of decomposition of solid solutions of central compositions, is the linear spinodal theory proposed by Cahn. This model has been remarkably successful in predicting morphologies and, when used with caution, quite useful in predicting early-stage kinetics.

At the other extreme of the decomposition spectrum one finds familiar kinetic models of nucleation, growth, and coarsening. An incipient instability is introduced in the metastable system by constructing artificially a critical nucleus in equilibrium with the surroundings. The nucleus profile is either calculated by the Cahn and Hilliard method or assumed, as in classical nucleation theory. The rate of nucleation is strongly dependent on the Boltzmann factor $e^{-W/kT}$, where W is the reversible work required for the formation of the critical nucleus. The nucleus is then allowed to become unstable, and its growth is described by the Avrami[71] and Johnson and Mehl[72] models. This produces the familiar sigmoid curves of overall growth kinetics. Coarsening, or Ostwald ripening, follows.

If the term "spinodal decomposition" is reserved for processes adequately described by Cahn's linear diffusion equation, it is found[16] that spinodal decomposition is an increasingly transitory stage as the average composition approaches the spinodal. It is quickly followed by growth and initial-stage coarsening, which are primarily governed by the nonlinear diffusion equation. Curves describing the overall kinetics inside the spinodal are also of sigmoid shape,[16] so that even "typical" decomposition processes in solid solutions cannot be readily distinguished from one another on the basis one-parameter kinetic measurements alone.

Experimental studies of the kinetics of spinodal decomposition have been difficult to interpret. It appears that the measurements have yielded an *effective* free energy second derivative f_0'' which does not have a clear thermodynamic meaning. In other words, investigators were setting themselves the difficult task of measuring basic thermodynamic properties from nonequilibrium processes. The spinodal itself, defined by $f'' = 0$, is not an equilibrium thermodynamic concept, and its position on the phase diagram is model dependent from a theoretical standpoint and path dependent from an experimental standpoint. The coherent solvus, on the other hand, is a well-defined equilibrium concept, at least in isotropic solids, and its experimental determination has met with considerable success as evidenced, for example, by Figure 3.

References

1. J. W. Gibbs, On the equilibrium of heterogeneous substances, in *Scientific Papers*, pp. 105 and 252, Dover, New York (1961).
2. M. Volmer and A. Weber, Nuclei formation in supersaturated states, *Z. Physik. Chem.* **119**, 277–301 (1925).
3. R. Becker and W. Döring, Kinetic treatment of grain formation in supersaturated vapors, *Ann. Physik.* **24**, 719–752 (1935).
4. D. Turnbull, Phase changes, in *Solid State Physics* (F. Seitz and D. Turnbull, eds.), pp. 226–308, Academic, New York (1956).

5. J. W. Christian, *The Theory of Phase Transformations in Metals and Alloys*, Pergamon, London (1965).
6. K. C. Russell, Nucleation in solids, in *Phase Transformations* (H. I. Aaronson, ed.), pp. 219–268, American Society for Metals, Metals Park, Ohio (1970).
7. M. Hillert, A theory of nucleation of solid metallic solutions, D.Sc. Dissertation, Massachusetts Institute of Technology, Cambridge, Mass. (1956).
8. M. Hillert, A solid-solution model for inhomogeneous systems, *Acta Met.* **9**, 525–535 (1961).
9. J. W. Cahn and J. E. Hilliard, Free energy of a nonuniform system, I. Interfacial free energy, *J. Chem. Phys.* **28**, 258–267 (1958).
10. J. W. Cahn and J. E. Hilliard, Free energy of a nonuniform system, II. Nucleation in a two-compartment incompressible fluid, *J. Chem. Phys.* **31**, 688–699 (1959).
11. J. W. Cahn, On spinodal decomposition, *Acta Met.* **9**, 795–801 (1961).
12. J. W. Cahn, On spinodal decomposition in cubic crystals, *Acta Met.* **10**, 179–183 (1962).
13. J. W. Cahn, Coherent fluctuations and nucleation in isotropic solids, *Acta Met.* **10**, 907–913 (1962).
14. H. E. Cook, D. de Fontaine, and J. E. Hilliard, A model for diffusion on cubic lattices and its application to the early stages of ordering, *Acta Met.* **17**, 765–773 (1969).
15. D. de Fontaine and H. E. Cook, Early-stage clustering and ordering kinetics in binary solid solutions, in *Critical Phenomena in Alloys, Magnets, and Superconductors* (R. E. Mills, E. Ascher, and R. I. Jaffee, eds.), pp. 257–275, McGraw-Hill, New York (1971).
16. D. de Fontaine, A computer simulation of the evolution of coherent composition variations in solid solutions, Ph.D. Dissertation, Northwestern University, Evanston, Illinois (1967).
17. D. de Fontaine, An analysis of clustering and ordering in multicomponent solid solutions—I. Stability criteria, *J. Phys. Chem. Solids* **33**, 297–310 (1972).
18. D. de Fontaine, An analysis of clustering and ordering in multicomponent solid solutions—II. Fluctuations and kinetics, *J. Phys. Chem. Solids* **34**, 1285–1304 (1973).
19. J. E. Morral and J. W. Cahn, Spinodal decomposition in ternary systems, *Acta Met.* **19**, 1037–1045 (1971).
20. D. de Fontaine, in *Solid State Physics* (H. Ehrenreich, F. Seitz and D. Turnbull, eds.), to be published.
21. J. W. Cahn, The later stages of spinodal decomposition and the beginnings of particle coarsening, *Acta Met.* **14**, 1685–1692 (1966).
22. J. S. Langer, Theory of spinodal decomposition in alloys, *Ann. Phys. (N.Y.)* **65**, 53–86 (1971).
23. L. H. Shendalman and J. T. O'Toole, Nucleation and coarsening in binary condensed phases, *J. Colloid. Interface Sci.* **27**, 145–160 (1968).
24. L. A. Swanger, P. K. Gupta, and A. R. Cooper, Jr., Computer simulation of one-dimensional spinodal decomposition, *Acta Met.* **18**, 9–14 (1970).
25. J. W. Cahn, Spinodal decomposition, *Trans. AIME* **242**, 166–180 (1968).
26. J. W. Cahn, Unmixing in binary critical systems, in *Critical Phenomena in Alloys, Magnets, and Superconductors* (R. E. Mills, E. Ascher, and R. I. Jaffee, eds.), pp. 41–64, McGraw-Hill, New York (1971).

Chapter 3

27. J. E. Hilliard, Spinodal decomposition, in *Phase Transformations* (H. I. Aaronson, ed.), pp. 497–560, American Society for Metals, Metals Park, Ohio (1970).
28. D. de Fontaine, Development of fine coherent precipitate morphologies by the spinodal mechanisms, in *Ultrafine Grain Metals* (Burke and V. Weiss, eds.), pp. 93–131, Syracuse Univ. Press (1970).
29. A. Bonfiglioli, La decomposition espinodal, Comision Nacional de Energia Atomica PMM/I-91, Buenos Aires, Argentina (1972).
30. A. G. Khachaturyan and R. A. Suris, Theory of periodic distributions of concentrations in a supersaturated solid solution, *Soviet Phys.—Crystallography* **13**, 63–67 (1968).
31. R. Kikuchi, Cooperative phenomena in the triangular lattice, *J. Chem. Phys.* **47**, 1664–1668 (1967).
32. P. C. Clapp, A critical examination of the validity of the pair-wise interaction model for ordered alloys, in *Ordered Alloys: Structural Applications and Physical Metallurgy* (B. H. Kear, ed.), pp. 25–35, Clator's Press, Baton Rouge, La. (1970).
33. D. W. Hoffman, Configurational entropy and solute correlation in disordered alloys, *Trans. AIME*, in press.
34. R. Cadoret, A statistical treatment of the free energy of binary nonhomogeneous solutions, *Phys. Stat. Sol. (b)* **46**, 291–298 (1971).
35. J. F. Nye, *Physical Properties of Crystals*, Oxford Univ. Press (1957).
36. H. E. Cook and D. de Fontaine, On the elastic free energy of solid solutions—I. Microscopic theory, *Acta Met.* **17**, 915–924 (1969).
37. A. G. Khachaturyan, Microscopic theory of diffusion in crystalline solid solutions and the time evolution of the diffuse scattering of X-rays and thermal neutrons, *Soviet Phys.—Solid State* **9**, 2040–2046 (1968).
38. H. E. Cook and D. de Fontaine, On the elastic free energy of solid solutions—II. Influence of the effective modulus on precipitation from solution and the order-disorder reaction, *Acta Met.* **19**, 607–616 (1971).
39. D. W. Hoffman, Concerning the elastic energy of dilute interstitial alloys, *Acta Met.* **18**, 819–833 (1970).
40. W. H. Zachariasen, *X-Ray Diffraction in Crystals*, Wiley, New York (1945).
41. A. Guinier, A new type of X-ray diagram, *Compt. Rend.* **206**, 1641–1643 (1938).
42. G. D. Preston, The diffraction of X-rays by age-hardening Al–Cu alloys, *Proc. Roy. Soc.* **A167**, 526–538 (1938).
43. M. Murakami, O. Kawano, and Y. Murakami, On the determination of the solvus temperature for G.P. zones in an Al–0.8 at.% Zn alloy, *J. Inst. Metals (London)* **99**, 160 (1971).
44. J. Lašek, Über die Einfluss der durchschnittlichen Zusammensetzung auf die Lage der kohärenten Mischungslücke von Al–Zn Legierungen, *Czech. J. Phys.* **15**, 848–857 (1965).
45. V. Gerold, Die Zonebildung in Al–Zn Legierungen, *Phys. Stat. Sol.* **1**, 37–49 (1961).
46. V. Gerold and W. Mertz, On the decomposition of an aluminum–zinc alloy, *Scripta Met.* **1**, 33–35 (1967).
47. G. J. C. Carpenter and R. D. Garwood, Hardness reversion and the metastable phase boundary for G.P. zones in Al–Zn alloys, *J. Inst. Metals (London)*, **94**, 301–304 (1966).
48. T. Niklewski, P. Spiegelberg, and K. Sunbulli, The solvus curve for G.P. zones in Al–Zn alloys: A diffuse X-ray study, *Metal Sci. J.* **3**, 23–25 (1969).

49. A. J. Ardell, K. Nuttall, and R. B. Nicholson, The decomposition of concentrated Al–Zn alloys, in *The Mechanism of Phase Transformation in Crystalline Solids*, pp. 22–26, The Institute of Metals, London, England (1968).
50. B. Golding and S. C. Moss, A recalculation of the gold nickel spinodal, *Acta Met.* **15**, 1239–1241 (1967).
51. D. de Fontaine, An approximate criterion for the loss of coherency in modulated structures, *Acta Met.* **17**, 477–482 (1969).
52. H. Goldstein, *Classical Mechanics*, p. 53, Addison–Wesley, Reading, Mass. (1950).
53. G. A. Korn and T. M. Korn, *Mathematical Handbook for Scientists and Engineers*, pp. 830–841, McGraw-Hill, New York (1968).
54. R. Kikuchi, A theory of cooperative phenomena, *Phys. Rev.* **81**, 988–1003 (1951).
55. R. Kikuchi, Boundary free energy in the lattice model—III Solution of the paradox, *J. Chem. Phys.* **57**, 787–798 (1972).
56. L. D. Landau and E. M. Lifshitz, *Statistical Physics*, pp. 366–369, Addison–Wesley, Reading, Mass. (1958).
57. P. C. Clapp and S. C. Moss, Correlation functions in disordered binary alloys I, II, *Phys. Rev.* **171**, 418–427; 754–763 (1968).
58. M. A. Krivoglaz, *Theory of X-Ray and Thermal-Neutron Scattering by Real Crystals*, Plenum, New York (1967).
59. S. Wilkins, Determination of long-range interaction energies from scattering of X-rays by disordered alloys, *Phys. Rev. B* **2**, 3935–42 (1970).
60. D. de Fontaine, Bose–Einstein condensation of concentration fluctuations in binary solid solutions, in *Critical Phenomena in Alloys, Magnets, and Superconductors* (R. E. Mills, E. Ascher, and R. I. Jaffee, eds.), pp. 277–287 McGraw-Hill, New York (1971).
61. P. Glansdorff and I. Prigogine, *Thermodynamic Theory of Structure, Stability and Fluctuations*, Wiley, New York (1971).
62. E. L. Huston, J. W. Cahn, and J. E. Hilliard, Spinodal decomposition during continuous cooling, *Acta Met.* **14**, 1053–1062 (1966).
63. K. B. Rundman and J. E. Hilliard, Early stages of spinodal decomposition in an aluminum–zinc alloy, *Acta Met.* **15**, 1025–1033 (1967).
64. A. Guinier, Nouvelle interprétation des diagrammes à "side-bands," *Acta Met.* **3**, 510–512 (1955).
65. A. Bonfiglioli and A. Guinier, La structure des zones G.P. dans les alliages aluminium–zinc au premier stade de leur formation, *Acta Met.* **14**, 1213–1224 (1966).
66. H. E. Cook, The kinetics of clustering and short-range order in stable solid solutions, *J. Phys. Chem. Solids* **30**, 2427–2437 (1969).
67. H. E. Cook, Brownian motion in spinodal decomposition, *Acta Met.* **18**, 297–306 (1970).
68. Y. Yamauchi, Doctoral Dissertation, Northwestern Univ., Evanston, Ill. (1973).
69. R. Acuña Laje, Transformaciones de fases coherentes en aleaciones de Al–Zn, Doctoral Dissertation, IMAF, Universidad Nacional de Córdoba (1971).
70. D. Turnbull and J. C. Fisher, Rate of nucleation in condensed systems, *J. Chem. Phys.* **17**, 71–73 (1949).
71. M. Avrami, Kinetics of phase change—I. General theory, *J. Chem. Phys.* **7**, 1103–1112 (1939).
72. W. A. Johnson and R. F. Mehl, Reaction kinetics in processes of nucleation and growth, *Trans. AIME* **135**, 416–458 (1939).

Chapter 3

73. C. Wagner, Theorie der Alterung von Niederschlägen durch Umlösen, *Z. Electrochem.* 581–591 (1961).
74. I. M. Lifshitz and V. V. Slyozov, The kinetics of precipitation from supersaturated solid solutions, *J. Phys. Chem. Solids* **19**, 35–50 (1961).
75. E. P. Butler and G. Thomas, Structure and properties of spinodally decomposed Cu–Ni–Fe alloys, *Acta Met.* **18**, 347–365 (1970).
76. R. Cadoret and P. Delavignette, Etude de la décomposition spinoidale au microscope electronique dans les alliages CuNiFe, *Phys. Stat. Sol.* **32**, 853–865 (1969).
77. J. W. Cahn, A correction to spinodal decomposition in cubic crystals, *Acta Met.* **12**, 1457 (1964).
78. R. B. Nicholson, unpublished work.
79. J. W. Cahn, Phase separation by spinodal decomposition in isotropic systems, *J. Chem. Phys.* **42**, 93–99 (1965).
80. J. W. Cahn and R. J. Charles, The initial stages of phase separation in glasses, *Phys. Chem. Glasses* **6**, 181–191 (1965).
81. J. Zarzycki and F. Naudin, Spinodal decomposition in the B_2O_3–PbO–Al_2O_3 system, *J. Non-Cryst. Solids* **1**, 215–234 (1969).

4

Relation of the Thermochemistry and Phase Diagrams of Condensed Systems

Larry Kaufman and Harvey Nesor
ManLabs, Inc.
Cambridge, Massachusetts

1. Introduction

The relation between the thermochemistry and the phase diagram of a condensed system is an identity. Although this identity was rather well known to the chemists, metallurgists, and ceramists who practiced in the early decades of this century, most practitioners educated after the second world war seem to be unaware of the equivalence. Indeed, the field of phase diagrams is almost exclusively tilled by metallurgists and ceramists, while thermochemical aspects of condensed systems is the exclusive province of chemists. This dichotomy is extremely wasteful since the integrated description of a system permits interrelation of phase diagram and thermochemical data yielding a more complete description of a given system. The description of the magnesium–gallium–aluminum system shown in Table 1 and Figure 1 illustrates the interrelation between thermochemical and phase diagram data and serves to review the whole problem for those who are accustomed to dealing only with the component parts. Table 1 summarizes the free energy of formation of Mg–Ga and Mg–Al compounds at 300°K. These free energies of formation have been calculated from measurements of the thermochemical properties of the Mg–Ga and Mg–Al systems[1,2] using techniques which have been described in detail[3] for combining phase diagram and thermochemical data. The Ga–Al system exhibits slightly positive deviations from ideality which result in limited

Chapter 4

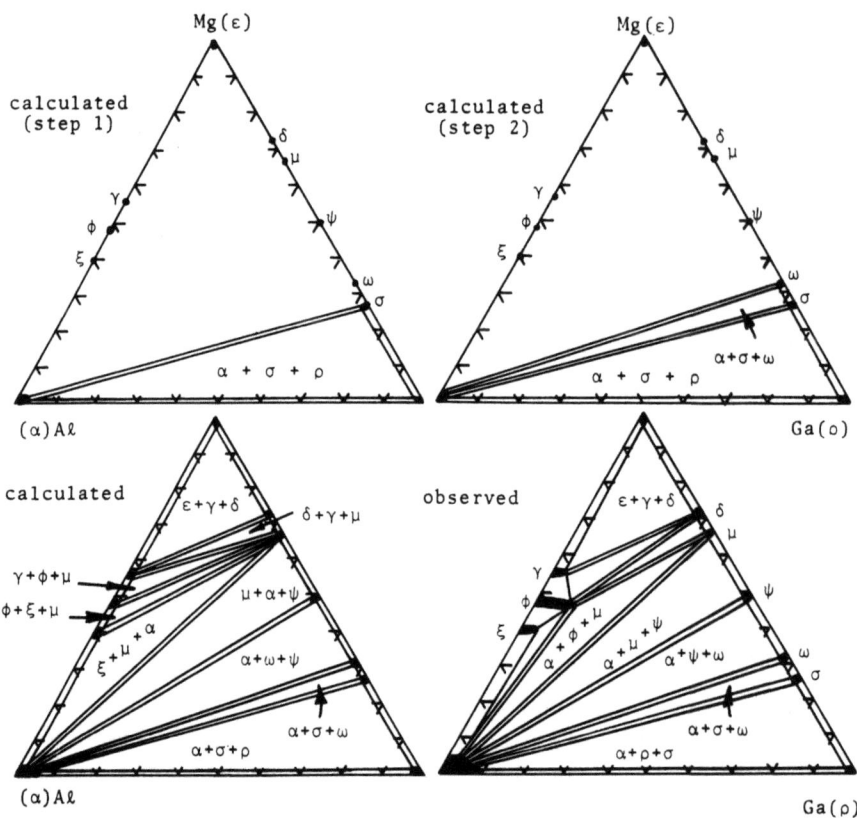

Fig. 1. Calculated and observed isothermal sections in the Mg–Ga–Al system at 300°K.

solubility of Al in Ga and of Ga in Al at 300°K. No stable Ga–Al compounds have been reported. It is now instructive to calculate an equilibrium diagram for Mg–Ga–Al at 300°K based on the assumption that all of the compounds (and pure metals) shown in Table 1 exist at fixed compositions without additional solubility of other components. The first step of the calculation is illustrated in the upper left corner of Figure 1. Here we consider the possible reactions

$$\xi + \rho \rightleftarrows \alpha + \sigma \tag{1a}$$

$$\phi + \rho \rightleftarrows \alpha + \sigma \tag{1b}$$

$$\gamma + \rho \rightleftarrows \alpha + \sigma \tag{1c}$$

In each case the reaction proceeds to the right based on Table 1. Thus Eq. (1) yields the first panel in Figure 1. It should be noted that

TABLE 1
Free Energy of Formation of Mg–Al and Mg–Ga Compounds at 300°K

Compound	Phase designation	ΔF_f, cal/g-atom
Mg_2Al_3	ξ	−1060
$Mg_{23}Al_{30}$	ϕ	−1130
$Mg_{17}Al_{12}$	γ	−1160
Mg_2Ga_5	σ	−2470
$MgGa_2$	ω	−2570
$MgGa$	ψ	−2620
Mg_2Ga	μ	−2650
Mg_5Ga_2	δ	−2300
Mg	ε	0
Ga	ρ	0
Al	α	0

if compounds were present in Ga–Al, the competition between ε and the Ga–Al compound on the one hand and $\alpha + \sigma$ on the other hand would require investigation in addition to the examples in Eq. (1). The next step is to investigate the stability of $\alpha + \omega$:

$$\rho + \xi \rightleftarrows \alpha + \omega \quad (2a)$$

$$\rho + \phi \rightleftarrows \alpha + \omega \quad (2b)$$

$$\rho + \gamma \rightleftarrows \alpha + \omega \quad (2c)$$

$$\xi + \sigma \rightleftarrows \alpha + \omega \quad (2d)$$

$$\phi + \sigma \rightleftarrows \alpha + \omega \quad (2e)$$

$$\gamma + \sigma \rightleftarrows \alpha + \omega \quad (2f)$$

Now, in addition to Ga and Mg–Al compound combinations competing with the α–ω pair, the combinations between σ and the Mg–Al compounds compete with the α–ω pair. However, in all cases the $\alpha + \omega$ pair yields the lowest free energy as shown in the upper right panel of Figure 1. Similar analyses are performed for $\alpha + \psi$, $\alpha + \mu$, and $\alpha + \delta$. In these cases it is found that $\alpha + \psi$ and $\alpha + \mu$ are stable but $\alpha + \delta$ is not. Finally ξ/δ versus ϕ/μ and ϕ/δ versus γ/μ competition are considered to yield the calculated section shown on the lower left part of Figure 1. This result is to be compared with the observed section shown in the lower right portion of Figure 1.[4] The main discrepancy is due to neglecting solubility in the compound phases. This approximation leads to the error in the ϕ/δ versus γ/μ calculation and the suppression of the ζ phase by the ϕ/α combination. Nevertheless, the calculated diagram bears a strong resemblance to the observed diagram. Moreover, this exercise illustrates many important aspects of the relation of thermochemistry and phase diagrams, not the least of which is how competition between phases (or pairs of phases) forms the basis of the phase diagram. Thus stability is not the property of a phase itself but is rather determined in combination with other phases or groups of phases.

The only important feature absent from the foregoing considerations is the aspect of variable composition. Thus as the temperature is increased a liquid phase will expand from the gallium corner until it covers the entire diagram. Although calculation of the phase diagram from thermochemical properties is more complicated when phases of variable composition are considered, the principles of computation are the same as outlined above.

Accordingly, subsequent sections of this chapter will review present methods for coupling thermochemical descriptions with phase diagram data in order to produce a more complete description of phase stability. In addition, examples will be presented to illustrate cases where the coupling techniques can be employed to predict phase stability in complex systems. Although the examples presented are drawn exclusively from metal and metal–carbide systems, the methods are general and have also been applied to oxide, sulfide, alkali halide, and III–IV compound systems. The basic relations between thermochemical and phase diagram characteristics are identical in all cases.

2. Current Efforts at Coupling Thermochemical and Phase Diagram Data

The example of the Mg–Ga–Al system presented above provides a convenient means for introducing the relation between the thermochemical and phase diagram characteristics of a system. However, the example noted is simplified by consideration of phases of fixed composition at one temperature. Naturally, a full consideration of the problem must deal with phases of variable composition over wide ranges of temperature (and pressure). Under such conditions it is immediately apparent that calculation of a complete diagram requires examination of many competing phases having a variety of complex structures and stoichiometries. Although such a task may seem impossible at first consideration, application of modern computer techniques has provided a means for reducing the problem to tractable limits.[3,5-29]

In particular, when dealing with solution phases which can exist over wide ranges of composition the important feature is definition of the excess free energy. Thus in dealing with a solution phase ϕ in the i–j system, the free energy F^ϕ is defined as

$$F^\phi = (1 - x)F_i^\phi + xF_j^\phi + RT[(1 - x)\ln(1 - x) + x \ln x] + F_E^\phi \quad (3)$$

where F_i^ϕ and F_j^ϕ are the free energies of pure i and pure j in the ϕ form, F_E^ϕ is the excess free energy of the phase ϕ, and x is the atom fraction of j. Here ϕ can be the liquid, bcc, hcp, fcc, or other solid-solution phases. At present there are no physical models which can predict the form or magnitude of F_E^ϕ although empirical schemes do exist for estimating F_E^ϕ.[3] All that can be said is that $F_E^\phi = 0$ for $x = 0$ and $x = 1$. Another important problem which arises in dealing with equilibria between two or more solution phases (say ϕ and α) is definition of $\Delta F_i^{\alpha \to \phi}$ and $\Delta F_j^{\alpha \to \phi}$. These quantities are dependent upon temperature and form an important link in coupling the thermochemical and phase diagram characteristics. Moreover, since they are properties of the pure metals i and j, they appear in the coupling equations of any system containing these metals. By contrast the F_E function is uniquely associated with a given i–j pair. Although some progress has been made recently in calculating $\Delta F_i^{\alpha \to \phi}$ from first principles,[6] the level of accuracy is not yet adequate for purposes of coupling thermochemical and phase diagram data. On the other hand, empirical methods have been successfully employed in establishing $\Delta F_i^{\alpha \to \phi}$ values (lattice stability terms) for many pure metals where α and ϕ refer to the liquid, bcc, hcp, and fcc structures.[3,5,17-19] Figure 2 shows recent comparisons[3,17] of predicted lattice stability values for chromium, nickel, and cobalt with those deduced from thermochemical measurements. The ordinate of the Cr–Ni diagram in Figure 2 is the activity coefficient of chromium in fcc Cr–Ni solid solutions relative to pure bcc chromium. The ordinate of the Co–Fe diagram in Figure 2 is the heat of formation of bcc cobalt–iron alloys from bcc iron and fcc cobalt at 1143°K divided by the atom fraction of cobalt. Table 2 displays numerical values for iron,[3,33] nickel,[5,19] and cobalt[19] for the bcc, hcp, fcc, and liquid forms. Figure 3 compares these results for nickel and cobalt with recent estimates by Miodownik,[34] who reconsidered the lattice stability of cobalt by

Chapter 4

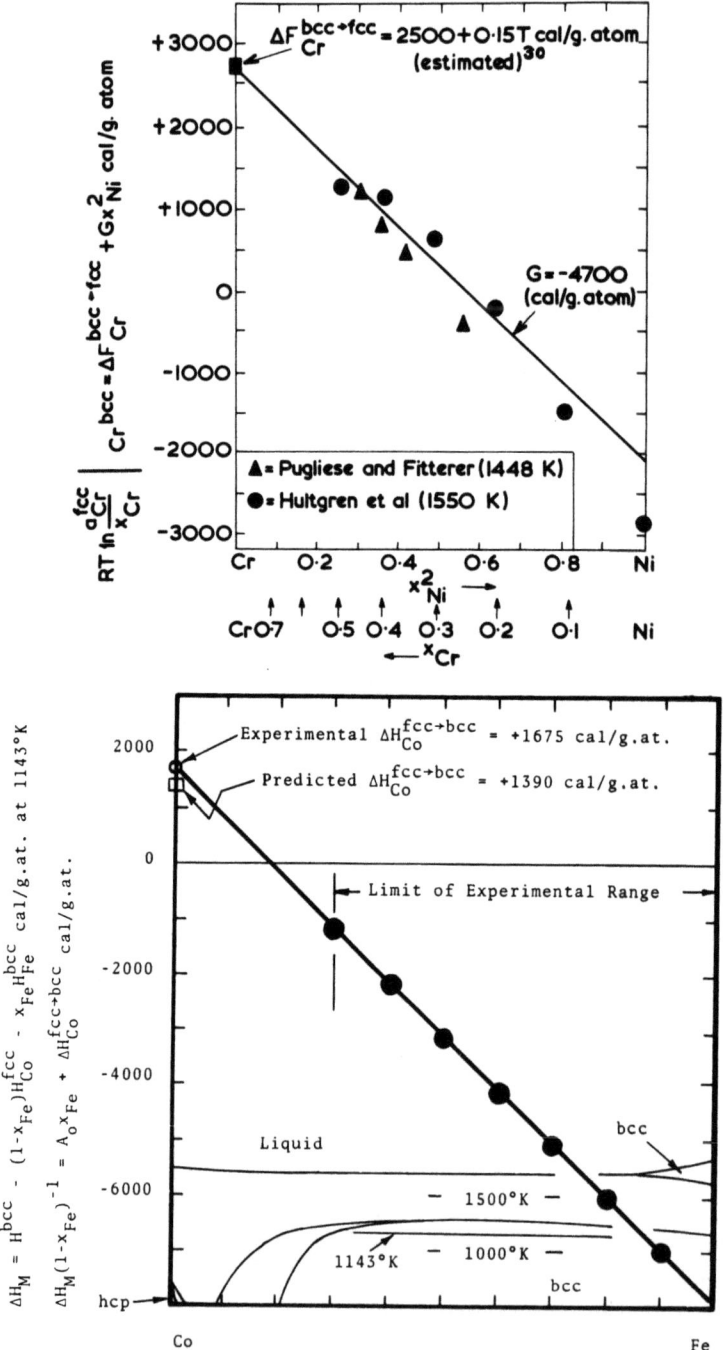

Fig. 2. Comparison of predicted and observed lattice stability for chromium[6] and cobalt.[19]

TABLE 2
Compilation of Free-Energy Difference Equations*

Iron	
L–fcc	$3524 - 1.959T$ above $1665°K$
L–bcc	$3300 - 1.824T$ above $1665°K$
bcc–fcc	$-1251.2 + 2.2468T - 0.12655 \times 10^{-2}T^2 + 0.2204 \times 10^{-6}T^3$ for $1100 \leq T \leq 1800$
fcc–bcc	$1460 - 0.8274T - 0.17858 \times 10^{-2}T^2 + 0.1225 \times 10^{-5}T^3$ for $300 \leq T \leq 1100$
fcc–bcc	$1303 + 1.78 \times 10^{-3}T^2 - 2.87 \times 10^{-5}T^3 + 4.91 \times 10^{-8}T^4$ below $300°K$
hcp–fcc	$-437 + 1.12T$ above $300°K$
Nickel	
L–fcc	$4210 - 2.44T$ above $1500°K$
fcc–bcc	$-940 - 0.982 \times 10^{-3}T^2 + 0.116 \times 10^{-5}T^3 - 0.337 \times 10^{-9}T^4$ for $0 \leq T \leq 1800$
fcc–bcc	$-940 - 0.74167 \times 10^{-3}T^2 + 0.5555 \times 10^{-6}T^3$ for $0 \leq T \leq 1200$
fcc–bcc	$-1330 + 0.25T$ for $T \geq 900$
Cobalt	
L–fcc	$3870 - 2.19T$ above $1500°K$
fcc–bcc	$-1662 + 0.1509 \times 10^{-2}T^2 - 0.6701 \times 10^{-6}T^3$ for $0 \leq T \leq 1800$
fcc–bcc	-560 for $1300 \leq T \leq 1800$
fcc–bcc	$-1840 + 1.0T$ for $300 \leq T \leq 1300$

* Units are cal/g-atom and °K.

including magnetic terms which were omitted earlier.[30,3] The lattice stability values for iron, nickel, chromium, and cobalt shown in Table 2 and Figures 2 and 3 were used in analyses of the Fe–Co, Fe–Ni, Ni–Cr, Fe–Cr, Ni–Co, and Cr–Co systems.[19] Utilization of these lattice stability descriptions in many systems provides a means for assessing their accuracy and obtaining better values. Such analyses should be pursued in the future. Although it is difficult to place absolute error limits on the current values, it is interesting to note that a recent compilation of the heats of fusion of iron, cobalt, and nickel[35] indicated uncertainty limits of up to ± 600 cal/g-atom in some measurements. Reference to Figure 3 suggests that the present uncertainty limits for nickel and cobalt might be ± 300 cal/g-atom over most of the temperature range of interest. In the case of iron the uncertainty limits are undoubtedly smaller.

Graphical illustration of the role played by the lattice stability terms in establishing phase equilibria is shown in Figures 4–6.[3] Figure 4 shows the comparison between calculated and observed versions of the Hf–Ta system.[3] Recent observations[36] are in better agreement with the calculations than the comparison in Figure 4.

Chapter 4

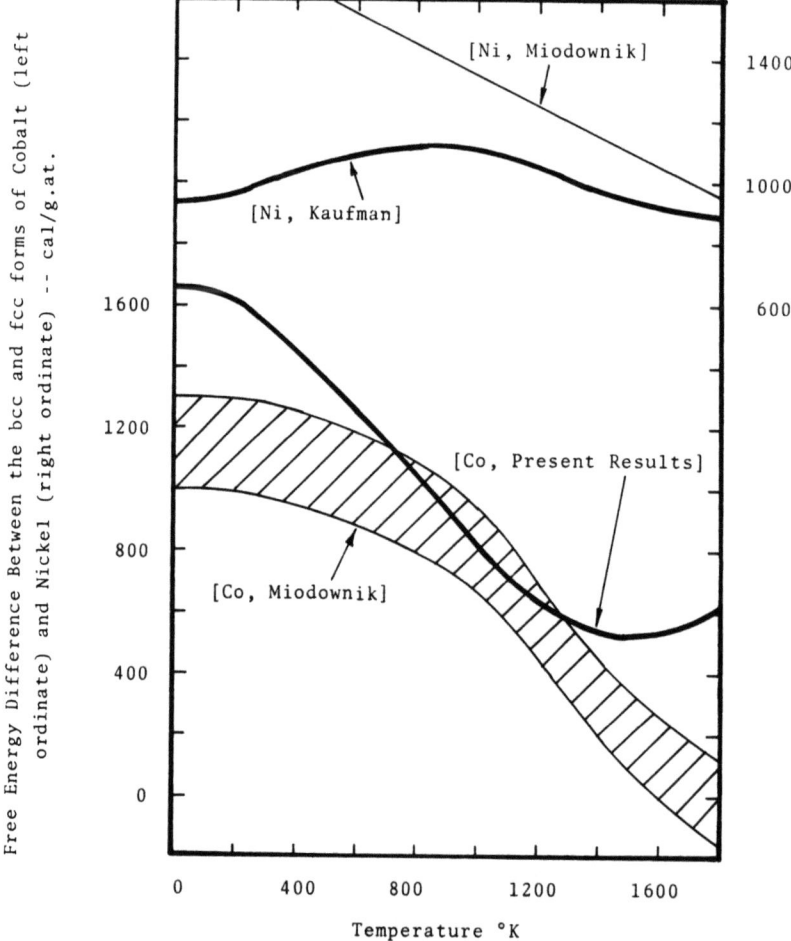

Fig. 3. Free energy of the bcc form minus the free energy of the fcc form of cobalt and nickel as a function of temperature.[19]

Figures 5 and 6 show calculated free energy–composition curves for L (liquid), β (bcc), ε (hcp), and α (fcc) phases in the Hf–Ta system at various temperatures. Several important features are to be noted. At 1200°K (Figure 6) the β and ε phases are more stable than the α and L phases since their free energies (F^β and F^ε) lie below the free energies of the fcc and liquid phases (F^α and F^L). The free energies of the β and ε phases are equal at $x_0^{\beta\varepsilon} = 0.23$ and 1200°K (Figure 5). However, the most stable situation is formation of a two-phase field bounded by the "common tangent rule" (equivalent to equilibration

Relation of the Thermochemistry and Phase Diagrams of Condensed Systems

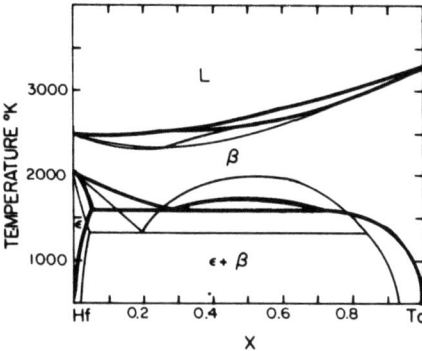

Fig. 4. Regular solution calculated (thick lines) and observed (thin lines) in hafnium–tantalum phase diagram.[3]

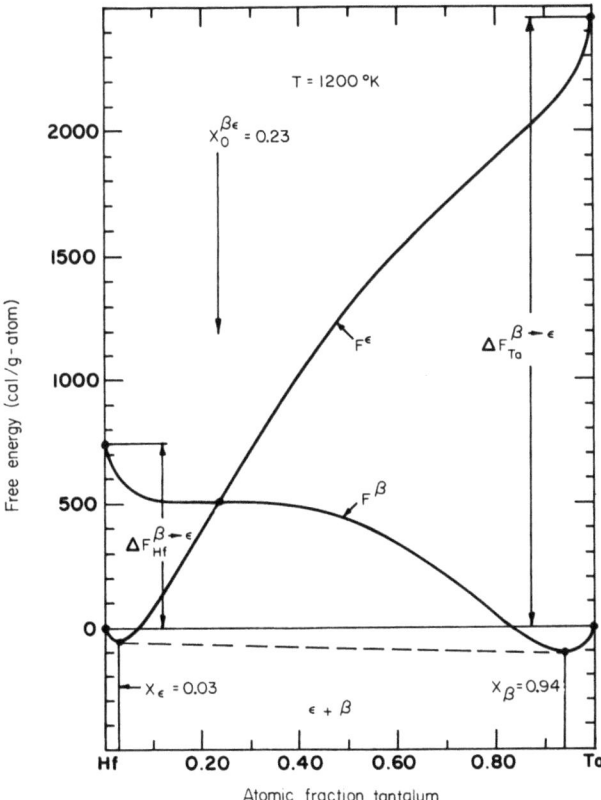

Fig. 5. Calculated free energy–composition relations for the bcc (β) and hcp (ε) solid solutions in the hafnium–tantalum system at 1200°K, illustrating common tangent location of the phase boundaries.[3]

Chapter 4

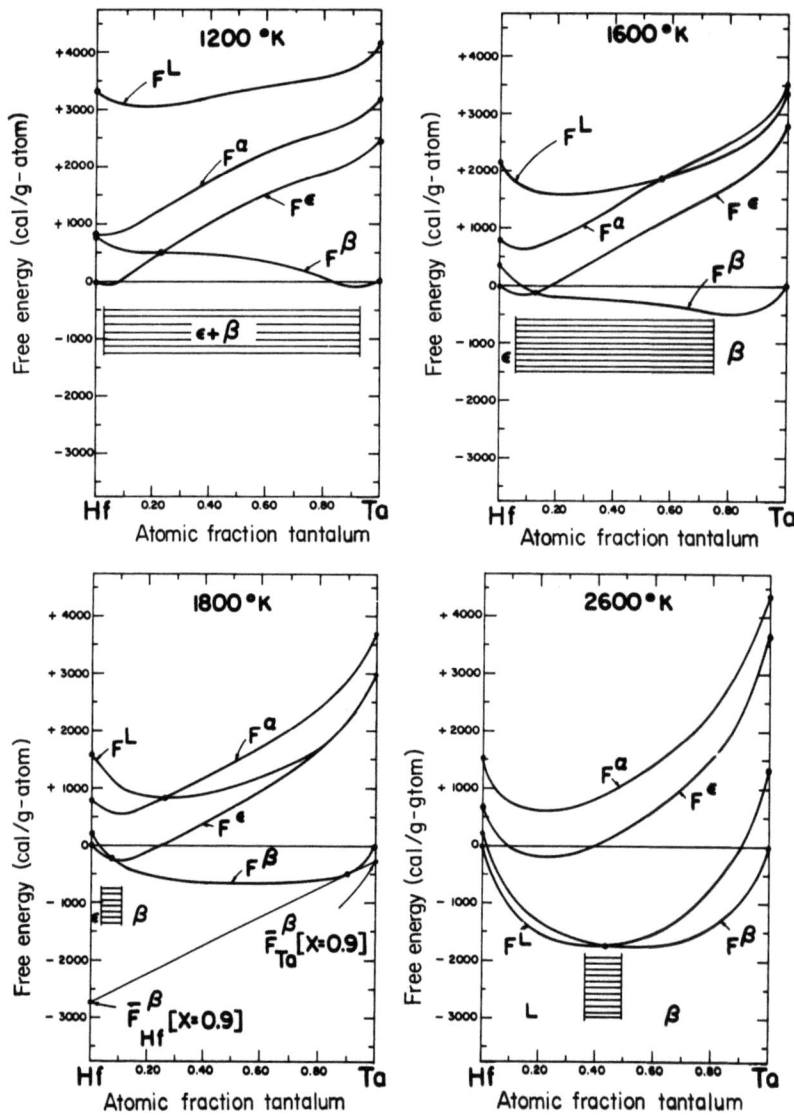

Fig. 6. Calculated free energy–composition curves for the liquid (*L*), fcc (α), β, and ε solutions in the hafnium–tantalum system. Panel at the lower left illustrates partial molar free energies (or chemical potentials).[3]

of the partial molar free energies) at $x_\varepsilon = 0.03$ and $x_\beta = 0.94$. Figure 6 illustrates the important role played by $\Delta F_{\text{Hf}}^{\beta \to \varepsilon}$ and $\Delta F_{\text{Ta}}^{\beta \to \varepsilon}$ in controlling phase stability. Moreover, Figure 6 shows that the other lattice stability differences (involving the α and *L* phases) determine the relative free energies and hence the stability of these phases. The partial molar free energies (or chemical potentials) of hafnium and tantalum in the β phase at $x = 0.9$ and $1800°K$ (i.e., $\bar{F}^\beta_{Hf}[x = 0.9]$ and $\bar{F}^\beta_{Ta}[x = 0.9]$) are illustrated in Figure 5. These examples show how the lattice stability terms contribute to the relative *position* of the free energy–composition curves. The *shape* of these curves is controlled by the free energy of mixing. The inflection in $F^\beta[x]$ leads to the miscibility gap.

Previous analyses of metallic systems[3,17] have made extensive use of symmetric functions to describe the excess free energy of mixing. However, in some cases the complex nature of the system of interest requires that more general forms be considered along the lines presented in Eqs. (209)–(212) of Ref. 3. Accordingly, the excess free energy of solution phases has been described by the following equation for the i–j system, in which x is the atomic fraction of j[19]:

$$F_E = x(1 - x)\{(1 - x)g[T] + xh[T]\} \tag{4}$$

where $g[T]$ and $h[T]$ are temperature-dependent functions. Thus the excess partial free energies of components i and j are given as[3]

$$\bar{F}_{Ei} = x^2\{g[T] + (2x - 1)(h[T] - g[T])\} \tag{5}$$

$$\bar{F}_{Ej} = (1 - x)^2\{g[T] + 2x(h[T] - g[T])\} \tag{6}$$

When $g[T] \equiv h[T]$ the excess free energy is symmetric; when $g[T] = h[T] = $ constant the solution is regular; when $g[T] = h[T] = 0$ the solution is ideal.

Two descriptions will be employed for $g[T]$ (and $h[T]$). The simplest consists of the approximation

$$g[T] = g_0 + g_1 T \tag{7}$$

where g_0 and g_1 are constants. Equation (7) will be employed over a limited temperature range well removed from the third-law range. On the other hand, description of $g[T]$ over a wider temperature range inclusive of $0°K$ will require the form of

$$g[T] = g_2 + g_3 T^2 + g_4 T^3 \tag{8}$$

where g_2, g_3, and g_4 are constants.

A variety of methods have been employed to formulate the free energy of compound phases,[3,17–19,26,36–46] including characterization of the deviations from stoichiometry.[26,38–46] In the present case the simplest description of a compound of fixed stoichiometry $i_{(1-x_*)}j_{x_*}$ will be considered[3] which has the structure designated ψ.

In this case the free energy of the compound F^ψ is given by

$$F^\psi = (1 - x_*)F_i^\theta + x_* F_j^\theta + x_*(1 - x_*)$$
$$\times \{[x_*(1 - x_*)]^{-1} F_E^L[x_*] - C^\psi\} \quad (9)$$

where $F_E^L[x_*]$ is the excess free energy of the liquid phase at composition x_*, C^ψ is the compound parameter of $i_{(1-x_*)}j_{x_*}$ (which can be temperature dependent), and F_i^θ and F_j^θ are the free energies of pure i and j in the θ form. The latter is designated as the base phase for the compound. Definition of the compound parameter in Eq. (9) is an alternative means for describing the free energy of formation of the compound, which at the same time fixes its melting point. Specification of θ, C^ψ, and the excess free energy of the liquid fixes the free energy of formation of the compound since

$$\Delta F_f = F^\psi - (1 - x_*)F_i^0 - x_* F_j^0 \quad (10)$$

where F_i^0 and F_j^0 are the free energies of the stable forms of i and j at the temperature (and pressure) of interest.

We will now consider a number of ternary systems to illustrate how an exact description of Eqs. (3)–(10) can be used to couple thermochemical and phase diagram data. These examples will be employed to illustrate how explicit definition of these equations can be employed to predict thermochemical data and/or phase diagram data which have not been measured.

3. Calculation of the Iron–Chromium–Nickel System[19]

The iron–nickel–chromium system is of great technical importance and offers an excellent opportunity for illustrating the coupling between phase diagram and thermochemical data. The analyses presented illustrate how a description of the phase stability of liquid, bcc, fcc, and sigma phases in the component binary systems are combined to calculate the stability of these phases in the ternary system.[19]

The excess free energy of the liquid, fcc, and bcc phases in the iron–nickel system are given in Table 3. In line with the definition of Eq. (4), the excess free energy of the liquid is

$$F_E^L = x(1 - x)[(1 - x)(-2000 + 0.65T)$$
$$+ x(-7700 + 2.20T)] \quad \text{cal/g-atom} \quad (11)$$

where x is the atom fraction of nickel. Thus $g[T]$ is given first and

TABLE 3
Compilation of Equations for Excess Free Energy of Mixing of Solution Phases*

Fe–Ni	
liquid	$-2000 + 0.65T$ for $1600 \leq T \leq 1900$
	$-7700 + 2.20T$ for $1600 \leq T \leq 1900$
fcc	$500 - 0.91573 \times 10^{-3}T^2 + 0.39029 \times 10^{-6}T^3$ for $0 \leq T \leq 1800$
	$-8320 + 0.58327 \times 10^{-2}T^2 - 0.24859 \times 10^{-5}T^3$ for $0 \leq T \leq 1800$
bcc	$320 + 0.31727 \times 10^{-3}T^2 - 0.37930 \times 10^{-6}T^3$ for $0 \leq T \leq 1800$
	$-3890 + 0.84637 \times 10^{-2}T^2 - 0.31528 \times 10^{-5}T^3$ for $0 \leq T \leq 1800$
Cr–Ni	
liquid	-2000 for $1500 \leq T \leq 2200$; -2000 for $1500 \leq T \leq 2200$
fcc	$-2000 + 0.11202 \times 10^{-2}T^2 - 0.18649 \times 10^{-5}T^3$ for $0 \leq T \leq 1800$
	$-6000 + 0.22651 \times 10^{-2}T^2 - 0.6231 \times 10^{-6}T^3$ for $0 \leq T \leq 1800$
bcc	$12{,}800 - 6.50T$ for $1000 \leq T \leq 2200$
	$-3200 - 0.50T$ for $1000 \leq T \leq 2200$
Fe–Cr	
liquid	$4970 - 2.5T$ for $1700 \geq T \leq 2200$; $4970 - 2.5T$ for $1700 \leq T \leq 2200$
fcc	$1700 - 1.5T$ for $800 \leq T \leq 1700$; $1700 - 1.5T$ for $800 \leq T \leq 1700$
bcc	$6000 - 2.5T$ for $300 \leq T \leq 2200$; $6000 - 2.5T$ for $300 \leq T \leq 2200$

* Units are cal/g-atom and °K.

$h[T]$ second in the tabular listing. These functions are selected to yield the best description of the thermochemical and the phase diagram data. Combination of these excess free energies with the lattice stability results shown in Table 2 yields the thermochemical properties which are compared with observations in Tables 4 and 5 and the phase diagram shown in Figure 7. The latter agrees with the experimental phase diagram to within 2 at.% and 20°K.[54–56] The main discrepancy lies in the calculated limit of solubility of nickel in the bcc phase, which is observed near 6 at.%[57,58] nickel, in contrast to the present calculated value near 4 at.%. Figure 7 shows the computed temperature–composition curve defining equal free energy of the bcc and fcc phases which is related to diffusionless martensitic transformation [labeled $T_0(\text{bcc/fcc})$].[59]

Figure 7 also contains the ordered FeNi$_3$ phase, which is represented as a "line compound" existing at one composition. This simplified description[3] is based on an approximate representation of the FeNi$_3$ phase as a fully ordered phase whose free energy is described by

$$F[\text{FeNi}_3] = 0.25 F_{\text{Fe}}(\text{fcc}) + 0.75 F_{\text{Ni}}(\text{fcc}) + 0.1875(L - C) \quad (12)$$

where the base phase is fcc and C is the compound parameter defined

Chapter 4

TABLE 4
Comparison of Calculated and Observed Thermochemical Properties of Fe–Ni Alloys*

Liquid (1873°K) Atom fraction Ni	H_E(obs) ± 200	H_E(calc)	F_E(obs) ± 90	F_E(calc)
0.1	−200	−226	−159	−91
0.2	−400	−502	−298	−215
0.3	−650	−778	−431	−340
0.4	−950	−926	−556	−456
0.5	−1400	−1212	−661	−545
0.6	−1420	1301	−721	−591
0.7	−1150	−1257	−700	−575
0.8	−820	−1049	−574	−483
0.9	−420	−641	−338	−296

fcc (1323°K) Atom fraction Ni	H_E(obs) ± 100	H_E(calc)	S_E(obs) ± 0.10	S_E(calc)
0.1	−50	−35	—	+0.01
0.2	−150	−187	−0.04	−0.03
0.3	−380	−286	−0.10	−0.10
0.4	−550	−631	−0.19	−0.17
0.5	−800	−840	−0.33	−0.25
0.6	−1000	−982	−0.40	−0.31
0.7	−1050	−1012	−0.45	−0.33
0.8	−950	−889	−0.39	−0.29
0.9	−650	−565	−0.25	−0.19

* H_E and F_E in cal/g-atom, S_E in cal/g-atom °K. Observed values for liquid from Refs. 2 and 47; for fcc from Refs. 48–50.

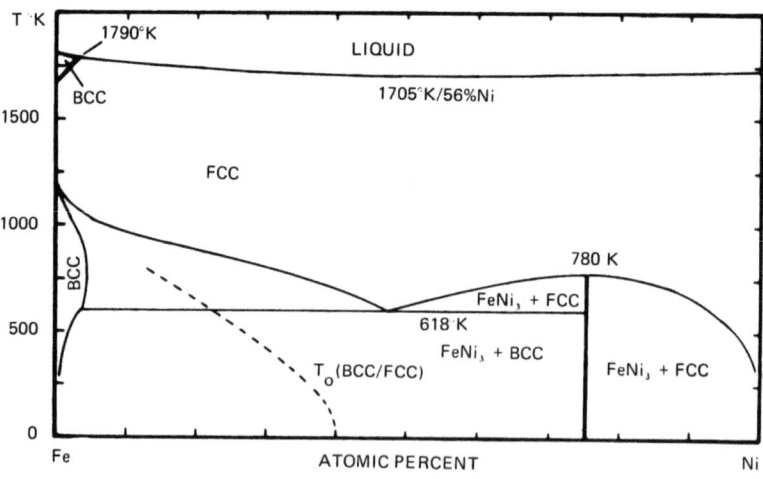

Fig. 7. Calculated iron–nickel phase diagram.[19]

TABLE 5
Comparison of Calculated and Observed Thermochemical Properties of Fe–Ni Alloys*

fcc (1200°K)			Activity of Fe (1273°K)	
Atom fraction Ni	H_E(obs) ± 100[2]	H_E(calc)	Obs.[51]	Calc.
0.1	−182	−31	0.900	0.910
0.2	−350	−200	0.800	0.826
0.3	−524	−442	0.700	0.732
0.4	−719	−712	0.600	0.622
0.5	−923	−957	0.495	0.491
0.6	−1056	−1125	0.380	0.348
0.7	−1061	−1147	0.224	0.215
0.8	−878	−1025	0.141	0.106
0.9	−503	−654	0.067	0.026

		H(fcc) − H(bcc)	
Atom fraction Ni	T°K	Obs.[52,53]	Calc.
0.15	700	910 ± 100	1018
0.20	600	810 ± 100	937
0.25	500	670 ± 100	790

* H in cal/g-atom.

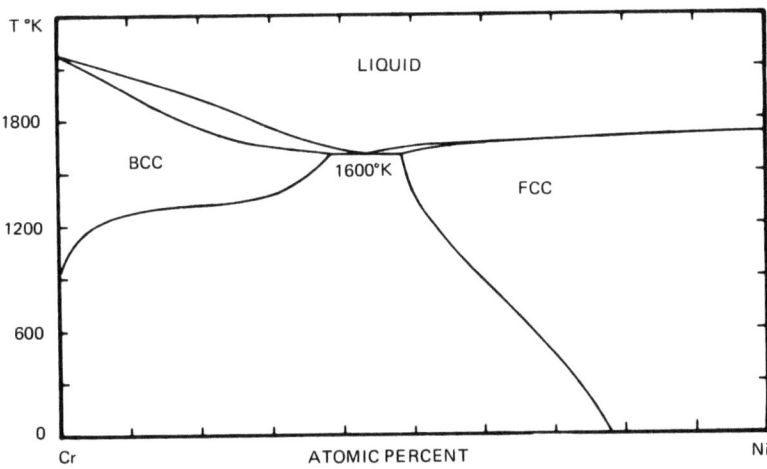

Fig. 8. Calculated chromium–nickel phase diagram.[19]

TABLE 6
Comparison of Calculated and Observed Thermochemical Properties of Cr–Ni and Fe–Cr Alloys*

Cr–Ni bcc (1550°K)† Atom fraction Ni	ΔH(obs) ± 75[2]	ΔH(calc)
0.10	1225	1141
0.20	1940	1802
0.30	2140	2079
0.35	2130	2104

Cr–Ni fcc (1550°K)‡ Atom fraction Ni	ΔH(obs) ± 75[2]	ΔH(calc)	ΔS(obs) ± 0.15[2]	ΔS(calc)
0.50	1540	1550	0.85	0.86
0.60	860	904	0.56	0.53
0.70	350	330	0.35	0.21
0.80	−40	−76	0.15	−0.04
0.90	−210	−218	0.00	−0.03

Cr–Ni liquid (1873°K) Atom fraction Ni	Activity coefficient of Cr	
	Obs.[60]	Calc.
0.460	0.903	0.892
0.568	0.927	0.840
0.724	0.957	0.754
0.725	0.887	0.753
0.768	0.797	0.728
0.891	0.862	0.653

Fe–Cr (1300°K)	Activity coefficient of Cr	
	Obs.[64]	Calc.
fcc 0–13% Cr	1.6–2.3	2.67–2.71
bcc 13% Cr	1.6	2.24

* ΔH in cal/g-atom, ΔS in cal/g-atom °K.
† $\Delta H = H[\text{bcc}] - (1 - x)H_{\text{Cr}}[\text{bcc}] - xH_{\text{Ni}}[\text{fcc}]$.
‡ $\Delta H = H[\text{fcc}] - (1 - x)H_{\text{Cr}}[\text{bcc}] - xH_{\text{Ni}}[\text{fcc}]$, $\Delta S = S[\text{fcc}] - (1 - x)S_{\text{Cr}}[\text{bcc}] - xS_{\text{Ni}}[\text{fcc}] - S(\text{Ideal})$.

in Eq. (12). In the present situation the value $C = 13{,}600 - 12.0T$ was used* to characterize FeNi$_3$ leading to a heat of formation of −3300 cal/g-atom and an entropy of formation of −2.17 cal/g-atom at 300°K.[19]

---

* The value $C = 13{,}600 - 12.0T$ corrects a typographical error in Table 6 of Ref. 19.

Table 3 shows the description of the excess free energy of the liquid, bcc, and fcc solutions in Cr–Ni. These quantities yield the phase diagram in Figure 8, which agrees with the observed diagram[54–56] to within 2 at.% and 20°K, and the thermochemical properties shown in Table 6. The latter compare favorably with observations.[2,60] Hall and Algie [61] have reviewed the stability of the sigma phase in ternary systems containing nickel and chromium. Although the sigma phase is not stable in the binary system, small additions of silicon and phosphorus are sufficient to stabilize this phase. Reference to Figure 2 of Hall and Algie's review[61] suggests that the most suitable composition of the Cr–Ni sigma phase is $Cr_{0.62}Ni_{0.38}$. Fixing the stoichiometry permits evaluation of the largest (positive) value of the compound parameter C which is consistent with the thermochemical description of Cr–Ni in which the sigma phase is *not stable*. On this basis it is found that $C = 2.5T$ for the Cr–Ni sigma phase based on the bcc structure at a composition $Cr_{0.62}Ni_{0.38}$. Thus the free energy of formation of the Cr–Ni sigma phase is equal to

$$-470 - 0.59T + 0.38(F_{Ni}[bcc] - F_{Ni}[fcc]) \quad \text{cal/g-atom}$$

Applying the results in Table 2 indicates that the free energy of formation of this phase is equal to $34 - 0.684T$ cal/g-atom above 900°K.

Thus the coupled description of phase diagram and thermochemical data permits characterization of the unstable Cr–Ni sigma phase. This description can be utilized in predicting the stability of the sigma phase in alloy systems containing chromium and nickel.

The description of the iron–chromium system[19] presented in Table 3 is based largely on previous analyses.[62,63] The simple "line compound" description of the sigma phase[3] represents the free energy of this phase as being equal to

$$0.53F_{Fe}(bcc) + 0.47F_{Cr}(bcc) + (0.53)(0.47)(L - C)$$

where Table 3 indicates $L = 4970 - 2.5T$ with $C = 300 + 4.3T$. The calculated heat and entropy of formation of +1140 cal/g-atom and 1.73 cal/g-atom °K for this compound agree with the observed values.[62] Moreover, the calculated thermochemical properties of the bcc and liquid phases are in good agreement with observed values. These equations generate the calculated phase diagram shown in Figure 9, which is in good agreement with observed and computed diagrams.[16,54–56,62] While the comparison of calculated and

Chapter 4

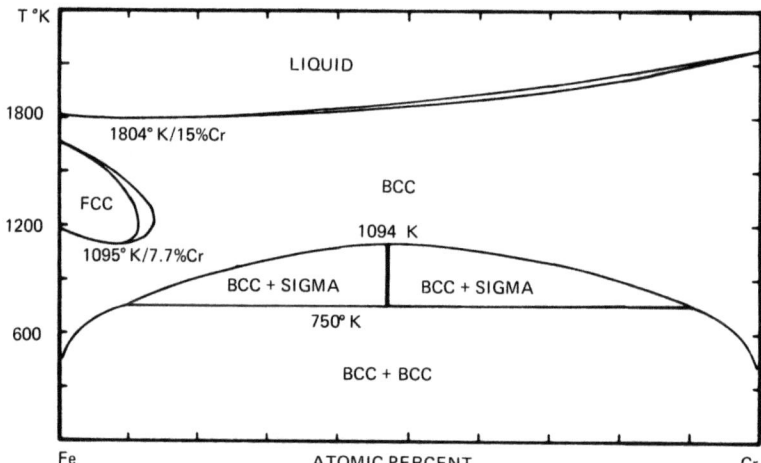

Fig. 9. Calculated iron–chromium phase diagram.[19]

observed activity coefficients for chromium in the fcc solid solution shown in Table 6 is not in as good agreement as might be desired, it should be noted that a similar discrepancy exists when the observed activity of chromium in the bcc phase at the edge of the bcc solution is compared with the calculated values. In this case the "calculated" values reflect a wide range of measured thermochemical data.[62]

Extension of the present description of asymmetric binary phases into the ternary systems has been carried out on the basis described earlier.[3,18,19] Thus the ternary analog of Eq. (4) formulates the excess free energy of a ternary alloy with x atom fraction of chromium, y atom fraction of nickel, and $1 - x - y$ atom fraction of iron as

$$F_E = x(1 - x - y)(1 - y)^{-1}\{(1 - x - y)g[T] + xh[T]\} \\ + y(1 - x - y)(1 - x)^{-1}\{(1 - x - y)l[T] + ym[T]\} \\ + xy(x + y)^{-1}(xn[T] + yp[T]) \tag{13}$$

where $g[T]$ and $h[T]$ refer to the Fe–Cr system, $l[T]$ and $m[T]$ refer to the Fe–Ni system, and $n[T]$ and $p[T]$ refer to the Cr–Ni system.

It should be noted that the formulation of the ternary excess free energy given above is identical* to the Kohler equation[7] which represents the ternary excess free energy in terms of the binary com-

---

* This equivalence has been pointed out by Dr. I. Ansara of CNRS, ENSEEG, Domain University, St. Martins d'Heres, France.

ponents as

$$F_E = (x_i + x_j)^2 (F_E[ij])_{x_i/x_j} + (x_i + x_k)^2 (F_E[ik])_{x_i/x_k}$$
$$+ (x_j + x_k)^2 (F_E[jk])_{x_j/x_k} \qquad (14)$$

where x_i, x_j, and x_k are the atom fractions of i, j, and k in the ternary system. Thus $x_i = 1 - x - y$, $x_j = x$, and $x_k = y$. On the basis of Eq. (4), the excess free energy of the ij binary system is given as

$$F_E[ij] = x_i^* x_j^* (x_i^* g[T] + x_j^* h[T]) \qquad (15)$$

where x_i^* and x_j^* are the concentrations of i and j in the ij binary system. If the ratio x_i/x_j is fixed and equal to x_i^*/x_j^*, then, since $x_i^* + x_j^* = 1$,

$$x_j^* = x_j/(x_i + x_j) \quad \text{and} \quad x_i^* = x_i/(x_i + x_j) \qquad (16)$$

As a consequence,

$$(F_E[ij])_{x_i/x_j} = (x_i x_j)(x_i + x_j)^{-2} \{x_i(x_i + x_j)^{-1}$$
$$\times g[T] + x_j(x_i + x_j)^{-1} h[T]\} \qquad (17)$$

Applying these definitions to the derivation of $(F_E[ik])_{x_i/x_k}$ and $(F_E[jk])_{x_j/x_k}$ followed by substitution into Eq. (14) yields Eq. (13).

The free energy of the ternary compound phases in the systems of interest is formulated along the lines of Eqs. (291) and (292) of Ref. 3, allowing for the variation in compound stoichiometry.[18,19] Thus consideration of a compound ψ with a base θ which runs between $i_{(1-x_*^1)} j_{x_*^1}$ and $j_{x_*} k_{(1-x_*)}$ and having a general composition $i_{z_\psi} j_{x_\psi} k_{(1-x_\psi-z_\psi)}$ is defined as follows:

$$p = (x_* - x_*^1)/(1 - x_*) \qquad (18)$$

$$\left. \begin{array}{l} x_\psi = x_*^1 + y_\psi p \\ y_\psi = 1 - x_\psi - z_\psi \\ z_\psi = 1 - x_*^1 - y_\psi(1 + p) \end{array} \right\} \qquad (19)$$

Thus the free energy of the compound ψ is

$$F^\psi = z_\psi F_i^\theta + x_\psi F_j^\theta + y_\psi F_k^\theta + [1 - (y_\psi/x_\psi)] \Delta F_A$$
$$+ [y_\psi/(1 - x_\psi)] \Delta F_B + RT[y_\psi \ln y_\psi + z_\psi \ln z_\psi$$
$$- (1 - x_\psi) \ln(1 - x_\psi)] \qquad (20)$$

---

* Equation (18) presents the correct definition for z_ψ. The earlier definition[18,19] contained a typographical error.

Chapter 4

where

$$\Delta F_A = x_*^1(1 - x_*^1)[(1 - x_*^1)L_{ij} + x_*^1 L_{ji} - C_{ij}^\psi] \quad \text{cal/g-atom} \quad (21)$$

$$\Delta F_B = x_*(1 - x_*)[x_* L_{jk} + (1 - x_*)L_{kj} - C_{jk}^\psi \quad \text{cal/g-atom} \quad (22)$$

where the L and C values in Eqs. (21) and (22) are defined for the component binary systems. These equations permit calculation of all compound/solution phase interactions. The solution phase interactions are carried out along the lines indicated earlier[3] on a pairwise basis. After each pairwise calculation has been completed the results are combined to yield the most stable configuration.

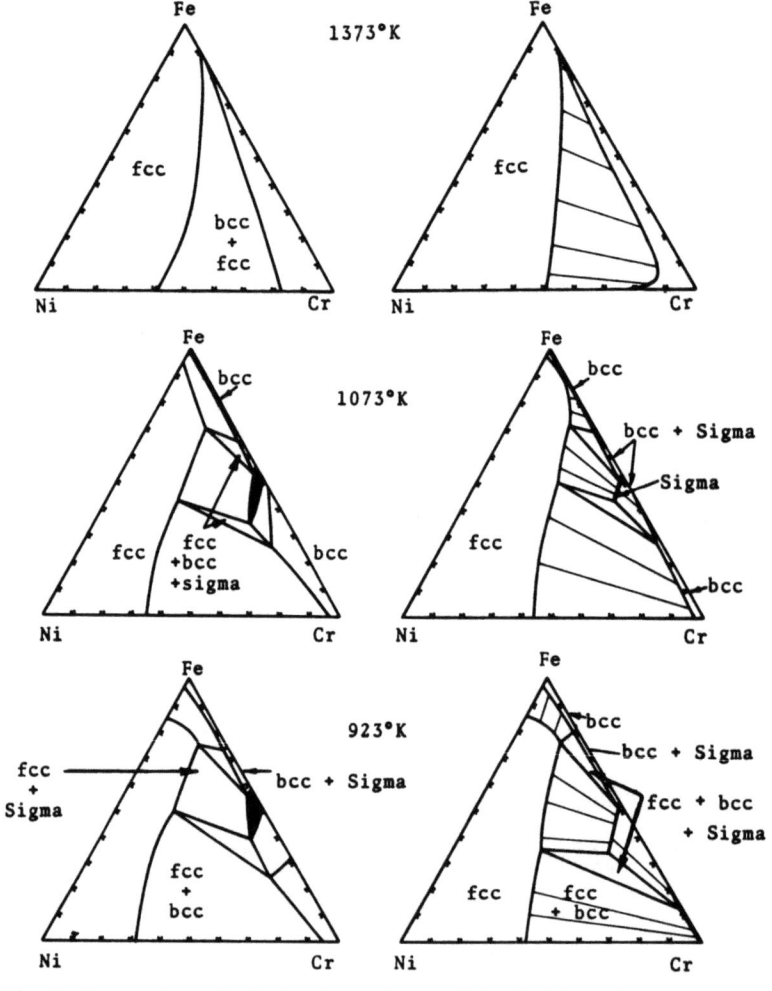

Fig. 10. Comparison of calculated and observed Fe–Cr–Ni sections (calculated sections on right).[19]

Reference to Figure 10 shows that the fcc and bcc phases are most stable at 1373°K in Fe–Cr–Ni. The computed section (which displays some of the calculated tie lines) is in good agreement with the observations of Pugh and Nisbet.[65] At lower temperatures the sigma phase appears in the 1073°K and 923°K sections where the calculations are compared with the diagrams suggested by Lindemer.[66] This comparison provides a direct check on the estimate of $C = 2.5T$ for the Cr–Ni sigma phase. The result is quite good since at 923°K the stability of the sigma phase in the calculated diagram is larger than that in the observed diagram, while at 1073°K the reverse is true. It should be recalled that the above-noted value of $C = 2.5T$ is the largest possible value consistent with the absence of a stable sigma phase in Cr–Ni.

An additional noteworthy result of the analysis of the iron, nickel, cobalt, and chromium binary systems[19] is the calculation of the pressure of the bcc/hcp transformation in iron–cobalt alloys. This transition is observed to occur in pure iron at room temperature near 130 kbar. If the effects of cobalt additions on the volume difference

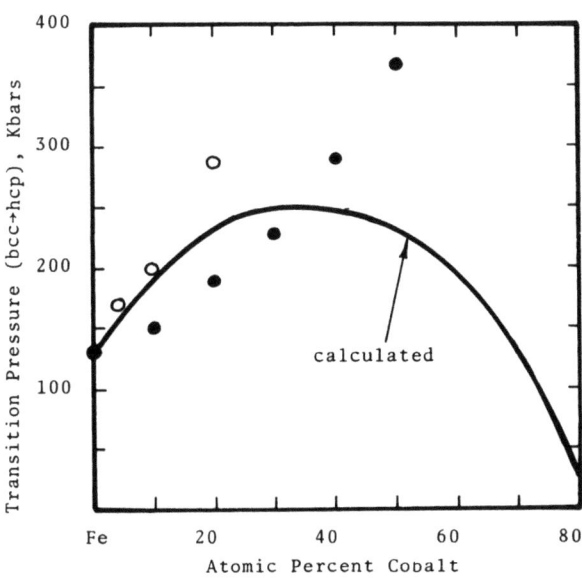

Fig. 11. Comparison of calculated and observed transition pressures in iron–cobalt alloys. Static transition pressure measurements are denoted by open circles.[67] Dynamic transition pressure measurements are denoted by filled circles.[68]

between the bcc and hcp phases is ignored as a first approximation,[3] then the transition pressure can be calculated as 130 kbar times the free energy difference between the bcc and hcp forms in iron–cobalt alloys at 300°K, divided by the corresponding free energy difference for pure iron (1010 cal/g-atom[3]). The results are shown in Figure 11 along with experimental observations.[67,68] It should be noted that the curve must bend over as indicated by the calculations since pure cobalt is stable in the hcp form. The difference between the dynamic transition pressures reported[68] near the equiatomic composition and the calculated results may be due to effects of ordering reactions in the experimental observations which are not taken into account in the calculations.

Comparison of similar calculations[69] with observed bcc/hcp transformations under pressure in iron–manganese alloys yields excellent agreement.[70]

In line with the analysis of the iron–chromium–nickel system presented above, it is of interest to note a recent analysis of the Mo–Ni binary system which can be employed in analyses of Fe–Mo–Ni or Ni–Mo–Cr. The results are displayed in Table 7 and Figure 12. The latter is in excellent agreement with observed phase diagram data for this system.[54–56] Figure 12 is calculated from the

TABLE 7
Summary of Excess Free Energy and Compound Parameters for the Mo–Ni System*

Liquid	$-4200 + 3.3T$ for $1500 \leq T \leq 2900°K$; $-4200 + 3.3T$ for $1500 \leq T \leq 2900°K$
bcc	$+5600 + 3.3T$ for $500 \leq T \leq 2900°K$; $+5600 + 3.3T$ for $500 \leq T \leq 2900°K$
fcc	$-3250 + 3.3T$ for $300 \leq T \leq 1800°K$; $-3250 + 3.3T$ for $300 \leq T \leq 1800°K$

Compound	Structure	Base	Compound parameter
$Mo_{0.53}Ni_{0.47}$	Complex orthorhombic (μ)	bcc	$-1700 + 4.27T$
$Mo_{0.25}Ni_{0.75}$	Ordered orthorhombic (ϕ)	fcc	$-615 + 5.68T$
$Mo_{0.20}Ni_{0.80}$	Ordered $D1_a$ (ξ)	fcc	$-805 + 6.09T$

Compound	ΔH_f(calc)	ΔS_f(calc)
$Mo_{0.53}Ni_{0.47}$	-220	$+0.05$
$Mo_{0.53}Ni_{0.75}$	-30	$+0.41$
$Mo_{0.20}Ni_{0.80}$	-30	$+0.42$

* Excess free energy and ΔH_f in cal/g-atom, ΔS_f in cal/g-atom °K, T in °K.

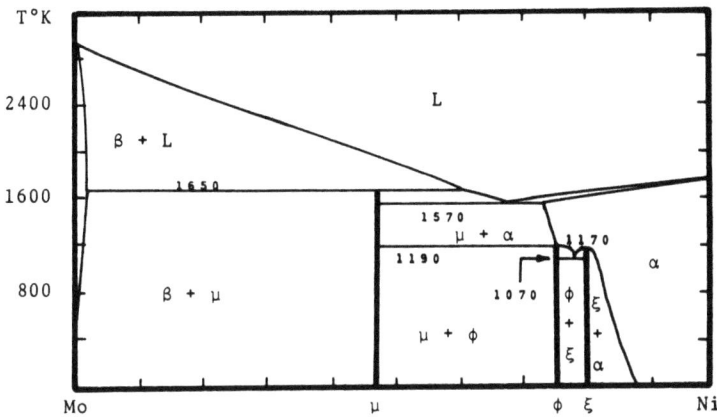

Fig. 12. Calculated molybdenum–nickel phase diagram.

excess free energy and compound parameters shown in Table 6 and the lattice stability of molybdenum presented earlier.[3] Although there are no measured thermochemical data available at present to compare with the calculations of Table 7, ternary phase diagram data for the Ni–Mo–C[71] and Ni–Mo–Al[72] systems observed near 1300°K can be employed to estimate limits for the free energy of formation of the Ni–Mo compound phases. This procedure is essentially the reverse of that applied in the introduction of this chapter to the Mg–Ga–Al system. Thus, if the $Mo_{0.53}Ni_{0.47}$ phase is approximated by $Mo_{0.5}Ni_{0.5}$, it is found[71] that

$$2MoNi + C \rightarrow Mo_2C + 2Ni \qquad (23)$$

The observed free energy of formation of Mo_2C of[2] -4300 cal/g-atom at 1300°K in combination with Eq. (23) indicates that the free energy of formation of $Mo_{0.50}Ni_{0.50}$ is more positive than -3200 cal/g-atom.

Similarly in the Ni–Mo–Al system[72]

$$3NiMo + Ni_3Al \rightarrow Ni_6Al + 3Mo \qquad (24)$$

since the observed free energy of formation of Ni_3Al is -8000 cal/g-atom and that of Ni_6Al is -5100 cal/g-atom at 1300°K,[2] Eq. (2) indicates that the free energy of formation of $Ni_{0.5}Mo_{0.5}$ is more positive than -600 cal/g-atom. This result is in keeping with the predictions of Table 7. In addition, the observed Ni–Mo–Al phase diagram suggests that

$$4Ni_3Al + 3Ni_4Mo \rightarrow 4Ni_6Al + 3Mo \qquad (25)$$

Chapter 4

at 1100°K. At this temperature the free energy of formation of Ni_6Al is -5150 cal/g-atom while the free energy of formation of Ni_3Al is -8200 cal/g-atom. Thus Eq. (25) implies that at 1100°K the free energy of formation of $Ni_{0.80}Mo_{0.20}$ is more positive than -1100 cal/g-atom. This result is also in keeping with the calculations shown in Table 7.

4. Prediction of Partial Titanium–Aluminum–Gallium and Titanium–Aluminum–Tin Isothermal Sections

Prediction of the effects of alloying elements on the field of hcp stability in titanium alloys is a problem of considerable practical interest.[17] Additions of Al, Sn, Ga, and Zr result in extensive fields of hcp stability. Simultaneous additions of Al and Sn and of Al and Ga yield high-strength alloys with attractive high-temperature properties. Thus prediction of the Ti–Ga–Al and Ti–Sn–Al systems are of considerable interest and present a means for illustrating the efficacy of the technique since limited phase diagram and thermochemical data are available.[1,2,54–56]

Analyses of the Ti–Zr, Ti–Al, Ti–Ga, and Ti–Sn systems[17] are summarized in Figures 13 and 14. In these cases the excess free energies of the bcc, hcp, fcc, and liquid phases are symmetric so that $g[T] = h[T]$ in Eq. (4). The $g[T]$ values for Ti–Al, Ti–Ga, and Ti–Sn[17] are in agreement with recent vapor pressure measurements.[73] The Ti–Zr

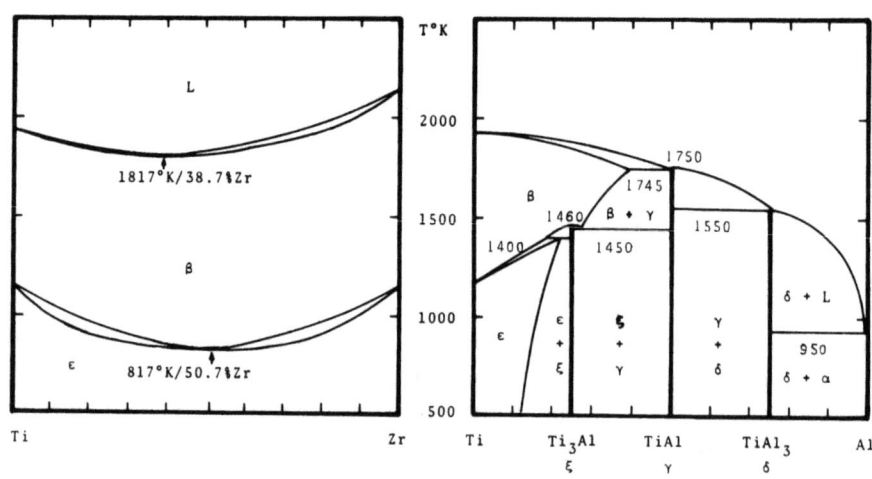

Fig. 13. Calculated Ti–Zr and Ti–Al phase diagrams.[17]

Relation of the Thermochemistry and Phase Diagrams of Condensed Systems

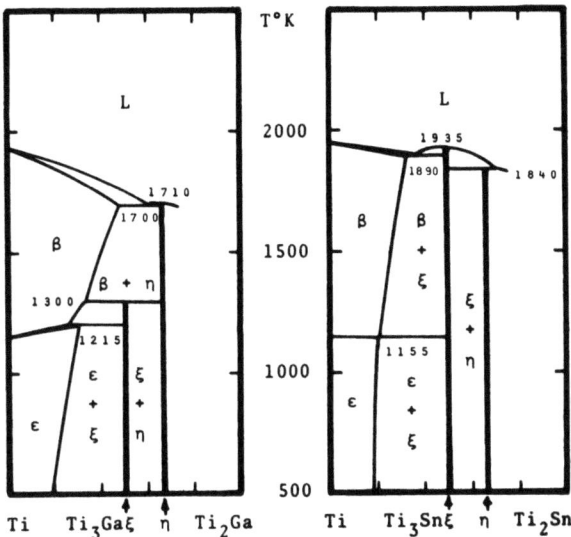

Fig. 14. Calculated partial Ti–Ga and Ti–Sn phase diagrams.[17]

system has been reanalyzed[17] to provide a slightly better description of the phase diagram than that provided by the earlier analysis.[3] The current solution parameters are -746, $+2090$, and $+894$ for the liquid, hcp, and bcc phases, respectively.[17] Lattice stability parameters for Ti, Zr, and Al have been presented earlier.[3] Corresponding values for Sn and Ga are given in Ref. 17. It is interesting to note that Predel and Stein's estimate of the enthalpy difference between the rhombohedral (ρ) and fcc forms of gallium is 2000 ± 1000 cal/g-atom.[74] This is of the same order of magnitude as the current estimate for the difference between the hcp and rhombohedral forms.[17]

The present Ti–Al analysis[17] is more complete than that presented earlier,[3] and describes the AuCu-type TiAl phase, γ, and the tetragonal TiAl$_3$ phase, δ, as well as the hexagonal DO_{19} Ti$_3$Al structure, designated by ξ, which has counterparts in the ξ Ti$_3$Ga and ξ Ti$_3$Sn structures. Considerable controversy exists concerning the constitution of the Ti–Al system in the vicinity of Ti$_3$Al.[54–56,75,78] The limitations of the present "line compound" description preclude critical analysis of the controversy which centers on ordering versus miscibility gap decompositions. The present description of Ti$_3$Al leads to a peritectoid bcc/hcp/Ti$_3$Al at 1400°K in keeping with the findings of Clark et al.[75] However, a small alteration of the Ti₃Al compound parameter to $7800 + 3.77T$ would reduce the stability of this phase sufficiently to eliminate the peritectoid, leading to the phase diagram configuration suggested by Blackburn. Such an alteration would decrease the entropy of formation of ξ Ti₃Al by 0.043 cal/g-atom °K. Thus the two alternatives are separated by a 4% entropy difference corresponding to 60 cal/g-atom or 0.0026 eV/atom at 1400°K!

Partial Ti–Ga and Ti–Sn diagrams[17] are shown in Figure 14, and include the Ti₂Ga and Ti₂Sn η phases, which exhibit the hexagonal Ni₂In structure. Recent phase equilibria and vapor pressure measurements[73] suggest that $B = -13,000$ for Ti–Ga and $B = -21,000$ for Ti–Sn, and that the composition of the $\beta/(\beta + L)$ boundaries in Ti–Ga at 1770 and 1700°K occur at 0.16 and 0.26 atom fractions of gallium, respectively. The present solution model is too simple to reflect the minimum in the bcc/hcp phase boundaries reported for Ti–Sn.[54–56]

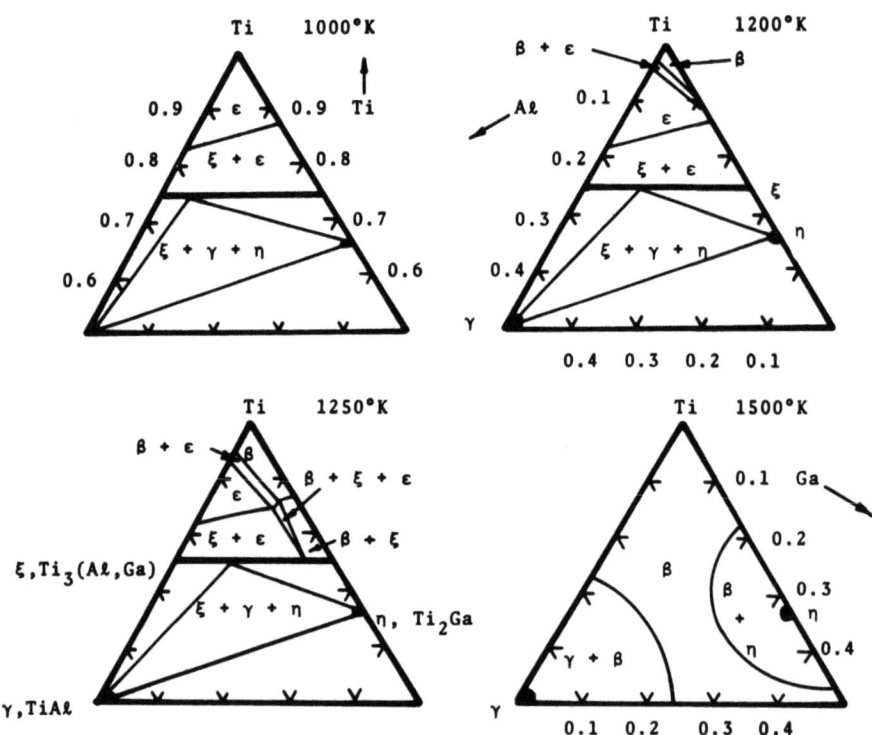

Fig. 15. Calculated partial isothermal sections in the Ti–Ga–Al system.[17]

Relation of the Thermochemistry and Phase Diagrams of Condensed Systems

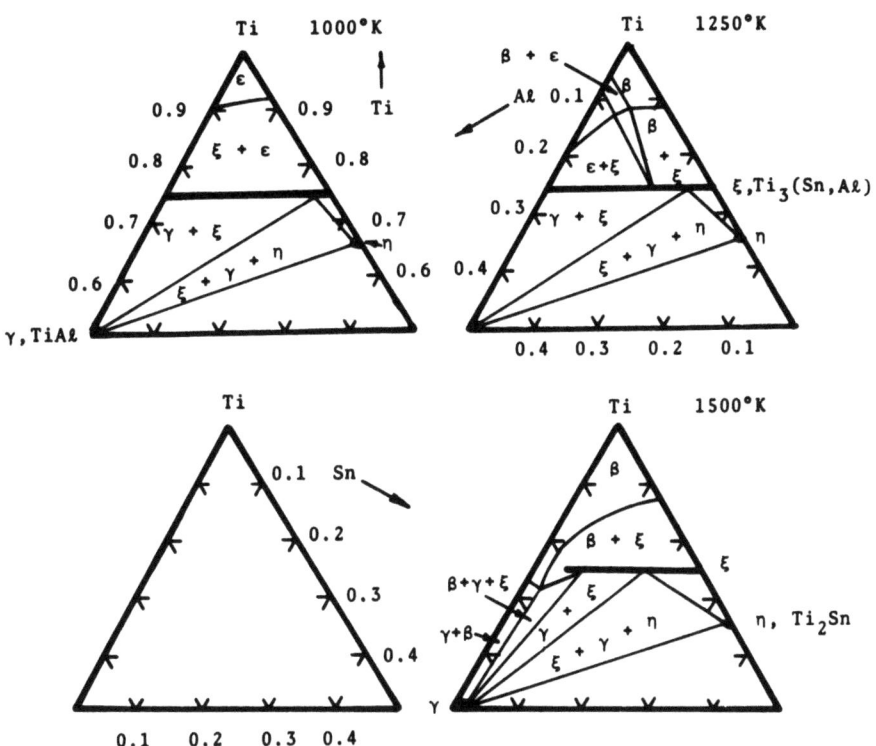

Fig. 16. Calculated partial isothermal sections in the Ti–Sn–Al system.[17]

Isothermal sections of the Ti–Ga–Al and Ti–Sn–Al systems have been computed as shown in Figures 15 and 16 based[17] on the foregoing description of the component binary systems employing the techniques described earlier.[3] In these calculations the TiGa (γ) and TiSn (γ) counterphases of TiAl (γ) were based on the α structure with $C = 0$, while the Ti_2Al (η) counterphase of Ti_2Sn (η) was based on the ε structure with $C = 0$. The Al–Sn and Al–Ga interaction parameters were estimated from thermochemical data for these binary systems[1,2] as being equal to $+28000$ and $+400$ cal/g-atom, respectively, for all the solution phases in these systems. The latter value for Al–Ga relates to characterizing the Al–Ga system by slightly positive deviations from ideality in the introductory section of this chapter.

5. Calculation of the Chromium–Columbium–Nickel System

Utilization of modern methods for controlled directional solidification has led to development of *in situ* composites. These

Chapter 4

structures, consisting of aligned two-phase materials in which the reinforcing phase is produced simultaneously, are usually formed by freezing eutectic or near-eutectic compositions. While rational application of this method depends on extensive phase diagram data (i.e., location of eutectic compositions, solid–liquid tie lines, curves of twofold saturation, etc.), such information is presently not available for most three-component systems. It is virtually nonexistent for systems containing four or more elements. Application of the present methods for calculating phase diagrams to the Cr–Cb–Ni system (which is a system of current technical interest) provides an additional illustration of the wide utility of this approach.[18]

The Cr–Ni system has been described in Section 3. The Cb–Cr system, which was analyzed on the basis of a regular solution model,[6] was reconsidered[18] in order to provide close agreement with recent phase diagram data[79] as shown in Figure 17. The latter agrees with observed phase diagrams[54–56,79] to within 2 at.% and 30°K. The lattice stability parameters for Cb and Cr are those presented earlier,[3] while the values for bcc nickel are given in Table 2.

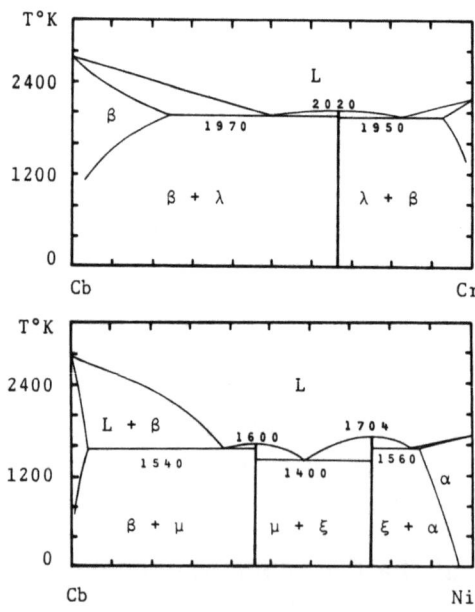

Fig. 17. Calculated binary phase diagrams for the Cb–Cr and Cb–Ni systems.[18]

Thus the simple set of parameters shown in Tables 2 and 3 and Ref. 18 provides a complete description of the phase diagrams and thermochemical properties of the component binary systems. The only distortion occurs in the Cb–Ni system in the vicinity of the μ phase, where the present description treats the phase as a line compound, $Cb_{0.539}Ni_{0.461}$, while the compound phase is observed to exist over a range of composition between 40 and 50 at. % nickel.

Calculation of the Cr–Cb–Ni ternary is carried out following the methods developed earlier.[3,18] The excess free energy of the ternary solution phase containing x atom fraction of columbium and y atom fraction of nickel is defined on the basis of Eq. (13). Thus, for example, the excess free energy of the fcc (α) phase is described by the matrix[18]

$$g[T] = 7000, \quad h[T] = 5400$$
$$l[T] = -2000 + 0.11202 \times 10^{-2}T^2 - 0.18649 \times 10^{-5}T^3$$
$$m[T] = -6000 + 0.22651 \times 10^{-2}T^2 - 0.6231 \times 10^{-6}T^3$$
$$n[T] = -11,500, \quad p[T] = -35,500$$

Analogous descriptions follow for the bcc and liquid phases.

Similarly, the compound phases are defined as in Eqs. (18)–(22). Thus, for example, the λ phase in the present case runs between $Cr_{0.667}Cb_{0.333}$ and $Ni_{0.75}Cb_{0.25}$, with θ = hcp. Hence $x_*^1 = 0.333$ and $x_* = 0.25$. Further, Ref. 18 shows that

$$L_{ij} = L_{ji} = +1400, \quad L_{jk} = -13,000, \quad L_{kj} = -37,000$$
$$C_{ij}^\psi = 17,200 + 1.0T, \quad C_{jk}^\psi = 9500$$

In this manner the free energies of the solution phases (α, β, L) and compound phases (λ, ξ, μ) are defined explicitly and the fifteen pairwise interactions can be calculated to yield the tie lines.[3] The results are then examined to determine the most stable configurations along the lines illustrated earlier.[3] Figures 18–20 show the resulting isothermal sections computed at temperatures between 1373 and 1973°K. Figure 18 contains two partial experimental sections[81] for comparison, while Figure 20 shows the computed liquidus projections and a partial experimental liquidus projection. Before considering the individual features of Figures 18–20, it should be noted that the stoichiometry and compound parameters attributed to the unstable counterphases CbCr (ξ), and Ni_3Nb (λ) were inferred[18] from the experimental section at 1373°K. Moreover, the experimental diagrams show two polymorphic forms of the Laves phase,

207

Chapter 4

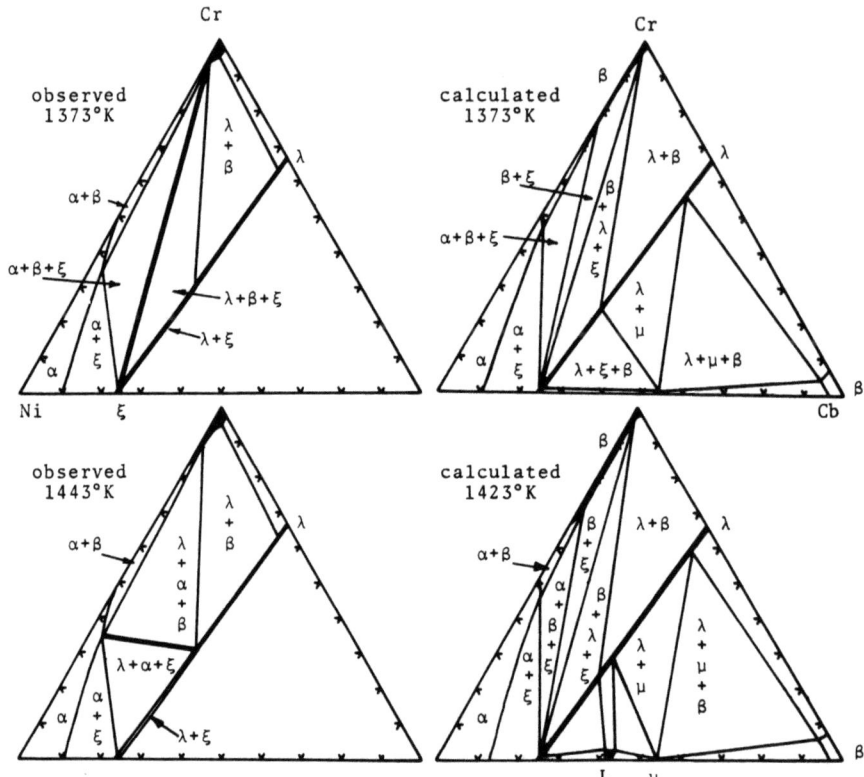

Fig. 18. Calculated and observed isothermal sections.[18]

one of which is stable above 1860°K and the other below 1860°K. Addition of nickel stabilizes the high-temperature form so that the latter is stable at 1373°K when the nickel content exceeds 5 at. %. This feature has been omitted from the experimental diagrams and the calculations for the sake of simplicity.[18] Some of the computed isothermal sections display tie lines in the two phase fields; these have been included where possible without undue complications.

Reference to Figure 18 shows relatively good agreement between the computed and experimental sections with the exception of one detail which is illustrated in the experimental section at 1443°K. The latter suggests that at 1443 the two-phase $\lambda + \alpha$ field is more stable than the two-phase $\beta + \xi$ field and that a 2/2 reaction occurs on cooling. This feature is completely absent in the calculations. In fact, if the free energy change for the 2/2 reaction depicted in the

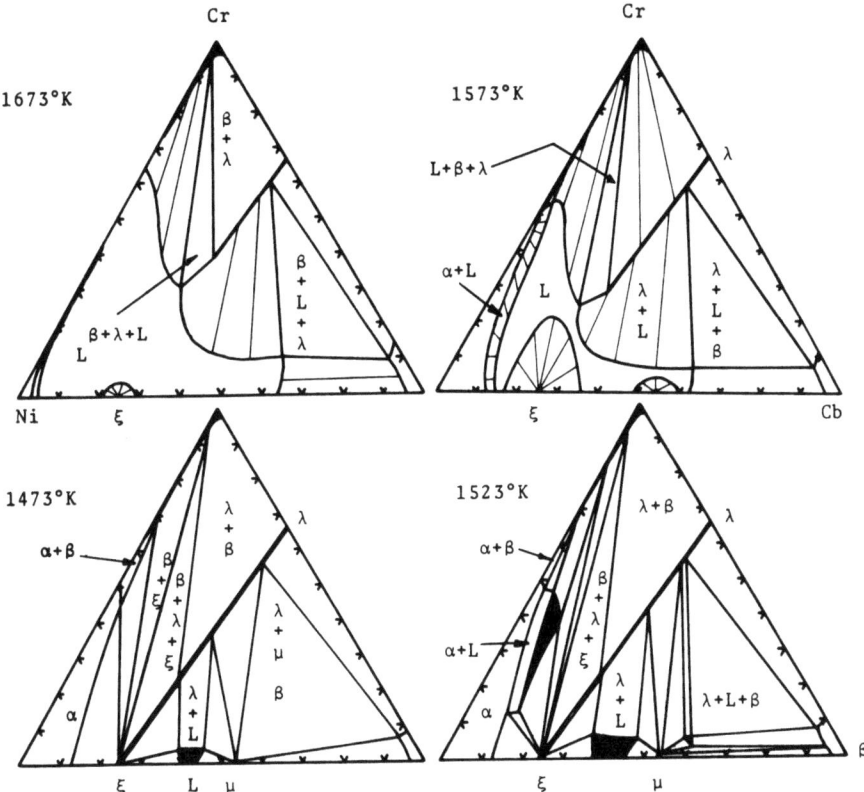

Fig. 19. Calculated isothermal sections.[18]

experimental diagram is considered as

0.367[90Cr, 10Ni](β) + 0.633[75Ni, 25Cb](ξ)

$\rightarrow$ 0.493[31Cr, 29Cb, 40Ni](λ) + 0.507[35Cr, 3Cb, 62Ni](α)

a positive free energy change equal to $1760 + 0.04T$ cal/g-atom is obtained. This result is mainly due to the large negative free energy of formation of Ni_3Cb (ξ) relative to Cr_2Cb (λ). Since the calculations agree with the thermochemical data concerning these compounds,[18] there is substantial reason to question the experimental section at 1443°K which displays the $\alpha + \lambda$ equilibrium as the stable configuration.

Figure 19 shows calculated isothermal sections at 50°K intervals between 1473 and 1673°K illustrating the appearance and growth of

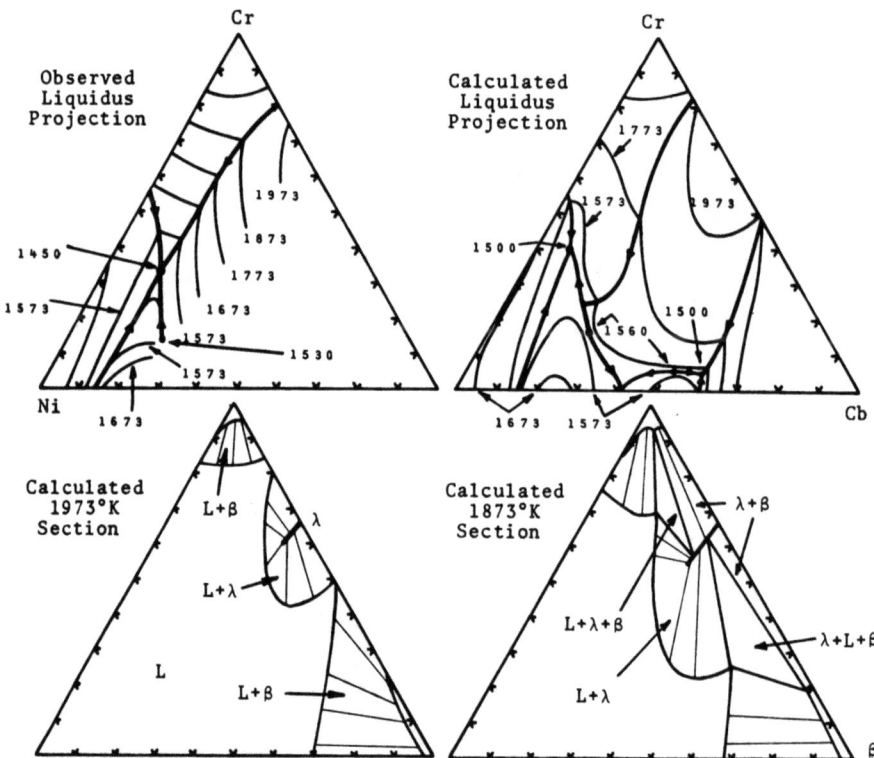

Fig. 20. Calculated and observed isothermal sections and liquidus projections.[18]

liquid-phase fields. Additional calculated sections at 1873 and 1973°K show further expansion of the liquid fields as well as the calculated and observed liquidus projections. These projections afford an additional comparison of calculation and observation. Thus the quasibinary eutectic between ξ and λ is reported[81] to occur at 1530°K at a composition corresponding to 13 at. % Cr, 25 at. % Cb, and 62 at. % Ni. The calculated conditions are 1560°K and 16 Cr, 25 Cb, and 59 Ni. In addition, the minimum freezing point near the Cr–Ni edge is reported at 1450°K near 32 Cr, 13 Cb, 54 Ni, as a result of the intersection of four curves of twofold saturation. The calculations put the minimum freezing point near 1500°K at 39 Cr, 9 Cb, 59 Ni, at the intersection of three curves of twofold saturation. This can be seen by examining the computed isothermal sections at 1573 and 1523°K in Figure 19. Thus this difference between the calculated and observed liquidus projections (three-curve versus four-curve intersection at the minimum freezing point) is related to the earlier 2/2 reaction discrepancy. The calculations suggest that the λ phase is not sufficiently stable to compete with the ξ, α, β, and liquid phases in the vicinity of the minimum freezing point. Apart from these differences in detail, the calculated and observed results are in relatively good agreement.

6. Prediction of Isothermal Sections in the Cb–Zr–Cr, Cb–Hf–Cr, Cb–Mo–Cr, and Cb–Ti–Zr Systems

Development of oxidation-resistant columbium-base alloys which can be used at high temperatures has been hampered by limited phase diagram and thermochemical data. Accordingly, analysis of the component binary systems of the Cb–Cr–Zr, Cb–Cr–Hf, Cb–Cr–Mo, and Cb–Ti–Zr phase diagrams were performed. The results, shown in Tables 8 and 9, were synthesized to yield the binary and ternary phase diagrams shown in Figures 21–28. The Cb–Cr binary system was discussed in Sectlon 5. The description of the Hf–Cb

TABLE 8
Summary of Excess Free Energy Parameters for the Mo–Cr System and Comparison of Calculated and Observed[2] Thermochemical Properties*

Liquid (L)	$3000 - 1.4T$ for $2000 \le T \le 3000°K$; $6100 - 2.7T$ for $2000 \le T \le 3000°K$
bcc (β)	$5100 - 1.4T$ for $300 \le T \le 2700°K$; $8200 - 2.7T$ for $300 \le T \le 2700°K$

Atom fraction chromium	H_E^β		S_E^β		F_E^β (1471°K)	
	Calc.	Obs. ± 100	Calc.	Obs. ± 0.10	Calc.	Obs. ± 100
0.10	487	495	0.138	0.103	284	343
0.20	915	963	0.266	0.236	524	615
0.30	1266	1345	0.376	0.356	713	821
0.40	1521	1600	0.461	0.451	843	958
0.50	1663	1725	0.513	0.463	908	1023
0.60	1670	1710	0.523	0.478	901	1021
0.70	1526	1555	0.485	0.475	813	922
0.80	1212	1244	0.390	0.342	638	741
0.90	710	745	0.231	0.203	370	446

Calculated miscibility gap summit = 1153°K, 0.617 Cr
Observed miscibility gap summit = 1150°K, 0.61 Cr

* H, F in cal/g-atom, S in cal/g-atom °K, T in °K.

Chapter 4

TABLE 9
Summary of Solution and Compound Parameters for Components of
Cb–Cr–Zn, Cn–Cr–Hf, Mo–Cr–Cb, and Zr–Ti–Cb Systems*

System	L	B	E	Compound	x_*	Base	Parameter
Cb–Zr	1500	4400	8000	—	—	—	—
	1500	4400	8000	—	—	—	—
Zr–Cr	−2000	6000	9000	Laves (λ)	0.667	ε	16,300
	−2000	6000	9000	—	—	—	—
Cb–Hf	3219	5137	5137	—	—	—	—
	3219	5137	5137	—	—	—	—
Hf–Cr	0	6000	9000	Laves (λ)	0.667	ε	17,200
	0	6000	9000	—	—	—	—
Mo–Cb	−7100	−5500	−5500	—	—	—	—
	−7100	−5500	−5500	—	—	—	—
Ti–Cb	3121	3125	3125	—	—	—	—
	3121	3125	3125	—	—	—	—
Ti–Zr	−746	+2090	+894	—	—	—	—
	−746	+2090	+894	—	—	—	—
Cb–Cr	1400	5400	5400	Laves (λ)	0.667	ε	$17,200 + 1.0T$
	1400	7000	7000	—	—	—	—
Mo–Cr	$(3000 - 1.4T)$	$(5100 - 1.4T)$	$(5100 - 1.4T)$	—	—	—	—
	$(6100 - 2.7T)$	$(8200 - 2.7T)$	$(8200 - 2.7T)$	—	—	—	—

* E in cal/g-atom, T in °K.

system given in Table 9 is exactly the same as that employed earlier[3,82] and yields the phase diagram shown in Figure 23, which is in excellent agreement with recent observations by Carpenter et al.[36]

The earlier description of $L = 4297$, $B = 6934$, and $E = 6934$ cal/g-atom[3] for the Zr–Cb system has been revised as shown in Table 9 to provide a more accurate calculation of the miscibility gap maximum and monotectoid temperatures.

Figures 24 and 25 show calculated isothermal sections and limited experimental results[83] for the Cb–Cr–Zr case. The $ZrCr_2$–$CbCr_2$ quasibinary system exhibits nearly linear behavior[84] in accordance with the present description. Figure 26 shows several calculated Cb–Cr–Hf sections which are qualitatively similar to the Cb–Cr–Zr cases in Figures 24 and 25.

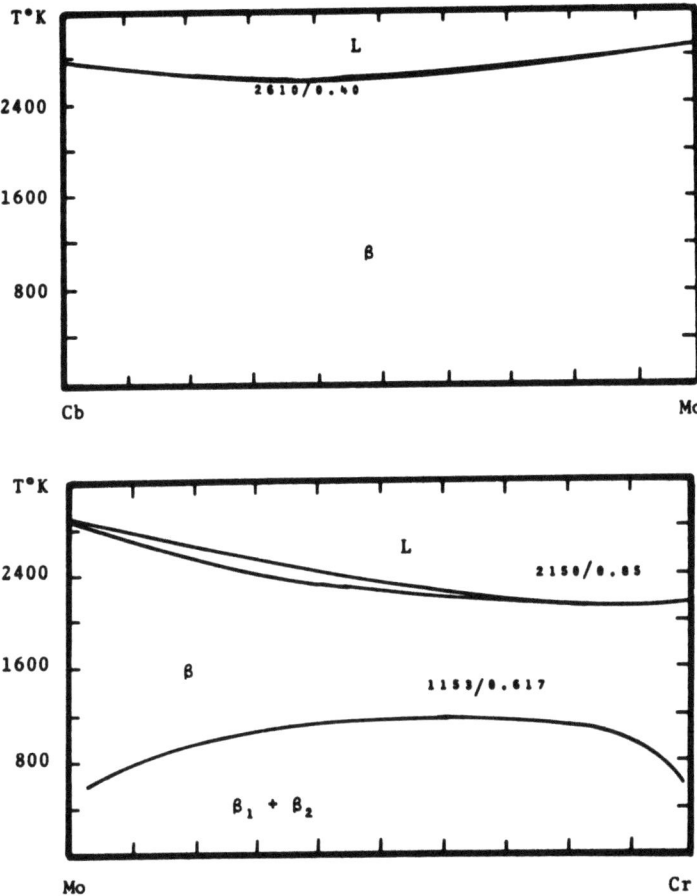

Fig. 21. Calculated Cb–Mo and Mo–Cr phase diagrams.

The Cb–Cr–Mo system provides an interesting case since extensive phase equilibria and oxidation data are available.[85–87]*
As indicated earlier, the Cb–Cr case is supported by experimental data on the heat of formation of $CbCr_2$.[80] In addition, the calculated miscibility gap in Mo–Cr with a maximum computed at 1153°K is in agreement with thermochemical data. It is not possible to infer interaction parameters for the Cb–Mo system directly from the phase diagram. Application of estimation procedures developed and applied uniformly to all the transition metals yields small, positive

---

* It should be noted that the compilation in Ref. 87 incorrectly reports the results at 1200°C[85] in weight per cent when the original results[85] were given in atomic per cent.

Chapter 4

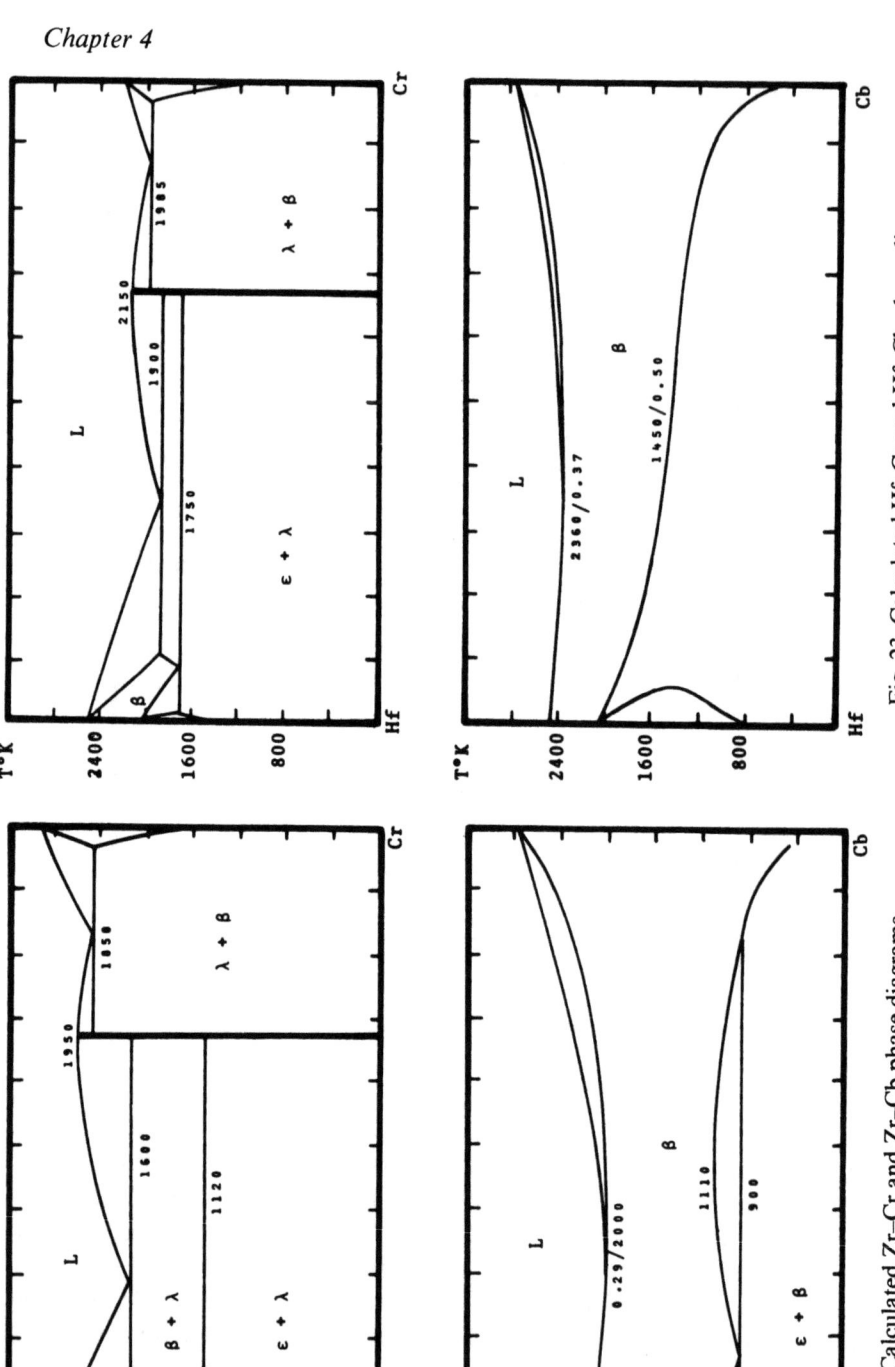

Fig. 23. Calculated Hf–Cr and Hf–Cb phase diagrams.

Fig. 22. Calculated Zr–Cr and Zr–Cb phase diagrams.

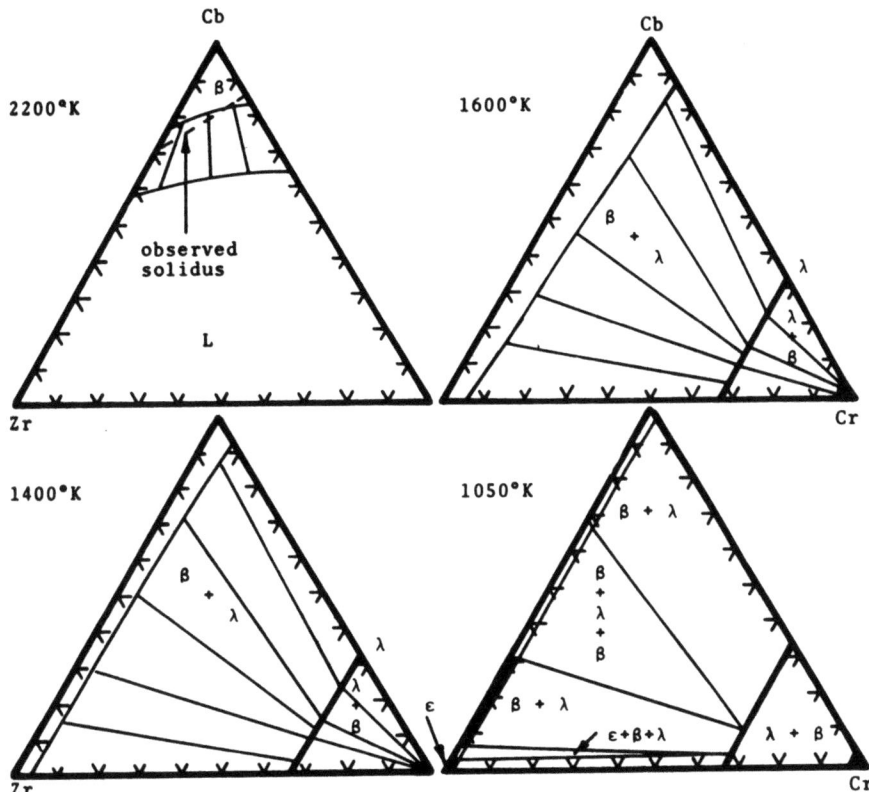

Fig. 24. Calculated and observed isothermal sections in the Cb–Cr–Zr system.

interaction parameters for the Cb–Mo case. Recently Singhal and Worrell[88] reported thermochemical data for Ta–W, Cb–Mo, and Ta–Mo at 1473°K. The bcc solutions in these systems can be approximated by regular solution parameters of -6400, -8800, and $-10,400$ cal/g-atom, respectively. In order to evaluate the Singhal–Worrell result, which is somewhat unexpected, an analysis of the bcc phase in the Mo–Cb–Cr system was performed as shown in Figures 26 and 27. The critical feature in this analysis is the miscibility gap which is observed at 1273°K[86,87] and 1473°K.[85,87] Reference to the experimental 1273 and 1473°K panels in Figure 27 shows that the miscibility gap extends far over to the Mo–Cr edge. This observation strongly suggests that the interaction parameter for the bcc phase in Mo–Cb is a large, negative quantity, supporting the Singhal–Worrell result. This conclusion stems from the fact that if $B \approx 4600$ for Mo–Cr, then the gap maximum occurs at 1157°K. In the Nb–Cr case $B = 5500$,

215

Chapter 4

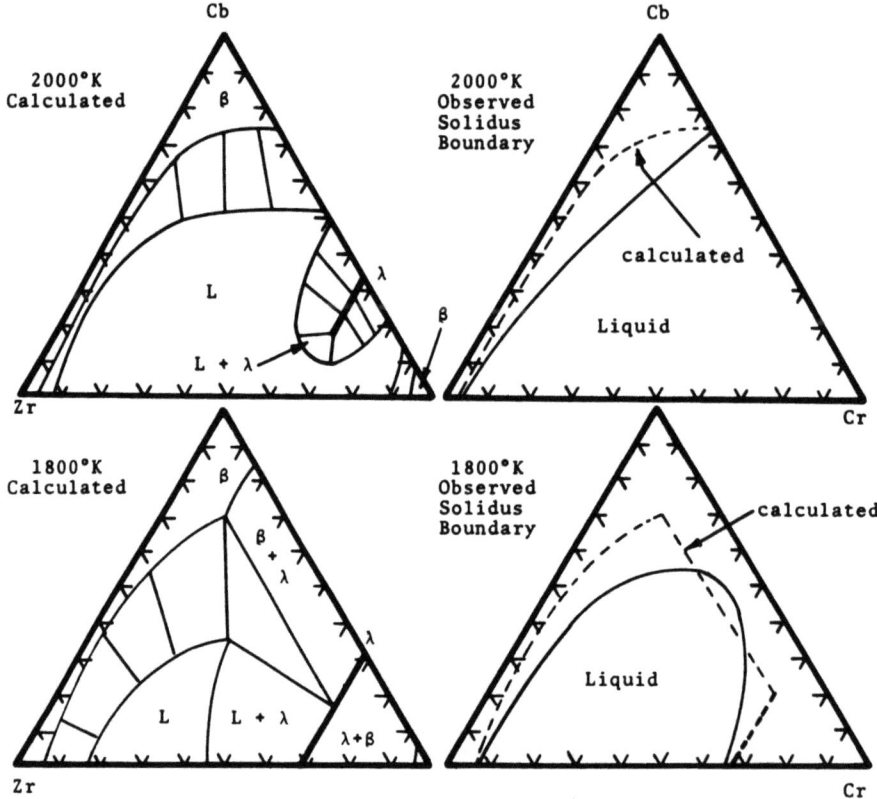

Fig. 25. Calculated and observed isothermal sections in the Cb–Cr–Zr system.

hence the gap maximum occurs at 1383°K. If the Mo–Cb system were characterized by $B = 0$, then no gap would be stable in Mo–Cr–Cb at 1473°K, and at 1273°K the gap would be covered by the $\beta + \lambda$ equilibria. Thus negative values of B for Mo–Cb are implied, leading to an isolated gap at high temperatures with a summit temperature above 1473°K. (See pp. 299–300 of Ref. 3.) If a value of $B = -8800$ is chosen for Mo–Cb (in keeping with the Singhal–Worrell result), the miscibility gap extends far into the ternary at 1273 and 1473°K. Reduction in magnitude to $B = -5500$ yields the results shown in Figures 26 and 27, which show conformity between the calculated and observed miscibility gaps at 1273 and 1473°K. In addition, it should be noted that $B = -3178$ for Cr–V, which is the analog of Mo–Cb (i.e., Table XXI, p. 188 of Ref. 3).

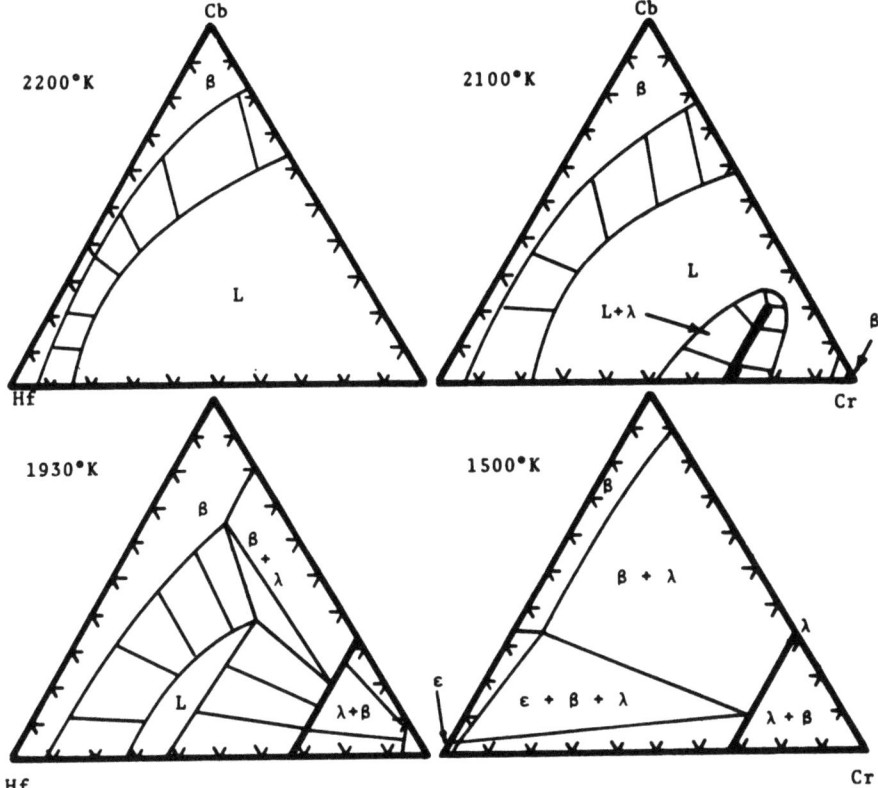

Fig. 26. Calculated isothermal sections in the Cb–Cr–Hf system.

On this basis the interaction parameter for the bcc phase in Mo–Cb was set equal to -5500 cal/g-atom. The value $L = -7100$ is based on the bcc/liquid equilibrium in Mo–Cb. Comparison of the calculated and observed bcc/liquid equilibria at 2650 and 2400°K shown in Figure 28 yields reasonable agreement. Thus the present description provides a means for calculating the entire ternary phase diagram and the thermochemical properties. The summit conditions corresponding to the peak of the isolated miscibility gap are 1626°K, 21 at.% Mo, and 50 at.% Cr. This description could be employed in performing a detailed analysis of the oxidation behavior of Cb–Cr–Mo alloys.[86] Figure 28 shows a final check on the Ti–Zr, Ti–Cb, and Cb–Zr values in Table 9. Here the calculated Zr–Ti–Cb section at 843°K is in excellent agreement with the observed section.[89]

Chapter 4

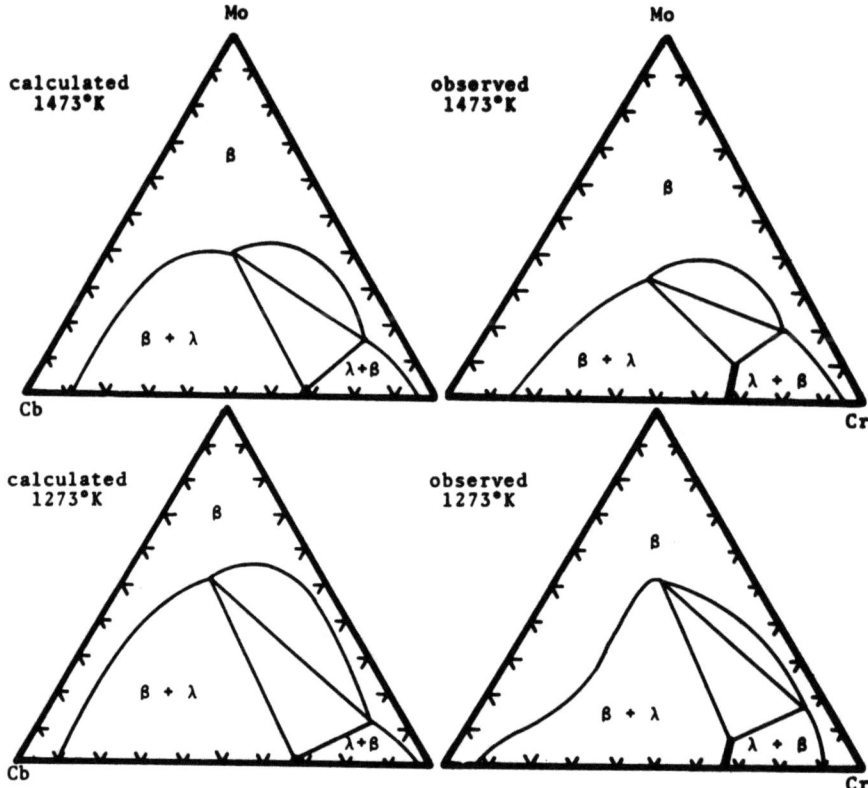

Fig. 27. Calculated and observed isothermal sections in the Mo–Cr–Cb system.

7. Analysis of the Quasibinary Eutectic Iron–Titanium Carbide

Metal–metal carbide composites which have been fabricated by means of controlled directional solidification techniques have been found to exhibit attractive high-temperature properties. Since little phase diagram data are available concerning these systems, it is of interest to apply the computational methods to calculation of the solidification characteristics. In order to illustrate the procedure, the iron–carbon–titanium system will be considered. In addition, the binary components of the nickel–carbon–titanium and chromium–carbon–titanium systems will be analyzed so that these systems could be readily considered. As a consequence, seven binary systems, Fe–Ti, Ti–Ni, Ti–Cr, Fe–C, Ti–C, Ni–C, and Cr–C, must be considered.

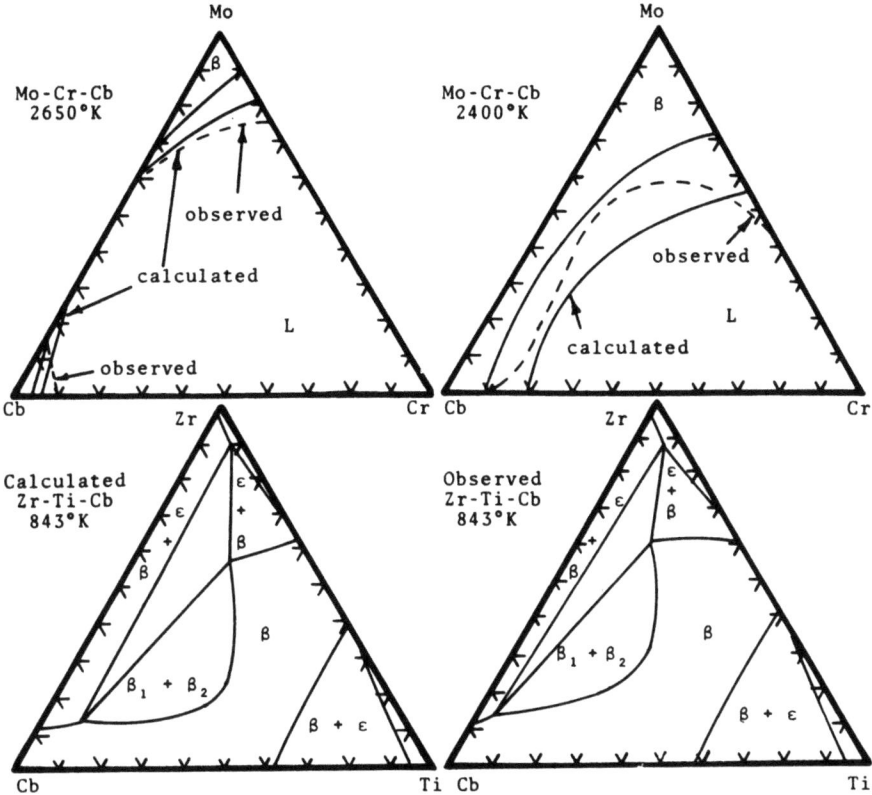

Fig. 28. Calculated and observed isothermal sections in Mo–Cr–Cb and Zr–Ti–Cb.

The description of the Fe–Ti system is summarized in Tables 10 and 11. Figure 29 shows the computed phase diagram which results from the lattice stability parameters[3] combined with the solution and compound parameters. The latter yield calculated thermochemical results which agree with observation.[2] The free energy difference between bcc and hcp iron is shown graphically in Ref. 3. Table 2 contains the free energy between the remaining structures of iron. The computed phase diagram is in good agreement with the observed diagram.[54–56] Moreover, the thermochemical description permits calculation of the M_s temperature for bcc → hcp transformations on cooling titanium-rich alloys.[90]

Tables 10 and 11 summarize the thermochemical properties of the Ti–Ni system. Lattice stability data for titanium and nickel were presented earlier.[3] The latter are modified by the results shown in

TABLE 10
Summary of Excess Free Energy Parameters*

System	Liquid	bcc	fcc	hcp
Fe–Ti	−15,000	−11,000	−8,000	+7,900
	−11,000	−5,500	−2,500	+7,900
Range, °K	1400–2000	700–2000	1000–1800	500–1200
Ti–Ni	−27,000 + 7.0T	−18,600 + 7.0T	−21,600 + 7.0T	−10,600 + 7.0T
	−45,000 + 13.0T	−39,200 + 13.0T	−43,600 + 13.0T	−10,600 + 7.0T
Range, °K	1200–2000	1000–2000	500–1200	500–1200
				Graphite
Ni–C	−26,600 + 10.3T	—	−19,000 + 1.03T	10,000
	−23,800 + 3.6T	—	−26,000 + 16.18T	10,000
Range, °K	1500–4000		500–2000	500–4000
Cr–C	−33,000	−32,000	—	10,000
	−33,000	−32,000	—	10,000
Range, °K	1600–4000	500–2300		500–4000
Ti–C	−42,000	−50,000	—	20,000
	−42,000	−50,000	—	20,000
Range, °K	1700–4000	1000–2000		500–4000
Fe–C	−21,900 + 2.52T	−8,400 − 9.04T	−22,600 − 0.70T	10,000
	−37,300 + 11.70T	−22,900 − 6.95T	−37,100 + 1.39T	10,000
Range, °K	1300–4000	300–2000	300–2000	500–4000

* In cal/g-atom. T in °K.

TABLE 11
Summary of Compound Parameters

Compound	Structure	Base	Compound parameter,* cal/g-atom
$Fe_{0.667}Ti_{0.333}$	Laves phase	hcp	14,000
$Fe_{0.50}Ti_{0.50}$	CsCl	bcc	$6,400 + 0.18T$
$Ti_{0.667}Ni_{0.333}$	FCC ($Fd3m$)	fcc	$-2,000 + 5.00T$
$Ti_{0.50}Ni_{0.50}$	CsCl	bcc	$2,200 + 4.05T$
$Ti_{0.25}Ni_{0.75}$	Hexagonal	hcp	$6,700 + 3.21T$
$Ni_{0.75}C_{0.25}$	Rhombohedral	fcc	$14,000 + 6.2T$
$Cr_{0.793}C_{0.207}$	Complex fcc	bcc	$27,400 - 0.78T$
$Cr_{0.70}C_{0.30}$	Hexagonal	bcc	$33,900 - 1.36T$
$Cr_{0.60}C_{0.40}$	Orthorhombic	bcc	$39,200 - 2.18T$
$Ti_{0.56}C_{0.44}$	NaCl	fcc	$104,000 - 13.4T$
$Fe_{0.75}C_{0.25}$	Orthorhombic	fcc	$14,800 + 3.67T$

* T in °K.

Table 2. Figure 29 shows the calculated phase diagram, which is in good agreement with observations.[54–56] A similar comparison results when calculated and observed[2,81] thermochemical properties are compared. In addition, the thermochemical description permits calculation of the M_s temperature in titanium-rich alloys.[90] A complete description of the Ti–Cr system has been presented recently.[90]

Tables 10 and 11 and Figure 30 summarize the present description of the Ni–C system which has been developed in accordance with the observed phase diagram,[54–56] recent results on carbon solubility in fcc nickel,[92] and the metastable fcc–liquid–Ni_3C eutectic.[93] Although the positional entropy of the interstitial fcc solid solution of carbon in nickel[94] should not be equated with that of a substitutional solid solution, which is the procedure employed here, the errors are small in the solution range of interest (i.e., less than 2 at. % C) and are compensated for by the excess free energy terms in Table 10. This computational procedure was employed to take advantage of existing computer software which is available for substitutional solution interactions.

The free energy difference between the graphitic (g) and liquid forms of carbon is estimated on the basis of the most recent measurements of the graphite/vapor/liquid triple point[95,96] and an analysis of the entropy difference between the liquid and graphitic forms of carbon[97] which has been updated to reflect the most recent values of the entropy of fusion of the diamond cubic forms of Si and Ge.[3] This procedure leads to an enthalpy of fusion of 27,300 cal/g-atom and

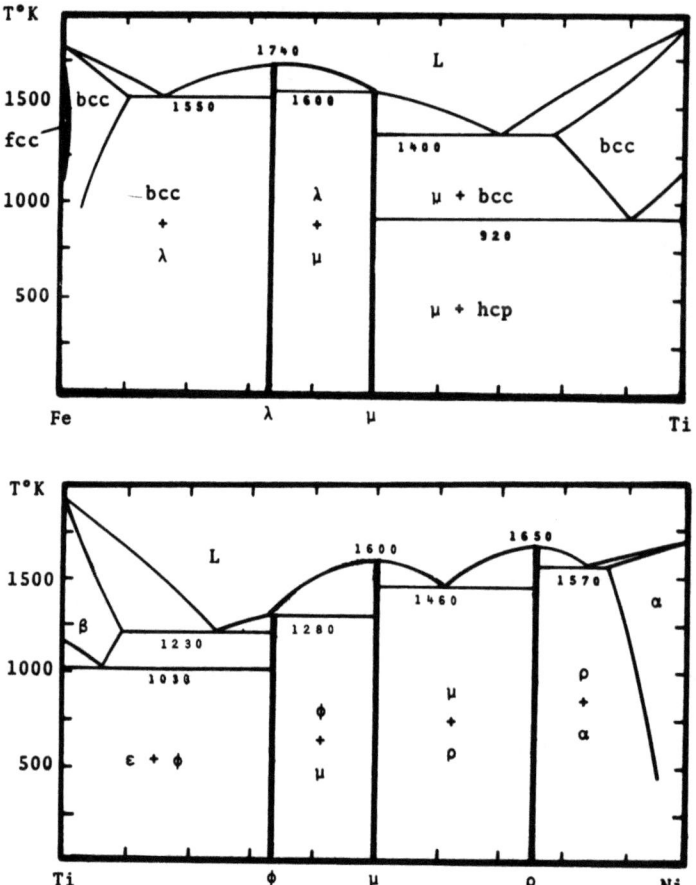

Fig. 29. Calculated phase diagrams for the Fe–Ti and Ti–Ni systems.

an entropy of fusion of 6.5 cal/g-atom °K for the graphitic form of carbon. The enthalpy of the fcc form of carbon is estimated to be 33,100 cal/g-atom greater than the graphitic form, while the enthalpy of the bcc form is estimated to be 35,100 cal/g-atom greater than the graphitic form. The entropy of the bcc and fcc forms of carbon is estimated to be 3.5 cal/g-atom °K greater than the graphitic form of carbon. Consideration of the Re–C system suggests that the hcp form is 5800 cal/g-atom more stable than the fcc form. Finally the free energy differences between the graphitic and liquid forms of Ni, Cr, Ti, and Fe have been estimated on the basis of the present analyses of Ni–C, Cr–C, Ti–C, and Fe–C as being equal to $-6.0T$ cal/g-atom.

Relation of the Thermochemistry and Phase Diagrams of Condensed Systems

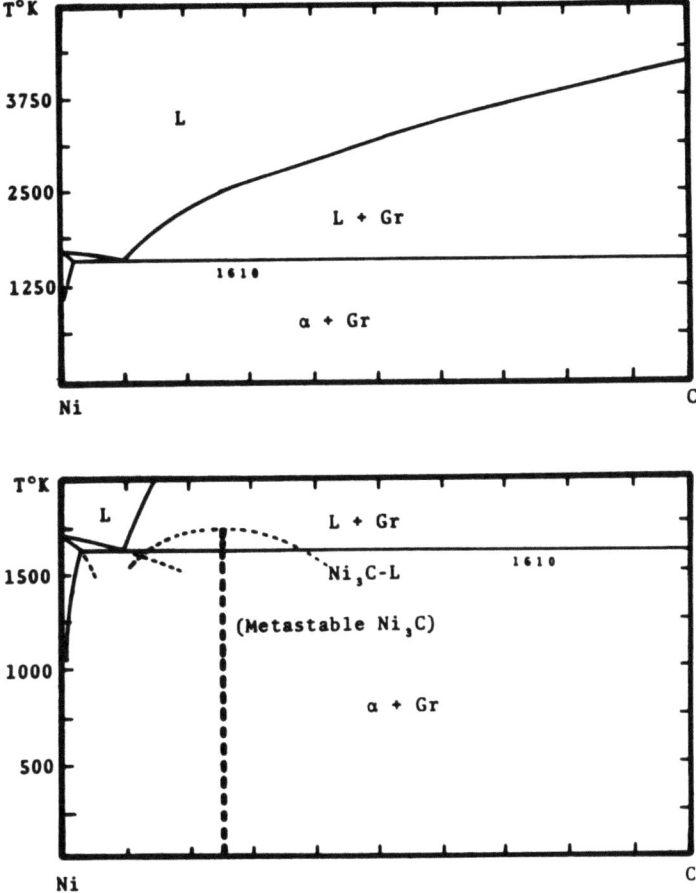

Fig. 30. Calculated phase diagram for the Ni–C system.

Tables 10 and 11 and Figure 31 summarize the current description of the Cr–C system which is treated on the basis of simple regular solutions. As in the Ni–C case, the positional entropy of the interstitial solution of carbon in bcc iron should not be calculated on the same basis as a substitutional solid solution. However, since the range of interest does not exceed 2 at.%, the resulting errors are not large. The computed results are in good agreement with the recent thermochemical data of Kulkarni and Worrell[98] and the phase diagram given by Rudy.[79]

Tables 10 and 11 and Figure 32 display the present description of the Ti–C system in which the NaCl-type titanium carbide phase

Chapter 4

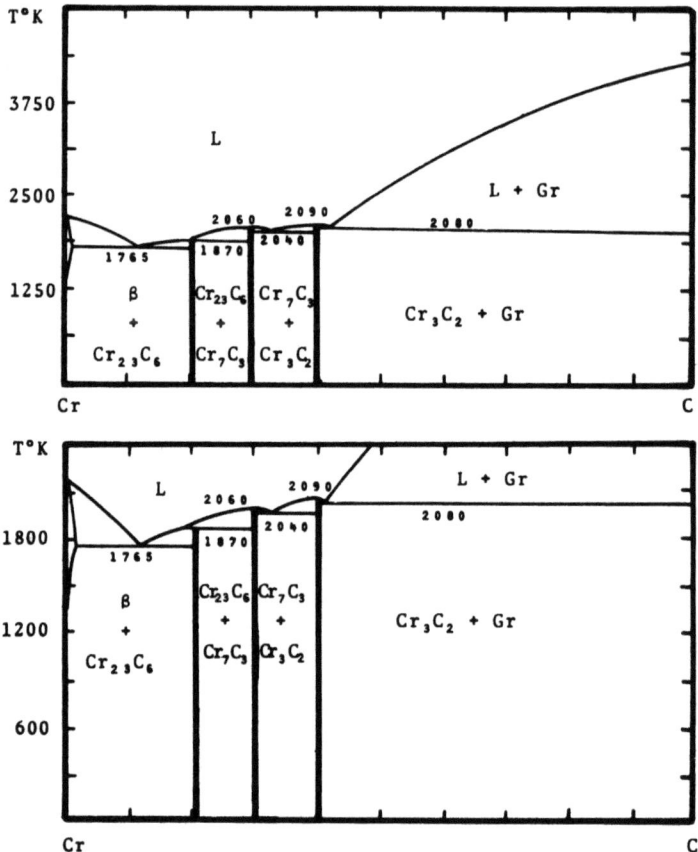

Fig. 31. Calculated phase diagram for the Cr–C system.

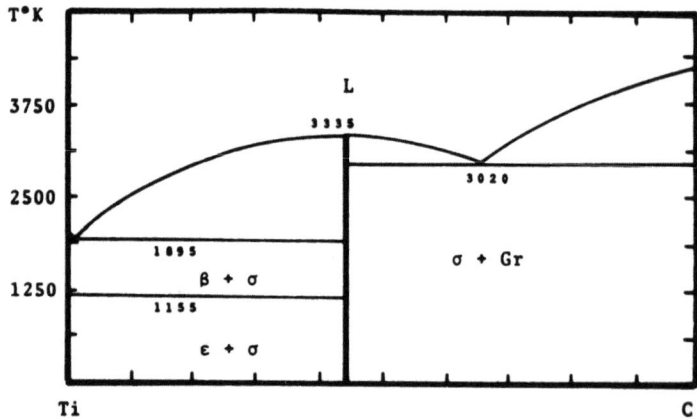

Fig. 32. Calculated phase diagram for the Ti–C system.

Relation of the Thermochemistry and Phase Diagrams of Condensed Systems

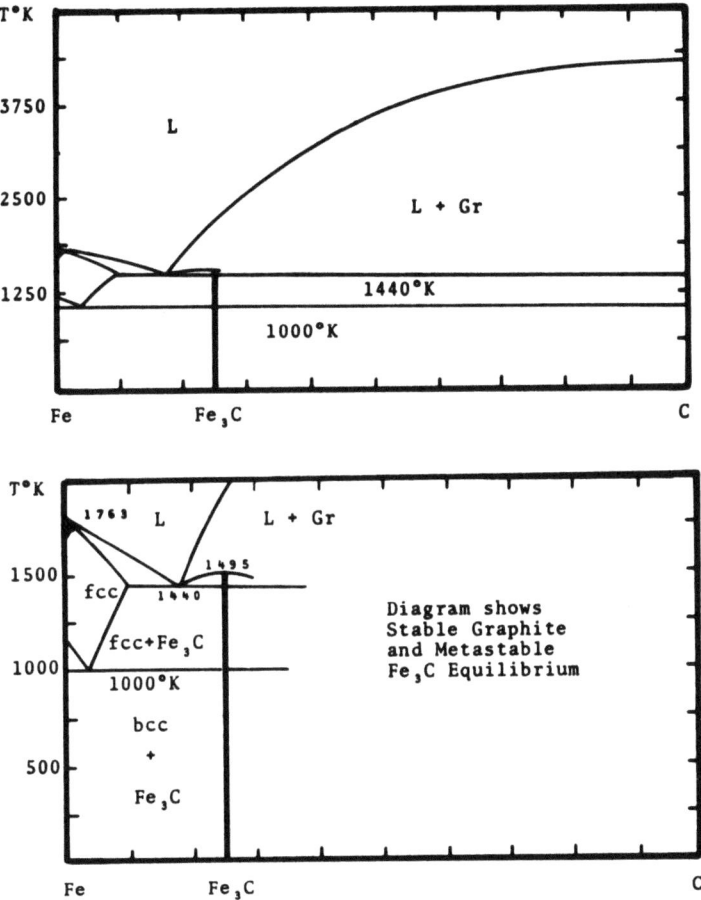

Fig. 33. Calculated phase diagram for the Fe–C system.

is stable over a range of composition between 40 and 50 at.%. The present simple formulation treats this phase as a line compound $Ti_{0.56}C_{0.44}$ (σ) (at which the maximum melting point is observed[79]) and characterizes the bcc, liquid, and graphite phases as regular solutions. The present description is in reasonable agreement with the observed thermochemical properties.[2] It should be noted that a more detailed description of the titanium carbide (σ) phase can be developed by applying the Schottky–Wagner model to allow for stoichiometric deviations. This procedure has been applied to the Ti–C system.[97] However, in the present case where interest is focused on the Fe–Ti–C liquidus surface, the present simple line compound formulation should suffice.

Chapter 4

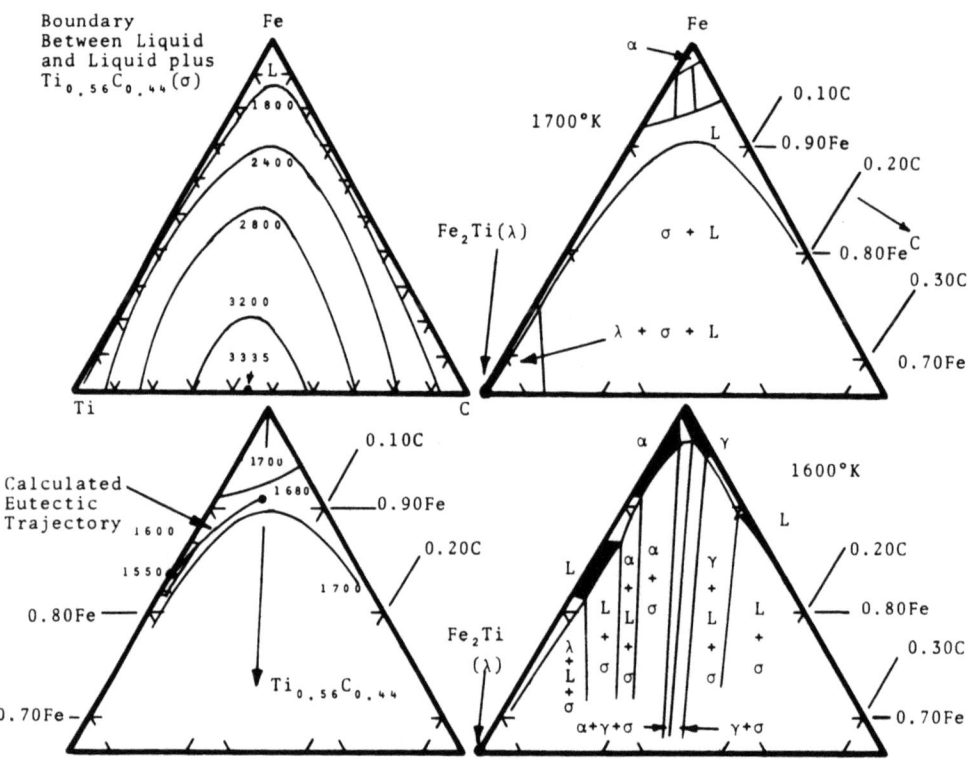

Fig. 34. Calculated liquidus in the Fe–Fe$_2$Ti–TiC region of the Fe–C–Ti system.

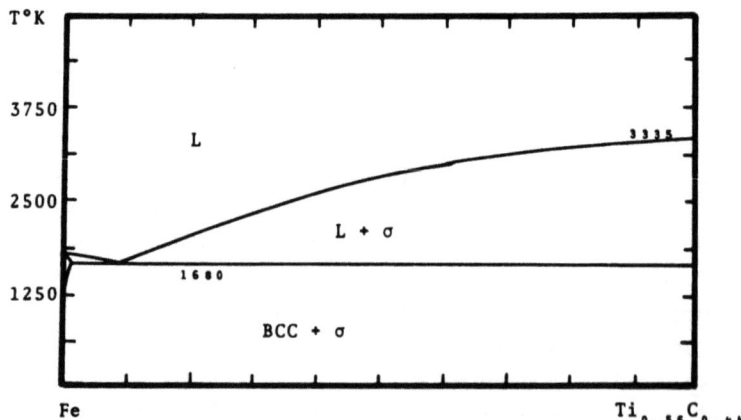

Fig. 35. Calculated quasibinary eutectic in the Fe–C–Ti system.

The description of the iron–carbon system given in Tables 10 and 11 yields the calculated Fe–C phase diagram shown in Figure 33 as well as thermochemical properties which compare well with observation.[99] The current description also provides an adequate calculation of the austenite/martensite reactions in iron–carbon alloys.[94] As in the case of the Ni–C, Cr–C, and Ti–C systems described earlier, the bcc, fcc, and liquid solutions are treated as substitutional solutions, to take advantage of existing computer software. Calculation of the activity of carbon in the fcc phase based on Table 10 shows that serious errors are not introduced by this procedure since the range of composition is restricted. In the case of liquid solutions calculation of the carbon activity from Table 10 shows that deviations occur at the higher carbon levels above 20 at. %. However, these deviations appear to be well within the experimental scatter of various individual investigations.[100]

The foregoing description of the binary phase diagram has been employed to compute the liquidus boundaries shown in Figure 34. In these calculations the quasibinary calculation Fe–$Ti_{0.56}C_{0.44}$ versus liquid yields the diagram in the upper left corner. Expanded compositional views at 1700 and 1600°K which include the bcc (α), fcc (γ), and Fe_2Ti (λ) phases suggest quasibinary eutectic formation between the $Ti_{0.56}C_{0.44}$ (σ) phase and the bcc phase with a eutectic near 1680°K and a composition of $0.91Fe + 0.09(Ti_{0.56}C_{0.44})$. Figure 32 shows a vertical section displaying the calculated eutectic temperature and composition.

Jellinghaus[101] has reported the results of an experimental study of the Fe–C–Ti system. Before considering his results with reference to the present ternary calculations shown in Figure 34, it is important to note that he reports the melting point of Fe_2Ti (λ) as 1540°C, or 1813°K. This temperature is 110°C higher than the value of 1430°C (1703°K) given in Elliot's compilation[55] and 73°K higher than the calculated melting point (1740°K) shown in Figure 29. In addition, the temperature of the liquid/bcc/Fe_2Ti eutectic reported by Jellinghaus[101] is approximately 100°K higher than that given by Elliot[55] and Figure 29. Jellinghaus reports a quasibinary eutectic between liquid, Fe_2Ti (λ), and titanium carbide at 1733°K at a composition of 53 at. % Fe, 37 at. % Ti, 10 at. % C, and a melting point minimum for liquid → bcc iron–carbon–titanium + Fe_2Ti + titanium carbide at 1600°K at 80 at. % Fe, 6 at. % C, 14 at. % Ti. These results are not in keeping with the calculations shown in Figure 34 due to the above-mentioned discrepancy in the melting point of Fe_2Ti. The present calculations place the quasibinary between Fe_2Ti and $Ti_{0.56}C_{0.44}$ near 1735°K and 66 at.% Fe, 33 at.% Ti, 1 at.% C. Moreover, Figure 31 *does not* indicate a melting point minimum for liquid → bcc iron–carbon–titanium + Fe_2Ti + titanium carbinde.

Jellinghaus[101] has heated mixtures of 90 at.% Fe–5 at.% C–5 at.% Ti, 85Fe–7.5C–7.5Ti, 80Fe–10C–10Ti, etc., to determine the minimum melting temperature. Microstructural results were employed to fix the temperature and composition of the liquid → bcc iron–carbon–titanium + titanium carbide reaction. The above-mentioned experiments indicate the critical composition between 90Fe–10($Ti_{0.5}C_{0.5}$) and 85Fe–15($Ti_{0.5}C_{0.5}$) with the critical temperature between 1400 and 1470°C. These findings are in reasonable agreement with the predicted results shown in Figure 35, which correspond to 91Fe(Ti, C)–9($Ti_{0.56}C_{0.44}$).

8. Summary

As indicated at the outset of this review, thermochemical and phase diagram descriptions of condensed systems provide alternative methods of characterization which are intimately related. The distinction is artificial and restrictive in cases where one formalism or the other is not considered. Thus the examples presented here illustrate that intimate coupling of both methods often leads to a more complete picture of the system. In retrospect, one might speculate that the origin of the diverse paths which thermochemists and "phase diagramists" have taken is a result of the systems of interest to these groups. Thus thermochemists, who were interested in equilibrium reactions involving products and reactants of well-defined stoichiometry, may have concluded that the multiplicity of reactants and products (all of variable compositions) making up a phase diagram could not be treated systematically. Moreover, metastable phases which persist in condensed systems provide additional headaches. The "phase diagramists," on the other hand, who were required to deal with complex systems containing variable composition and metastable phases, concluded intuitively that a simple thermochemical description could never provide meaningful guidance. Although these conclusions may have been true in the past, present computational methods which permit examination of competition between a wide variety of phases have altered the situation. Since these operational barriers have been largely eliminated, coordinated thermochemical and phase diagram analyses should find increased applications in the manipulation of condensed systems. Although the examples presented here deal exclusively with metal and metal–carbide systems, they illustrate two-phase solution equilibria, miscibility gap formation, and compound–compound and compound–solution phase interactions which form the basis of all phase diagrams. Consequently, the methods applied to the systems considered could readily be applied to oxide, sulfide, alkali halide, and III–V compound systems and to organic chemical systems as well.

Acknowledgments

This work has been sponsored by the Metals and Ceramics Division, Air Force Materials Laboratory, Wright–Patterson Air Force Base, Ohio, under contract AF 33(615)-71-C-1471 and the Engineering Materials Program, Division of Materials Research of the National Science Foundation, Washington, D.C., under Grant GH-36121.

References

1. R. Hultgren, R. L. Orr, P. D. Anderson, and K. K. Kelley, *Selected Values of Thermodynamic Properties of Metals and Alloys*, Wiley, New York (1963).
2. R. Hultgren *et al.*, *Supplement to Selected Values of Thermodynamic Properties of Metals and Alloys*, University of California, Berkeley (1966–1971).
3. L. Kaufman and H. Bernstein, *Computer Calculation of Phase Diagrams*, Academic, New York (1970).
4. K. A. Bolbshakov, P. I. Fedorov, E. I. Smarina, and I. N. Smirnova, *Zh. Neorgan. Khim.* **9**, 1883 (1964).
5. C. Bodsworth and O. Kubaschewski, *Metallurgical Chemistry*, Preface (O. Kubaschewski, ed.), HMSO, London (1972).
6. L. Kaufman, in *Metallurgical Chemistry* (O. Kubaschewski, ed.), p. 373, HMSO, London (1972).
7. I. Ansara, in *Metallurgical Chemistry* (O. Kubaschewski, ed.), p. 403, HMSO, London (1972).
8. H. Harvig, T. Nishizawa, and B. Uhrenius, in *Metallurgical Chemistry* (O. Kubaschewski, ed.), p. 431, HMSO, London (1972).
9. J. F. Counsell, E. B. Lees, and P. J. Spencer, in *Metallurgical Chemistry* (O. Kubaschewski, ed.), p. 451, HMSO, London (1922).
10. J. F. Counsell and O. Kubaschewski, in *Metallurgical Chemistry* (O. Kubaschewski, ed.), p. 649, HMSO, London (1972).
11. C. H. P. Lupis and H. Gaye, in *Metallurgical Chemistry* (O. Kubaschewski, ed.), p. 469, HMSO, London (1972); *Scripta Met.* **4**, 685 (1970).
12. G. Kirchner, H. Harvig, and B. Uhrenius, *Met. Trans.* **4**, 1059 (1973).

13. H. Harvig, G. Kirchner, and M. Hillert, *Met. Trans.* **3**, 329 (1972).
14. G. Kirchner, G. Larbo, and B. Uhrenius, *Praktische Metallographie* **8**, 641 (1971).
15. G. Kirchner, H. Harvig, K.-R. Moquist, and M. Hillert, *Arch. Eisenhütt.* **44**, 1973 (in press).
16. G. Kirchner, T. Nishizawa, and B. Uhrenius, *Met. Trans.* **4**, 167 (1973).
17. L. Kaufman and H. Nesor, in *Titanium Science and Technology* (R. I. Jaffee and H. M. Burte, eds.), pp. 773–800, Plenum, New York (1973).
18. L. Kaufman and H. Nesor, *Conf. on In Situ Composites*, National Academy of Science, Washington, D.C., NMAB 308 III, pp. 21–29.
19. L. Kaufman and H. Nesor, *Z. Metallk.* **64**, 249 (1973).
20. P. J. Spencer and F. H. Putland, *J. Iron and Steel Inst.* **211**, 293 (1973).
21. I. Ansara, E. Bonnier, and J. Mathieu, *Z. Metallk.* **64**, 258 (1973).
22. N. J. Olson and G. N. Toop, *Trans. Met. Soc. AIME* **245**, 905 (1969).
23. S. M. Carmio and J. L. Meijering, *Z. Metallk.* **64**, 170 (1973).
24. E. Rudy and Y. A. Chang, in *Plansee Proceedings, 1964* (F. Benesovsky, ed.), p. 786, Metallwerke Plansee AG, Reutte/Tyrol.
25. Y. A. Chang and D. Naujock, *Met. Trans.* **3**, 1693 (1972).
26. R. F. Brebrick, *Met. Trans.* **2**, 1657, 3377 (1971).
27. Y. K. Rao, in *Phase Diagrams* (A. Alper, ed.), Volume 1, p. 1, Academic, New York (1970).
28. J. F. Breedis and L. Kaufman, *Met. Trans.* **2**, 2359 (1971).
29. J. B. Gilmore, G. R. Purdy, and J. Kirkaldy, *Met. Trans.* **3**, 1455 (1972).
30. L. Kaufman, *Phase Stability in Metals and Alloys* (P. S. Rudman, J. Stringer, and R. I. Jaffee, eds.), p. 125, McGraw-Hill, New York (1967).
31. L. A. Pugliese and G. R. Fitterer, *Met. Trans.* **1**, 1997 (1970).
32. F. Mueller and F. H. Hayes, *J. Chem. Thermo.* **3**, 599 (1971).
33. L. Kaufman, E. V. Clougherty, and R. J. Weiss, *Acta Met.* **11**, 323 (1963).
34. A. P. Miodownik, in *Int. Symp. on Metallugical Chemistry—Application in Ferrous Metallurgy* (B. B. Argent and M. W. Davies, eds.), HMSO, England (1972).
35. B. Predel and R. H. Mohs, *Arch. Eisenhütt.* **41**, 61 (1970).
36. R. W. Carpenter, C. T. Liu, and P. G. Mardon, *Met. Trans.* **2**, 125 (1971).
37. M. B. Panish, in *Phase Diagrams* (A. Alper, ed.), Volume III, p. 53, Academic, New York (1970).
38. C. Wagner, *Acta Met.* **6**, 309 (1958).
39. J. J. Vieland, *Acta Met.* **11**, 137 (1963).
40. G. G. Libowitz and J. B. Lightstone, *J. Phys. Chem. Solids* **28**, 1145 (1967).
41. R. F. Brebrick, in *Progress in Solid State Chemistry* (H. Reiss, ed.), Volume III, p. 213, Pergamon (1966); *J. Solid State Chem.* **1**, 88 (1969).
42. L. Kaufman and E. V. Clougherty, in *Metallurgy at High Pressure and High Temperature* (K. A. Gschneider, M. T. Hepworth, and N. A. D. Parlee, eds.), Gordon and Breach, New York; *Met. Soc. Conf. AIME* **22**, 322 (1964).
43. E. Rudy, *Thermodynamics of Nuclear Materials*, p. 243, IAEA, Vienna (1962).
44. L. Kaufman, in *Compounds of Interest in Nuclear Reactor Technology* (J. T. Waber, P. Chiotti, and W. N. Miner, eds.), pp. 193, 267, AIME, New York (1964).
45. L. Kaufman and E. V. Clougherty, in *Metals for the Space Age Plansee Proceedings, 1964* (F. Benesovsky, ed.), p. 722, Metallwerke Plansee, Reutte/Tyrol.
46. L. Kaufman and G. Stepakoff, *Third Conf. on the Performance of High Temperature Systems* (G. Bahn, ed.), p. 33, Gordon and Breach, New York (1964).

47. B. Predel and R. H. Mohs, *Arch. Eisenhütt.* **41**, 143 (1970).
48. A. P. Miodownik, *Acta Met.* **18**, 51 (1970).
49. O. Kubaschewski and W. Slough, *Progr. Mat. Sci.* **14**, 3 (1969).
50. O. Kubaschewski and L. E. H. Stuart, *J. Chem. Eng. Data* **12**, 418 (1967).
51. D. Henriette, C. Gatellier, and M. Ollette, in *Int. Symp. on Metallurgical Chemistry* (O. Kubaschewski, ed.), p. 97, HMSO, London (1972).
52. E. Scheil and W. Normann, *Arch. Eisenhütt.* **30**, 751 (1959).
53. E. Scheil and E. Saftig, *Arch. Eisenhütt.* **31**, 623 (1960).
54. M. Hansen and K. Anderko, *Constitution of Binary Alloys*, McGraw-Hill, New York (1958).
55. R. P. Elliot, *Constitution of Binary Alloys*, First supplement, McGraw-Hill, New York (1965).
56. F. A. Shunk, *Constitution of Binary Alloys*, Second supplement, McGraw-Hill, New York (1968).
57. M. M. Rao, R. J. Russell, and P. G. Winchell, *Trans. TMS—AIME* **239**, 641 (1967).
58. J. I. Goldstein and R. E. Ogilvie, *Trans. TMS—AIME* **233**, 2083 (1965).
59. L. Kaufman and M. Cohen, *Trans. AIME* **206**, 1393 (1956).
60. R. Fruehan, *Trans. Met. Soc. AIME* **242**, 2007 (1968).
61. E. P. Hall and S. H. Algie, *Metallurgical Rev.* **11**, 61 (1966).
62. F. Mueller and O. Kubaschewski, *High Temperature–High Pressure* **1**, 543 (1969).
63. L. Kaufman, *Scripta Met.* **4**, 437 (1969).
64. R. H. Moore, *Chemical Metallurgy of Iron and Steel*, Iron and Steel Inst. of London (1973), p. 360.
65. J. W. Pugh and J. D. Nisbet, *Trans. Met. Soc. AIME* **178**, 268 (1950).
66. T. Lindemer, Oak Ridge National Laboratory, Oak Ridge, Tennessee, Private communication (December 1971).
67. F. P. Bundy, *J. Appl. Phys.* **38**, 2446 (1967).
68. T. R. Loree, C. M. Fowler, E. G. Zukas, and F. S. Minshall, *J. Appl. Phys.* **37**, 1918 (1966).
69. J. F. Breedis and L. Kaufman, *Met. Trans.* **2**, 2359 (1971).
70. P. M. Giles and A. Marder, *Met. Trans.* **2**, 1371 (1971).
71. A. Fraker and H. H. Stadelmaier, *Trans. Met. Soc. AIME* **245**, 847 (1969).
72. V. Ya. Markiv, V. V. Burnashova, L. I. Pryakhina, and K. P. Myasnikova, *Metally* **1969**, 180.
73. H. L. Gegel and M. Hoch, in *Titanium Science and Technology* (R. I. Jaffee and H. M. Burte, eds.), pp. 923–934, Plenum, New York (1973).
74. B. Bredel and D. W. Stein, *Acta Met.* **20**, 515 (1972).
75. D. Clark, K. S. Jepson, and C. J. Lewis, *J. Inst. Metals* **91**, 197 (1962–1963).
76. M. J. Blackburn, *Trans. Met. Soc. AIME* **239**, 1200 (1967).
77. F. A. Crossley, *Trans. Met. Soc. AIME* **242**, 726 (1968).
78. M. J. Blackburn, *Trans. Met. Soc. AIME* **242**, 728 (1968).
79. E. Rudy, Compendium of Phase Diagram Data, AFML-TR-65-2, Part V, Wright-Patterson Air Force Base, Ohio, 1969.
80. J. F. Martin, F. Muller, and O. Kubaschewski, *Trans. Faraday Soc.* **66**, 1065 (1970).
81. V. N. Svechnikov and V. M. Pan, *Summaries Sci. Work, Inst. Metal Phys., Acad. Sci. Ukr SSR* **1962**(8), 46–57; **1962**(15), 156–162.

Chapter 4

82. L. Kaufman and H. Bernstein, *Phase Diagrams* (A. Alper, ed.), Volume 1, p. 45, Academic, New York (1970).
83. M. M. Savel'eva and N. V. Grum-Grzhimaylo, *Izv. Akad. Nauk SSSR, Metally* **1**, 224 (1969).
84. I. I. Kornilov, S. P. Alisova, and P. B. Budberg, *Izv. Akad. Nauk SSSR, Neorg. Mater.* **1**, 2205 (1965).
85. M. I. Zakharova and D. A. Prokoshkim, *Izv. Akad. Nauk SSSR Otd. Tekh. Nauk Met. i Topliva* **4**, 59 (1961).
86. H. J. Goldschmidt and J. A. Brandt, *J. Less-Common Metals* **3**, 44 (1961).
87. Binary and Ternary Phase Diagrams of Cb, Mo, Ta and W, DMIC Report 183, February 7, 1963, Battelle Memorial Institute, Columbus, Ohio.
88. S. C. Singhal and W. L. Worrell, in *Int. Symp. on Metallurgical Chemistry* (O. Kubaschewski, ed.), p. 149, HMSO, London (1971).
89. Toshio Doy, Masahiro Kitada, and Fumihiko Ishida, *J. Japan. Inst. Metals* **32**, 684 (1968).
90. L. Kaufman and H. Nesor, in *Ann. Rev. of Material Sci.* (R. Huggins, ed.), Vol. 3, Annual Reviews, Palo Alto, California (1973).
91. R. M. German and G. R. St. Pierre, *Met. Trans.* **3**, 2819 (1972).
92. J. Swartz, *Met. Trans.* **2**, 2318 (1971).
93. H. Strong, *Trans. Met. Soc. AIME* **233**, 643 (1965).
94. L. Kaufman, S. V. Radcliffe, and M. Cohen, in *Decomposition of Austenite by Diffusional Processes* (V. F. Zackay and H. I. Aaronson, eds.), p. 313, AIME—Interscience, New York (1962).
95. G. J. Schoessow, Graphite Triple Point and Solidus–Liquidus Interface Experimentally Determined to 1000 Atmospheres, NASA CR-1148, July 1968.
96. N. S. Diaconis, E. R. Stover, J. Hook, and G. J. Catalano, Graphite Melting Behavior, AFML-TR-71-119, July 1971.
97. L. Kaufman and E. V. Clougherty, Metallurgy at high pressures and high temperatures, in *Metallurgical Society Conferences* (K. A. Gschneider, M. T. Hepworth, and N. A. P. Parlee, eds.), Volume 22, p. 322, Gordon and Breach, New York (1964).
98. A. D. Kulkarni and W. L. Worrell, *Met. Trans.* **3**, 2363 (1972).
99. J. Chipman, *Met. Trans.* **3**, 55 (1972).
100. H. Schenk, E. Steinmetz, and M. Gloz, in *Metallurgical Chemistry* (O. Kubaschewski, ed.), p. 445, HMSO, London (1972).
101. W. Jellinghaus, *Arch. Eisenhütt.* **40**, 843 (1969).

5

Theory of Crystal Growth

Kenneth A. Jackson
Bell Laboratories
Murray Hill, New Jersey

1. Introduction

Crystal growth processes are determined by mass and heat transfer (by diffusion and convection) and by the intrinsic kinetic processes associated with the attachment of atoms or molecules to the crystal. These processes determine not only the growth rate and morphology of the crystal growth process, but also the composition, homogeneity, and perfection of the crystals produced.

Diffusion processes play an important role during the growth of crystals. The growth front represents a source of both heat and matter. Both the latent heat generated during freezing and the changes in composition generated by the selective incorporation of various species into the crystal are transported from the immediate vicinity of the crystal by diffusion. In many cases these diffusion processes dominate the growth process, and both the growth rate and growth morphology are determined by diffusion. Convection can also play an important role. The rate of attachment of atoms or molecules to the crystal depends on the material, on the phase into which it is growing, and on the growth conditions. Diffusion can be either fast or slow compared to the attachment kinetics, and the way these processes interact determines the progress of growth.

Chapter 5

2. Mass and Heat Transfer

2.1. Diffusion during Crystal Growth

The diffusion process obeys the usual diffusion equation

$$D\nabla^2 u = \partial u/\partial t \qquad (1)$$

This same equation applies for both thermal and compositional diffusion, with D being the appropriate diffusion coefficient and u being the temperature or composition. The boundary condition at the interface takes the form

$$D(\nabla u)_I = -Kv \qquad (2)$$

where $(\nabla u)_I$ is the gradient at the interface, v is the local interface velocity, and K is the temperature rise associated with the latent heat from crystallizing a unit volume $(L/C\rho)$, or, for the mass diffusion case, the amount of the component being considered which is rejected by the crystal per unit volume frozen.

In Eq. (2) the importance of the diffusion length D/v is apparent. It is this length which determines the scale of diffusion segregation during crystal growth. It is the distance over which the temperature or chemical potential fields can be made uniform in the vicinity of the moving interface by diffusion. The magnitude of the diffusion length varies sufficiently from one growth operation to another that qualitatively different effects are produced. A few examples will illustrate the point. For thermal diffusion in metals D is the order of 0.1–1 cm²/sec. For slow growth, say $v \sim 10^{-3}$ cm/sec, $D/v = 100$–1000 cm; the heat flow is macroscopic, and depends on heat loss from the crucible. For rapid growth, $v \sim 10^2$ cm/sec, typical for dendritic growth in pure materials, the diffusion distance is a fraction of a millimeter, and heat flow occurs on a local scale near the dendrite tip. For growth from aqueous solution the thermal diffusivity is about 10^{-3} cm²/sec, but the growth rate is usually very small (1 mm/day $\sim 10^{-6}$ cm/sec), so that the heat flow is again macroscopic, and is controlled by the thermal environment of the container. For vapor-phase growth the growth rate is usually sufficiently slow so that heat generated by the growth process is not a problem. However, at elevated temperatures radiation effects can produce important temperature gradients.

For mass flow the diffusion coefficients are usually considerably smaller. For diffusion in liquids $D \sim 10^{-5}$ cm/sec, so that at slow growth rates, $v \sim 10^{-3}$, the diffusion length is 10^{-2} cm; the species rejected by the growing crystal accumulate in a thin boundary layer adjacent to the crystal. Under conditions where the average growth rate is controlled by heat flow, instabilities can occur due to mass diffusion which produce segregation on the scale of the diffusion length, due to cellular or dendritic growth. This effect will be discussed in more detail below. For rapid growth the compositional segregation occurs on an even finer scale. For example, during rapid dendritic growth, at, say, 100 cm/sec, the diffusion length is $\sim 10^{-7}$ cm, and the boundary layer is so thin that the validity of the macroscopic diffusion equation is questionable. The result is indistinguishable from a trapping process, where the composition of the rapidly growing crystal is the same as the composition of the melt. For growth from solution, at the other extreme, the diffusion distance can be several centimeters, because of the slow growth rate. The bath will then have uniform composition solely as a result of diffusion. In the more usual case the diffusion distance for solution growth is the order of a few millimeters, so that the bath needs to be stirred in order to keep its composition uniform. In vapor-phase growth the concept of diffusion does not apply to molecular beam methods, but does apply to growth from the gas phase at moderate pressures, where the motion of atoms in the gas phase is diffusive. The diffusion coefficients in this case are small, $\sim 10^{-5}$, but the growth rates are also slow, so that the diffusion distance is usually fairly long. For rapid growth the diffusion distance is a fraction of a millimeter and instability effects associated with boundary layers have been observed.[1] In hydrothermal growth, where the solvent phase is at high pressure, diffusion distances of the order of a few millimeters are typical.

Diffusion in the crystalline phase can also be important. Thermal diffusivities in crystals are generally similar to those in the melt of the same material, so that thermal diffusion distances are long except for very rapid growth. Compositional diffusion in the crystal phase is much slower: $D \sim 10^{-10}$ for substitutional impurities. Only composition striations with a very short period, which have been grown in by irregular growth, will be smoothed out by subsequent annealing.

These examples serve to illustrate the point that diffusion can produce qualitatively different effects under various conditions of growth. Slow growth with rapid diffusion produces macroscopic effects, with heat loss from the container to its surrounding being important. Fast growth with slow diffusion results in atomic-scale segregation and trapping at the crystal growth surface. The most interesting effects occur at neither of these extremes, but where the diffusion length is smaller than the dimensions of the crystal, and so diffusion has a complex influence on the growth process.

2.2. Convection

Convection often plays an important role during the crystal growth process. Convection arises from three principle sources: (1) density differences arising from temperature gradients, (2) density differences arising from composition gradients, (3) convection caused by rotation of the crystal and/or crucible (as in Czochralski or floating-zone growth). Convection resulting from the first two of these is gravity driven.

The convective mass flow of the liquid usually has some irregular character, even if the flow is lamellar. If the flow is turbulent, the irregularities are greater. The convection tends to reduce composition or temperature gradients in the system, and can result in mass flows over distances which are long compared to those obtained by diffusion.

The convective pattern depends on the shape of the container and on the existing density gradients. Horizontal gradients readily produce convective flow, but a vertical gradient without horizontal gradients must be great enough to exceed the critical Raleigh number to produce convection. Often there are cells set up in the convecting liquid, with little intermixing between the cells. There is some irregularity produced by irregular motion within the cells, and major irregularities are produced as the cell pattern shifts due to changing geometry as growth proceeds. In horizontal boats, temperature fluctuations due to convection driven by the horizontal gradients are common.[2] At low temperature gradients the flow pattern produces periodic temperature fluctuations which can resemble the complex temperature fluctuations associated with turbulent flow. Gravity-driven convection can be controlled in some cases by orienting the temperature gradient with respect to the gravitation field so that the dense fluid is at the bottom (i.e., growing upward). This is the usual geometry in Bridgman growth. Gravity-driven convection in this case is eliminated if the species rejected into the melt have a greater density than the melt far from the interface. However, if the rejected species are less dense, then temperature-induced density gradients may be overcome by the composition-induced density gradients.

For a conducting melt, convection can be reduced and often eliminated entirely by a rather weak magnetic field.[3]

The principal effect of convection is to produce a fluctuating temperature at the interface, which results in an irregular growth rate. It is not unusual to have alternating growth and remelting, with the average growth rate being the net difference between the two.[4] The fluctuating growth rate causes corresponding fluctuations in the composition of the solid due to the presence of the compositional boundary layer at the interface. The presence of these composition fluctuations is well known in as-grown crystals and manifests itself as swirls or banding.

The effect of rotation of the crystal or the crucible during Czochralski growth is to cause the liquid in contact with the rotating surfaces to rotate, but also to flow radially outward due to centrifugal force. If both the crucible and crystal rotate at the same rate, the liquid can rotate at a constant rate. The flow is outward at both the top under the crystal and at the bottom of the container. In the top half of the liquid there is flow down the walls and up the center. In the bottom half there is flow up the walls and down the center. There is little fluid motion at the center of the crystal. Changing the relative rotation speeds results in a more complex pattern of convection cells. Part of the melt rotates with the crucible and part rotates with the crystal, producing cells with individual flow patterns. These are known as Taylor–Proudman cells.[5] There is very little mixing between the fluid in the various cells. Counterrotation of the crystal and the crucible produces even more complicated flow patterns.

It is clear from recent studies of convection[5] that convection patterns are complex and that detailed study of the particular geometry and thermal conditions is necessary to optimize the homogeneity and uniformity of the growing crystals.

3. Interface Morphology

The morphology of the solid–liquid interface depends primarily on diffusion if the intrinsic growth kinetics are sufficiently rapid, as is the case for metals and other materials having low entropy of fusion.[6] Figure 1 shows a planar interface in a low-entropy-of-fusion material (CBr_4). The interface is parallel to an isotherm. A small amount of impurity causes the interface to become cellular, as in Figure 2. This transition from planar to cellular growth will be discussed in more detail below. This phenomenon has been observed in many metal systems growing from the melt, and also during vapor growth. With several percent impurity, dendritic growth will normally occur

Chapter 5

Fig. 1. Planar interface in carbon tetrabromide ($L/kT_E = 0.8$) growing from the melt between glass cover slides approximately 25 μm apart. Photograph taken using a temperature gradient microscope state (110×).

(Figure 3). These three growth forms—planar, cellular, and dendritic—are found universally in the low-entropy transformations, where diffusion processes dominate the growth morphology.

In the cellular and dendritic growth forms the impurity or heat diffuses both forward and laterally away from the tip and collects in the interdendritic spaces. Growth continues until the interdendritic liquid is no longer undercooled. The dendrite growth morphology subdivides the liquid so that it is all within the diffusion distance (D/v) of an interface. The lateral composition or thermal gradients rapidly disappear from the interdendritic or intercellular liquid. The final stage of growth depends on the rate at which heat is removed from the system. For thermal dendrites further growth occurs as the latent heat is removed. For the impurity dendrites further growth requires progressive lowering of the temperature, with the interdendritic liquid remaining near the liquidus composition for the local temperature. The early stages of dendritic growth (the growth near

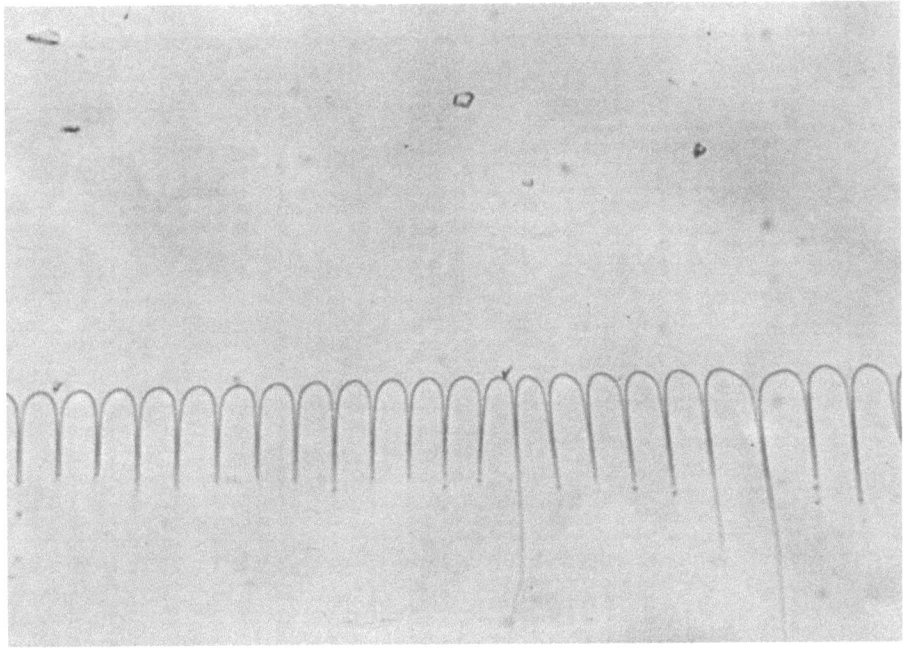

Fig. 2. Same conditions as Fig. 1. Cellular growth in carbon tetrabromide containing a small amount of impurity (110×).

the dendrite tip) occur with large thermal or concentration gradients in the liquid as the heat or second components are distributed laterally into the undercooled liquid. The later stages of growth depend on the structure inherited from the initial growth but proceed as heat is evacuated at much slower rates.

3.1. Constitutional Supercooling

As discussed previously, there is a compositional boundary layer in the liquid, so that the freezing temperature of the liquid increases away from the interface, as shown in Figure 4. The liquid at the interface is at or near the local liquidus temperature. The liquid far from the interface is usually at a temperature above the liquidus temperature. By superimposing the temperature and liquidus temperature (Figure 4), it can be seen that there can be a region near the interface which is supercooled, since the liquid is at a temperature

Chapter 5

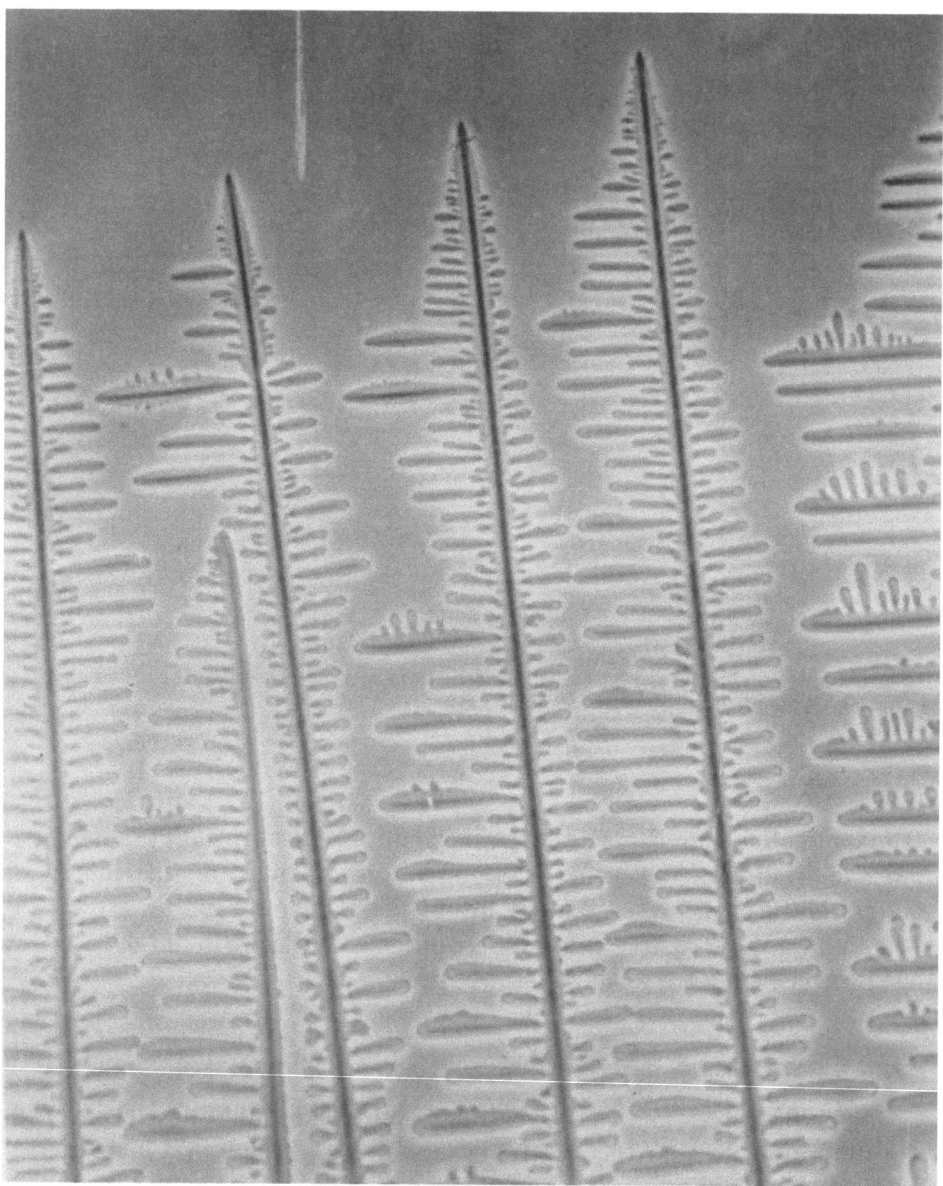

Fig. 3. Same conditions as Fig. 1. Dendrite growth in carbon tetrabromide with several percent impurity (110×).

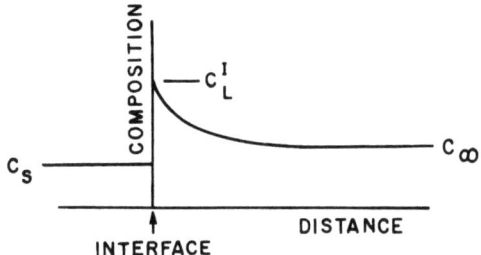

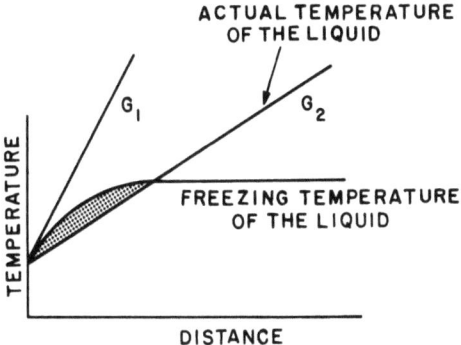

Fig. 4. Top: Composition profile ahead of a moving interface. Bottom: Constitutional supercooling due to the buildup of impurity at the interface. Constitutional supercooling exists in the shaded region for temperature gradient G_2, but none exists for temperature gradient G_1.

below its liquidus temperature. If the temperature increases rapidly into the liquid, then the temperature line will not cross the liquidus temperature line, and there will be no constitutional supercooling. The flatter the temperature gradient, the further the undercooled region extends into the liquid.

The condition for constitutional supercooling to exist is[7]

$$G \leq \frac{mvC_\infty}{D}\frac{1-k}{k} \qquad (3)$$

where G is the temperature gradient, m is the slope of the liquidus line on the phase diagram, k is the distribution coefficient, D is the diffusivity, and C_∞ is the composition in the liquid far from the interface. The equation for constitutional supercooling can be rewritten as[6]

$$G \leq \Delta T_f/(D/v) \qquad (4)$$

Chapter 5

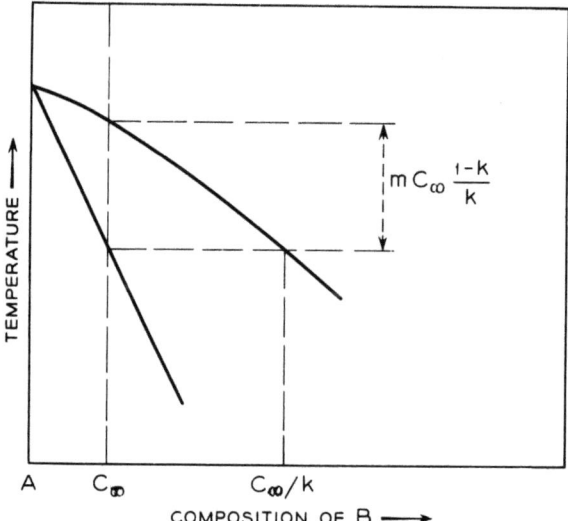

Fig. 5. Portion of a phase diagram. The freezing range ΔT_f of a mixture of composition C_∞ is given by $\Delta T_f = mC_\infty(1-k)/k$.

where $\Delta T_f = mC_\infty(1-k)/k$ is the temperature difference between the solidus and liquidus lines on the phase diagram at composition C_∞, as illustrated in Figure 5. This temperature interval, divided by the diffusion length, D/v, defines the critical temperature gradient which must be exceeded if constitutional supercooling is to be eliminated.

It has been demonstrated experimentally[8] that the occurrence or nonoccurrence of cellular growth in metals can be predicted from Eq. (3).

The condition for constitutional supercooling was derived for the case of diffusion and no convection. A similar condition can be derived for the case where there is both convection and diffusion in the melt. The flux of solute away from the interface is a boundary condition on the diffusion problem: and diffusion is the only mechanism for mixing right at the interface. Convection changes the composition at the interface by changing the far-field composition, but the gradient at the interface must satisfy the diffusion boundary condition [Eq. (2)].

The amount of solute rejected by the crystal can be stated in terms of the composition of the solid, regardless of the convection occurring in the liquid.[9] Equation (2) can therefore be written as

$$\left(\frac{dc}{dx}\right)_I = \frac{1-k}{k}\frac{V}{D}C_s \qquad (5)$$

And so the constitutional supercooling condition becomes

$$G \leq \frac{mV}{D}C_s\frac{1-k}{k} \qquad (6)$$

This equation is similar to Eq. (3) for the case where there is only diffusion causing mixing in the liquid. (For diffusion only, $C_s = C_\infty$.) Equation (6) is, however, quite general, since the composition of the solid C_s can vary in any arbitrary way as the crystal grows. C_s can be determined by making measurements on the solid after solidification. The constitutional supercooling equation [Eq. (6)] has been verified experimentally on gallium-doped germanium by Bardsley et al.[10]

Fig. 6. Hydrothermally grown quartz crystal showing several faceted slow-growing faces, but instability on the fastest-growing face.

Chapter 5

The scale of the substructure which forms because of the constitutional supercooling is usually of the order of the diffusion distance D/v. For metals this is typically 0.1 mm or so. The onset of constitutional supercooling typically occurs with one part in 10^4 or so impurity in metals. In semiconductors the small distribution coefficients ($k \sim 0.001$) can result in constitutional supercooling at even smaller impurity contents.

There are many cases, however, where supercooling of the liquid or solution exists, but where the interface does not become cellular. For example, Figure 6 shows quartz grown hydrothermally. The quartz was grown on a seed cut so that the fastest growing direction is parallel to the thin dimension of the seed. Only the fastest growing face exhibits a cellular structure; the slower growing faces do not. In this case the formation of cells depends on the crystallographic direction. It is clear that although the constitutional supercooling condition correctly predicts the onset of cellular growth in metals, it does not correctly predict it for all forms of crystal growth. For materials which grow with facets bounding them the crystal faces are stable despite the fact that the crystals are growing into a supercooled medium. The observations on quartz indicate that the fast-growing faces of a crystal can become unstable, while with the same growth conditions the slow-growing faces remain faceted.

3.2. Interface Stability

The stability of a planar interface in the presence of constitutional supercooling can be analyzed using stability theory. The constitutional supercooling condition is similar to the condition for the growth of instabilities on a plane interface if interface kinetics can be ignored. These instabilities grow to form the cellular substructure (Figure 2) which has been used to verify the constitutional supercooling condition. The stability analysis can also take into account factors such as surface energy and interface kinetics which will influence the onset of instability.

The analysis of the stability of a planar interface was first done by Mullins and Sekerka[11] and we will outline their analysis. We consider a unidirectional solidifying dilute alloy. Imagine a sinusoidal perturbation of very small amplitude on the interface:

$$Z = \delta(t) \sin wx \qquad (7)$$

where Z is a coordinate fixed in the interface in the direction of growth, the x coordinate is parallel to the interface, and $\delta(t)$ is the amplitude of the perturbation, which may be changing in time. The temperature of the perturbed interface is given by

$$T_\Phi = T_E - mC_\Phi - (T_E\sigma/L)\delta w^2 \sin wx \tag{8}$$

where T_Φ and C_Φ are the local temperature and composition of the interface, respectively, m is the slope of the liquidus line, T_E is the melting point, σ is the interfacial free energy, and L is the latent heat of fusion. The growth velocity of the interface is made up of a steady-state part V, which corresponds to the translation of the interface, and a part which depends on whether the perturbation is changing in amplitude,

$$v(x) = V + \dot\delta \sin wx \tag{9}$$

The boundary condition for the removal by diffusion of the impurity at the interface is given by

$$v(x) = \frac{D}{C_\Phi(k-1)}\left(\frac{\partial c}{\partial z}\right)_\Phi \tag{10}$$

where D is the diffusion coefficient and k is the distribution coefficient. The temperature of the interface can be written as

$$T_\Phi = T_0 + a\delta \sin wx \tag{11}$$

where a will be chosen to satisfy the thermal diffusion equation.

We will look for a solution which reduces to

$$C_\Phi = C_0 + b\delta \sin wx \tag{12}$$

on the interface and which satisfies the diffusion equation

$$D\nabla^2 C + V(\partial c/\partial z) = 0 \tag{13}$$

and the boundary condition Eq. (10). The appropriate solution to the diffusion field is given by

$$C(x, z) = C_0 + \frac{G_c D}{V}\left[1 - \exp\left(\frac{-vz}{D}\right)\right]$$
$$+ \delta(b - G_c)(\sin wx)\exp(-w^*z) \tag{14}$$

which satisfies Eq. (14) and reduces to Eq. (13) for $Z = \Phi$, where

$$G_c = (V/D)C_0(k - 1) \tag{15}$$

Chapter 5

and

$$w^* = (V/2D)[(V/2D)^2 + w^2]^{1/2} \qquad (16)$$

It is convenient to introduce reduced temperature gradients

$$\mathcal{G} = (K_L/K)G, \qquad \mathcal{G}' = (K_s/K)G' \qquad (17)$$

where K_s and K_L are the thermal diffusivities in the solid and liquid, respectively, and G' and G are the corresponding thermal gradients in the unperturbed system.

The composition gradient can be calculated from Eq. (14) and inserted into the boundary condition Eq. (10). The parameter b in Eq. (14) can be obtained from the continuity condition, Eq. (8). Collecting the terms which are linear in δ gives

$$\frac{\dot{\delta}}{\delta} = vw\left[-\frac{T_E\sigma}{L}w^2 - \frac{\mathcal{G}' + \mathcal{G}}{2} - mG_c\frac{w^* - (v/D)}{w^* - (1-k)v/D}\right]$$
$$\times \left[(\mathcal{G}' - \mathcal{G})\left(2 - \frac{wmG_c}{w^* - v(1-k)/D}\right)\right]^{-1} \qquad (18)$$

The denominator of this equation is positive, so that the perturbation grows or shrinks depending on the sign of the numerator. It turns out that for most experimental conditions the surface free-energy term is unimportant. The principal destabilizing term is due to solute diffusion and the principal stabilizing effect is heat flow. The condition for stability is approximately

$$-mG_c < \mathcal{G} \qquad (19)$$

which in view of the definitions of G_c and $\mathcal{G}$ [Eqs. (16) and (12)] is very similar to the condition for constitutional supercooling. The interface becomes unstable under almost the same conditions for the onset of constitutional supercooling.

During melting a similar analysis can be applied. The corresponding condition for constitutional superheating of the solid [cf. Eq. (4)] is

$$G \leq \Delta T_f/(D_s/v) \qquad (20)$$

which contains the diffusion coefficient in the solid rather than in the liquid.

Diffusion in the solid is so slow that it is almost impossible to eliminate constitutional superheating during melting, because the critical temperature gradient is so large. However, the instability condition involves the liquid diffusion coefficient, because instabilities in the interface involve both phases. The condition for instability on melting is similar to that on freezing (except for a factor k multiplying the composition). Chen and Jackson[12] have demonstrated experimentally that the melting instability is correctly predicted by instability theory. This means that the solid can be significantly superheated without the onset of interface instability.

These experiments demonstrate that the stability theory correctly predicts the onset of instability, whereas the superheating or supercooling condition does not, unless the two criteria happen to coincide, as they do for the freezing case.

The analysis which has been presented above does not take into account the growth kinetics. Equation (8) assumes that the interface temperature is at the local equilibrium temperature. The isotropic growth rate kinetics can be added to the analysis[13] by adding a kinetic undercooling term

$$\Delta T_k = \beta v = \beta V + \beta \delta \sin wx \qquad (21)$$

to Eq. (8) to give

$$T_\Phi = T_E - mC_\Phi - (T_E \sigma / L)\delta w^2 \sin wx - \beta V - \beta \delta \sin wx \qquad (22)$$

and Eq. (18) then becomes

$$\frac{\dot\delta}{\delta} = vw \left[-\frac{T_E \sigma}{L} w^2 - \frac{\mathscr{G}' + \mathscr{G}}{2} - mG_c \frac{w^* - v/D}{w^* - (1-k)v/D} \right]$$
$$\times \left[(\mathscr{G}' - \mathscr{G})\left(2 - \frac{wmG_c}{w - v(1-k)/D}\right) + \beta wv \right]^{-1} \qquad (23)$$

In this expression the numerator is unchanged from Eq. (18) and both terms in the denominator are positive, so that the stability condition is not affected by the addition of an isotropic kinetic term. The kinetic term does affect the rate of growth of the perturbation, but not whether the perturbation will grow or decay. Let us examine its consequences in two different crystal growth geometries. In the first case let us consider linear growth such as in the Bridgman method where the growth rate and temperature gradient are controlled by the experimenter. In this case the isotropic kinetic term simply means that the interface temperature is lower than it would be with fast kinetics. The interface will be in a cooler part of the apparatus. The gradients may also be affected by this shift but the change should not

Chapter 5

be large. In this case the onset of instability in terms of the measured growth rate and the measured gradient will be identical with or without the isotropic kinetic factor. The second case to consider is a crystal growing in an isothermal bath. In this case the bath temperature is specified by the experimenter but the growth rate and temperature gradient are not. The large kinetic term will mean that the interface temperature is well below the equilibrium temperature and that the growth rate is slow and is limited by the intrinsic kinetics. The diffusion distance will be large and the diffusion boundary layer will be thick. The onset of instability will therefore be much different with a large kinetic coefficient than with a small kinetic coefficient, and we would expect that if the kinetic coefficient were large enough, the instability would not occur.

3.3. Anisotropic Growth Kinetics

When the growth kinetics of a crystal are strongly dependent on orientation the anisotropy will influence the instability. For some materials, such as the metals growing from the melt, the anisotropy in the growth rate is small, but the kinetic term is small in this case. For most cases where the kinetic coefficient is a significant factor in the growth process the growth rate is strongly anisotropic. The stability analysis for this case has recently been treated by Chernov.[14] The anisotropic part of the growth kinetics tends to make the valleys fill in by lateral growth rather than by the growth or reduction of the amplitude of the sine perturbation. In terms of the formal analysis this lateral growth corresponds to the generation of harmonic terms, whereas the analysis of Mullins and Sekerka[11] considers the only response to be a change in amplitude of the various wavelengths in the original perturbation. Because the perturbation produces harmonics, two different wavelengths will not evolve similarly in time. For purposes of analysis the evolution of an interface perturbation of some general form cannot be Fourier-decomposed. The growth of the perturbation always depends on its initial shape. The Chernov analysis assumes an initial perturbation and follows its evolution in time. The evolving shape depends both on the height and the width of the initial perturbation. Fortunately, the stability criterion while it does depend on these details, does not seem to be sensitive to them, so that a general statement can be made about the onset of instability: Although a slowly growing face is more stable due to the anisotropic growth rate, the stability criterion with anisotropic growth kinetics is similar to that with isotropic growth kinetics [Eq. (22)] for the particular growth direction.

The theory of instability has been cast[15] into a form where the time evolution of perturbations of the interface can be followed. This has been done using a Green's function method. The perturbation of the interface as a function of position and time can be written as

$$Z = 2\int_{-\infty}^{\infty} \Phi_0(x')G(x - x';t)\,dx' \qquad (24a)$$

where

$$G(x;t) = \exp(t/\tau)[\cos(2\pi x/\lambda_0)][\exp(-x^2/4\mathscr{D}t)]/(4\pi\mathscr{D}t)^{1/2} \qquad (24b)$$

is a Green's function which describes the time evolution of Φ_0; τ, λ_0, and $\mathscr{D}$ are the characteristic time, length, and diffusion coefficient associated with the problem. The time evolution of a perturbation is illustrated schematically in Figure 7.

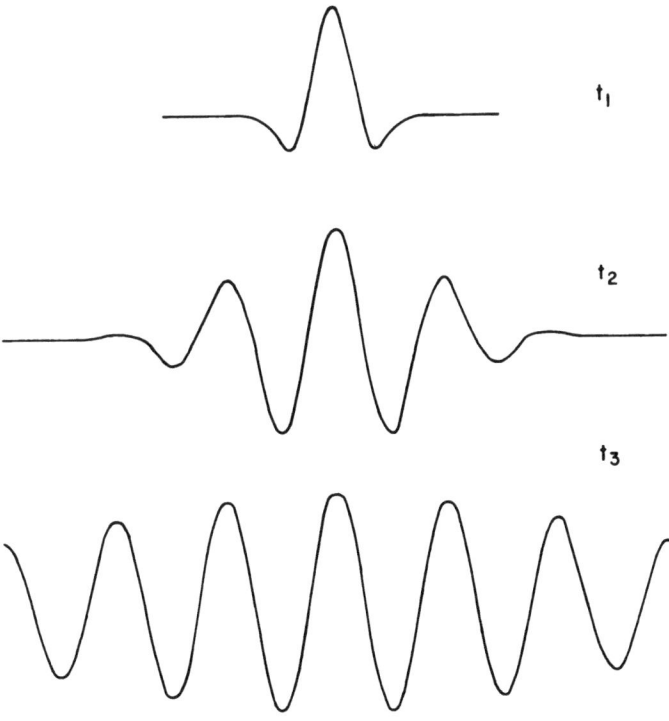

Fig. 7. The time evolution of an interface perturbation.

Chapter 5

Detailed confirmation of several aspects of the time-dependent stability theory has been made by Hardy and Coriell[16] in experiments on cylinders of ice in undercooled water. The onset and evolution of the instability agree in detail with the predictions of the theory for several different impurities, although the results for some impurities are not understood.

3.4. Banding

Banding is the term used to describe composition variations in a crystal which are parallel to its growth front. These bands are variations in composition which are usually produced by fluctuations in the growth rate of the crystal, which can be due to convection, the rotation of the crystal, the operation of a temperature controller, etc. Bands can take the form of swirl patterns in Czochralski or float-zone crystals. Banding is due to the presence of the diffusion boundary layer at the interface during crystal growth. Under given steady-state growth conditions there is a constant amount of solute in the boundary layer at the crystal interface. If the growth rate of the interface changes for any reason, then the steady-state boundary layer contains a different amount of impurity. This adjustment is made by incorporating more or less of the impurity into the crystal, which appears as a band of composition which is different than the average composition. Using the diffusion-limited distribution of impurity in the liquid as given by

$$C = [(1 - k_0)/k_0]C_\infty e^{-v/Dx} + C_\infty \tag{25}$$

the amount of impurity in the boundary layer at the interface is given by

$$\int_0^\infty (C - C_\infty)\, dx = \frac{D}{v} \frac{1 - k_0}{k_0} C_\infty \tag{26}$$

A sudden change in growth rate from v_1 to v_2 means that an amount of solute given by

$$D \frac{1 - k_0}{k_0} C_\infty \left(\frac{1}{v_1} - \frac{1}{v_2} \right) \tag{27}$$

will appear as a band in the crystal. The dynamics of this process has been described in detail by Smith et al.[17] Any variation of growth rate of a crystal will thus produce variations in composition parallel to the growth front.

Banding determines the distribution of residual impurities during the growth of high-purity single-component semiconductors. The effect is particularly pronounced during growth of intermediate phases which are not of a congruently melting composition as, for example, during the growth of a solid solution phase. Here the banding results in composition variations of the major constituents. Many compounds do not melt congruently at the stoichiometric composition. These compounds must be grown from noncongruently melting compositions in order to obtain a stoichiometric solid. The banding in this case results in departures from stoichiometry.

The effects of banding are minimized for congruently melting compositions, where there is no boundary layer due to redistribution of the major constituents. For congruent melting, the distribution of minor impurities will be determined by banding, just as for a single-component primary phase.

4. Heat Flow

Considerable effort has been expended on solutions to macroscopic heat flow problems such as those encountered in castings. The problems are known as "moving source" problems. For the pure material the interface temperature is taken to be the melting point, and the heat of fusion is assumed to evolve at the interface between solid and liquid. For other than steady-state conditions the solution of such problems is not trivial. Alloy solidification is even more complex. Alloys are usually treated by ignoring the details of dendritic growth, and making the simplifying assumption that the liquid freezes over a temperature range as given by the phase diagram. The fraction of solid at any temperature is obtained from the lever rule: The latent heat of fusion is assumed to evolve over the temperature interval between the liquidus and solidus, with the amount being proportional to the volume fraction frozen. The model agrees well with experimental measurements,[18] and provides a computational scheme for predicting the course of solidification in complex castings.

5. Diffusion-Limited Growth

The diffusion-limited growth of crystals into supercooled melts has received considerable attention. The most significant work in this area is the classical work of Ivantsov,[19] who developed a mathematical method for treating these problems. Ivantsov showed

Chapter 5

that a paraboloid is the appropriate shape for a growing dendrite, and also treated the problem of the temperature or composition distribution around a polyhedral crystal, an important problem which has application not only to solution growth, but also to the formation of facets during Czochralski growth of semiconductors.

The formation of facets during growth results from anisotropy in the growth rate. If the crystal is growing into a supersaturated solution, then the isoconcentrates around the crystal tend to be spherical because diffusion in the liquid phase is isotropic. This means that the concentration gradients will be largest at the corners of the crystal and the supersaturation will be least at the center of the faces. This effect leads to the well-known "hopper" morphology of bismuth crystals growing from the melt (Figure 12), and to hollow cylinders of ice growing from the vapor phase. In these cases the supercooling or supersaturation at the center of the face has become so low that the crystal stopped growing there. The same phenomenon has been observed in the growth of silicon, where concentration of impurities at the center of a facet has resulted in a depression forming at its center.

5.1. Dendritic Growth

The treatment of the parabolic dendrite tip by Ivantsov[19] assumed that the surface was isothermal. Horvay and Cahn[20] extended this treatment to paraboloids having elliptical cross sections. It was realized by Temkin[21] that the surface of the dendrite could not be isothermal. He added the effect of surface tension due to curvature of the dendrite tip and the effect of finite molecular attachment kinetics. Several recent treatments[22] have modified and refined the analysis of dendritic growth.

A basic problem in dendritic growth, as with other unconstrained growth forms, is that a valid mathematical solution to the heat flow problem may not be a stable solution.

The solution presented by Ivantsov[19] contained a relationship between the growth rate of the dendrite and the bath temperature given by

$$C \Delta T/L = -\tfrac{1}{2} P e^{P/2} E_i(-\tfrac{1}{2}P) \qquad (28)$$

where E_i is the exponential integral function, C is the specific heat, L is the latent heat, ΔT is the difference between the interface temperature and the bath temperature, and P is the Peclet number (dimensionless) given by $P = v\rho/K$, where v is the growth velocity of the dendrite tip, ρ is the radius of curvature of the paraboloidal tip, and K is the thermal diffusivity.

Only one value of P is allowed for a particular temperature. But P does not uniquely define the growth velocity, only the product ρv. Experimentally, a unique growth rate is observed for a particular undercooling, so the growing crystal selects one combination of v and ρ for a particular set of growth conditions. A growing dendrite tip is shown in Figure 8. The instabilities which lead to dendrite branching can be seen just back from the tip. In this photograph it does not appear that the instability occurs close enough to the tip to influence its shape. In other dendrites, however the shape and the growth rates vary cyclically. The tip broadens, slows, then splits; the two branches grow rapidly, then slow, broaden, and split again. The average growth rate and tip radius are clearly dependent on instabilities in such a case. For the dendrite shown in Figure 8 the instabilities may also be playing a similar, though more subtle role. It is also possible that the slightly anisotropic growth rate stabilizes the tip shape in the fast-growth direction and that instabilities occur away from the tip region because the stabilizing influence is absent further from the tip. A satisfactory mechanistic treatment of the dendrite tip has not been done yet, taking into account the anisotropic nature of the tip. It is of course clear that the anisotropy, though small, is important, since dendrites, when growing freely, always grow in crystallographic directions.

Temkin[21] circumvented the problem by noting that if the dendrite tip were too small, it would slow down because of its surface tension, and it would slow down because of heat flow if it were too large. He therefore suggested that the dendrite should grow at the maximum rate represented by a compromise between these extremes. He found a relationship between the growth rate v and the undercooling ΔT that can be written as

$$v = A\, \Delta T^n \qquad (29)$$

where A is a constant and n is an exponent which lies between 1.2 and 2.4 depending on the relative importance of surface tension and atomic attachment kinetics. Most of the experimental data on dendritic growth can be fitted to Eq. (29) with reasonable values for A and n.

Refinements[22] of this treatment changed the way in which surface tension and the intrinsic growth kinetics are introduced into

Chapter 5

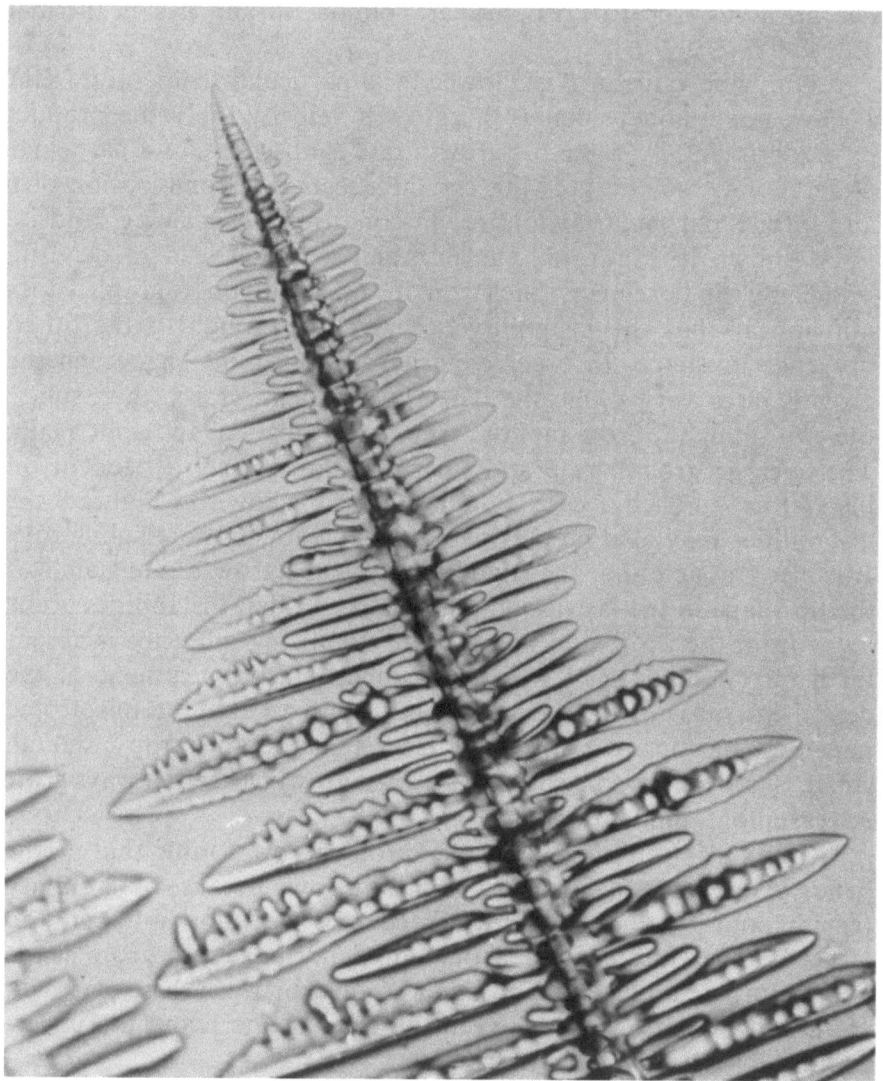

Fig. 8. Dendritic growth in succinonitrile (110 ×).

the problem, and are also in agreement with the available data. Recent work[23] uses the parabolic dendrite tip as a first approximation, and the changes in shape due to the dendrite surface not being isothermal are treated as corrections. This analysis is a rigorous treatment of the nonisothermal dendrite and gives very good agreement for the data on ice.

6. Intrinsic Growth Kinetics

The classical theory for the intrinsic growth rate of a crystal from its melt was developed by Wilson[24] and independently by Frenkel.[25] The basic premise of the theory is that crystal growth is a reversible process in which atoms are both leaving and joining the crystal.[26] Above the equilibrium temperature the leaving process predominates, so the crystal shrinks; below the equilibrium temperature the joining process predominates, so the crystal grows. The net rate of growth is given by

$$v = R_A - R_D \qquad (30)$$

where R_A is the rate of arrival of molecules at the crystal surface and R_D is the rate of departure. We assume that R_A and R_D are simply activated processes,

$$R_A = R_A^0 \exp(-Q_A/kT), \qquad R_D = R_D^0 \exp(-Q_D/kT)$$

We note that $Q_D - Q_A = L$, the latent heat of fusion. Since $R_A = R_D$ at equilibrium ($T = T_E$), then

$$R_A^0/R_D^0 = \exp(-L/kT_E) \qquad (31)$$

So that

$$v = R_A^0[\exp(-Q_A/kT)][1 - \exp(-L\,\Delta T/kT_E T)] \qquad (32)$$

where $\Delta T = T_E - T$. For small ΔT (i.e., $L\,\Delta T/kT_E T \ll 1$)

$$v = R_A^0(L\,\Delta T/kT_E T)\exp(-Q_A/kT) \qquad (33)$$

For numerical calculations $R_L^0 \approx av$ (the interatomic distance times the atomic vibration frequency) is normally used, so that $R_A^0 \approx av \exp(-L/kT_E)$. In the Wilson[24] theory Q_A was associated with the activation energy for diffusion in the liquid phase, whereas Frenkel[25] associated Q_A with the temperature dependence of viscosity. Both are, in fact, usually rather similar, although neither of these activation energies, which are bulk properties, necessarily correlates precisely with jump probabilities at an interface.

Equations (32) and (33) incorporate two of the major factors determining the growth rate. The growth rate depends on the temperature difference, or, equivalently, on the chemical potential difference between the two phases, with the dependence being linear at small undercooling. The growth rate also depends on the mobility of the molecules in the liquid, through the factor $\exp(-Q_A/kT)$. Thus the growth rate for a viscous glass-forming material is much smaller

Chapter 5

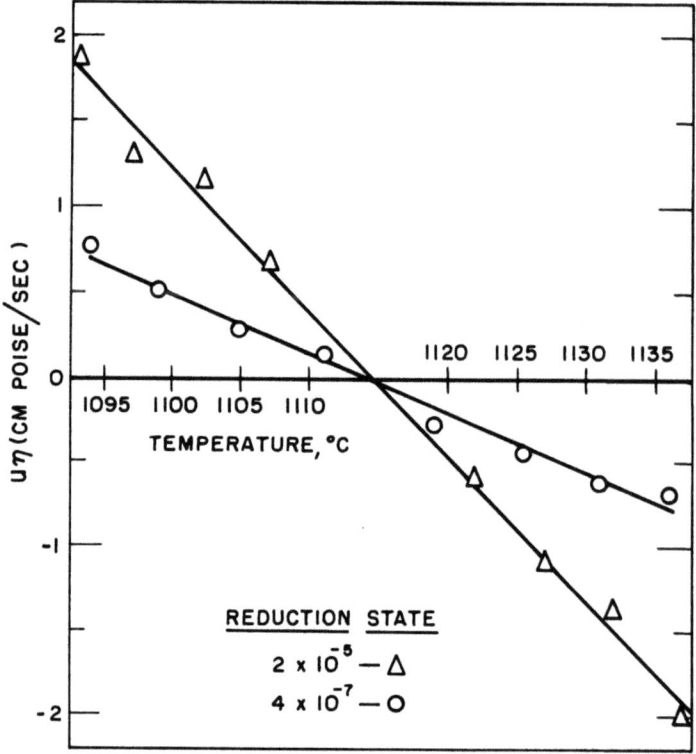

Fig. 9. Reduced growth rate vs. undercooling for GeO_2 ($L/kT_E = 1.3$).

than that of a melt which is highly fluid at the growth temperature, for equivalent undercooling.

The applicability of this equation is demonstrated (for example) in Figure 9. The reduced growth rate of GeO_2 is linear with reduced undercooling through the melting point. In this temperature range the viscosity changes by several orders of magnitude, but its temperature dependence is seen to be the same as the temperature dependence of the growth rate. In this instance the liquid is tetrahedrally bonded, so that the bond shifts required for viscous flow are likely to be the same as those for crystal growth. The temperature dependence of these two processes is thus more likely to be similar than for a "close-packed" melt.

However, the theory of Wilson and Frenkel takes no account of the structure of the interface. It implicitly assumes that all the sites on the crystal surface are equivalent, and are active growth sites. In general this is not the case. The Wilson–Frenkel growth law can be generalized[27] by multiplying by a factor f which is the fraction of surface sites which are growth sites:

$$v = av[\exp(-L/kT_E)][\exp(-Q_A/kT)] \\ \times [1 - \exp(-L\,\Delta T/kT_ET)]f \qquad (34)$$

In general f will be a function of temperature. There are two traditional models of growth for which f can be estimated.[27] Both assume that growth occurs only at steps on a surface and that f is given by the step density. One of the models assumes that the steps are created by surface nucleation, while the other model assumes that steps are created by screw dislocations.

In the surface nucleation model the number of critical nuclei on the surface is given by

$$N_n^* \approx N \exp(-g\sigma^2 T_E/kTL\,\Delta T) \qquad (35)$$

where σ is the edge free energy of the disk-shaped nucleus and g is a geometric factor which depends on the shape of the nucleus. The average distance between disks is approximately $1/(N_n^*)^{1/2}$, so that the total step density is of the order of $(N_n^*)^{1/2}$. For surface nucleation, then,

$$f_{SN} = N^{1/2} \exp(-g\sigma^2 T_E/kTL\,\Delta T) \qquad (36)$$

The ΔT in the denominator of the exponent dominates the temperature dependence of f_{SN}. For small ΔT the growth rate is very small. At some critical undercooling ΔT^*, f_{SN} increases abruptly. There are several measurements of growth rate which exhibit the functional form of Eq. (36), but neither the preexponential factor nor the coefficient in the exponent are predicted correctly. The discrepancy can be accounted for if the steps on the surface are not geometric steps, but roughened by thermal energy. This would decrease the edge free energy of a disk, and thereby decrease the critical undercooling for growth (a decrease of a factor of ten in σ is necessary). A theoretical model for this can be based on interface roughening, which will be discussed below.

Screw dislocations generate a spiral such that the spacing between the turns of the spiral is related to the critical nucleus radius, $(n^*)^{1/2} = g^{1/2}\sigma T_E/(L\,\Delta T)$, so that

$$f_D = L\,\Delta T/g^{1/2}\sigma T_E \qquad (37)$$

Using the approximation of Eq. (33), this gives $v \propto \Delta T^2$ for small undercooling.

Chapter 5

There are several experimental measurements of growth rate which have this functional form, but the magnitude of the growth rate is not correctly predicted. Once again the effects of step and surface roughness can account for the differences qualitatively.

6.1. Surface Roughness

The importance of surface roughness during crystal growth[27,28] can be seen by considering Figure 10, which shows a rough and a smooth edge for both a close-packed and non-close-packed edge of a two-dimensional crystal. It will be shown below that a small entropy change gives rise to rough faces, whereas a large entropy change results in smooth faces. Rough surfaces provide many sites for growth, as illustrated by the two top drawings, for both the close-packed and non-close-packed faces. The factor f should be close to one for both, and so the growth rate should be rapid and isotropic. This has already been illustrated in Figures 1–3. However, when a close-packed face is smooth (lower left) the formation of new layers is difficult, so that f for such a face is small. On the other hand, on the non-close-packed face (bottom right) the geometry of the surface provides many growth sites, even though the surface is atomically smooth. Atoms on the surface do not depend on atoms in the same layer to provide enough bonding. Nucleation of new layers is not necessary. The factor f is close to one. So a crystal which has smooth surfaces grows anisotropically, rapidly on the non-close-packed faces and slowly on the close-packed faces. For geometric reasons such a crystal will lose its fast-growing faces, and be bounded by the slowest-growing

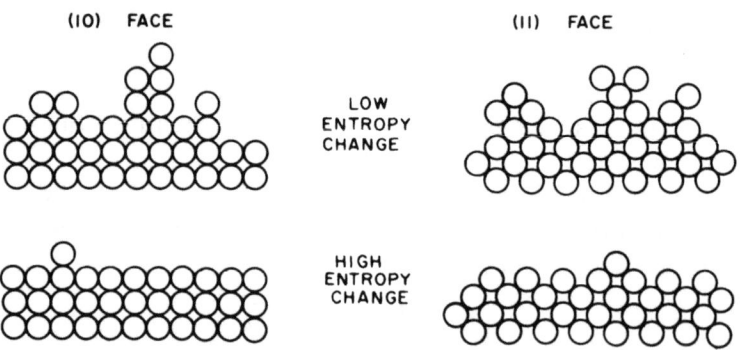

Fig. 10. The structure (rough vs. smooth) of the edge of a two-dimensional crystal for a close-packed (10) and non-close-packed (11) edge.

faces which form a closed figure, so that a slow growth rate will be observed for such crystals.

The first treatments of surface roughness as a cooperative phenomenon were based on two-dimensional one-layer models. Langmuir's[29] treatment of adsorption is one of these. This treatment included the entropy from the random distribution of molecules on the surface, but assumed no interaction energy between adsorbed molecules. Fowler and Guggenheim[30] treated the interaction energy in a simple approximation. Burton et al.[31] applied the exact solution to the two-dimensional Ising model to treat the problem exactly in this approximation. Jackson[27,28] used a model similar to that of Fowler and Guggenheim to treat the roughness of the crystal melt interface. This model, which can be treated analytically, considers the change in free energy of an initially plane interface due to the addition of molecules at random from the phase in contact with the crystal. The change in free energy ΔF_s can be written as[27]

$$\frac{\Delta F_s}{NkT_E} = -\frac{L\,\Delta T}{kT_E T}x + \alpha x(1-x) + \frac{T}{T_E}[x \ln x + (1-x)\ln(1-x)] \tag{38}$$

where N is the total number of sites on the surface, x is the fraction of sites which are occupied, T_E and T are the temperature of equilibrium and the actual temperature of the surface, respectively, L is the heat of transformation, and α is given by[27]

$$\alpha = (L/kT_E)\xi = (\Delta S/R)\xi \tag{39}$$

where $\xi = \eta_s/\eta$, the number of nearest-neighbor sites in a layer parallel to the surface, divided by the total number of nearest-neighbor sites. The factor L/kT_E is the ratio of the binding energy at the interface (taking into account the other phase) to the thermal energy, and may also be expressed as $\Delta S/R$, the entropy difference between the two phases divided by the gas constant.

The first term on the right-hand side of Eq. (38) is the change in free energy due to filling a fraction x of the surface sites: the chemical potential change per site times the number of sites filled. But these added molecules do not have the full number of nearest neighbors. The second term is the interaction energy term, which comes from the average number of neighbors in the layer, assuming the molecules are randomly distributed. The last term is the entropy of mixing,

Chapter 5

representing the decrease in free energy due to randomness in the partially filled layer.

Equation (3) is plotted in Figure 11 for various values of α, for the equilibrium temperature. [For nonequilibrium temperatures a linear term of slope $(-L\,\Delta T/kT_E T)$ is added.] The minima in Figure 11 give the equilibrium configuration of the surface. For large α the minimum free energy occurs for x small or close to one (i.e., $x = e^{-\alpha}$ and $x = 1 - e^{-\alpha}$) corresponding to a fraction $e^{-\alpha}$ of the sites filled with admolecules or a fraction $e^{-\alpha}$ vacant sites in the layer.[27] This corresponds to a smooth surface. For small α the minimum occurs at $x = 1/2$, which corresponds to a rough surface. We have already seen that the growth rate is rapid, isotropic, and, in general, diffusion limited for low entropy of fusion (Figures 1–3).[6] Figure 12

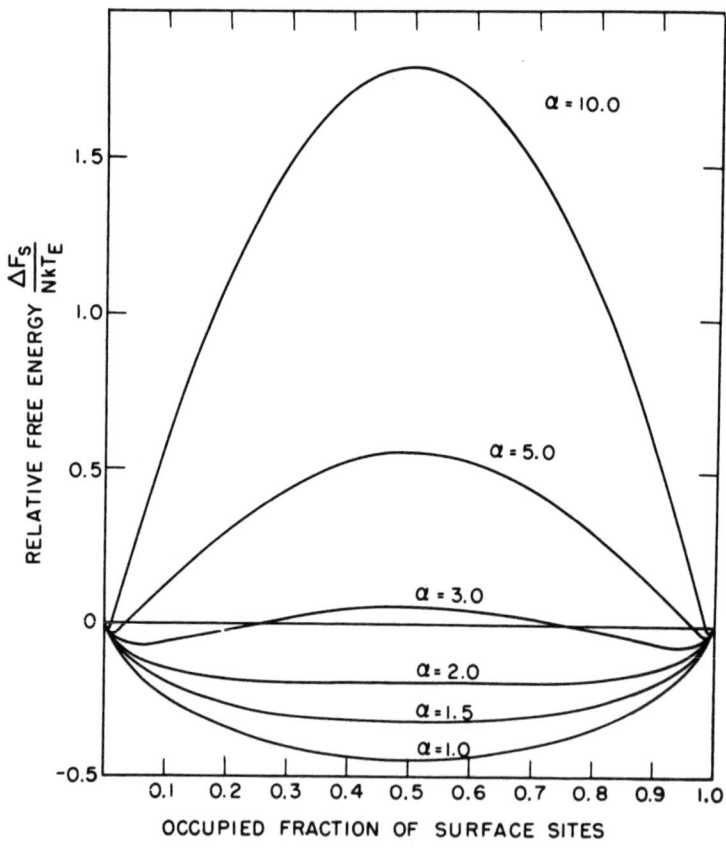

Fig. 11. Free energy of a surface as a function of coverage for various values of α.

Fig. 12. Decanted bismuth crystals, showing hopper growth (3×).

shows a decanted bismuth crystal which had been grown from the melt. Figure 13 shows a growing crystal of trichloroacetic acid, which has the same entropy of fusion as bismuth, growing from the melt. Figure 14 shows benzil, for which $\Delta S/R \approx 6$, growing from the melt. Figure 15 shows polyethylene growing from the melt. Each of these morphologies is typical for a wide variety of materials growing with similar entropy changes. The entropy of transformation thus provides a convenient classification scheme for various crystal growth morphologies.[6] Indeed, the small-entropy-of-fusion materials grow rapidly and the growth rate is isotropic. The larger the entropy

Chapter 5

Fig. 13. Trichloroacetic acid ($L/kT_E = 2.14$) growing under the same conditions as Fig. 1 (90×).

of fusion, the slower the growth rate (for the same undercooling) and the more anisotropic is the growth.

The entropy of fusion has also been used successfully to classify the morphologies observed as a result of eutectic solidification, where two solid phases grow simultaneously.[32,33]

6.2. Multilevel Interface

In the simplest model of a multilevel interface[37] atoms in each layer are assumed to be arranged randomly on top of the atoms in the layer below, as illustrated in Figure 16. The change in free energy due

Theory of Crystal Growth

Fig. 14. Benzil ($L/kT_E = 6$) growing under same conditions as Fig. 1 (90×).

to roughening of the surface is given by an equation similar to Eq. (38), but each of the terms is now summed over all the layers i parallel to the interface:

$$\frac{\Delta F_s}{NkT_E} = \frac{L\,\Delta T}{kT_E T}\left[\sum_{i=-\infty}^{0}(1-x_i) - \sum_{i=1}^{\infty} x_i\right] + \alpha \sum_{i=-\infty}^{\infty} x_i(1-x_i)$$

$$+ \frac{T}{T_E}\sum_{i=-\infty}^{\infty}(x_{i-1}-x_i)\ln(x_{i-1}-x_i) \quad (40)$$

There is no contribution from the planes far from the interface, since $x_i = 1$ in the solid and $x_i = 0$ in the liquid far from the

Chapter 5

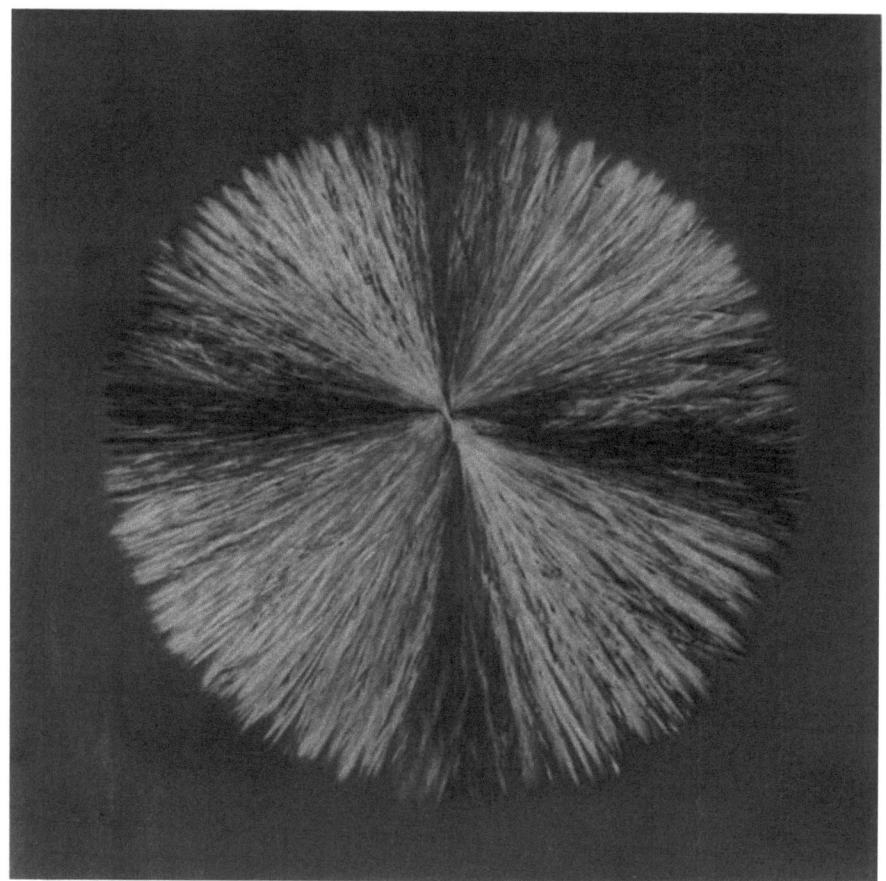

Fig. 15. Spherulitic growth of polyethylene.

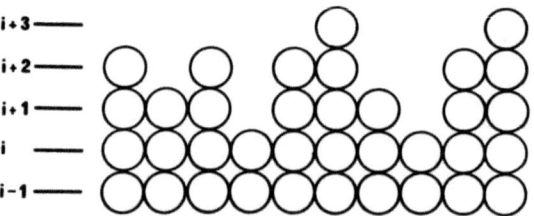

Fig. 16. Schematic drawing of a multilevel interface.

interface. The equilibrium configuration corresponds to the minimum in Eq. (40), and is given by the recursion relationship

$$\frac{x_i - x_{i+1}}{x_{i-1} - x_i} = \exp\left[\alpha \frac{T_E}{T}(2x_i - 1) + \frac{L\Delta T}{kT_E T}\right] \quad (41)$$

The values of x_{i+2} and x_{i-1} can be calculated from this equation if x_i and x_{i+1} are known, so that all the x_i can be calculated from any two, say x_0 and x_1. The values x_0 and x_1 can be chosen uniquely so that $x_{-\infty} = 1$ and $x_\infty = 0$. This can be done by successive approximations in a computer.

The interface profiles calculated[35] from Eq. (41) are shown in Figure 17. For large α the interface is smooth, with few solid atoms in layer one and few liquid atoms in layer zero. For large α the solution to Eq. (41) is $x_{-1} = 1$, $x_0 = 1 - e^{-\alpha}$, $x_1 = e^{-\alpha}$, $x_2 = 0$, in agreement with the one-level model, Eq. (38). For small α the interface is quite thick, extending over several atomic layers.

The lowest free-energy position of the interface is as shown in Figure 17, with the number of solid atoms in layer one equal to the number of liquid atoms in layer zero, i.e., $x_1 = 1 - x_0$. As the crystal grows, the mean position of the interface moves from plane to plane. A sequence of such positions is shown in Figure 18. Using Lagrange multipliers, the minimum free energy of the interface can be determined for any value of n,[35] where

$$n = N\left[\sum_{i=1}^{\infty} x_i - \sum_{i=-\infty}^{0}(1 - x_i)\right] \quad (42)$$

is the number of atoms which have been added since the mean interface was midway between layers zero and one. The minimum free energy of the interface is periodic, with the period of the spacing

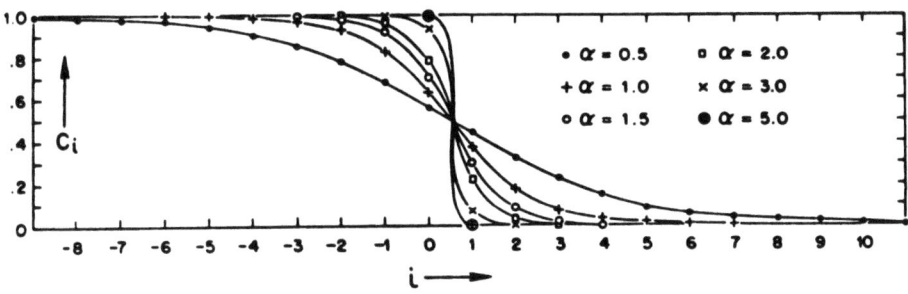

Fig. 17. Interface profiles calculated from Eq. (41).

Chapter 5

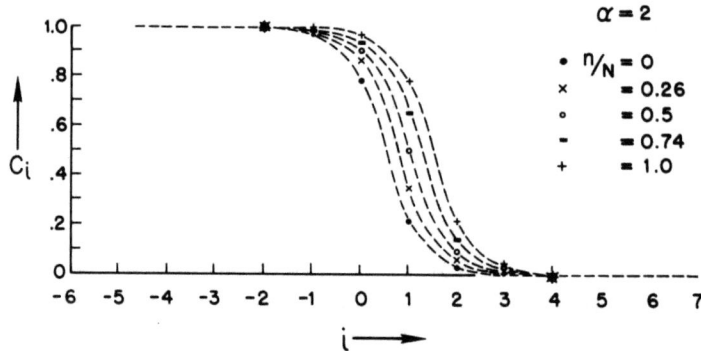

Fig. 18. Sequence of interface positions for interface motion.

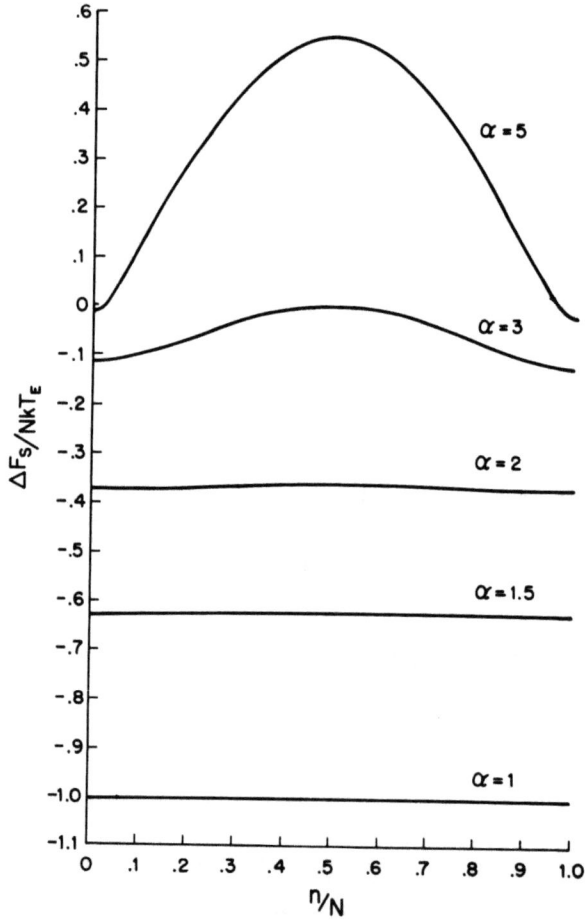

Fig. 19. Surface free energy (compared to a flat interface) as a function of position for various values of α.

between planes, that is, $\Delta F_s(n + N) = \Delta F_s(n)$. The free energy as a function of position of the interface is shown in Figure 19 for various values of α.[35] For large α there is a significant maximum in free energy, which occurs when the interface is centered on an atom plane, that is, for $x_1 = \frac{1}{2}$, for example. The maximum in free energy represents a barrier to growth. For small α the barrier is small.

6.3. The Growth Rate Equation

During growth the interface passes from plane to plane. In the usual case the configuration of the interface will be periodic, passing through an equivalent sequence of configurations each time it passes from one plane to the next. Any step in this sequence will be labeled by n, the number of atoms added since the last minimum in free energy. Consider a large number of equivalent interfaces undergoing steady-state growth at the same rate. At any instant in time the fraction of such surfaces which are labeled n is $\langle n \rangle$. The term $\langle n \rangle$ may also be thought of as the fraction of time which a given interface spends in configurations labeled n. If P_n^+ and P_n^- are the rates at which configurations labeled n gain and lose atoms, respectively, then the net rate at which configurations n pass to configurations $n + 1$ is[36]

$$R' = \langle n \rangle P_n^+ - \langle n+1 \rangle P_{n+1}^- \tag{43}$$

R' must be independent of n for $\langle n \rangle$ to be independent of time. This does not imply that $\langle n \rangle$ is independent of n, that is, it does not imply that the interface spends an equal amount of time at each n. It means that the interface is likely to spend the same amount of time in configuration n each time is passes through it. From the definition of $\langle n \rangle$

$$\sum_{n=0}^{N-1} \langle n \rangle = 1 \tag{44}$$

Since the configuration labeled 0 is equivalent to the configuration labeled N, Eqs. (43) and (44) form a closed set from which all the $\langle n \rangle$ can be eliminated to express the net growth rate of a crystal in terms of the P's. For the case where P_n^+ is independent of n we can write $p_n = P_n^-/P^+$, and the growth rate can be written[32]

$$\frac{v}{P^+} = \frac{N\left(1 - \prod_{n=0}^{N-1} p_n\right)}{N + \sum_{n=0}^{N-1} p_n + \sum_{n=0}^{N-1} p_n p_{n+1} + \cdots + \sum_{n=0}^{N-1} p_n p_{n+1} \cdots p_{n-2}} \tag{45}$$

Chapter 5

where $v = NR'$. We can write p_n as

$$p_n = \exp(-\Delta f_n/kT) \qquad (46a)$$

where Δf_n is the change in free energy on going from configuration n to $n + 1$, that is,

$$\Delta f_n = (d/dn)(\Delta F_s) \qquad (46b)$$

Thus the growth rate can be determined for the model of the interface described in the last section from the curves of ΔF_s vs. n (Figure 19), or indeed for any model for which ΔF_s vs. n is obtainable. The Δf_n can be calculated from ΔF_s, p_n calculated using Eq. (46a), then the normalized growth rate can be calculated by inserting the p_n into Eq. (45).

The simplest form of ΔF_s vs. n, namely ΔF_s independent of n, is implicitly assumed in the Wilson–Frenkel equation (32), and occurs in the multilevel model for small α (Figure 19). In this case $p_n = \exp(-\Delta f_0/kT)$, where $\Delta f_0 \approx L\,\Delta T/T_E$, the change in free energy on adding a molecule to the crystal. Equation (45) becomes

$$\frac{v}{P^+} = \frac{N(1 - e^{-N\Delta f_0/kT})}{N + Ne^{-\Delta f_0/kT} + Ne^{-2\Delta f_0/kT} + \cdots + Ne^{-(N-1)\Delta f_0/kT}}$$
$$= 1 - e^{-\Delta f_0/kT} \qquad (47)$$

Identifying $P^+ = R_A = R_A^0 \exp(-Q_A/kT)$ as in Eqs. (30)–(32), Eq. (47) is identical to Eq. (3), the Wilson–Frenkel growth rate. This result is in fact a direct consequence of assuming that the interface configuration does not change with undercooling.

Similarly, for screw dislocation growth p_n is independent of n, because of the spiral nature of the surface. This makes v/P^+ linearly dependent on undercooling (for small undercooling) as in Eq. (47) or (33). However, for the screw dislocation model it is assumed that atoms can be added only on the steps of the spiral, so that P^+ depends on the density of loops of the spiral, which is linearly proportional to undercooling. So $v \propto \Delta T^2$, as in Eq. (38). Similarly, surface nucleation can be incorporated into this scheme by choosing appropriate ΔF_s vs. n. The general form of the growth equation (45), thus incorporates the Wilson–Frenkel law as a limiting case, and other models for the growth rate can also be derived using it.

The growth rate for the zeroth-order model of the surface (Section 6.2) can be calculated as outlined above. An example of the results of such a calculation are presented for $\alpha = 2$ in Figure 20.

Theory of Crystal Growth

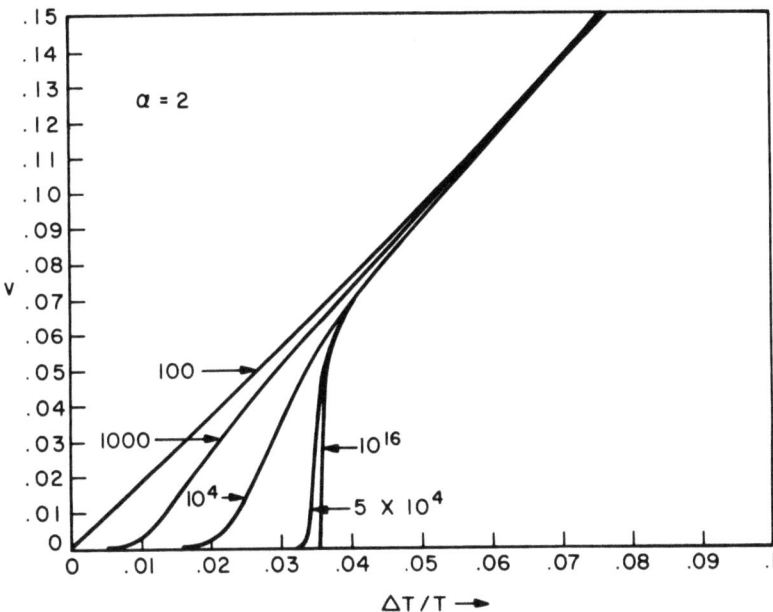

Fig. 20. Growth rate calculated from Eq. (45) for the zeroth-order model of the interface for various surface sizes.

The growth rates have been calculated for various sizes of surface N. For a large interface, $N = 10^{16}$, there is no growth up to some critical undercooling. At undercoolings greater than the critical value the growth rate rises rapidly. The critical undercooling ΔT_c depends on the height of the free-energy barrier in Figure 19, or, more precisely, on the maximum slope in the ΔF_s vs. n curve:

$$L \, \Delta T_c / T = (d \, \Delta F_s / dn)_{\max} \qquad (48)$$

ΔT_c will be large for large α and small for small α, as illustrated by Figure 19.

The various curves in Figure 20 were calculated from the same ΔF_s, changing only N in Eq. (45), and therefore the number of terms in the sum. The simple interpretation of the result is that a large number of terms is needed to make a discontinuity in the growth rate. The corresponding physical significance can be seen from the term in the numerator of Eq. (45):

$$\left[1 - \prod_{n=0}^{N-1} p_n \right]$$

269

Chapter 5

which can be written as

$$\left[1 - \exp\left(-\frac{1}{kT}\prod_{n=0}^{N-1}\frac{d\Delta F_s}{dn}\right)\right]$$

Because of the periodic nature of ΔF_s,

$$\sum_{n=0}^{N-1}(d\Delta F_s/dn) = NL\,\Delta T/kT_E T.$$

So the numerator term is

$$[1 - \exp(-NL\,\Delta T/kT_E T)]$$

The N in the exponent underscores the well-known fact that the sharpness of the melting point depends on the size of the system. A small crystal cannot have a melting point which is sharply defined thermodynamically. This is the reason for the size dependence of the growth rate. For a small surface the interface temperature is not sharply defined. The small growing crystal thus has finite probability of surmounting the barrier, which cannot happen for the large crystal surface, whose energy is sharply defined. For a sufficiently small surface the barrier might as well not exist.

But this raises the question of surface nucleation. A large surface is made up of segments of small surface. The calculation for a large surface constrains the surface (by implicit assumption) to move as a plane. Small segments of the surface can clearly grow faster than the whole surface. The barrier exists only for the large surface constrained to move parallel to itself. Small segments of the surface can grow without going over the barrier, provided there is not too much of a penalty involved with the additional roughening of the surface. The problem then reduces to determining the edge free energy of a step on the surface, in order to calculate the increased free energy associated with making steps on the surface. For a rough surface this will be less than the edge free energy of a sharp step on a smooth surface, but how much less has not been determined satisfactorily. For example, if the edge free energy were known, then each of the curves in Figure 20 could be shifted an appropriate amount by assuming that the melting point of the small segment of surface was shifted, according to the Gibbs–Thomson relation, by an amount proportional to the edge free energy times $b/\sqrt{N}$, where b is a geometric factor, depending on the shape of the disk. This has been done in Figure 21, where it has been assumed that the edge free energy is one-tenth its value for a sharp

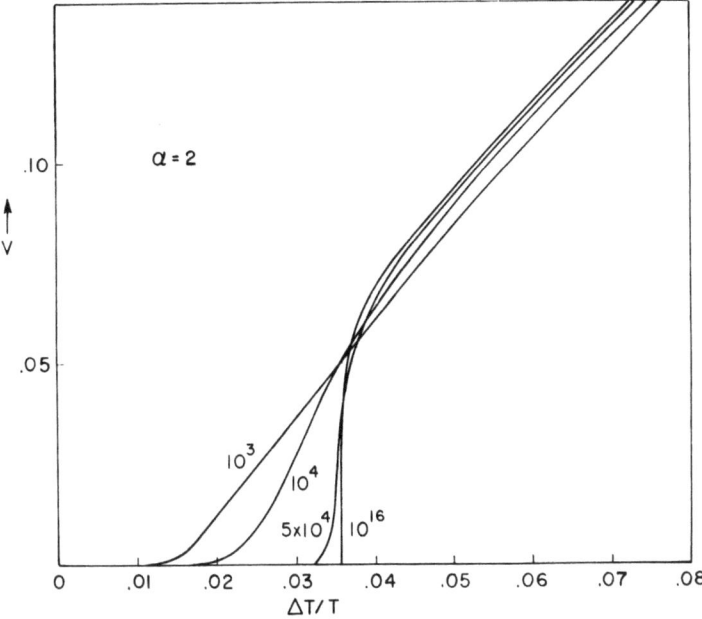

Fig. 21. Replotting of the growth rate data of Fig. 20, assuming a small ledge free energy.

step. The resultant growth rate, given by the upper envelope of the curves, indicates considerable growth at undercoolings less than T_c. Presumably, this sort of treatment, done correctly, would merge with surface nucleation theory, using the corresponding modified edge free energy.

6.4. Rate Equations

The growth rate can also be calculated for the large-surface case using rate equations.[34] By considering the evaporation and condensation (or melting and freezing) rates in each layer parallel to the interface, the net rate of increase of the number of atoms in layer i is given by

$$\frac{1}{P^+}\frac{dx_i}{dt} = (x_{i-1} - x_i) - (x_i - x_{i+1})\exp\left[\alpha\frac{T_E}{T}(2x_i - 1) + \frac{L\,\Delta T}{kT_E T}\right] \quad (49)$$

and the growth rate of the crystal by

$$v = \sum_{n=-\infty}^{\infty} \frac{dx_i}{dt}$$

Chapter 5

Note that for equilibrium, $dx_i/dt = 0$, Eq. (49) reduces to the recursion relationship, Eq. (41). The growth rate can be calculated by a computer by starting with some arbitrary initial configuration of the interface, say $x_i = 1, n < 0, x_i = 0, n \geq 0$. The time evolution of the occupancy of each layer x_i can then be calculated from Eq. (49). After an initial transient it was found that the growth rate settled down either to zero if $T < T_c$, or to a finite growth rate which was periodic with the repeat distance of the lattice, for large undercoolings. The growth rates calculated were similar to those presented in Figure 20 (calculations were not made for identical parameters). This method of calculation should give the same results as the previous calculation using the rate equation (45), since the model is the same, equivalent assumptions have been made, and only the method of analysis differs.

6.5. Clustering

The multilevel model of the interface considered above does not include the possibility of clustering of molecules in the various layers. It corresponds in spirit to a Bragg–Williams or zeroth-order approximation. Clustering can be taken into account to first order by considering a Bethe–Pieirls or first-order model of the interface. In this treatment an order parameter β_i is introduced for each layer, the value of which is determined by a self-consistency condition in each layer. The number of configurations which are considered increases from the zeroth-order treatment, and the complexity of the analysis increases considerably. However, a final expression for the free energy of the interface can be derived in a form similar to Eq. (40), and can be treated in the same way to obtain a calculated growth rate.

Figure 22 shows $\Delta F_s/NkT$ for the first-order model compared with the corresponding surface free energy calculated for the zeroth-order model [Eq. (40)]. It is evident that the maximum is much lower for the first-order model than for the zeroth-order model. The ratio of the two was found to be approximately a factor of three. The reason for this can be seen by considering a large-α, smooth interface. The zeroth-order model overestimates the free energy of a half-filled layer (corresponding to the maximum) by requiring the atoms to be randomly distributed. The first-order model permits some clustering, and so is a better approximation. Successively higher-order cluster expansions would give lower maxima, but these are increasingly difficult to analyze. The growth rates calculated from the zeroth- and first-order models are shown in Figure 23 for

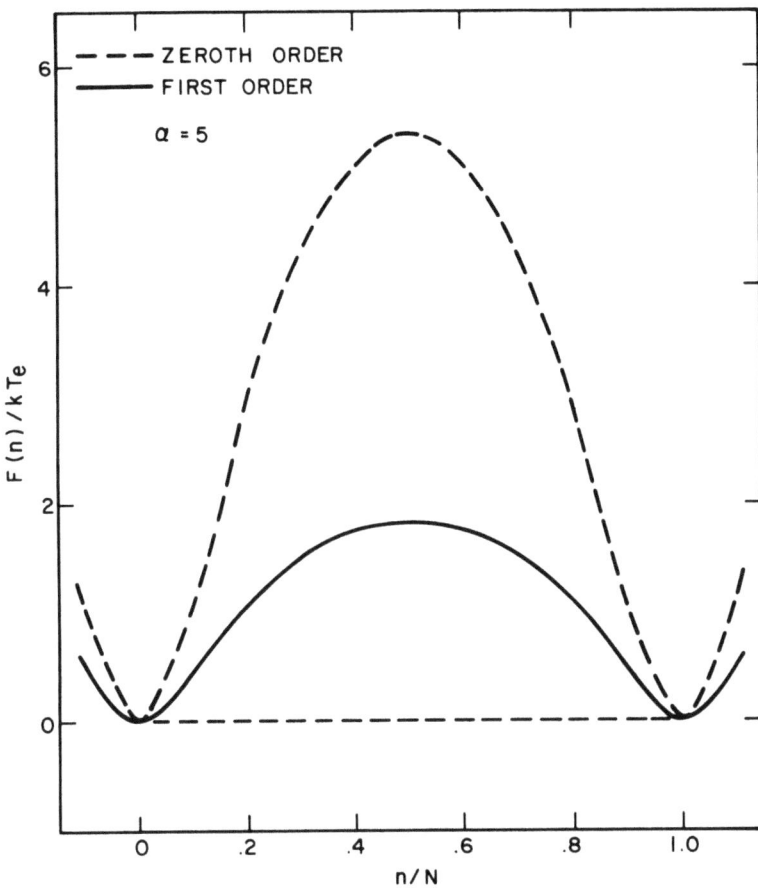

Fig. 22. Comparison of the free energy of an interface as a function of position for the zeroth- and first-order models.

large N and for various values of α. The critical undercooling for growth, ΔT_c, for a particular α is less for the first-order model than for the zeroth-order model by about a factor of three. Presumably, ΔT_c would be even smaller for higher-order models. Indeed, it is evident that the barrier in this form is a direct consequence of the implicit assumption that the interface must move as a plane, which is a nonphysical assumption. The nucleation barrier which actually exists is not described by these models.

6.6. Small Crystals Using Periodic Boundary Conditions

The growth rate can be calculated in closed form for a small crystal surface using periodic boundary conditions.[37] The method

Chapter 5

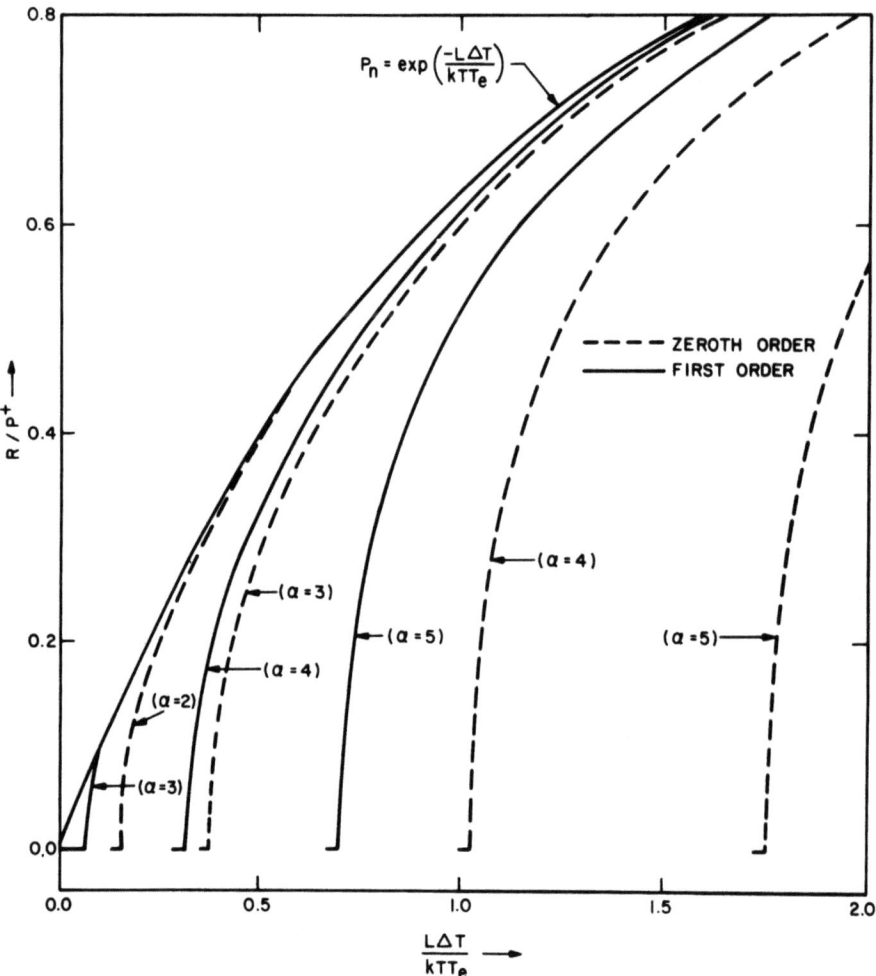

Fig. 23. Comparison of growth rates calculated from Eq. (45) for the zeroth- and first-order models.

used is to enumerate the possible configurations of the surface and to determine all the transitions between the configurations. The probability of finding the surface in a particular configuration depends on the relative rates of the transitions. The growth rate is the net rate at which atoms are added to the configurations, weighted for their probability of occurrence.

If we denote the rate at which atoms join the crystal per site as R^+, then the rate at which atoms leave an n-bonded site on the surface is given by

$$R_n^- = R^+ P_n \qquad (50)$$

where

$$P_n = \exp[(L/kT_E) - 2nL/vkT] \quad (51)$$

where n is the number of nearest neighbors of a particular surface atom, and v is the total possible number of nearest neighbors.

The growth rate of the crystal is given by

$$R = R^+ \left[\sum_i N_i - \sum_j N_j P_n^j \right] / N \quad (52)$$

where the i summation is over all sites at which atoms can arrive, and the j summation is over all surface atoms which can depart. P_n^j is the departure probability for atom j, given by Eq. (51), and N is the total number of sites on the surface.

As an example of the method, let us consider a three-site edge of a two-dimensional crystal. We will, for simplicity, exclude double steps on the surface. The possible configurations of the surface are shown in Figure 24. The three configurations labeled 1 are equivalent because of the periodic boundary conditions, as are the three labeled 2.

The possible transitions for a type 1 site are

$$2\langle 1 \rangle + \langle 1 \rangle P_1 = 3\langle 0 \rangle + 2\langle 2 \rangle P_2 \quad (53a)$$

where $\langle i \rangle$ is the probability of finding the surface in configuration i. The terms on the left involve atoms arriving at or leaving a type 1 site, thus destroying it, and the terms on the right involve atoms arriving at or leaving other sites to create a type 1 site. Similarly,

$$\langle 2 \rangle (1 + 2P_2) = 2\langle 1 \rangle + 3\langle 0 \rangle P_3 \quad (53b)$$

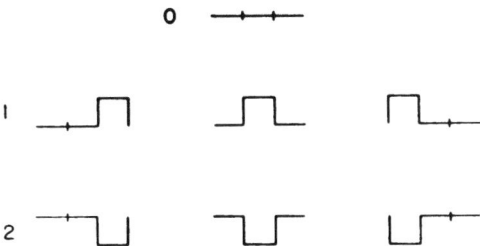

Fig. 24. Possible configurations for a three-site model of the growth of a two-dimensional crystal.

Chapter 5

and
$$3\langle 0\rangle(1 + P_3) = \langle 1\rangle P_1 + \langle 2\rangle \tag{53c}$$
These three equations can be rewritten as

$$\frac{3\langle 0\rangle}{P_1 + 2P_1P_2 + 2} = \frac{\langle 1\rangle}{1 + 2P_2 + 2P_2P_3} = \frac{\langle 2\rangle}{2 + 2P_3 + P_1P_3} \tag{54}$$

For $T = T_E$ and large α these reduce to
$$\langle 1\rangle = \langle 2\rangle = 3\langle 0\rangle e^{-\alpha} \tag{55}$$
where $\alpha = \frac{1}{2}L/kT_E$, i.e., the fraction of extra atoms (type 1 sites) and holes (type 2 sites) is $e^{-\alpha}$ times the number of surface sites, a result previously obtained from Eq. (38).

The growth rate is given by

$$R = R^+ \frac{3(1 - P_3)\langle 0\rangle + (2 - P_1)\langle 1\rangle + (1 - 2P_2)\langle 2\rangle}{3(\langle 0\rangle + \langle 1\rangle + \langle 2\rangle)}$$

$$= R^+ \frac{6(1 - P_1P_2P_3)}{11 + P_1 + 6P_2 + 6P_3 + 2P_1P_2 + 3P_1P_3 + 6P_2P_3} \tag{56}$$

For $L\,\Delta T/kTT_E \ll 1$ this is approximately

$$R = R^+ \frac{L\,\Delta T}{kT_E T}\left[\frac{18}{20 + 3e^\alpha + 12e^{-\alpha}}\right]$$

which is in the form of Eq. (34), with

$$f = \frac{18}{(20 + 3e^\alpha + 12e^{-\alpha})}$$

For small α, $f \approx \frac{1}{2}$; for large α, $f \approx 6e^{-\alpha}$. The factor f is close to one for rough surfaces, so that the predicted growth rate is close to the Wilson–Frenkel value, whereas for large α the predicted growth rate is significantly smaller, as is observed experimentally.

Calculations such as this can be made for edges of various lengths. The results of such calculations[37] are summarized in Table 1.

The number of configurations which must be enumerated increases rapidly with repeat distance. It is apparent from Table 1 that the f factors are not changing rapidly with surface size. Indeed, f_{11} is given by $f_{11} = N/(2N - 1)$, where N is the repeat distance, which converges to $\frac{1}{2}$ for large N.

Table 1 also indicates a large anisotropy in the growth rate for large L/kT_E, as is observed experimentally.

TABLE 1
Calculated f Factors for the Growth Rate of Two-Dimensional Square Crystals in the [10] and [11] Directions for Two-, Three-, and Four-Site Repeat Distances

L/kT_E	f_{10}			f_{11}		
	2-Site	3-Site	4-Site	2-Site	3-Site	4-Site
1	0.68	0.56	0.54	2/3	3/5	4/7
5	0.26	0.3	0.34	2/3	3/5	4/7
10	0.026	0.039	0.050	2/3	3/5	4/7

Similar calculations can be made for three-dimensional crystals. For 2×2 surfaces of a simple cubic crystal the growth rate factors f resulting from such a calculation[37] are presented in Table 2.

For large L/kT_E the growth rate is calculated to be much slower than for small L/kT_E on the closest packed face. The growth rate for large L/kT_E is very anisotropic, but for small L/kT_E is essentially independent of direction. These results are in qualitative agreement with experiment.

The calculated growth rate factors are proportional to $e^{-\alpha}$, where α is defined in Eq. (39). For small L/kT_E, α is small, and α is also small for non-close-packed faces. For growth on close-packed faces with large L/kT_E, α is large.

These calculations on crystals with small repeat distances have obvious limitations: Nucleation behavior is excluded by the small repeat distance. However, the results give a remarkably good qualitative picture of the growth of real crystals. Some physical insight can be obtained from a detailed examination of the equations. For example, the rate of arrival of atoms on the surface has been factored out of the f factor. Effectively, atoms arrive at the same rate on the (100) face for $L/kT_E = 10$ as on any other face. And yet the growth

TABLE 2
Growth Rate Factors for 2×2 Faces of a Simple Cubic Crystal

L/kT_E	f_{100}	f_{110}	f_{111}
1	0.53	0.55	0.51
5	0.2	0.35	0.51
10	0.01	0.09	0.51

Chapter 5

rate is smaller by a factor of 50. The large arrival rate is cancelled by the large departure rate of single adatoms. These have a large departure rate (proportional to e^α), and so the small number of them (the density of adatoms is proportional to $e^{-\alpha}$) cancels out the large arrival rate to a first approximation, leaving only a small net growth rate proportional to the density of adatoms, $e^{-\alpha}$. For a non-close-packed surface a large fraction of the surface atoms are not single adatoms, and these contribute directly to the growth rate. For small L/kT_E the factor $e^{-\alpha}$ is close to one, and so the net growth rate due to the adpopulation is large anyway. But in addition there is a large number of adatoms, which is another way of saying the surface is rough, and so many of the surface atoms do not exist as single adatoms, and so contribute directly to the growth rate.

This mechanism operates for all growth rate processes, and indicates clearly why the Wilson–Frenkel rate is not achieved for materials growing with large L/kT_E.

6.7. Monte Carlo Calculations

The structure of the interface can be simulated by computer by using Monte Carlo methods. A small segment of a crystal is stored in the computer memory and atoms are added to the crystal or removed from it according to jump probabilities which are calculated from binding energies. The binding energy of a surface atom depends on its number of nearest neighbors. An atom is permitted to leave the crystal if its jump probability exceeds a random number. The crystal surface achieves some steady-state configuration, and depending on the jump probabilities, can be made to grow. The Monte Carlo method treats clustering properly, but in view of the size dependence of the effective barrier to growth (Figure 20), growth rates calculated by this method must be viewed with some caution. This method is to be distinguished from the use of the computer to solve statistical mechanics equations as in the methods discussed above. In Monte Carlo simulation and surface configurations are treated directly and clustering effects are built in to the simulation automatically.

Figure 25 compares the interface roughness calculated for equilibrium using zeroth-order, first-order, and Monte Carlo calculations. The surface roughness R is defined as the number of unsatisfied bonds parallel to the surface. All three give remarkably similar results. The curve labeled $8e^{-\alpha}$ is the roughness assuming $e^{-\alpha}$ adatoms and $e^{-\alpha}$ holes in the layer below (the solution for large α), and counting four

Theory of Crystal Growth

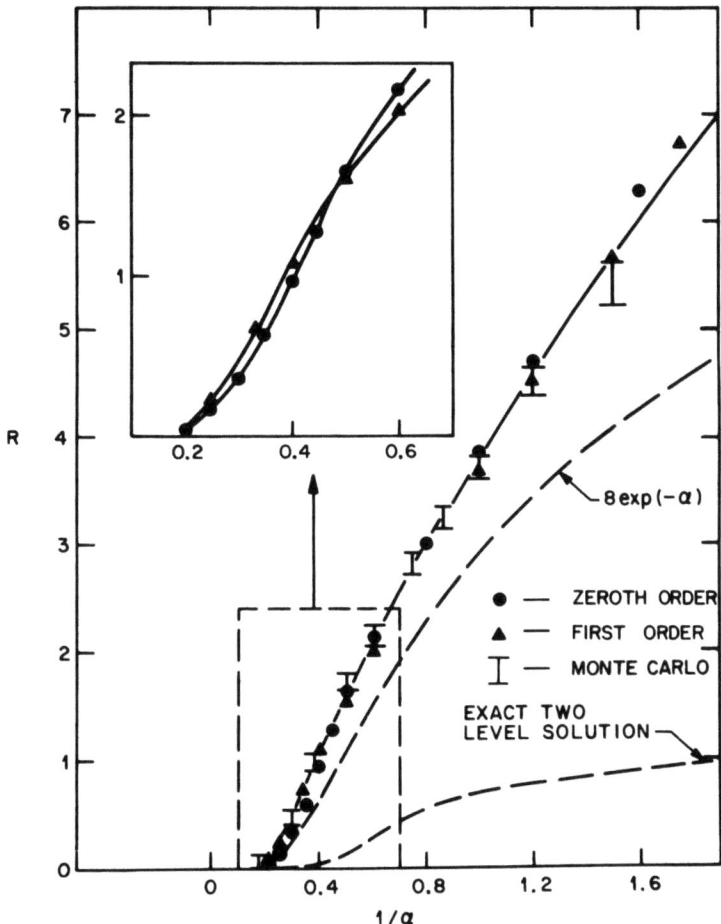

Fig. 25. Interface roughness R for zeroth- and first-order models, and from Monte Carlo simulation.

bonds for each. The surface roughness is remarkably insensitive to the details of clustering on the surface.

In view of the difference between the calculated growth rates for the zeroth- and first-order model (Figure 23) the roughness does not correlate with the growth rate.

Gilmer and Bennema[38] have determined growth rates for a limited range of α (3–5) by direct Monte Carlo simulation. Some of their results are shown in Figure 26. The critical undercooling for growth as shown in Figure 23 is not present. Their growth rates can be fitted to nucleation theory using an edge free energy about one-

Chapter 5

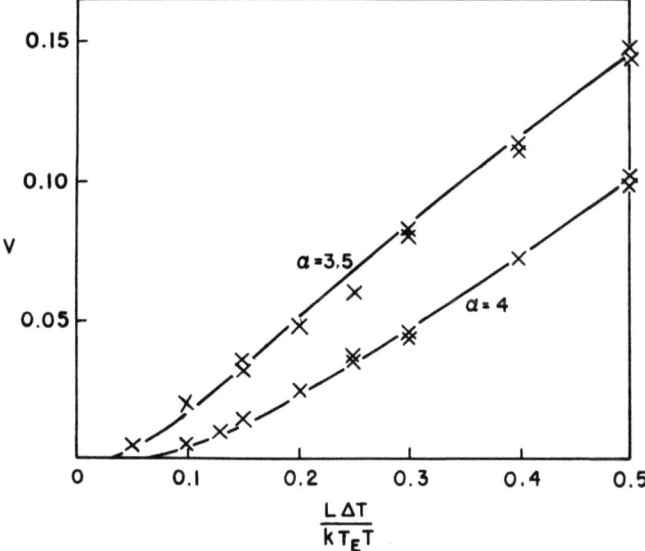

Fig. 26. Growth rates from Monte Carlo simulation for
α = 3.5 and 4.

tenth the bulk surface free energy. They observe clusters in the configurations of the surface which give the appearance of nucleation behavior. The size of these clusters is small compared to the size of the surface found by computer, so the size effects shown in Figure 20 are presumed to be absent. Introducing surface diffusion into the simulation increases the growth rate, which implies that the surface configurations were not equilibrium configurations, but rather nonequilibrium "growth" configurations, as implied by the growth rate calculations with small repeat distances.

The growth rates obtained by direct Monte Carlo computer simulation are similar (not identical) to the growth rates calculated for small repeat distances. Unfortunately, the Monte Carlo calculations to date are restricted to a small range of L/kT_E, but it is anticipated that more extensive calculations will show the same dependence on orientation and L/kT_E as the small crystal calculations.

The overall picture for the intrinsic growth process can be summarized as follows: For small entropy change the surface of the crystal is above its roughening temperature. There is for these cases no lattice resistance to the motion of the interface. Since the surface is rough, the growth rate is described by the modified Wilson–Frenkel equation (34), where the factor f may depend on the degree of surface roughness. The intrinsic growth rate will be rapid and almost isotropic. The growth and morphology will be dominated by diffusion processes (e.g., Figures 1–3).

For large entropy change the close-packed surfaces are smooth (corresponding to the surface being below its roughening temperature) and the intrinsic growth rate is slow. On the non-close-packed faces, which are above their roughening temperature, the growth rate is rapid. The intrinsic growth rate is therefore very anisotropic, and limited by the close-packed faces which bound the crystal. The intrinsic growth rate on the close-packed faces is nonlinear with chemical potential difference (see Figure 26) and depends on the entropy of fusion approximately as showh in Table 2, that is, the factor f is approximately proportional to $e^{-\alpha}$. For large entropy of change the growth morphology is dominated by the intrinsic growth kinetics, the crystals assume "crystalline" shapes, and diffusion processes do not dominate the morphology, although they are still important in controlling the overall conditions of growth (e.g., benzil, Figure 14).

There is a transition region from small to large entropy change, where the diffusion processes and the intrinsic growth processes both contribute strongly to the growth morphology (e.g., bismuth hopper growth, Figure 12).

For very large entropy change (e.g., the polymers, Figure 15) such large chemical potential differences are needed to drive the growth process, and the anisotropy in growth rate is so large, that only small defective crystals grow at quite slow rates, and the growth process is dominated by the intrinsic growth kinetics.

References

1. T. B. Reed, W. J. LaFleur, and A. J. Strauss, *J. Cryst. Growth* **3/4**, 115 (1968).
2. J. R. Carruthers, in *Proparation and Properties of Solid State Materials* (R. A. Lefever and W. R. Wilcox, eds.), to be published.
3. H. P. Utech and M. C. Flemings, in *Crystal Growth* (H. S. Peiser, ed.), Pergamon Press, Oxford (1967) p. 651.
4. A. F. Witt and H. C. Gatos, *J. Electrochem. Soc.* **113**, 808 (1966); **114**, 413 (1967); **115**, 70 (1968); **116**, 511 (1969).
5. J. R. Carruthers and K. Nassau, *J. Appl. Phys.* **39**, 5205 (1968).
6. K. A. Jackson, in *Progress in Solid State Chemistry*, Vol. 4 (H. Reiss, ed.), Pergamon Press, Oxford (1967), p. 53.
7. W. A. Tiller, K. A. Jackson, J. W. Rutter, and B. Chalmers, *Acta Met.* **1**, 428 (1953).

Chapter 5

8. D. Walton, W. A. Tiller, J. W. Rutter, and W. C. Winegard, *Trans. AIME* **203**, 1023 (1955); E. C. Holmes, J. W. Rutter, and W. C. Winegard, *Can. J. Phys.* **35**, 1223 (1957).
9. D. T. J. Hurle, *Solid State Electronics* **3**, 37 (1961).
10. W. Bardsley, J. M. Callan, H. A. Chedze, and D. T. J. Hurle, *Solid State Electronics* **3**, 142 (1961).
11. W. W. Mullins and R. F. Sekerka, *J. Appl. Phys.* **34**, 323 (1963); **35**, 444 (1964).
12. H. S. Chen and K. A. Jackson, *J. Crystal Growth* **8**, 184 (1971).
13. J. W. Cahn, *Crystal Growth* (H. S. Peiser, ed.), *J. Phys. Chem. Solids (Suppl.)* **1967** 681.
14. A. A. Chernov, in *Proc. IV All-Soviet Conf. on Crystal Growth*, MK II (1972), p. 168.
15. R. F. Sekerka, *J. Crystal Growth* **3/4**, 71 (1968).
16. S. C. Hardy and S. R. Coriell, *J. Crystal Growth* **3/4**, 569 (1968); **5**, 329 (1969); **7**, 147 (1970).
17. V. G. Smith, W. A. Tiller, and J. W. Rutter, *Can. J. Phys.* **33**, 723 (1955).
18. V. Koump and R. H. Tien, *Met. Trans.* **1** (1970), 1231.
19. G. P. Ivantsov, *Doklady Akad. Nauk SSSR* **58**, 567 (1947); *Growth of Crystals*, Vol. 1, Consultants Bureau, New York (1958), p. 59.
20. G. Horvay and J. W. Cahn, *Acta Met.* **9** (1961), 695.
21. D. E. Temkin, *Soviet Phys.—Doklady* **5** (1960), 609.
22. M. E. Glicksman and R. J. Schaefer, *J. Crystal Growth* **1**, 297 (1967); G. R. Kotler and L. A. Tarshis, *J. Crystal Growth* **3/4**, 603 (1968); E. Holtzman, *J. Appl. Phys.* **41**, 1460 (1970); R. Trivedi, *Acta Met.* **18** (1970), 287.
23. G. E. Nash and M. E. Glicksman, *Proc. IV All-Soviet Conf. on Crystal Growth*, MK II (1972), p. 143.
24. H. A. Wilson, *Phil. Mag.* **50**, 238 (1900).
25. J. Frenkel, *Physik Sowjet union* **1**, 498 (1932).
26. K. A. Jackson and B. Chalmers, *Can. J. Phys.* **34**, 473 (1956).
27. K. A. Jackson, *Liquid Metals and Solidification*, ASM, Cleveland (1958), p. 174.
28. K. A. Jackson, *Growth and Perfection of Crystals* (Doremus et al., eds.), Wiley, New York (1958), p. 319.
29. I. Langmuir, *J. Am. Chem. Soc.* **40**, 1361 (1918).
30. R. Fowler and E. A. Guggenheim, *Statistical Thermodynamics*, Cambridge Univ. Press (1965), p. 430.
31. W. K. Burton, N. Cabrera, and F. C. Frank, *Phil. Trans. Roy. Soc. (London)* **234A**, 299 (1951).
32. J. D. Hunt and K. A. Jackson, *Trans. Met. Soc. AIME* **236**, 843 (1966).
33. R. S. Fidler, M. N. Croker, and R. W. Smith, *J. Crystal Growth* **13/14**, 739 (1972).
34. D. E. Temkin *Crystallization Processes* (Sirota et al., eds.), Consultants Bureau, New York (1966), p. 15.
35. H. J. Leamy and K. A. Jackson, *J. Appl. Phys.* **42**, 2121 (1971).
36. K. A. Jackson, *J. Crystal Growth.* **5**, 13 (1969).
37. K. A. Jackson, *J. Crystal Growth* **3/4**, 507 (1968).
38. G. H. Gilmer and P. Bennema, *J. Crystal Growth* **13/14**, 148 (1972).

6

Growth from the Vapor

Morton E. Jones and Don W. Shaw
Central Research Laboratories
Texas Instruments Inc.
Dallas, Texas

1. Introduction

Growth from the vapor for preparing materials, particularly in the form of thin films, has become an extremely important technique. Probably the best example is its extensive use in the fabrication of silicon semiconductor devices and integrated circuits. Chemical vapor deposition is used to prepare the high-purity polycrystalline silicon, which is then melt-grown into single crystals. Thin silicon films, in which the actual devices and circuits are formed, are grown on slices of these crystals by chemical vapor deposition. Thin layers of silicon dioxide and silicon nitride are vapor-deposited for insulation and surface passivation. Finally, metal interconnect patterns are deposited, generally by vacuum evaporation. Chemical vapor deposition is also used for preparing high-purity metals, such as titanium, zirconium, hafnium, thorium, and chromium, and for a wide variety of other materials. Vacuum evaporation is used for depositing thin layers of many materials for surface coatings. The important present and potential uses of vapor deposition have spurred extensive research on these processes during the past decade. This chapter will discuss the general principles of growth from the vapor. Particular emphasis will be placed on the relative contributions of thermodynamic and kinetic factors to vapor deposition processes. Application of these principles will be illustrated using vapor-phase epitaxial growth as an example.

Chapter 6

1.1. Principles of Vapor Deposition

Vapor growth processes involve formation of the vapor species, transport of the species to the deposition region, and condensation or deposition to form a solid. The vapor species may be introduced into the system as the gases themselves or may be formed by heating a solid or liquid to high temperatures (evaporation), bombarding a solid with energetic ions (sputtering), creating an electrical discharge (plasma), or through a chemical reaction to give volatile products. Transport may occur through a vacuum impelled by the kinetic energy from evaporation or sputtering, in a flowing gas stream, or by diffusion in a concentration gradient. Deposition may occur by condensation or as a result of a chemical reaction and may take place in the vapor itself (homogeneous deposition) or on the surface of a solid substrate (heterogeneous deposition). Most practical vapor growth processes are heterogeneous since they have the capability of yielding single-crystal epitaxial layers and, in any case, give much more adherent deposits than does homogeneous deposition. Hence we will restrict our discussion to the heterogeneous case.

The deposition step usually involves a series of events. One or more of the reactants must be adsorbed on the solid substrate surface. These reactants may desorb back into the vapor, may diffuse to nucleation sites on the surface, or may undergo chemical reactions with other adsorbed or gaseous reactants to form a solid product. If gaseous products also result, they must desorb back into the vapor.

Vapor deposition processes are generally classified into two categories, physical and chemical. Physical processes are those for which the composition of the deposit is the same as that of the vapor. Examples are vacuum evaporation and sputtering in an inert ambient. Chemical processes are those in which a chemical reaction occurs on or near the deposition surface and the composition of the deposit and vapor are different. Some processes are difficult to classify, for example, reactive sputtering or evaporation of binary compounds which dissociate into elemental constituents in the vapor and then recombine in the deposit. Since these are variations of the simple physical processes, they are usually also classified as physical even though chemical reactions occur.

Since the subject of this chapter is growth from the vapor, we will concentrate on events that occur in the vicinity of the deposition surface. The source of the vapor species and transport to the deposition region will not be treated in detail. Physical deposition processes generally occur very far from equilibrium. Atoms or molecules arrive at a surface whose temperature is usually much lower than the source from which they were evaporated or ejected by ion bombardment. They have kinetic energies on the order of 0.1 eV for evaporation and tens or hundreds of electron volts for sputtering. The probability that such impinging species will be adsorbed on the surface, thermally equilibrate, and be incorporated into the deposit is expressed by a condensation coefficient which is determined empirically for various materials and conditions. Since the subjects of condensation and nucleation are covered in detail elsewhere in this volume, they will not be treated here. Rather, we shall restrict the discussion to chemical vapor deposition (CVD) processes.

1.2. Chemical Vapor Deposition

In general the design and complexity of a CVD system is determined by the physical states of the starting materials. In the simplest case all reactants exist as gases at room temperature. Pyrolysis of a gaseous hydride such as silane is a good example:

$$SiH_4(g) \stackrel{\text{carrier gas}}{=\!=\!=} Si(s) + 2H_2(g)$$

Deposition occurs by simply passing the gaseous mixture of the hydride and its carrier gas over a heated substrate.

In some cases one or more of the reactants exists as a volatile liquid or solid at room temperature. This complicates the system somewhat because provisions must be made for vaporization and transfer of the material into the deposition zone. Some reactants are unstable at low temperature and must be generated at high temperatures in a separate region of the apparatus. They must then be transported through a hot region to the deposition area. This situation exists for deposition of some III–V materials, such as GaAs, where gallium monochloride, which is unstable at low temperature, is a principal reactant leading to deposition.

Transport of the reactants from their source or generation location may be achieved by free convection or diffusion in a closed tube system or by forced convection with a carrier gas in an open tube system. Although closed tube processes are very simple, they are severely limited in versatility. Most large-scale CVD systems are of the open tube, gas flow variety. Figure 1 shows the basic elements in the deposition region of such a flow system. The carrier gas may itself be one of the reactants; for example, hydrogen when silicon is

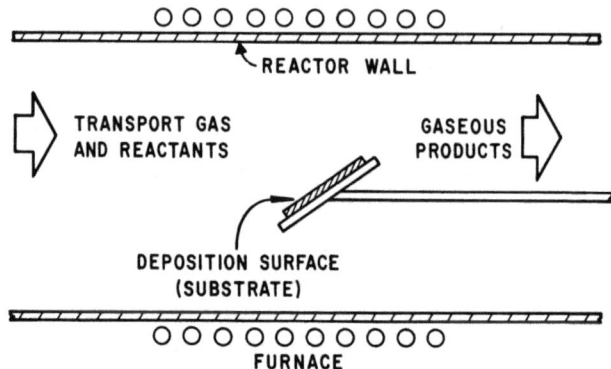

Fig. 1. Major elements of the deposition region in an open tube flow system.

deposited by reduction of silicon tetrachloride according to the reaction

$$\text{SiCl}_4(g) + 2\text{H}_2(g) = \text{Si}(s) + 4\text{HCl}(g)$$

Conversely, the carrier gas may be inert and serve only for transport of the reactants to the deposition region. Usually the carrier gas is present in excess and the reactant partial pressures can be controlled externally. With open tube systems temperature gradients are not necessary for transport of the reagents and the deposition can proceed in an isothermal environment, a feature which promotes uniformity in the deposits.

2. Thermodynamics of Gas–Solid Reactions

A thermodynamic analysis of a chemical vapor deposition process is important for determining to what degree the pertinent reactions will occur in the absence of kinetic limitations. Such an analysis also provides information on the species that are expected to be present in the deposition region.

There are several sources for thermochemical data applicable to chemical vapor deposition processes. Stull and Sinke[1] have compiled data for many of the elements. The JANAF tables[2] and National Bureau of Standards Circular 500[3] and its revision[4] contain data on many molecular species. Glassner[5] has obtained information on oxides, fluorides, and chlorides. The U.S. Bureau of Mines has published data on several inorganic compounds.[6-8] When the necessary thermodynamic quantities are not available, they may be estimated by methods described by Kubaschewski and Evans.[9] Finally, the *Journal of Chemical Thermodynamics* and *Chemical Abstracts* contain new data not included in earlier publications and tables. Since the usefulness of the results of thermochemical calculations may depend strongly on the accuracy of the initial data employed, it is essential that the best available values be used.

The difficulty of carrying out the desired thermodynamic analysis depends on the complexity and number of chemical reactions and species involved. For the simplest processes, involving only a single reaction and a few chemical species, a calculation of the equilibrium constant as a function of temperature allows a thorough analysis to be made. Where several reactions occur, and these are known, a set of equations involving the equilibrium constants and partial pressures of each species must be solved. Fortunately, methods[10-12] have been developed which require only a knowledge of the possible species present and their thermochemical data. No knowledge or assumptions of the actual reactions which occur are necessary. These methods make use of computer programs which minimize the free energy of the overall system and calculate the amount of solid species deposited and the partial pressures of all gaseous species at equilibrium.

In order to perform the desired thermodynamic calculations, either the equilibrium constants of the reactions or the free energies of formation of the chemical species at the reaction temperature are needed. However, many of the published compilations give only the heats and entropies of formation and the heat capacities, from which the desired quantities must be calculated; hence a brief review of the pertinent equations will be presented.

If the only data available for the species are the heats and entropies of formation at a single temperature, usually 298°K, and the heat capacities, the values at the deposition temperature can be calculated using

$$\Delta H°(T) = \Delta H°(T_0) + \int_{T_0}^{T} C_p \, dT$$

and

$$\Delta S°(T) = \Delta S°(T_0) + \int_{T_0}^{T} (C_p/T) \, dT$$

Chapter 6

The Gibbs free energies of formation can then be calculated from

$$\Delta G°(T) = \Delta H°(T) - T\Delta S°(T)$$

The free energy of each of the reactions is then the sum of the free energies of formation of the products minus that of the reactants,

$$\Delta G°(T) = \sum_p \Delta G_p°(T) - \sum_r \Delta G_r°(T)$$

The equilibrium constants can then be calculated from

$$\ln K(T) = -\Delta G°(T)/RT$$

Let us now consider some hypothetical deposition reactions to demonstrate the application of these principles to predicting deposition parameters. The simplest case would be the chemical decomposition of a binary gas AB to yield solid A(s) and gaseous B(g) as products. The reaction occurring at the solid substrate surface is

$$AB(g) = A(s) + B(g)$$

and, assuming ideal gas behavior, the equilibrium constant is

$$K = p_B/p_{AB}$$

Generally the reacting species, in this case AB, is carried to the deposition region in a stream of a carrier gas and the partial pressure in the entering stream p_{AB}^0 is known. Let us assume for this example that the carrier gas is inert and does not enter into the deposition reaction. From the ideal gas law

$$p = nRT/V$$

the equilibrium constant can be written

$$K = n_B/n_{AB}$$

However, since each mole of AB that decomposes yields one mole of A and one mole of B, $n_B = n_A$, $n_{AB} = n_{AB}^0 - n_A$, and

$$K = n_A/(n_{AB}^0 - n_A)$$

Hence

$$n_A = n_{AB}^0 K/(1 + K)$$

Normally the quantities are divided by time to give the quantity

$$\eta_A = \eta_{AB}^0 K/(1 + K) \tag{1}$$

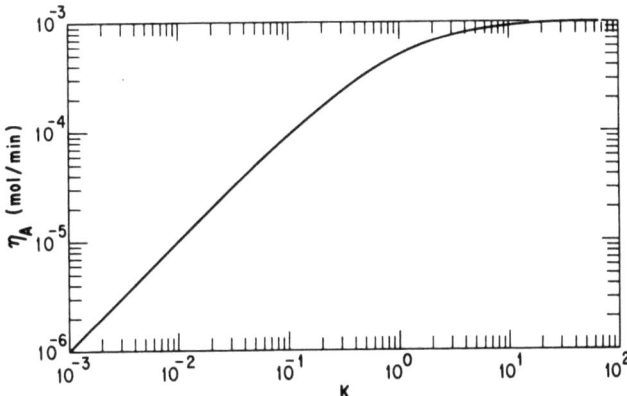

Fig. 2. Dependence of the decomposition rate on the equilibrium constant K for a simple decomposition reaction $AB(g) = A(s) + B(g)$ occurring under equilibrium conditions. $\eta_{AB}^0 = 10^{-3}$ mole/min.

where η_A is the deposition rate of A and η_{AB}^0 the rate of input of AB into the gas deposition region.

Equation (1) gives the maximum rate of deposition that can occur if equilibrium is obtained. Figure 2 shows the behavior predicted for a fixed input rate η_{AB}^0. For large values of the equilibrium constant we see that the deposition rate approaches the input rate and essentially all of the entering reactant will be decomposed. At low values of the equilibrium constant the deposition rate may be proportional to the input rate and to K. In actual systems kinetic factors generally prevent achieving equilibrium and the rates are less than would be predicted by Eq. (1).

For our next example, let us consider a somewhat more complex decomposition reaction

$$AB_x(g) = A(s) + xB(g)$$

with the equilibrium constant

$$K = (p_B)^x/p_{AB_x}$$

Again from the ideal gas law this becomes

$$K = (n_B)^x(RT/V)^{x-1}/n_{AB_x}$$

Since the reaction creates one mole of A and x moles of B for each mole of AB_x that decomposes,

$$n_{AB_x} = n_{AB_x}^0 - n_A \quad \text{and} \quad n_B = xn_A$$

Chapter 6

such that

$$K = (xn_A)^x(RT/V)^{x-1}/(n^0_{AB_x} - n_A)$$

or, rearranging and converting to moles per unit time,

$$(x/RT/f)^{x-1}(x/K)\eta_A{}^x + \eta_A = \eta_{AB_x}$$

where f is the volumetric flow rate. Let us next consider a simple recombination reaction such as

$$A(g) + B(g) = AB(s)$$

for which the equilibrium constant is

$$K = 1/p_A p_B$$

From the ideal gas law this becomes

$$K = 1/n_A n_B (RT/V)^2$$

Since

$$n_A = n_A{}^0 - n_{AB} \quad \text{and} \quad n_B = n_B{}^0 - n_{AB}$$

then

$$K = 1/(n_A{}^0 - n_{AB})(n_B{}^0 - n_{AB})(RT/V)^2$$

Solving for the rate of formation of AB, $\eta_{AB} = n_{AB}/t$, gives

$$\eta_{AB} = \tfrac{1}{2}\{\eta_A{}^0 + \eta_B \pm [(\eta_A{}^0 - \eta_B{}^0)^2 + 4(f/RT)^2(1/K)]^{1/2}\}$$

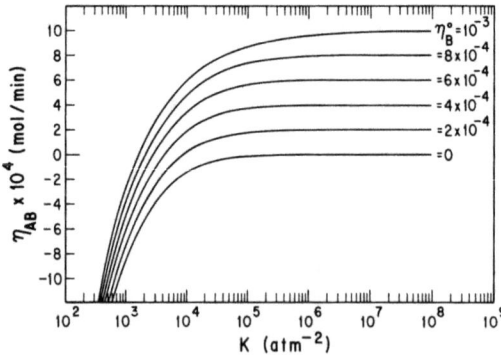

Fig. 3. Dependence of the deposition rate on the equilibrium constant K for the reaction $A(g) + B(g) = AB(s)$ occurring under equilibrium conditions.

Since the extent of deposition must increase with increasing K, only the negative sign gives a meaningful solution; thus

$$\eta_{AB} = \tfrac{1}{2}\{\eta_A{}^0 + \eta_B{}^0 - [(\eta_A{}^0 - \eta_B{}^0)^2 + 4(f/RT)^2(1/K)]^{1/2}\}$$

Figure 3 shows the equilibrium growth rate predicted by this equation for a fixed input rate of A and several input rates of B. For large values of K this equation reduces to

$$n_{AB} = \tfrac{1}{2}(\eta_A{}^0 + \eta_B{}^0 - |\eta_A{}^0 - \eta_B{}^0|)$$

Hence if $\eta_A{}^0 \gg \eta_B{}^0$, then $\eta_{AB} = \eta_B{}^0$; and if $\eta_B{}^0 \gg \eta_B{}^0$, then $\eta_{AB} = \eta_A{}^0$. Therefore the rate of deposition is equal to the input rate of A or B, whichever is smaller.

For very small values of K,

$$\eta_{AB} = -(f/RT)(1/K)^{1/2}$$

The negative value indicates that not only does no AB(s) deposit, but if any is already present, it will be removed by etching. As shown, the behavior of such simple decomposition or recombination reactions may be easily predicted; however, more complex reactions involving several species require computer techniques such as described in Refs. 10–12.

3. Kinetics of Gas–Solid Reactions

The degree to which most chemical vapor deposition reactions occur, particularly in flowing gas or open tube systems, is often determined more by kinetic limitations than by thermodynamics. Such processes occur through a series of sequential steps:

1. Transport of the reactive species to the deposition regions.
2. Transfer of the species from the main gas stream to the deposition surface.
3. Adsorption of one or more reactants, and possibly of non-reactive species.
4. Chemical reaction, either between adsorbed species or adsorbed and gaseous species.
5. Desorption of reaction products.
6. Transfer of products to the main gas stream.
7. Transport of products away from the deposition region.

Whichever step is inherently the slowest will determine the overall rate of the deposition process.

Steps 1, 2, 6, and 7 involve material transport in the vapor. For steps 1 and 7, transport is usually in a flowing gas, although free convection is important in a sealed tube system. Reactants are normally introduced into the system diluted in a carrier gas, which itself may also be a reactant. Steps 2 and 6, which represent transfer between the main gas stream and the deposition surface, occur by physical processes such as diffusion or convection. Processes with rates determined by any of these four steps are described as being controlled by mass transport.

Steps 3–5 occur at or on the substrate surface. Processes whose rates are limited by one of these steps are described as chemically, kinetically, or surface controlled. Frequently the rate-limiting process for a particular deposition reaction will change as the process parameters are varied. For example, if the input reactants are in low concentration and the gas velocity is low, the initial mass transport step might control the process. This is particularly true if the deposition temperature is high such that surface reactions proceed rapidly. However, if the concentrations and gas velocity are high, such that the rate of reactant input to the deposition surface is no longer limiting, surface kinetics might take over as the controlling step. Low deposition temperatures tend to favor kinetic control.

Let us now consider the behavior to be expected for the different rate-limiting steps by determining, for each case, how the deposition rate will depend on the following experimental parameters:

1. The total mass or volume flow rate of gas.
2. The linear gas velocity in the deposition region.
3. The hydrodynamic nature of the gas flow in the deposition region.
4. The concentrations or partial pressures of the various vapor species.
5. The temperature of the deposition surface.
6. Temperature gradients in the system.
7. The nature of the deposition surface, such as the crystallographic orientation in the case of epitaxial deposition.

3.1. Mass Transport Limiting Steps

Deposition reactions where the rate-determining step is transport of the reactive species to the deposition region are frequently referred to as mass input limited. For this case a near-equilibrium condition exists and a thermodynamic analysis is valid. The deposition rate will increase with gas velocity due to the increase in mass input of the reactants. It will be independent of the linear gas velocity in the deposition region, the hydrodynamic nature of the gas flow, temperature gradients in the system, and the crystallographic orientation of the substrate. The rate will have a temperature dependence which can be predicted by thermodynamics. In this regime deposition reactions that are endothermic, such as the reduction of $SiCl_4$ by H_2 to give $Si(s)$, will show an increasing deposition rate with increasing temperature. Those processes that are exothermic, such as the deposition of GaAs from As_4 and GaCl, will show decreasing deposition rates with increasing temperature. The rate will increase with increasing concentration of those vapor species that limit the amount of deposition, but will be independent of the concentration of those species that are present in excess. For example, as shown in Section 2, for the reaction $A(g) + B(g) = AB(s)$, the limiting species will be either $A(g)$ or $B(g)$, whichever is present at the lower concentration.

Reactions where transfer of the reactive species between the main gas stream and the deposition surface is rate limiting are frequently called diffusion controlled, since diffusion is often the primary transfer mechanism. Such systems are often treated by considering a main gas stream, a deposition surface, and a stagnant boundary layer between the two, as shown in Figure 4. The partial pressures of the species in the main stream are assumed to be those at the input, p_i^0. If the surface reactions are sufficiently rapid, the partial pressures of the species at the deposition surface are the equilibrium values p_i^{eq} calculated from thermodynamic data. For the simple binary case of a species i in a carrier gas, the diffusion flux is given by

$$J_i = D_i \, dC_i/dx$$

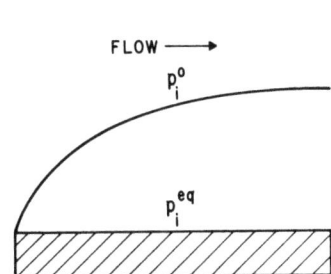

Fig. 4. Boundary layer between the main gas stream and the deposition surface in a flowing gas system.

Chapter 6

which, for a uniform concentration gradient across the thickness L, becomes

$$J_i = D_i(C_i^0 - C_i^{eq})/L$$

However, the concentrations may be converted to partial pressures using the ideal gas law,

$$C_i = n_i/V = p_i/RT$$

such that the flux becomes

$$J_i = D_0(p_i^0 - p_i^{eq})/RTL \tag{2}$$

The deposition rate r will then be proportional to J_i.

Let us now examine the effect of the experimental variables on the deposition rate for diffusion-controlled systems. The rate will be independent of the total mass or volume flow rate if the partial pressures and linear gas velocity are held constant. On the other hand, the gas velocity and the hydrodynamic nature of the flow will both influence the thickness and shape of the stagnant boundary layer and, through this, the deposition rate. For the flow rates and geometries employed in practical deposition systems, the flow regime is typically in the laminar region. The thickness of the stagnant layer is proportional to the inverse square root of the linear velocity in the gas stream, $L \propto v^{-1/2}$, hence increasing the velocity will decrease the thickness and increase the deposition rate correspondingly.

Temperature will affect the deposition rate for diffusion-controlled systems through several factors. The diffusion coefficient is temperature dependent, as are the viscosity, gas velocity, and density, which determine the thickness of the boundary layer. The diffusion coefficient varies approximately as

$$D/D_0 \sim (T/T_0)^{1.75} \tag{3}$$

The density is inversely proportional to temperature,

$$d/d_0 = T_0/T \tag{4}$$

the viscosity varies approximately as

$$\mu/\mu_0 \sim (T/T_0)^m \tag{5}$$

where m is between 0.6 and 1.0, and the gas velocity is proportional to temperature,

$$v/v_0 = T/T_0 \tag{6}$$

The boundary layer thickness depends on these quantities as[13]

$$L/L_0 = [(\mu/\mu_0)(d_0/d)(v_0/v)]^{1/2} \qquad (7)$$

Hence, from Eqs. (2)–(7) we obtain the growth rate dependence

$$r/r_0 \sim (T/T_0)^{0.75-(m/2)}$$

or the rate varies between $(T/T_0)^{0.45}$ and $(T/T_0)^{0.25}$.

This dependence would give an apparent activation energy at 1000°C of 3.5–4.5 kcal/mole, which is typical of a diffusion-controlled process.

The concentrations or partial pressures of the vapor species will affect the deposition rate through the concentration gradient across the boundary layer. As shown by Eq. (2), the flux, and hence the rate, is proportional to the difference between the concentration of the species in the main gas stream and at equilibrium at the surface. If the reaction is one that goes nearly to completion such that p_i^{eq} approaches zero, the deposition rate will be proportional to the input partial pressure of species i.

Since equilibrium is assumed to exist at the deposition surface, the rate will not be dependent on the nature of the surface, such as its crystallographic orientation.

3.2. Surface Rate-Limiting Steps

Next let us consider the kinetics for those cases involving a rate-limiting process at the deposition surface. We will first treat the simple case of the decomposition of a compound AB. Gaseous AB is adsorbed,

$$AB(g) \xrightleftharpoons{k_{AB}} AB*$$

where k_{AB} is an adsorption rate constant. Hence the rate of adsorption is given by

$$r_{AB} = k_{AB}\theta p_{AB} \qquad (8)$$

where θ is the fraction of surface sites not already occupied by adsorbed species. After adsorption, two parallel events can occur: (1) AB* can desorb without undergoing decomposition,

$$AB* \xrightleftharpoons{k_{-AB}} AB(g)$$

at a rate

$$r_{-AB} = k_{-AB}\theta_{AB} \qquad (9)$$

Chapter 6

proportional to the desorption rate constant k_{-AB} and to the fraction of surface sites θ_{AB} containing adsorbed AB; or (2) AB* can decompose,

$$AB* \overset{k_r}{=} A(s) + B*$$

at a rate

$$r = k_r \theta_{AB} \tag{10}$$

where k_r is the reaction rate constant, with B* subsequently desorbing,

$$B* \overset{k_{-B}}{=} B(g)$$

at a rate

$$r_{-B} = k_{-B} \theta_B \tag{11}$$

Under steady state conditions the concentrations of AB* and B* on the surface become constant and so

$$r_{AB} = r_{-AB} + r \tag{12}$$

and $r_{-B} = r$. From Eqs. (8)–(13) and recognizing that $\theta = 1 - \theta_{AB} - \theta_B$ we obtain

$$\theta_{AB} = k_{AB} p_{AB} / \{k_{-AB} + k_r + [1 + (k_r/k_{-B})] k_{AB} p_{AB}\}$$

giving the deposition rate

$$r = k_r k_{AB} p_{AB} / \{k_{-AB} + k_r + [1 + (k_r/k_{-B})] k_{AB} p_{AB}\}$$

Let us examine the behavior for various rate-limiting conditions. If the inherent rate of adsorption is slow such that

$$k_{AB} \ll k_r, k_{-AB}, k_{-B}$$

Eq. (14) reduces to

$$r = k_r k_{AB} p_{AB} / (k_{-AB} + k_r)$$

and the rate becomes first order in the partial pressure of AB for all values of pressure, as shown in Figure 5(a). Since the mass transport steps have been assumed to be rapid, the rate will be independent of the flow rate and gas velocity. Since the process being treated involves chemisorption, it is an activated process and should exhibit an exponential dependence on temperature. The deposition rate for cases where the adsorption step is rate limiting should show a dependence on the nature of the deposition surface; for example, on crystallographic orientation when a single crystal is used for the substrate.

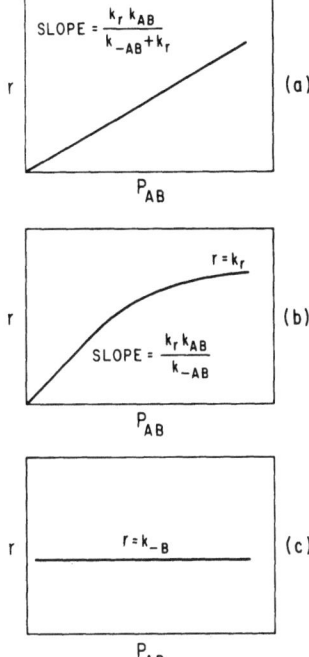

Fig. 5. Dependence of the deposition rate for the decomposition reaction $AB(g) = A(s) + B(g)$ for the rate-limiting step being (a) the adsorption of $AB(g)$, (b) the decomposition reaction on the surface, or (c) desorption of $B(g)$.

This is due to the fact that the composition and nature of the surface at an atomic level are dependent on orientation.

If the rate-limiting step is the surface chemical decomposition reaction, then $k_r \ll k_{AB}, k_{-AB}, k_{-B}$ and Eq. (14) becomes

$$r = k_r k_{AB} p_{AB}/(k_{-AB} + k_{AB} p_{AB})$$

or

$$r = k_r \beta_{AB} p_{AB}/(1 + \beta_{AB} p_{AB})$$

where $\beta_{AB} = k_{AB}/k_{-AB}$ is the Langmuir adsorption coefficient. The same independence of flow rate and gas velocity and dependence on temperature and crystallographic orientation will be observed as for absorption; however, the rate will be first order in p_{AB} only at low pressures. At high pressures it will become a constant, independent of p_{AB}. This behavior is shown in Figure 5(b).

If the desorption step of the reaction product B is rate limiting, such that

$$k_{-B} \ll k, k_{AB}, k_{-AB}$$

Chapter 6

then Eq. (14) becomes

$$r = k_{-B}$$

Hence, when desorption of the gaseous product is rate limiting, the deposition rate is independent of the partial pressure of the reactant species at all pressures, as shown in Figure 5(c). Again the rate will be independent of the flow rate and gas velocity, increase exponentially with temperature, since desorption is also an activated process, and can depend upon the crystallographic orientation of the deposition surface. Thus for a simple decomposition reaction, if it is first determined that the process is kinetically controlled, for example, by the independence of deposition rate on the input flow rate and its dependence on the crystallographic orientation of the deposition surface, then the specific rate-limiting process on the surface may be deduced from the pressure dependence of the deposition rate.

Next let us consider deposition reactions of the form $A(g) + B(g) = AB(s)$, where reactive vapor species A and B with partial pressures p_A and p_B combine to form solid AB on the surface. There are two ways for such a process to occur. Either A or B may adsorb on the surface and subsequently react with the vapor form of the other species, or both may adsorb, followed by a reaction between the adsorbed species. Let us first consider the case where only A adsorbs and then reacts with B(g). The sequence of steps will be adsorption,

$$A(g) \stackrel{k_A}{=} A* \qquad (13)$$

followed by either desorption,

$$A* \stackrel{k_{-A}}{=} A(g) \qquad (14)$$

or reaction,

$$A* + B(g) \stackrel{k_r}{=} AB(s) \qquad (15)$$

The rates of the three steps are given, respectively, by

$$r_A = k_A \theta p_A, \qquad r_{-A} = k_{-A} \theta_A, \qquad r = k_r \theta_A p_B$$

When a steady state is reached

$$r_A = r_{-A} + r \qquad \text{and} \qquad \theta = 1 - \theta_A$$

and we obtain

$$\theta_A = k_A p_A / (k_{-A} + k_r p_B + k_A p_A)$$

and
$$r = k_r k_A p_A p_B / (k_{-A} + k_r p_B + k_A p_A) \tag{16}$$

Now let us look at the implications of Eq. (16) for various rate-limiting steps. If adsorption of A is the slow step, $k_A \ll k_{-A}, k_r$, then

$$r = k_r k_A p_A p_B / (k_{-A} + k_r p_B)$$

The deposition reaction is first order in p_A and is first order in p_B at low partial pressure, becoming independent of p_B at high partial pressure. If the chemical reaction (15) is the slow step, then

$$r = k_r k_A p_A p_B / (k_{-A} + k_A p_A)$$
$$= \beta_A k_r p_A p_B / (1 + \beta_A p_A)$$

and the deposition is first order in both p_B and p_A at low partial pressures, but becomes independent of p_A at high partial pressure. If desorption is also a negligibly slow step, the expressions simplify to $r = k_A p_A$ for the case when adsorption of A is limiting and $r = k_r p_B$ when the reaction is limiting. Observation of the first-order dependence on the partial pressure of only one of the species for such a reaction leads to an unambiguous interpretation of the rate-limiting process only when the nature of the reaction itself is known, i.e., which species is being adsorbed. Thus the same dependence on the partial pressures of the reactants occurs when adsorption of B is rate limiting as for when A is adsorbed and the chemical reaction is rate limiting.

If the reaction between A and B occurs by both being adsorbed and subsequently reacting on the surface, the steps are

$$A(g) \overset{k_A}{=} A^*$$
$$B(g) \overset{k_B}{=} B^*$$
$$A^* \overset{k_{-A}}{=} A(g)$$
$$B^* \overset{k_{-B}}{=} B(g)$$
$$A^* + B^* \overset{k_r}{=} AB(s)$$

with rates

$$r_A = k_A \theta p_A, \quad r_B = k_B \theta p_B$$
$$r_{-A} = k_{-A} \theta_A, \quad r_{-B} = k_{-B} \theta_B, \quad r = k_r \theta_A \theta_B$$

Chapter 6

When a steady state is reached,

$$r_A = r_{-A} + r, \qquad r_B = r_{-B} + r$$

and

$$\theta = 1 - \theta_A - \theta_B$$

if the adsorption is competitive for surface sites. This leads to an implicit expression relating the surface coverage of A and B.

$$k_{-A}k_B p_B \theta_A - k_{-B}k_A p_A \theta_A + (k_B p_B - k_A p_A)k_r \theta_A \theta_B = 0 \qquad (17)$$

If the rate of the surface reaction is small compared to those of adsorption and desorption, hence equilibrium exists between the adsorbed and gaseous species, Eq. (17) is simplified and the surface coverages are given by the Langmuir isotherms,

$$\theta_A = \beta_A p_A/(1 + \beta_A p_A + \beta_B p_B)$$
$$\theta_B = \beta_B p_B/(1 + \beta_A p_A + \beta_B p_B)$$

and the rate is given by

$$r = k_r \beta_A \beta_B p_A p_B/(1 + \beta_A p_A + \beta_B p_B)^2$$

If the partial pressure of either species is increased, the deposition rate will first increase, then pass through a maximum, and, finally, decrease, as shown in Figure 6. This behavior results from the fact that at high partial pressures of one species, that species will

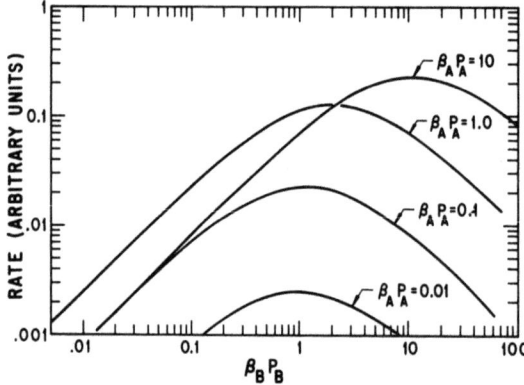

Fig. 6. Dependence of the deposition rate on the partial pressure p_B for the reaction $A(g) + B(g) = AB(s)$ when both gases adsorb and react on the surface.

dominate the adsorption sites, limiting their availability to the other reactant.

If the two species are adsorbed on different sets of surface sites, such that there is no competition involved, then the process behaves as if there were two separate surfaces involved and the coverage for each species is given by

$$\theta_i = \beta_i p_i/(1 + \beta_i p_i)$$

and the rate becomes

$$r = k_r \theta_A \theta_B = k_r \beta_A \beta_B p_A p_B/(1 + \beta_A p_A + \beta_B p_B + \beta_A \beta_B p_A p_B)$$

which at low partial pressures is first order for each species and at high pressures approaches a constant value (zeroth order).

In many chemical vapor deposition systems the reactive species are transported to the deposition region by a carrier gas which does not participate in the subsequent chemical reaction. However, this carrier gas may adsorb on the deposition surface, reducing the number of sites available to the reactants and acting as an inhibitor for the reaction. For the case of $A(g) + B(g) = AB(s)$ with A and B adsorbing competititively and with an inhibiting carrier gas i present, the rate can be shown to be

$$r = k_r \beta_A \beta_B p_A p_B/(1 + \beta_A p_A + \beta_B p_B + \beta_i p_i)^2$$

At high partial pressures of the carrier gas the rate decreases as the square of that pressure.

The examples which have been treated so far assume that the species adsorb without dissociation. If dissociation does occur, for example,

$$A_2(g) = 2A^*$$

then the adsorption involves two surface sites and

$$r_A = k_A \theta^2 p_A$$

and the surface coverage at equilibrium is

$$\theta_A = \beta_A^{1/2} p_A^{1/2}/(1 + \beta_A^{1/2} p_A^{1/2})$$

The rate of the reaction

$$A^* + B^* = AB(s)$$

where A_2 adsorbs dissociatively and B adsorbs competitively, is given by

$$r = k_r \beta_A^{1/2} \beta_B^{1/2} p_A p_B / (1 + \beta_A^{1/2} p_A^{1/2} + \beta_B p_B)^2$$

The expressions illustrated for simple vapor deposition reactions can be extended to more complex systems. For the general case of a reaction involving n gaseous species adsorbing competitively and the ith species dissociating on adsorption into m_i subspecies, the rate becomes

$$r = k_r \left[\prod_{i=1}^{n} (\beta_i^{1/m_i} p_i^{1/m_i}) \right] \left[1 + \sum_{i=1}^{n} (\beta_i^{1/m_i} p_i^{1/m_i}) \right]^{-n}$$

These examples illustrate the basic approaches for kinetic data analysis and provide guidance for the most efficient experimental approaches. In actual vapor growth systems the initial effort must be devoted to identification of the basic rate-limiting regime. If experimental conditions can be defined which correspond to kinetically limited growth rates, then the rate–partial pressure data can be analyzed in the above manner and rate expressions derived for plausible mechanisms can be tested with the experimental rate data. In every case, however, care must be taken to eliminate mass transport contributions to the rate data used in the mechanism evaluation. On the other hand, if the process appears to be essentially mass transport limited, no mechanistic information is expected. Instead, the emphasis is placed on process control through evaluation of the transport variables, i.e., fluid dynamics, convection, etc. In the next section many of the previously described principles will be illustrated with the well-developed vapor growth processes of silicon and gallium arsenide epitaxial deposition.

4. Epitaxial Growth

Epitaxial growth refers to the deposition of oriented single-crystal layers of a material on the surface of a single-crystal substrate. If the substrate is of the same material as the deposit, the process is referred to as homoepitaxy, and if the compositions are different, it is called heteroepitaxy. The deposition of silicon on silicon is a good example of homoepitaxy, and silicon on sapphire or spinel, of heteroepitaxy.

Epitaxial growth is important for both scientific and practical reasons. For studying kinetic or surface-controlled deposition reactions it provides a surface of known composition and structure.

The ability to change the nature of this surface by varying the crystallographic orientation is useful for obtaining data for deducing the rate-limiting step and the deposition mechanism. The effects of crystallographic orientation will be examined later for the homoepitaxial growth of gallium arsenide. On the practical side, epitaxial growth is a major reason for the revolutionary developments in electronics due to silicon integrated circuits. In many such circuits the individual devices are formed in epitaxial layers with highly controlled properties. We will examine in detail the thermodynamics and kinetics of some of the processes used to grow epitaxial layers of silicon and gallium arsenide.

The most important factors for obtaining good epitaxial growth are lattice match between the substrate and layer, near equality of thermal expansion coefficients, chemical compatibility, and surface cleanliness. For homoepitaxy the first three factors are, by definition, the same for substrate and deposit and surface cleanliness is the important factor.

4.1. Silicon Deposition

Chemical vapor deposition of silicon is important not only for growing epitaxial layers on silicon substrates, but also for preparing high-purity polycrystalline silicon, which is subsequently melt-grown into single crystals for substrates. Several chemical systems

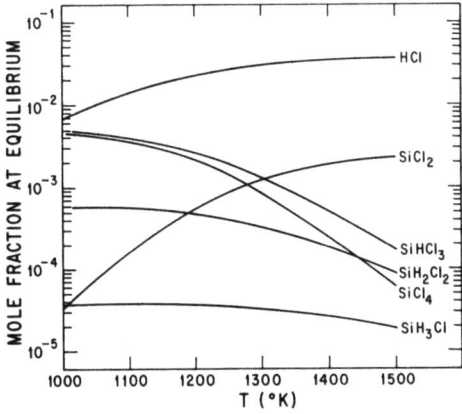

Fig. 7. Concentrations of the vapor species in equilibrium with solid silicon for an input gas with a $SiCl_4$ partial pressure of 37.5 mm in hydrogen and a total pressure of 1 atm.

Chapter 6

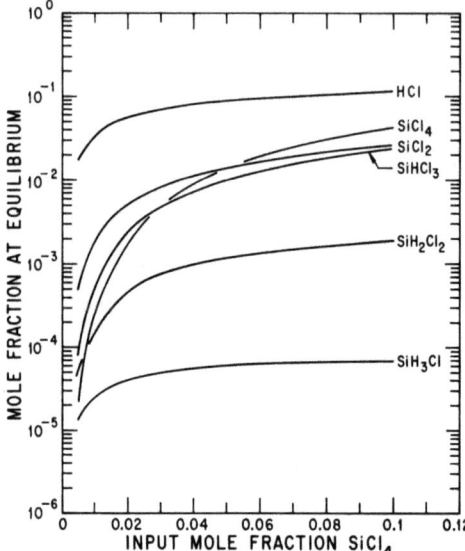

Fig. 8. Concentrations of the vapor species in equilibrium with solid silicon at 1400°K and a total pressure of 1 atm.

have been used for silicon deposition, the most common being the thermal decomposition of silane, SiH_4, and the hydrogen reduction of volatile chlorosilanes, e.g., $SiCl_4$, $SiHCl_3$, and SiH_2Cl_2. The $SiCl_4/H_2$ system has been the most thoroughly investigated and will be emphasized in the following discussion.

4.1.1. Thermodynamics of Silicon Deposition

Excellent thermochemical data are available for the Si/Cl/H systems. This has been used to produce several thermodynamic analyses.[14–17] The principal species in equilibrium with solid silicon in the temperature range 1000–1600°K are $SiCl_4$, $SiHCl_3$, SiH_2Cl_2, SiH_3Cl, $SiCl_2$, HCl, and H_2. Figure 7 shows the concentration of these species as a function of temperature for an initial $SiCl_4$ partial pressure of 37.5 mm and a total pressure of 1 atm using thermochemical data of Hunt and Sirtl.[17] A striking feature is the rapid increase in the equilibrium concentration of $SiCl_2$ with temperature, becoming the dominant silicon-containing species at about 1350°K. Figure 8 shows the dependence of the concentrations of the species as a function of the initial $SiCl_4$ partial pressure at 1400°K. A greater

Growth from the Vapor

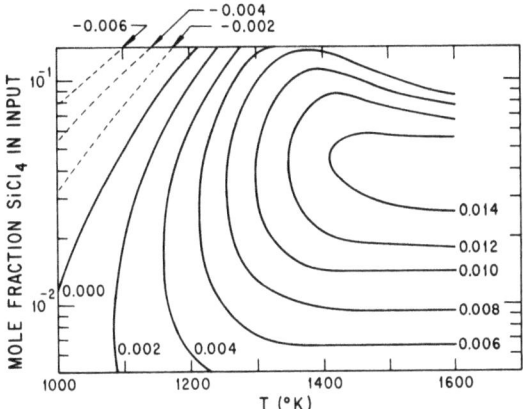

Fig. 9. Contours of constant deposition rate (g/min) as a function of input gas composition and temperature for an input flow rate of 1 liter/min at 1 atm. Calculated from the thermodynamic data of Hunt and Sirtl.[17]

total input concentration of Si-containing species favors formation of $SiCl_4$.

The equilibrium data may be converted to the form shown in Figure 9, where contours of equal deposition rate are plotted as a function of temperature and input vapor composition for a total input flow rate of 1 liter/min at atmospheric pressure with all of the input gas assumed to reach equilibrium with the substrate. For such a case, or for one proposed by Sedgwick[18] where a fixed fraction of the gas stream equilibrates, the curves predict, for temperatures greater than 1200°C, an initial increase in growth rate with increasing concentration of $SiCl_4$ followed by a decrease in rate and eventually etching. This is the qualitative behavior observed for $SiCl_4$–H_2 systems, as shown in Figure 10. This behavior results from the etching reaction

$$SiCl_4(g) + Si(s) = SiCl_2(g)$$

which becomes important at high $SiCl_4$ concentrations.

Unlike thermochemical data, the quantity of experimental rate data for the $SiCl_4/H_2$ deposition system is distressingly low. Though kinetic measurements have been made by several investigators, none of these studies has systematically covered all of the important parameters and the ranges of these parameters necessary to define the possible mechanisms. This would require unambiguous identification of a range of deposition conditions, if any, where the growth rate

Chapter 6

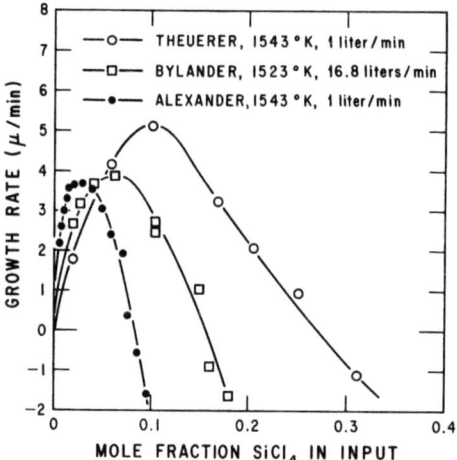

Fig. 10. Dependence of silicon growth rate on the input gas composition as determined by three investigators.[22-24] (From Ref. 27 with permission of John Wiley & Sons.)

is kinetically limited. One factor which hinders such an identification is the basic endothermic nature of the deposition process. This can result in a deposition rate which increases with temperature due to the shift in equilibrium alone. If the process were controlled by the rate of mass transfer, the rate–temperature behavior would be similar to that expected for a surface-limited or kinetically controlled process. This problem has been discussed in some detail from a theoretical point of view[19] and Harper and Lewis[15] have shown that a thermodynamic analysis can predict values of an "apparent activation energy" (assuming complete equilibrium at the crystal surface) which are similar to the observed values. Thus an activated rate–temperature behavior does not permit unambiguous identification of kinetic control in the $SiCl_4/H_2$ deposition system. Moreover, there are large discrepancies in the activation energy values reported in the literature, which vary from 23 kcal/mole[20] up to 44 kcal/mole.[21] This may be contrasted to the very good agreement between various literature values in the GaAs deposition system, which will be discussed later.

There is qualitative agreement on the effect of the $SiCl_4$ input partial pressure on the deposition rate. Figure 10 shows such data as determined by three investigators.[22-24] The curves all have the same shape, with the rate increasing as a function of the $SiCl_4$

concentration, ultimately passing through a maximum, and then decreasing until etching rather than growth occurs. However, the concentrations at which the maximum and the onset of etching occur differ considerably.

Disagreement exists concerning the extent to which the substrate surface orientation influences the silicon deposition rate. Some authors[25] have reported essentially no variation with orientation, while others find differences by as much as a factor of two.[26] Generally the rate is observed to be a function of the gas velocity in the deposition region, although this may tend to saturate at high flow rates.[24,27] Future progress with the silicon deposition system requires a thorough, self-consistent set of experimental rate data. Regions must be specified with respect to the dependence of the rate on flow velocity, substrate orientation, and temperature. Then from a knowledge of the effects of reactant partial pressures within each region insight can be gained with respect to the growth mechanism.

4.1.2. Proposed Mass Transport Models for Silicon Deposition

Many of the experimental and theoretical studies of silicon epitaxial deposition have been confined to the mass transfer characteristics of these systems. One of the goals of such efforts is to improve the overall deposition efficiency and uniformity by identifying the critical mass transfer parameters, e.g., boundary layer size and shape, convection effects, etc. Most of the resulting mass transfer models have two features in common. They assume that the surface processes are so rapid that the growth rate is completely mass transfer limited (although usually this is not experimentally verified) and that the partial pressures of the gaseous species are equal to the equilibrium values at the growing crystal surface. In general, the more sophisticated models depend less on empirically derived constants, but at the expense of being limited to a particular apparatus geometry. One of the more general treatments is the quasi-equilibrium model described by Sedgwick.[18] It is assumed that the gas stream may be divided into two parts, one of which passes by the substrate without opportunity to react and the other of which completely equilibrates with it. The fraction that does reach equilibrium is determined empirically and then used to predict the growth rate under other conditions. With this and most other mass transfer treatments it is essential to compute the equilibrium partial pressures from a set of inlet partial pressures using a thermodynamic analysis. It can be shown that Sedgwick's model is equivalent to the basic diffusion approach [Eq. (2)] if one assumes that all diffusing species have the same diffusion coefficients and that the rate is determined by differences between the bulk stream partial pressures of the species and their equilibrium values at the crystal surface.

Among the more detailed mass transfer models for silicon deposition is the boundary layer approach of Bradshaw.[28,29] The dimensions of the stagnant region are derived using a streamline approximation. The problem of boundary layer thicknesses has been approached experimentally by injecting TiO_2 particles into the reactor and photographing the streamline patterns in the substrate regions.[30] Some approaches do not consider a well-developed boundary layer with inlet partial pressures outside the layer; instead, assumptions are made with respect to the gas velocity distributions. A good example is the work of Shepherd,[21] where a parabolic velocity profile is assumed between the substrate and the reactor walls. The gas velocity distributions and boundary layer thicknesses can be accurately determined at the surface of a rotating disk. This has been a fruitful approach to studying mass-transfer-limited deposition.[31] Recently, free convection has received considerable theoretical and experimental attention as a means of mass transfer in silicon epitaxial growth systems.[32,33]

4.1.3. Proposed Surface Reaction Model for Silicon Deposition

Although the experimental rate data are far from sufficient to verify possible mechanisms, we can consider the species present at the silicon-vapor interface and speculate on the most probable events which may occur. The silicon surface itself consists of (a) silicon atoms bonded to from one to three underlying silicons in the crystal, with the excess electrons creating dangling bonds, and (b) absorbed species from the vapor. The principal silicon-containing species in the vapor, with the exception of $SiCl_2$, are fully coordinated molecules with four atoms bonded in a tetrahedral configuration to the silicon. If we assume these molecules adsorb without dissociation, then they must do so with the peripheral hydrogens or chlorines in contact with the substrate surface. Subsequent reduction must then involve removal of the peripheral atoms to bring the silicon atom into direct contact with the surface, where it can be incorporated into the crystal. However, for the nonlinear $SiCl_2$ molecule the silicon atom is not sterically blocked from the atoms in the surface layer and can adsorb in essentially the same position it will later occupy in the solid. Moreover, the presence of residual unbonded electrons on the silicon atom in this molecule should greatly enhance its adsorption by interacting with the dangling bonds on the surface. Hence the most probable adsorbed species for providing silicon atoms to the crystal should be $SiCl_2$. Assume the following elemental processes, beginning with adsorption of $SiCl_2$,

$$SiCl_2(g) \overset{k_{SiCl_2}}{=} SiCl_2{}^*$$

After adsorption, reduction can occur by gaseous hydrogen,

$$SiCl_2{}^* + H_2 \overset{k_1}{=} Si(s) + 2HCl(g)$$

or by gaseous $SiCl_2$

$$SiCl_2{}^* + SiCl_2(g) \overset{k_2}{=} Si(s) + SiCl_4(g)$$

In addition to these deposition reactions, the reverse reactions which lead to etching can also occur. These would involve $SiCl_4$,

$$SiCl_4(g) + Si(s) \overset{k_{-2}}{=} SiCl_2{}^* + SiCl_2(g)$$

and HCl,

$$Si(s) + 2HCl(g) \overset{k_{-1}}{=} SiCl_2{}^* + H_2(g)$$

The rates of these reactions are given by

$$r_1 = k_1 \theta_{SiCl_2} p_{H_2}, \qquad r_2 = k_2 \theta_{SiCl_2} p_{SiCl_2}$$
$$r_{-1} = k_{-1} \theta p_{HCl}^2, \qquad r_{-2} = k_{-2} \theta p_{SiCl_4}$$

and the net rate by

$$r = r_1 + r_2 - r_{-1} - r_{-2}$$

Since

$$\theta_{SiCl_2} = \beta_{SiCl_2} p_{SiCl_2}/(1 + \beta_{SiCl_2} p_{SiCl_2})$$

and $\theta = 1 - \theta_{SiCl_2}$, we have

$$r = (k_1 \beta_{SiCl_2} p_{SiCl_2} p_{H_2} + k_2 \beta_{SiCl_2} p_{SiCl_2}^2 - k_{-1} p_{HCl}^2 - k_{-2} p_{SiCl_4})$$
$$\times (1 + \beta_{SiCl_2} p_{SiCl_2})^{-1}$$

To test this mechanism would require a considerable amount of experimental data. Measurement of the vapor composition in the vicinity of the deposition surface would show whether homogeneous gas-phase equilibrium was being reached, so that the partial pressures of the species near the surface could be calculated thermodynamically. The region of parameters in which the rate is kinetically controlled would have to be determined. Within this region the deposition rate would be measured as the concentrations of the input species, $SiCl_4$ and HCl, and the temperature are varied. If homogeneous gas-phase equilibrium has been verified, these input data can be used to calculate the gas composition in the deposition region. If the proposed mechanism is valid, the deposition rate should be highest when the concentration ratio of $SiCl_2$ to $SiCl_4$ is high, tending toward etching when the opposite condition exists.

The importance of the hydrogen reduction of chlorosilanes to preparation of both polycrystalline and epitaxial single-crystal silicon for electronic devices makes an understanding of the mechanism, and hence the suggested thorough kinetic study, highly desirable.

4.2. Gallium Arsenide Deposition

Epitaxial growth of a compound is considerably more complicated than growth of an elemental material such as silicon. At least two volatile species must be formed and transported to the deposition region, where their concentrations must be within the ranges required for deposition of a stoichiometric compound. With GaAs the required partial pressure of the arsenic species at the epitaxial growth temperature can be obtained from the element; however, the vapor pressure of elemental gallium is far too low, so it must be converted to a volatile compound. With most GaAs systems, gallium monochloride is used for this purpose. Since GaCl is unstable at room temperature, it must be formed *in situ* within the growth apparatus by reaction of a chlorine-containing species, typically HCl with elemental Ga (or possibly with gallium arsenide). This requirement results in epitaxial growth systems with a high-temperature source region separate from the deposition or growth region. Several deposition systems have evolved which basically differ only in the choice of an original arsenic source, e.g., $Ga/HCl/As/H_2$, $Ga/HCl/AsH_3/H_2$, and $Ga/AsCl_3/H_2$. With all systems the basic reaction leading to epitaxial deposition can be written as

$$4GaCl(g) + As_4(g) + 2H_2(g) = 4GaAs(s) + 4HCl(g)$$

The gallium arsenide growth systems are fundamentally different from silicon from a thermodynamic point of view. The silicon processes described earlier are endothermic, and the expected extent of deposition increases with temperature. This permits deposition in an induction-heated system whose container walls can be maintained at a temperature lower than the actual deposition area. On the other hand, the gallium arsenide process is exothermic. Increases in temperature are expected to shift the equilibria toward less deposition. Since cooler areas promote deposition, the walls of the apparatus must be maintained at a temperature greater than the substrate in order to maximize epitaxial deposition. As we will see below, this exothermic nature of gallium arsenide epitaxial growth becomes very significant in the theoretical and experimental evaluation of the growth kinetics.

As mentioned in the previous section on kinetics, during growth of a compound material the possibility of competitive adsorption must be given serious consideration when evaluating possible deposition mechanisms. Although geometrically similar to the diamond-type lattice of silicon, the zinc blende lattice, which gallium arsenide assumes, is noncentrosymmetric. This results in two unique $\{111\}$ surfaces which, while having the same basic atomic geometry, differ in their chemical composition. One surface consists predominately of gallium and the other of arsenic atoms. Examination of the relative growth rates of these different surfaces under various deposition conditions can provide important clues to the basic mechanism of the growth process.

As might be expected, the quality of thermochemical data available for gallium arsenide in the literature is inferior to that for silicon. Often values must be estimated or derived from secondary or even tertiary sources. Consequently, thermodynamic analyses are not expected to be as accurate as for silicon deposition. Nevertheless, several useful analyses[34-36] have been reported. These analyses define the gas-phase species that are expected to exist at the deposition temperature. Furthermore, the results have been supported by Ban,[37] using on-stream mass spectroscopic analysis.

In contrast to the thermochemical data, there is a relative abundance of experimental kinetic data. The quantity and quality of this data are greatly superior to those for silicon systems. Most of the data were taken with an epitaxial system ($Ga/HCl/As/H_2$) equipped for continuous rate monitoring.[38] This apparatus, which is illustrated in Figure 11, consists of a continuously recording microbalance coupled to the epitaxial growth system. The substrate crystal is suspended from one arm of the microbalance, and the associated weight changes can be monitored *in situ* during the actual

Chapter 6

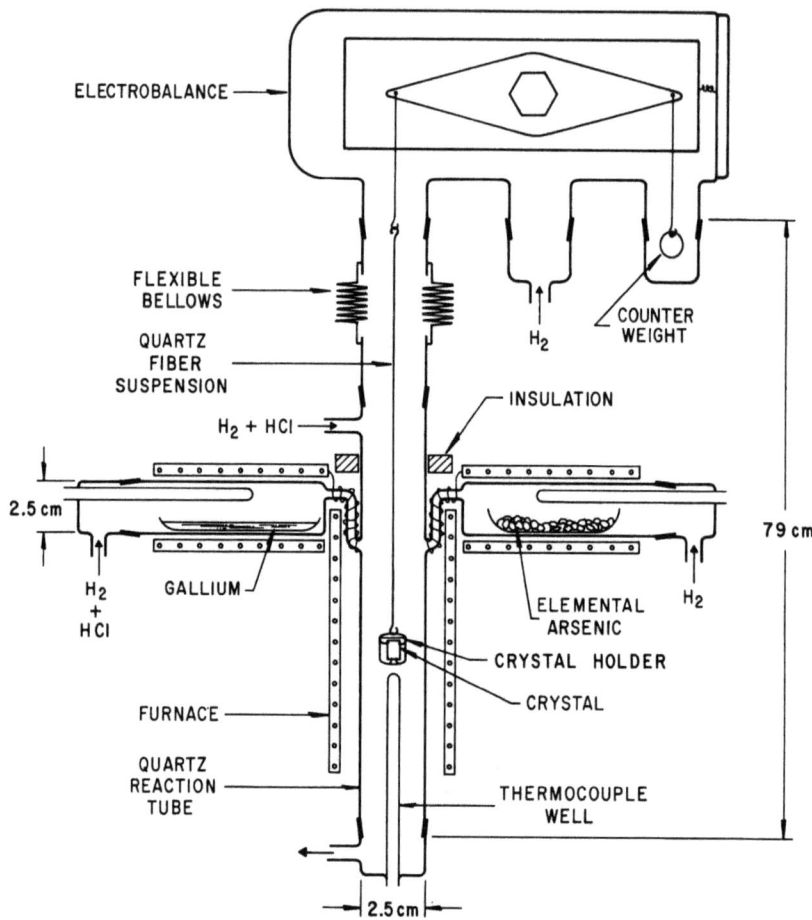

Fig. 11. Apparatus for monitoring the growth rate of GaAs.[38] (With permission of the Electrochemical Society.)

deposition while the epitaxial layer is growing. Furthermore, the electrical output from the microbalance can be electronically differentiated with respect to time, providing a continuous measurement of the actual deposition rate in addition to the weight change. The system is sensitive to changes in growth rate of approximately 0.004 μm/min. The substrate predeposition preparation can be eliminated to some extent as a variable since growth parameters can be altered and a new steady state rate can be immediately measured on the same substrate crystal. As will be shown, this approach has provided a good set of kinetic data for the gallium arsenide system even as a function of more than one variable, i.e., three-dimensional plots of rate as a function of temperature and arsenic partial pressure are available.

Another feature which has contributed to the reliability of the experimental GaAs rate data is the absence of extraneous deposition on the walls of the apparatus or substrate holder. Under many conditions GaAs will deposit only on the substrate even though the other, surrounding portions of the reactor are at the same temperature. This indicates that three-dimensional nucleation is a slow process. Extraneous deposits, if present, would compete with the substrate for reactants, thus reducing the partial pressures near the substrate from their expected values.

Let us now examine some of the experimental kinetic data for GaAs deposition, with the intention of determining the rate-limiting regime. The effect of the deposition temperature on the epitaxial growth rate[39] is shown in Figure 12. In generation of these data, four substrates of different crystallographic orientation were simultaneously subjected to deposition at each temperature. With respect to the rate–temperature behavior, there are two basic regions present in Figure 12. One is a low-temperature region, where the deposition

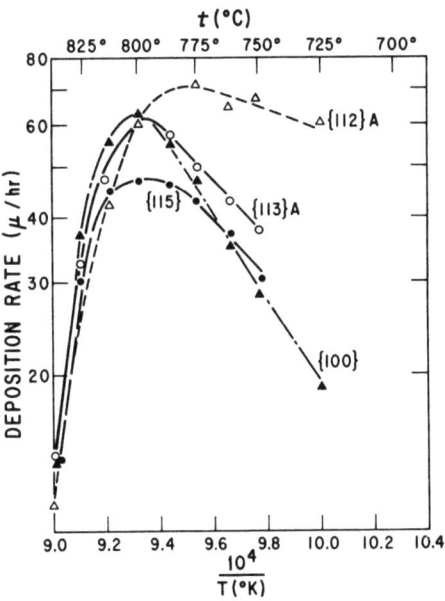

Fig. 12. Temperature dependence of GaAs deposition rate for growth on different crystallographic orientations.[39] (With permission of the Electrochemical Society.)

Chapter 6

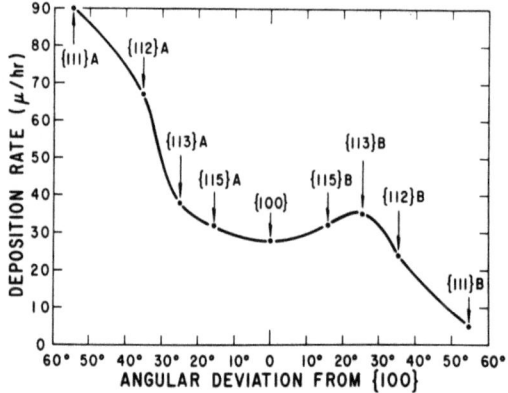

Fig. 13. Orientation dependence of GaAs deposition rate.[39] (With permission of the Electrochemical Society.)

rate increases with increasing temperature. In this region the rate is also sensitive to the substrate orientation. At higher temperatures the second region is encountered. Here the rate decreases as the temperature is increased and the rates of all orientations tend to converge.

The sensitivity of the growth rate to the substrate surface orientation at low temperatures becomes more obvious as several orientations are compared. This is illustrated in Figure 13 for deposition under the same experimental parameters. The rate for

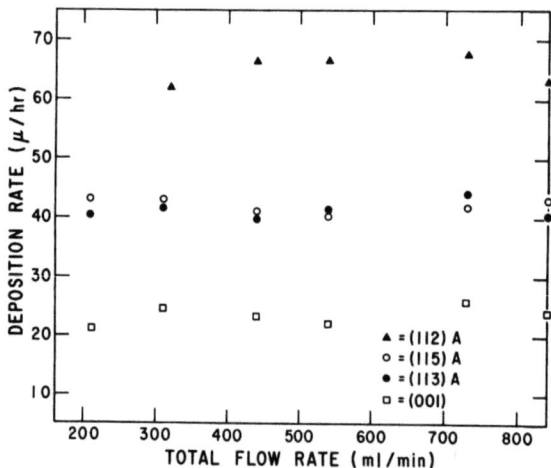

Fig. 14. Deposition rate of GaAs as a function of gas flow rate.[39] (With permission of the Electrochemical Society.)

the {111}A orientation is 16 times greater than that of the {111}B. These orientations are those referred to previously, which, in principle, have identical surface atomic geometries but different surface chemical compositions. The {111}A surface is gallium rich, while the {111}B is composed principally of arsenic atoms. Here we have an important clue to the growth mechanism. Specifically, under these conditions growth on an arsenic-rich surface appears to be inhibited; thus gallium addition appears to be a more difficult process than arsenic addition to surfaces of the same geometry and packing.

Data on the sensitivity of the growth rate to the total gas flow rate for the low-temperature region are shown in Figure 14. As is evident, the rate is essentially independent of the total flow rate.

The above data can be interpreted in terms of the fundamental rate-limiting regime. We conclude that there are two basic regions, depending on temperature. At high temperatures the rate varies inversely with the deposition temperature and is essentially independent of the substrate surface orientation. This temperature variation is qualitatively similar to the thermodynamically predicted behavior for an equilibrium process.[39] Quantitatively, however, the measured rate is considerably below that expected for complete equilibration of the incoming gas stream with the substrate. Such behavior could be due to a mass-transfer-limited process in which only a portion of the gas phase equilibrates with the substrate. Alternatively, the inverse temperature dependence could be the result of a slow etching or reverse reaction whose rate increases with increasing temperature. The latter explanation is unlikely because of the independence of the rate on substrate orientation. Thus we conclude that the high-temperature region represents a mass-transfer-controlled regime. Note that an unambiguous interpretation would not have been possible from the temperature behavior alone. Instead, it was essential to examine the rate as a function of two variables (temperature and substrate orientation).

At low temperatures the rate increases with increasing temperature. Such behavior is completely opposed to that predicted thermodynamically for this exothermic process. This fact alone is strong support for an activated, kinetically limited process. As mentioned previously for silicon deposition, endothermic processes whose rates are observed to increase with increasing temperature may be either kinetically or mass transport limited, but such behavior for an exothermic process is inconsistent with simple mass transport control. An apparent activation energy derived from the low-temperature data

Chapter 6

for the GaAs system should be representative of a chemical kinetic event rather than a thermodynamically determined heat of reaction. The conclusion of kinetic control at low temperatures is confirmed by the large variations in rate as a function of substrate surface orientation and by the independence of the rate on the total flow rate.

At this point we can summarize the experimental results and conclusions regarding rate limitations on the gallium arsenide system. The deposition process is mass transfer controlled at high temperatures. Here the rate varies inversely with temperature and, as discussed earlier, is expected to be sensitive to the flow dynamics in the vicinity of the substrate. At lower temperatures the process becomes insensitive to the flow dynamics and exhibits wide rate variations with different crystallographic orientations. Here a surface process controls the overall deposition rate. Since such processes are limited by a chemical event, we expect and observe a relatively large activation energy. The rate increases rapidly with increasing temperature. It is in this kinetically controlled region that any experimental data useful for deciphering the actual surface mechanism must be obtained. As alluded to in the previous section, such a well-defined kinetically limited region has not been demonstrated for silicon deposition. This remains the principal limitation to mechanism evaluation with the silicon system.

Further evaluation of the growth process requires knowledge of the dependence of the growth rate on the partial pressures of all

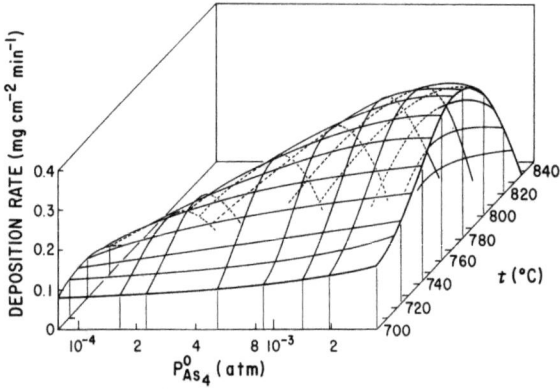

Fig. 15. Dependence of GaAs deposition rate on temperature and initial arsenic partial pressure. $p^0_{GaCl} = 3.46 \times 10^{-3}$ atm. $p^0_{HCl} = 9.75 \times 10^{-4}$ atm. (From Ref. 19 with permission of Plenum Press.)

reactants. In addition, since two growth regimes are possible with this system, some means must be provided for defining which limiting regime dominates at a given set of reactant partial pressures. In other words, one must ensure that the process remains kinetically limited over the entire range of reactant partial pressures if the data are to be applied to a mechanism determination. Fortunately, the aforementioned continuous rate monitoring apparatus provides a solution to this problem. The rate can be measured as the temperature is varied over a wide range for a given set of reactant partial pressures. The kinetic region can be identified as that portion of a rate–temperature plot where the rate is increasing with temperature. The process can then be repeated for a new set of reactant partial pressures to provide sufficient data to construct a complete rate–partial pressure–temperature diagram such as shown in Figure 15. The rate–partial pressure data on the low-temperature side of Figure 15 are taken in the kinetically limited region. Similar data can be taken to define the dependence of the rate on the partial pressure of the other reactant (GaCl).

There is good agreement for the apparent activation energy with GaAs deposition. For example, the following values have been found for $\{100\}$ deposition: 48.7 ± 3.5 kcal/mole for the $Ga/HCl/As/H_2$ system[38] and 49.1 ± 4.4 kcal/mole for the $Ga/AsCl_3/H_2$ system (the latter value was determined from 44 experimental curves).[40] These close values indicate that the same rate-limiting surface process is active in the two growth systems. Furthermore, no dependence of the activation energy on gas-phase composition has been found, indicating that the same rate-limiting process exists over the entire range of reactant partial pressures. On the other hand, different surface orientations have different activation energies. This is to be expected, since surfaces with different geometric arrangements and chemical compositions should produce different energy barriers even if the same basic reaction were limiting.

In general, curves such as Figure 15 show that the rate exhibits a fractional order dependence on the initial As_4 partial pressure. With respect to the gallium monochloride partial pressure an interesting feature is present. As shown by Figure 16, the rate actually decreases with increasing GaCl partial pressure at high values. In other words, the kinetics of the gallium arsenide deposition process is actually inhibited by one of the reactants. This inverse dependence indicates that a competitive adsorption process is involved in the growth mechanism, with GaCl competing with

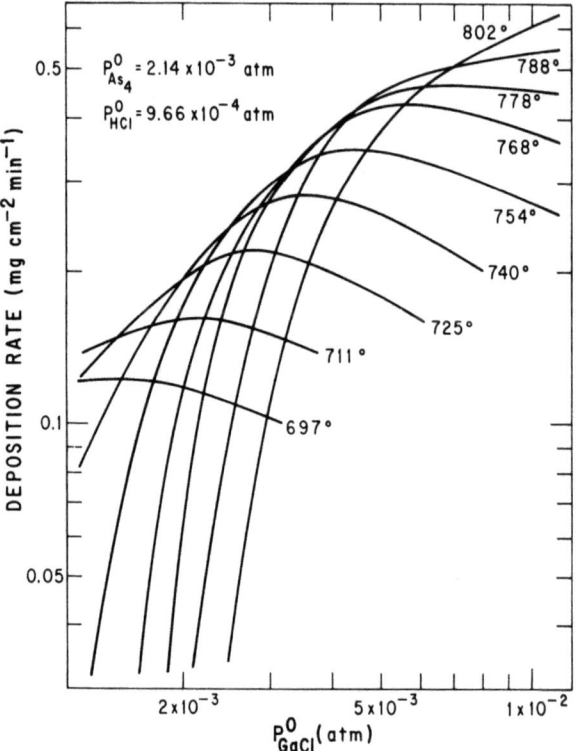

Fig. 16. Dependence of GaAs deposition rate on initial gallium monochloride partial pressure at several temperatures.[38] (With permission of the Electrochemical Society.)

arsenic for adsorption sites. For example, the As_4 molecules may be in gas-phase equilibrium with As_2, which in turn adsorb dissociatively on the {100} surface. Then, if GaCl should compete for these sites, say by attraction of the chloride portion of the molecule to the surface, the amount of surface sites available for arsenic adsorption would be reduced. Under these conditions the fraction of the surface sites occupied by arsenic atoms is given by

$$\theta_{As} = (\beta_{As_2}^{1/2} K^{1/4} p_{As_4}^{1/4})/(1 + \beta_{As_2}^{1/2} K^{1/4} p_{As_4}^{1/4} + \beta_{GaCl} p_{GaCl})$$

where K is the equilibrium constant for the reaction

$$As_4(g) = 2As_2(g)$$

and β_{As_2} and β_{GaCl} are the Langmuir adsorption coefficients for adsorption of As_2 and GaCl, respectively, at the arsenic sites. Incorporation of arsenic in the lattice requires that it be bonded to gallium atoms beneath. If we assume that the arsenic adsorption site consists of one or more lattice gallium atoms, then competitively adsorbed gallium monochloride must also bond to gallium. Such an adsorbed complex is not unreasonable since Ga–Cl–Ga bridges are well known.[41]

In view of the previous equation for θ_{As}, if the surface rate-limiting process involves a reaction of adsorbed arsenic atoms, then the rate of deposition will be expected to decrease with increasing gallium monochloride partial pressure as

$$\beta_{GaCl} p_{GaCl} > 1 + \beta_{As_2} K^{1/4} p_{As_2}^{1/4}$$

Several mechanisms may be envisioned which involve adsorbed arsenic atoms. From these, general rate expressions can be derived in the manner discussed in Section 3.2. Many of the rate expressions agree with the experimental results, but in view of the unknown adsorption coefficients and rate constants, it has so far been impossible to conclude which expression, if any of these, is the most valid. In any case, the competitive adsorption of GaCl on arsenic sites is most likely a key feature. As yet, suitable kinetic data exist only for the {001} orientation, but as data for other orientations becomes available, particularly the polar {111} orientations, the understanding of the GaAs growth mechanism should be further improved. Information on the mechanism for one orientation can be useful for deciphering mechanisms with other orientations.

So far we have discussed only the low-temperature, kinetically limited regime for GaAs deposition. At higher temperatures, as the process becomes mass transport limited, the same surface mechanism may be operative. However, the reactions are now sufficiently rapid in comparison with the gas-phase mass transport rate that reactants reaching the surface can equilibrate with it. This produces a concentration gradient within the gas phase, with partial pressures at the surface and in the bulk gas stream being the equilibrium and inlet values, respectively. This situation can be treated by the simple diffusion model described by Eq. (2). However, unambiguous computation of the boundary layer parameters is difficult in many systems. One practical approach consists in determination of the value of the combined physical constants in Eq. (2) empirically from knowledge of the growth rate at some reactant initial partial pressure and the computed equilibrium partial pressure.[38] Then, by assuming that this constant is a function only of the flow dynamics and that the reactant partial pressure is the only variable, the rate for all inlet

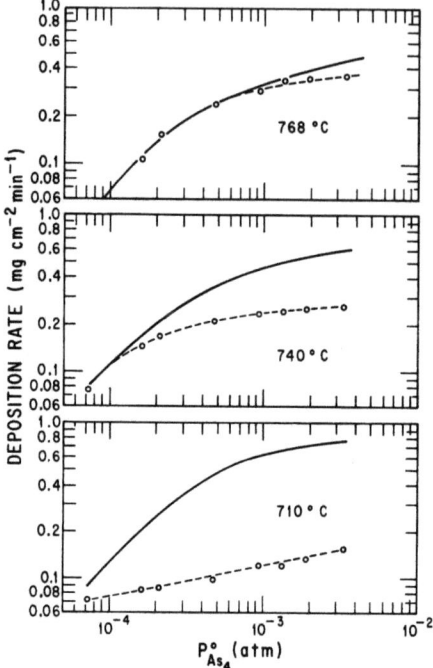

Fig. 17. Calculated[38] curves of GaAs deposition rate as a function of initial arsenic pressure for three deposition temperatures. Open circles represent experimental data.

partial pressures can be predicted. An example of this approach is illustrated in Figure 17, where the deposition rate, computed under the assumption of complete mass transfer control, is compared with experimental data. It is evident that poor agreement exists between the calculated values at low deposition temperatures; however, as the temperature is increased, the agreement becomes excellent. When the data are carefully examined, we find that the deviations between the experimental and calculated values are significant only when the data fall into the kinetic regime where the mass transfer model should not apply. At high temperatures very good agreement exists for all partial pressures.

The finite element method has been applied to mass-transfer-limited epitaxial growth.[42] This approach permits calculation of the deposition rate without resort to empirically derived constants. No assumptions of boundary layer thicknesses or velocity distributions are required. Figure 18 illustrates a comparison between the

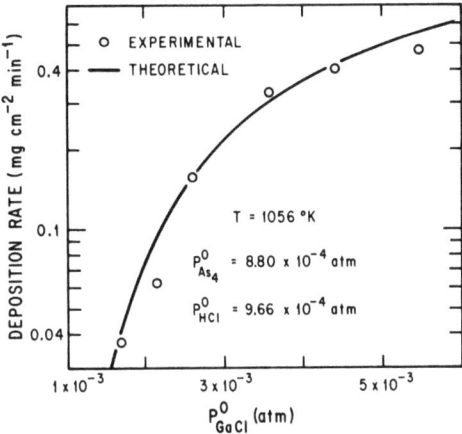

Fig. 18. Comparison of the GaAs deposition rate as a function of initial gallium monochloride partial pressure calculated by the finite element method[42] with the experimental results. (With permission of the *Journal of Crystal Growth*.)

computed and experimentally determined GaAs epitaxial deposition rates as a function of the GaCl inlet partial pressure. The model is not limited to a special system geometry. Unfortunately, the calculations are very involved and require a large-scale computer solution.

5. Summary

An attempt has been made to describe the important aspects of the growth of solids from a vapor, with particular emphasis on chemical vapor deposition. For simple systems and reactions a knowledge of the kinetics of the deposition as a function of the input variables allows determination of the rate-controlling step and, for surface-limited processes, the actual chemical mechanisms involved. For real systems where the number of vapor species can be large and the reactions complex, such as the reduction of chlorosilanes by hydrogen to form silicon or the deposition of gallium arsenide by the chloride process, an unambiguous definition of the mechanism may not be possible. To even deduce the rate-limiting regimes or to narrow the group of possible mechanisms requires the systematic gathering of a large amount of data. This has been done for the case of gallium arsenide deposition but not for the much more common and widely used silicon deposition processes.

Chapter 6

References

1. D. R. Stull and G. C. Sinke, *Thermodynamic Properties of the Elements*, American Chemical Society, Washington, D.C. (1956).
2. D. R. Stull and H. Prophet, *JANAF Thermochemical Tables*, 2nd ed., National Bureau of Standards, No. NSRDS-NBS 37 (June 1971).
3. F. Rossini, D. Wagman, W. Evans, S. Levine, and I. Jaffe, National Bureau of Standards Circular 500 (1952).
4. D. D. Wagman, W. H. Evans, V. B. Parker, I. Halow, S. M. Bailey, and R. H. Schumm, "Selected Values of Thermochemical Properties," National Bureau of Standards Technical Note 270-3 (1968).
5. A. Glassner, "Thermochemical Properties of the Oxides, Fluorides, and Chlorides to 2500 K," U.S. Atomic Energy Commission ANL-5750 (1960).
6. K. K. Kelley, U.S. Bureau of Mines Bulletin 477 (1950).
7. K. K. Kelley, U.S. Bureau of Mines Bulletin 584 (1960).
8. K. K. Kelley and E. G. King, U.S. Bureau of Mines Bulletin 592 (1961).
9. O. Kubaschewski and E. L. Evans, *Metallurgical Thermochemistry*, Pergamon Press, London (1958).
10. W. B. White, S. M. Johnson, and G. B. Danzig, Chemical equilibrium in complex mixtures, *J. Chem. Phys.* **28**, 751–755 (1958).
11. W. D. Madeley and J. M. Toguri, Computing chemical equilibrium compositions in multiphase systems, *Ind. Eng. Chem. Fund.* **12**, 261–2 (1973).
12. F. P. Boynton, Chemical equilibrium in multicomponent polyphase systems, *J. Chem. Phys.* **32**, 1880–1881 (1959).
13. R. B. Bird, W. E. Stewart, and E. N. Lightfoot, *Transport Phenomena*, Wiley, New York (1960).
14. R. F. Lever, The equilibrium behavior of the silicon–hydrogen–chlorine system, *IBM J. Res. Develop.* **8**, 460–465 (1965).
15. M. J. Harper and T. J. Lewis, "Thermodynamics of the Chlorine–Hydrogen–Silicon System," Great Britain Explosives Research and Development Establishment Report ERDE 6/M/66, Waltham, England (June 1966).
16. L. P. Hunt and E. Sirtl, in *Chemical Vapor Deposition* (J. M. Blocker, Jr. and J. C. Withers, ed.), Electrochemical Society, Princeton, New Jersey (1970), pp. 3–24.
17. L. P. Hunt and E. Sirtl, A thorough thermodynamic evaluation of the silicon–hydrogen–chlorine system, *J. Electrochem. Soc.* **119**, 1741–1745 (1972).
18. T. O. Sedgwick, Analysis of the hydrogen reduction of silicon tetrachloride process on the basis of a quasi-equilibrium model, *J. Electrochem. Soc.* **111**, 1381–1383 (1964).
19. D. W. Shaw, in *Crystal Growth—Theory and Techniques* (C. H. L. Goodman, ed.), Vol. 1, Plenum, London (1974), pp. 1–48.
20. R. W. Andrews, D. M. Rynne, and E. G. White, Effect of reactor geometry on growth rate of epitaxial silicon, *Solid State Tech.* **1969**, 61–66.
21. W. H. Shepherd, Vapor phase deposition and etching of silicon, *J. Electrochem. Soc.* **112**, 988–994 (1965).
22. H. C. Theuerer, Epitaxial silicon films by the hydrogen reduction of $SiCl_4$, *J. Electrochem. Soc.* **108**, 649–653 (1961).
23. E. G. Bylander, Kinetics of silicon crystal growth from $SiCl_4$ decomposition, *J. Electrochem. Soc.* **109**, 1171–1175 (1962).

24. E. G. Alexander, A surface reaction approach to the growth kinetics of epitaxial silicon from $SiCl_4$, *J. Electrochem. Soc.* **114**, 65C (1967).
25. W. Runyan, in *Semiconductor Silicon* (R. R. Haberecht and E. L. Kern, eds.), Electrochemical Society, Princeton, New Jersey (1969), pp. 169–188.
26. S. K. Tung, The effects of substrate orientation on epitaxial growth, *J. Electrochem. Soc.* **112**, 436–438 (1965).
27. M. E. Jones, in *Reactivity of Solids* (J. W. Mitchell, R. C. Devries, R. W. Roberts, and P. Cannon, eds.), Wiley, New York (1969), pp. 433–451.
28. S. E. Bradshaw, The kinetics of epitaxial silicon deposition by the hydrogen reduction of chlorosilanes, *Int. J. Electronics* **21**, 205–227 (1966).
29. S. E. Bradshaw, The effects of gas pressure and velocity on epitaxial silicon deposition by the hydrogen reduction of chlorosilanes, *Int. J. Electronics* **23**, 381–391 (1967).
30. F. C. Eversteyn, P. J. W. Severin, C. H. J. v. d. Brekel, and H. L. Peek, A stagnant layer model for the epitaxial growth of silicon from silane in a horizontal reactor, *J. Electrochem. Soc.* **117**, 925–391 (1970).
31. K. Sugawara, Silicon epitaxial growth by rotating disc method, *J. Electrochem. Soc.* **119**, 1740–1760 (1972).
32. R. Takahashi, Y. Koga, and K. Sugawara, Gas flow pattern and mass transfer analysis in a horizontal flow reactor for chemical vapor deposition, *J. Electrochem. Soc.* **119**, 1406–1412 (1972).
33. J. P. Dismukes and B. J. Curtis, in *Semiconductor Silicon 1973* (H. R. Huff and R. R. Burgess, eds.), Electrochemical Society, Princeton, New Jersey (1973), pp. 258–270.
34. R. R. Fergusson and T. Gabor, The transport of gallium arsenide in the vapor phase by chemical reaction, *J. Electrochem. Soc.* **111**, 585–592 (1964).
35. H. Seki, K. Moriyama, I. Askakawa, and S. Horie, Thermodynamics study of transport and epitaxial growth of GaAs in an open tube, *Japan J. Appl. Phys.* **7**, 1324–1331 (1968).
36. D. T. J. Hurle and J. B. Mullin, Thermodynamics of gas-phase equilibria: the Ga:As:Cl:H system, *J. Phys. Chem. Solids, Suppl.* 1, p. 241 (1967).
37. V. S. Ban, Mass spectrometric studies of vapor phase crystal growth, *J. Electrochem. Soc.* **118**, 1473–1478 (1971).
38. D. W. Shaw, Epitaxial GaAs kinetic studies: {001} orientation, *J. Electrochem. Soc.* **117**, 683–687 (1970).
39. D. W. Shaw, Influence of substrate temperature on GaAs epitaxial deposition rates. *J. Electrochem. Soc.* **115**, 405–408 (1968).
40. D. W. Shaw, Kinetics of transport and epitaxial growth of GaAs with a Ga–$AsCl_3$ system, *J. Cryst. Growth* **8**, 117–128 (1971).
41. I. A. Sheka, I. S. Chaus, and T. T. Mityureva, *Chemistry of Gallium*, Elsevier, Amsterdam (1966).
42. B. G. Secrest, W. W. Boyd, and D. W. Shaw, Application of the finite element method to mass transport limited epitaxial growth processes, *J. Cryst. Growth* **10**, 251–259 (1971).

7

Crystal Growth from the Melt

J. R. Carruthers
Bell Laboratories
Murray Hill, New Jersey

1. Introduction

Growth from the melt by solidification is the most widely used method for the preparation of large single crystals. Increasingly large numbers of materials have been prepared as single crystals by melt growth techniques for many diverse applications in solid state and quantum electronics as well as for basic studies in solid state physics and chemistry. These applications have, in turn, yielded much fundamental information on the relation of growth variables to the physical properties of the various crystals. These relationships, together with independent studies of melt growth, have provided a very detailed understanding of the dynamics of the processes and, in recent years, have led to an increased degree of control of properties during crystal growth. This chapter will emphasize these recent advances in the control of crystal structure, composition, shape, and dimensions during melt growth because many future advances in electronics and optics will rely heavily on the ability to control properties over dimensions ranging from the atomic scale to the macroscopic size of the crystal itself.

An understanding of melt growth processes involves some knowledge of phase equilibria, fluid dynamics, and defect structures in addition to a familiarity with the applications of the various materials. Consequently, studies of these phenomena relevant to crystal growth will be included in this chapter.

Other accounts of melt growth may be found in the books by Laudise,[1] Pfann,[2] Chalmers,[3] and Brice.[4] Excellent tutorial descriptions of crystal growth from melts are given by Hurle,[5] Mullin,[6] and Cockayne.[7]

This chapter will not include a description of crystal growth kinetics and the reader should consult the chapter in this volume by Jackson[8] for details. Also relevant to melt growth is the chapter by Wernick[9] in another volume of this treatise, on the relation of structure and composition to properties. The topic of polyphase growth, such as eutectic solidification from melts, also will not be included in this chapter.

2. Basic Principles and Techniques

2.1. Principles of Solidification Processes

Materials which do not decompose or dissociate uncontrollably upon melting may be prepared by unidirectional solidification as single crystals. Such materials should possess components with low or controllable vapor pressures and should melt congruently so that the crystal and melt have similar compositions and the growing interface conforms to the solidification isotherm in the system. A nucleus or seed is usually provided to initiate growth.

The interface is maintained in a stable position and spurious nucleation prevented by imposing a positive temperature gradient across the liquid (temperature increasing in the melt away from the interface). The growth interface is destabilized and becomes dendritic if the temperature gradient is negative in the liquid at the growth interface in a pure (one-component) system or if the equilibrium liquidus temperature becomes less than the actual melt temperature in a polycomponent system. The latter condition is known as constitutional supercooling and recent studies will be discussed in more detail later.

Growth is made to occur by moving the crystal–melt interface relative to the solidification isotherm of the system. This condition provides the thermal undercooling required to drive the growth process and ensures that the interface remains morphologically stable, with the generated latent heat of fusion being dissipated by conduction through the crystal. Materials possessing complex structural rearrangements at the growth interface require large undercoolings and generate large latent heats of fusion. Such materials, as, for example, polymers, cannot be grown easily with a stable, planar growth interface and so will not be considered further here.

2.1.1. Segregation during Steady-State Crystal Growth

In most growth systems of interest, the melt possesses dilute concentrations of components which are present either as intentionally added dopants, as impurities, or as nonstoichiometric components of the crystal itself. These extra components are segregated at the growth interface in accordance with a phase diagram relating the equilibrium solubilities of the components in the solid and liquid. The growth interface will be assumed to be advancing uniformly in a time-independent manner. The incorporation kinetics of the components can be analyzed in terms of mathematical solutions of the pertinent diffusion equations.

The equilibrium segregation behavior associated with the solidification of multicomponent systems can be determined from the corresponding phase diagram, as shown schematically in Figure 1 for a binary system. A melt with an initial composition C_L at a temperature T_i is in equilibrium with a solid phase of composition C_s.

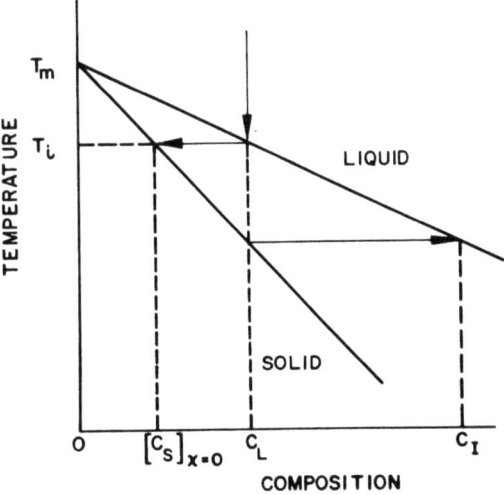

Fig. 1. Schematic equilibrium phase diagram of a binary or pseudobinary system in the dilute composition range (where liquidus and solidus curves are linear and decreasing, and $T_m - T_i$ is very small).

Chapter 7

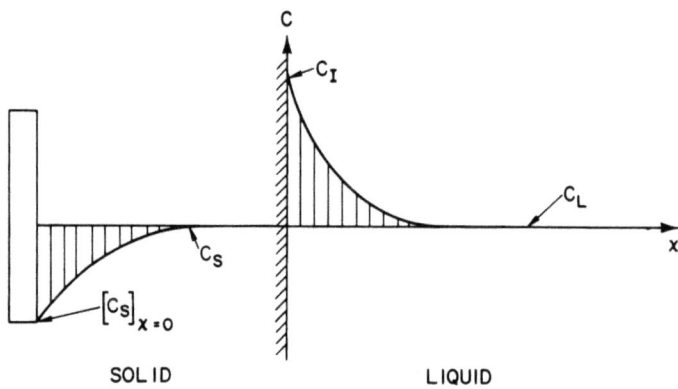

Fig. 2. Compositional solute profiles in the solid and liquid generated by uniform interface advance from $x = 0$ into a convectionless melt.

It will be assumed throughout this chapter that the solute concentrations are so dilute that the temperature T_i is nearly equal to the melting temperature of the pure solvent T_m and that T_i does not change significantly during growth due to compositional changes. Thus this treatment is not applicable to top-seeded solution growth.

With the initiation of growth at a given seed–melt interface, segregation takes place and the rejected component accumulates at the "liquid side" of the growth interface and diffuses in the direction of the bulk liquid. In the absence of convective influences, the melt composition at the interface reaches, with continuing growth, a steady-state value C_I at which the composition of the solidifying crystal C_s is the same as the composition C_L of the bulk liquid at a distance from the interface. This situation is indicated schematically in Figure 2.

If it is assumed that segregation at the growth interface is an equilibrium process, then the interface segregation coefficient k_I is identical with k_0, the equilibrium segregation coefficient, as obtained from the corresponding phase diagram:

$$k_I = C_s/C_I = k_0 \qquad (1)$$

Since the determination of C_I is associated with experimental difficulties, it is customary to use an "effective segregation coefficient" defined as

$$k_e = C_s/C_L \qquad (2)$$

where C_L is the composition of the bulk melt. It can be seen from Figure 2 that in convectionless melts for steady state conditions k_e assumes the value of one:

$$k_e = C_s/C_L = 1 \qquad (3)$$

There is experimental evidence in some systems that interface segregation is not an equilibrium process $(k_I \neq k_0)$.[10-16] Nonequilibrium segregation at the interface should, in principle, be expected at any finite growth rate and deviations from equilibrium should accordingly be proportional to the growth rate[10-13]; the extent of deviation from equilibrium is furthermore affected by the interface configuration on the atomic scale (for example, the "facet effect") and by specific adsorption–desorption kinetics which are sensitive to the chemical and physical nature of the system.[14-16] One particular model of segregation at growth interfaces which considers such nonequilibrium behavior has been proposed by Hall.[10,11] In this model, growth is assumed to occur by nucleation and rapid lateral growth of atomic-sized ledges across the interface. The lateral growth velocities are of such a magnitude that some solute atoms, which are adsorbed on the interface, may be trapped and incorporated into the solid. Under these conditions, a limited degree of equilibration is achieved by solute diffusion through the freshly grown solid back to the growth interface. If new layers are nucleated at a high rate, the originally trapped solute does not reach the equilibrium concentration.

In most melt growth systems, convective processes are present which influence solute mass transport in the melt close to the interface. Consequently, the contribution of convective diffusion to the segregation processes must be considered. One simple method of treating this problem is to assume that a diffusion boundary layer exists adjacent to the growth interface within which mass transport is diffusion controlled and outside of which liquid mixing is infinitely fast. Such a layer δ can be defined as shown in Figure 3 by the linear approximation[17]

$$(C_I - C_L)/\delta = [\partial C/\partial x]_{x=0} \qquad (4)$$

There are many other descriptions of convective diffusion processes which have more physical significance than Eq. (4). However, Burton, Prim, and Slichter (BPS)[17] used the simple relation to obtain solutions to the diffusion equations and obtained the following

Chapter 7

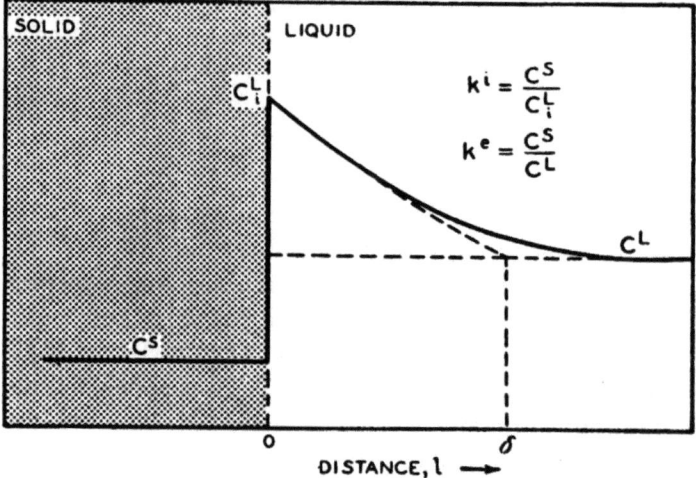

Fig. 3. Concentration profile at an advancing growth interface of the solute shown in Figure 1 when diffusion transport is limited to the region δ by convection in the melt. (From Ref. 17.)

relation between k_e and k_I:

$$k_e = \frac{k_I}{k_I + (1 - k_I)\exp(R\delta/D)} \qquad (5)$$

where k_I is the interface distribution coefficient, frequently assumed to be identical with k_0, R is the microscopic growth rate, D is the diffusion constant, and δ is the thickness of the diffusion boundary layer. [In systems subjected to forced convection (through seed rotation) it is customary to use the Cochran analysis[18] for the determination of δ.] As can be seen from the BPS analysis, any changes in growth rate R, diffusion boundary layer thickness δ, or bulk liquid composition will result in a change of k_e and thus a change in solid composition C_s. The relationship between k_e and $R\delta/D$ is plotted in Figure 4 for various values of k_I. It can be seen that convective processes exert a major influence on k_e when $R\delta/D \sim 1$ and that k_e varies from k_I under convection-controlled solute transport conditions ($D/R \gg \delta$) to unity for diffusion-controlled conditions ($D/R \ll \delta$). The latter condition implies that although the melt may possess convection, it will not influence solute segregation if the diffusion boundary layer thickness is greater than the diffusion distance D/R.

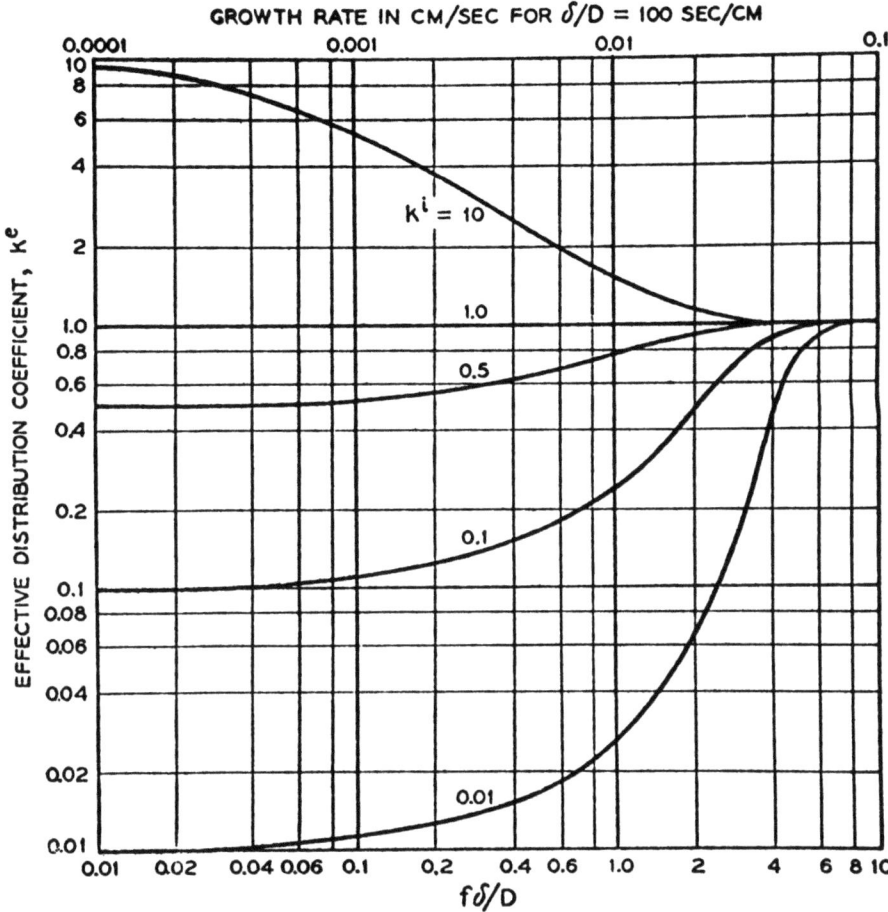

Fig. 4. The effective diffusion coefficient k^e (or k_e), as a function of the growth rate parameter $f\delta/D$ (or $R\delta/D$) for various values of the interface distribution coefficient k^i (or k_I). (From Ref. 17.)

Although the BPS model is useful for a well-behaved growth system, it has very limited utility for most melt growth, for the following reasons:

1. The diffusion boundary layer thickness δ is seldom, if ever, constant across the growth interface.[19–21]

2. The mechanisms of the transport processes are not a part of this model, so that effects such as density-driven flows due to compositional gradients and also boundary layer separation effects as observed by Carruthers and Winegard[22] cannot be assessed (see

Chapter 7

Levich[23] for a complete discussion of such convective diffusion analyses).

3. The analysis is basically a steady-state solution to the diffusion equation as used by most workers, although transient solutions for small-amplitude sinusoidal growth rate oscillations were worked out by Lewis (see Ref. 17; also see Ref. 16). However, most melt growth systems exhibit very *large*-amplitude growth rate oscillations and even remelting, for which the analysis does not apply.[24–26] In addition, the analysis cannot be used for situations where the mass transport processes are unsteady.[27]

4. The model neglects the nonequilibrium contributions to solute segregation, such as those discussed previously in this section. It appears that for all melt growth systems where such effects have been studied, the time-dependent behavior of the growth rate is the basic contributor to the nonequilibrium.[28]

Another type of compositional profile in melt-grown crystals is the normal freezing distribution of solute which results from the unidirectional solidification of stirred melts, as opposed to the quiescent melt shown in Figure 2. Since the solute rejected at the interface finds its way back into the bulk liquid under these conditions, then the melt concentration is continually changing during growth and this results in a continual variation in solute composition along the growth direction of the crystal. The solid composition C_s is then related to the initial melt composition C_0 and fraction solidified g by

$$C_s/C_0 = k_0(1 - g)^{k_0 - 1} \qquad (6)$$

A typical normal freezing variation is shown in Figure 5 as curve n.f.

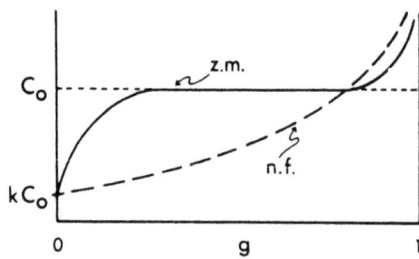

Fig. 5. Variation of the solute concentration along the growth direction of a unidirectionally solidified liquid (n.f.) and zone melted charge (z.m.) in the presence of convection in the melt. (From Ref. 5.)

for $k_0 = 0.2$. It can be seen that a continual change in solid composition occurs along the growth direction, as opposed to the case of a quiescent melt (Figure 2), where, after the initial transient, the composition remains constant. It is possible to achieve a longitudinal solute distribution in crystals grown from stirred melts similar to that obtained for quiescent melts by melting only a small zone of length l rather than the entire charge material. The resulting solute profile in the growth direction x is

$$C_s/C_0 = 1 - (1 - k_0)\exp(-kx/l) \tag{7}$$

The resulting profile is shown in Figure 5 as curve z.m. for $k_0 = 0.2$. Such a process is known as zone melting and such techniques will be

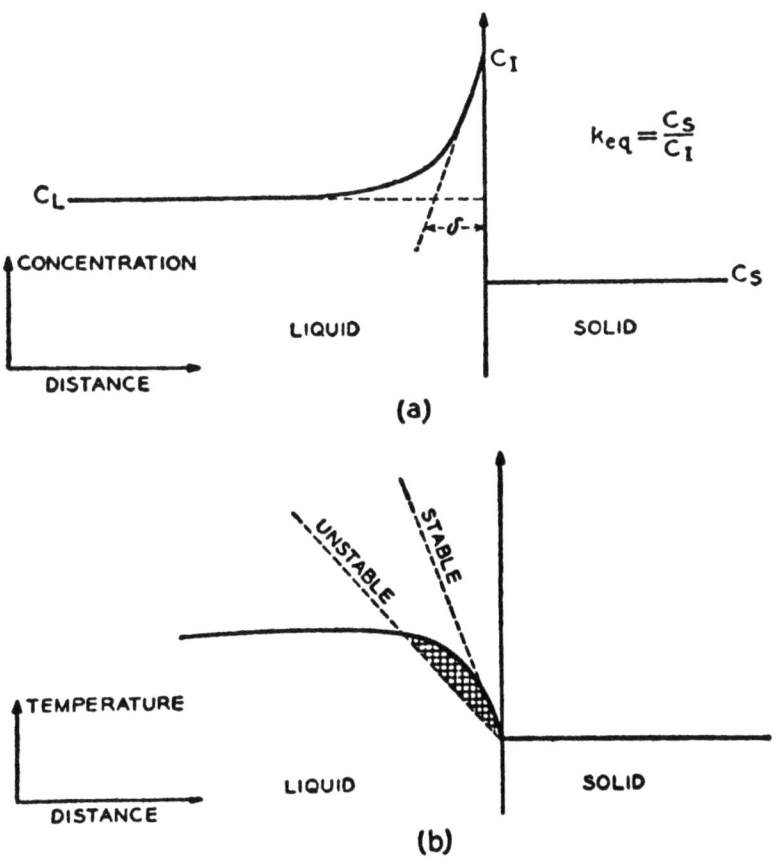

Fig. 6. Illustration of conditions leading to constitutional supercooling in the melt ahead of an advancing growth interface; (a) solute concentration profile, (b) liquidus profile associated with (a) and the actual liquid temperature (dashed line) for stable and unstable growth. (From Ref. 5.)

discussed in more detail later. The longitudinal solute distributions in most crystals occur under intermediate conditions of melt convection where $k_0 < k_{eff} < 1$. Equations (6) and (7) can be used in such circumstances by replacing k_0 with k_{eff}, providing $R\delta/D < 1$ (see Figure 4). Such profiles are often used to determine k_0 by extrapolating the growth rate dependence of k_{eff} to $R = 0$.

The existence of a region of solute enhancement (or depletion for $k_0 > 1$) ahead of the advancing growth interface because of incomplete mixing in the melt can lead to the onset of constitutional supercooling and associated cellular interface morphology under some circumstances. The situation is depicted in Figure 6, where the enhanced solute concentration profile is shown in (a) and the associated equilibrium liquidus temperature profile in (b). If the actual temperature gradient at the growth interface $[\partial T/\partial x]_{x=0}$ is greater than the liquidus temperature gradient, then the interface is stable. However, if the actual temperature gradient is lower than the liquidus temperature gradient, then a region of "constitutional supercooling" exists (shaded region in Figure 6b) in the presence of which the crystal–melt interface is unstable. The criterion for the existence of a region of constitutional supercooling in the melt is

$$[\partial T/\partial x]_{x=0} < m[\partial C/\partial x]_{x=0} \tag{8}$$

where m is the slope of the liquidus line on the appropriate phase equilibrium diagram. The compositional profile associated with quiescent melts is

$$C/C_L = 1 + [(1 - k_0)/k_0]e^{-X} \tag{9}$$

and for stirred melts is

$$C/C_L = k_{eff} + (1 - k_{eff})e^{\Delta - X} \tag{10}$$

where $X = Rx/D$ and $\Delta = R\delta/D$. Thus the critical temperature gradient for the onset of constitutional supercooling can be written as

$$[\partial T/\partial x]_{x=0} < -mC_L k_{eff}[(1 - k_0)/k_0]R/D \tag{11}$$

where $k_{eff} = 1$ for quiescent melts and $k_0 \leq k_{eff} < 1$ for stirred melts. These relations show that fast growth rates and small values of k_0 promote constitutional supercooling. Also, increased stirring can be seen to discourage constitutional supercooling through the lowering of the compositional gradient at the interface by a factor of k_{eff} over the quiescent case. Further developments regarding morphological stability will be dealt with in Section 4.1.

The considerations of solute segregation profiles dealt with in this section have indicated that crystals with uniform compositions over macroscopic dimensions may be prepared from large melts by eliminating all convective influences on solute mass transport during growth ($D/R \ll \delta$). In this way, uniform solute distributions can be obtained in the crystal both along the growth direction (Figure 2) as well as across the growth interface. However, these same conditions may also lead to constitutional supercooling, so that the effective growth rate and convection range for such composition control is

$$mC_L k_{\text{eff}}[(1 - k_0)/k_0]/G_L^0 < D/R \ll \delta \qquad (12)$$

where $G_L^0 = [\partial T/\partial x]_{x=0}$. Further considerations of composition control thus involve a more detailed understanding of the fluid dynamics of crystal growth melts and will be discussed in Sections 4.2.2 and 5.2.

2.1.2. Transient Segregation Phenomena

The time-dependent incorporation of solute during crystal growth occurs in the presence of growth rate fluctuations, diffusion boundary layer fluctuations, and changes in melt composition. These effects will all produce a striated or banded solute pattern parallel to the crystal–melt interface at that particular position. Such a distribution is often referred to as microsegregation because the spatial periods may range from microns to millimeters.

Growth rate fluctuations most commonly arise from the temperature fluctuations associated with unstable thermal convection as well as the rotation of the growth interface through an asymmetric thermal field. Vibrations and pressure fluctuations have also been identified as sources of growth rate variations. Transient solutions for the amplitude of compositional striations resulting from small sinusoidal growth-rate oscillations were first given by Lewis (see Ref. 17 with a full solution reproduced in Ref. 16). These solutions include the effects of convection on mass transport and apply reasonably well to small-amplitude rotational striations in Czochralski growth. Similar work was also reported by Hurle et al.,[25,26] who investigated the effect of temperature oscillations in the melt on segregation behavior during growth under conditions of convective transport. Their theoretical treatment indicated that for certain frequencies of temperature oscillation, resonant coupling should occur between thermal and solute fields to give large compositional

Chapter 7

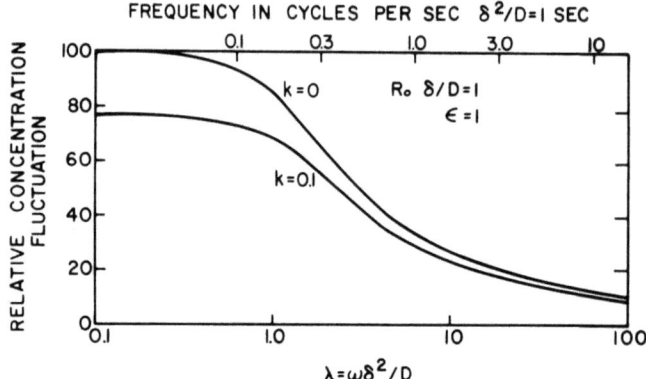

Fig. 7. Amplitude of compositional changes across striations introduced by small-amplitude sinusoidal growth-rate changes in stirred melts as a function of dimensionless frequency λ for different equilibrium distribution coefficients. The growth-rate amplitude ε and parameter $R\delta/D$ are unity for the curves shown. (From Ref. 17.)

amplitudes. This effect should be particularly pronounced in metals at low frequencies. They also reported that whenever the amplitude of the resulting growth-rate fluctuations, ΔR, is equal to R, the effective distribution coefficient k_e is reduced to one-half the value given by Eq. (5). For growth-rate fluctuations with small amplitudes ($\Delta R/R < 0.5$), the theories of Lewis (see Ref. 17) and Hurle et al.[25,26] give identical results. Figure 7 shows the relative amplitude of the composition variation with dimensionless frequency. The segregation behavior associated with growth-rate fluctuations of low frequencies (fluctuation periods τ larger than $2\pi\delta^2/D$) can be obtained from the steady state solution given in Eq. (5) of the diffusion equation. Thus for typical values of $\delta^2/D \approx 1$ sec and fluctuation periods longer than about 10 sec

$$\frac{\Delta C}{C} = \frac{\Delta R}{k_e} \frac{\partial k_e}{\partial R} \tag{13}$$

from which

$$\frac{\Delta C}{C} = \frac{\Delta R}{R} \frac{(1 - k_0)(R\delta/D) \exp(-R\delta/D)}{k_0 + (1 - k_0) \exp(-R\delta/D)} \tag{14}$$

In Figure 8 the amplitudes $\Delta C/C$ are plotted against $R\delta/D$ for three values of k_0. It is obvious from this plot that in order to optimize segregation homogeneity, experimental conditions associated with

Crystal Growth from the Melt

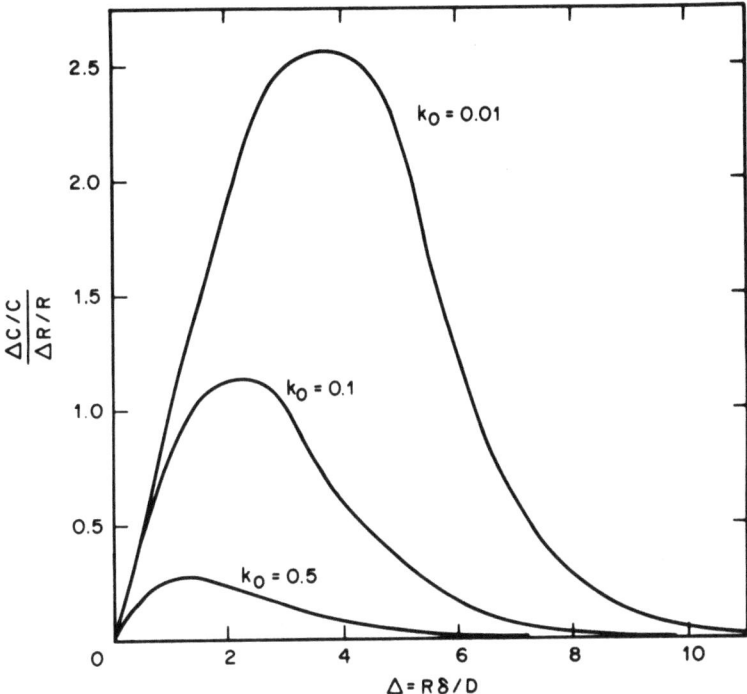

Fig. 8. Steady state compositional changes across striations (normalized by growth-rate amplitude) as computed for stirred melts, plotted as a function of the growth parameter $R\delta/D$ for various equilibrium distribution coefficients. This behavior is observed at low growth-rate fluctuation frequencies only [see Eq. (14)].

the maxima of the three curves are to be avoided. Under conditions where $R\delta/D \ll 1$, Eq. (14) assumes the form

$$\frac{\Delta C/C}{\Delta R/R} = (1 - k_0)\frac{R\delta}{D} \qquad (15)$$

This expression (corresponding to the linear portion in Figure 8—lower left corner) is a useful approximation. It should, however, be recalled that its applicability is restricted to conditions involving only low-frequency and low-amplitude growth-rate fluctuations. Alternately, minimization of $(\Delta C/C)/(\Delta R/R)$ to achieve optimum segregation homogeneity is also possible by establishing experimental conditions corresponding to $R\delta/D \gg 1$ (lower right corner in Figure 8). However, these conditions generally give rise to constitutional supercooling and therefore are not extensively applied. No analytical solutions to the diffusion equations are as yet available for diffusion-controlled solute transport in the absence of convection with sinusoidal growth-rate perturbations.

Under conditions where the growth-rate change is large ($\Delta R/R \geq 1$), theoretical analysis indicates that the concentration amplitude decreases to one-half the value predicted from Eq. (13).[25,26] Concentration profiles resulting from growth rate fluctuations which involve back melting have not yet been theoretically derived.

For high-frequency growth rate fluctuations ($\omega \gg D/\delta^2$), the amplitude of the associated compositional fluctuations has been shown to be below the value given in (15) and assumes the following form[26]:

$$\frac{\Delta C/C}{\Delta R/R} = (1 - k_0)\frac{R\delta/D}{(2\omega\delta^2/D)^{1/2}} \qquad (16)$$

Approximate solutions to the diffusion equations for compositional amplitudes produced by growth rate fluctuations of intermediate frequency can be found in Ref. 26.

Many convective flow phenomena must be expected to display a time-dependent behavior in terms of velocity perturbations. During crystal growth such perturbations may influence the diffusion boundary layer thickness and consequently affect the incorporated solute concentration in much the same manner as the growth rate. In fact the amplitude of compositional fluctuations resulting from periodic changes in diffusion boundary layer thickness can be estimated at low frequencies from Eq. (14) if $\Delta R/R$ is replaced by $\Delta\delta/\delta$. No theoretical work for high-frequency perturbations has been reported. It is, however, possible to estimate a cutoff frequency ω_m for which velocity perturbations penetrating through the momentum boundary layer no longer affect the diffusion boundary layer thickness:

$$\omega_m = v/[\delta_m(\delta_m - \delta)] \qquad (17)$$

where v is the kinematic viscosity, δ_m is the momentum boundary layer thickness, and $\delta = (D/v)^{1/3}\delta_m$ is the diffusion boundary layer thickness.

Taking representative values for semiconductor melts and commonly used rates of seed rotation ($v = 2.5 \times 10^{-3}$ cm^2/sec, $D = 10^{-4}$ cm^2/sec, $\delta_m = 0.1$ cm), then $\omega_m \cong 0.4$ sec^{-1}. From the low value of the cutoff frequency it can be seen that only relatively slow velocity changes will influence the diffusion boundary layer thickness. It is of interest to note that this cutoff frequency is of the same general

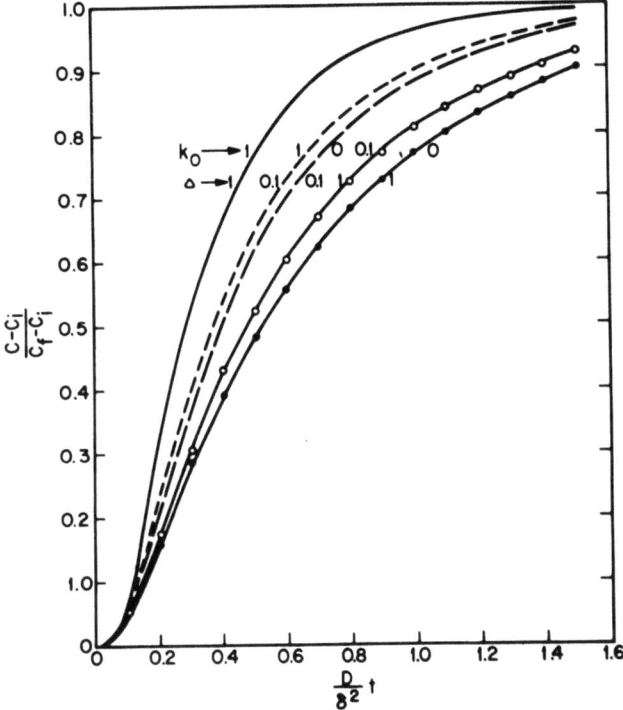

Fig. 9. Compositional change in the solid produced by an abrupt change of dopant concentration in a stirred melt, plotted as a function of dimensionless time for various equilibrium distribution coefficients k_0 and dimensionless growth parameter $\Delta = R\delta/D$. (From Ref. 17.)

magnitude as the cutoff frequency at the growth interface for thermal oscillations originating in the melt.

The third source of compositional striations is the abrupt change in the melt composition such as produced, for example, by dopant additions during growth. Theoretical solutions to the corresponding transient diffusion equations have been reported for the convective transport case.[17] These results are shown as a function of normalized time in Figure 9. For a wide range of conditions (k_0, $R\delta/D$), near-steady-state composition is reached after a time t given by

$$t \simeq 2\delta^2/D \tag{18}$$

For commonly employed semiconductor growth conditions in the presence of forced convection, the equilibration time t after any melt composition change is approximately 20 sec. It is assumed here that the magnitude of the liquid composition change is too small to measurably affect the liquidus (interface) temperature. Much longer times will be required to reach steady-state interface concentrations in convectionless growth:

$$t \simeq 2D/R^2 \qquad (19)$$

Again, no theoretical solutions are available for this case as yet.

The overwhelming interest in compositional control and segregation processes is reflected in the classification scheme often used for melt growth systems, the recent highlights of which are described next. Basic details of these techniques can be found in Refs. 1–7.

2.2. Normal Freezing Techniques

In normal freezing techniques an initially molten charge is directionally solidified. The three major techniques vary in the orientation of the melt and associated temperature gradients with respect to gravity.

2.2.1. Czochralski Growth

In order to understand the detailed interactions between crystal growth variables and compositional distribution in crystals, the Czochralski growth process is described in detail because it has been studied extensively.

The Czochralski growth geometry is shown schematically in Figure 10. The material to be grown is first melted by induction or resistance heating under a controlled atmosphere in a suitable nonreacting container. The melt temperature is then adjusted to be slightly above the melting point and a seed crystal is lowered to the melt surface. After thermal equilibration, the seed is contacted with the melt and the melt temperature is raised to establish the desired growth interface configuration. By activating a pulling mechanism, the seed is withdrawn from the melt (with simultaneous temperature adjustment) at such a rate that the crystal diameter is gradually increased to its desired value (frequently a "necking" process is used initially to reduce the dislocation density in the grown material). The growth interface in this configuration is expected to conform at all times to the location and shape of the solidification isotherm within the constraints imposed by the liquid meniscus (of curvature ρ and height h). Growth at constant crystal diameter is achieved by

Crystal Growth from the Melt

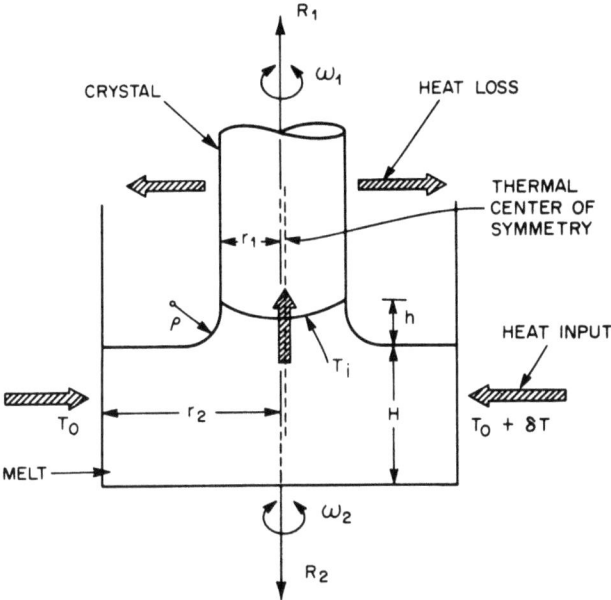

Fig. 10. Schematic illustration of growth parameters in the Czochralski process.

maintaining the solidification isotherm in a vertical position intersecting the meniscus at the point where the isotherm becomes perpendicular to the melt surface. This condition is maintained by adjustments of any one or more of the following parameters: the rate of pulling, the rate of liquid level drop (determined by the relative diameters of crystal and crucible), the heat fluxes into and out of the system, and, to a lesser extent, by seed and/or crucible rotation. Full dimensional control is frequently achieved by automatic control techniques to be described later, employing both analog and digital systems.

It should be realized that the crystal growth rate under the above conditions cannot be expected to be equal to the imposed pulling rate. The macroscopic growth rate is always greater than the pulling rate since the melt level drops during growth. The instantaneous microscopic growth rate can deviate substantially from the macroscopic growth rate, as will be discussed later.

The basic melt flow in the nonrotating growth system is a thermal convection shear flow generated by the imposed radial (horizontal) temperature gradient required for growth.[19] In growth systems where the thermal center coincides with the geometric center of the

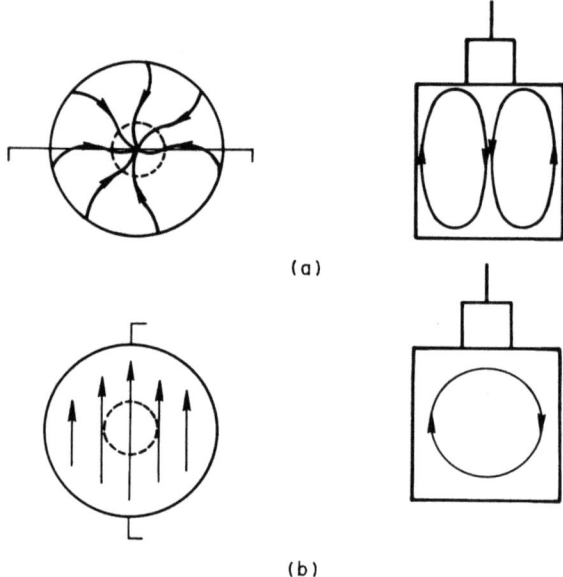

Fig. 11. Thermal convection flow patterns in Czochralski melts: (a) axisymmetric flow, (b) antisymmetric flow.

crucible, symmetric melt flows are established where hot liquid rises at the crucible wall and moves radially inward at the top surface to the cooler thermal center, where it descends to the crucible bottom.[29] This basic flow pattern is shown in Figure 11(a). The exact pattern established is very sensitive to the boundary conditions imposed on temperature and melt flow velocity, the melt aspect ratio (height to radius), and the thermal and viscous properties of the liquid. The melt flow pattern believed to be predominantly established in Czochralski growth systems is indicated in Figure 11(b). A detailed understanding of these flows is not yet available. However, there are four basic features which are important to crystal growth processes:

1. The nonuniform nature of the thermal convection flow at different points of the interface results in a correspondingly nonuniform diffusion boundary layer thickness δ and thus is responsible for radial variations in segregation at the freezing interface.

2. The unsteady (time-dependent) behavior of some thermal convection flows leads to temporal (transient) variations in R and δ, which in turn both generate compositional striations.

3. The lack of coincidence between the thermal center of symmetry and the geometric center of the melt leads to a drifting of the growing crystal from its original position.

4. The thermal convection flows may adversely affect the shape of the solidification isotherm (growth interface morphology).

These features of thermal convection flows are undesirable and therefore make it necessary to provide control over melt flow velocities and temperature distributions at the growth interface by other means. In Czochralski growth this approach has led to the introduction of crystal rotation. Such rotation provides a forced convection flow which tends to "isolate" the growth interface from thermal convection in the bulk liquid. Furthermore, under rotation, the diffusion boundary layer thickness should become more uniform (theoretically, δ is constant over a flat, rotating interface in an otherwise quiescent liquid) and the solidification isotherm shape becomes flatter. However, the effects of crystal rotation are not exclusively beneficial. If the growing crystal precesses about the rotation axis or if the rotation axis and thermal center of symmetry do not coincide, or if the thermal field is asymmetric, the growth interface during crystal rotation will be subject to a continuously varying temperature field over each rotational cycle. Consequently the instantaneous growth rate must be expected to vary continuously and thus produce compositional inhomogeneities with a period equal to the amount of crystal grown during each revolution (rotational striations).[15,30,31]

In melts where thermal asymmetries are pronounced it has become a common practice to slowly rotate the crucible in addition to rotating the seed in order to reduce the extent of thermal asymmetry. It was found, however, that this approach, while reducing the magnitude of rotational segregation effects, results in increased radial composition gradients. This effect is attributed to the establishment of a nonuniform diffusion boundary layer due to the introduction of additional convective flows which occur whenever the crystal and crucible do not rotate with the same velocity in the same direction.[32] Some of the basic forced flow patterns associated with the Czochralski configuration are shown schematically in Figure 12. During rotation of the crystal and crucible in the same direction (isorotation) two separate regions are established in the melt; the outer body of melt rotates with the crucible and possesses no relative motion, while vigorous secondary flows develop in a column beneath the crystal. In this region, called the Taylor–Proudman cell after its discoverers, the melt moves toward the faster rotating surface in the form of a

Chapter 7

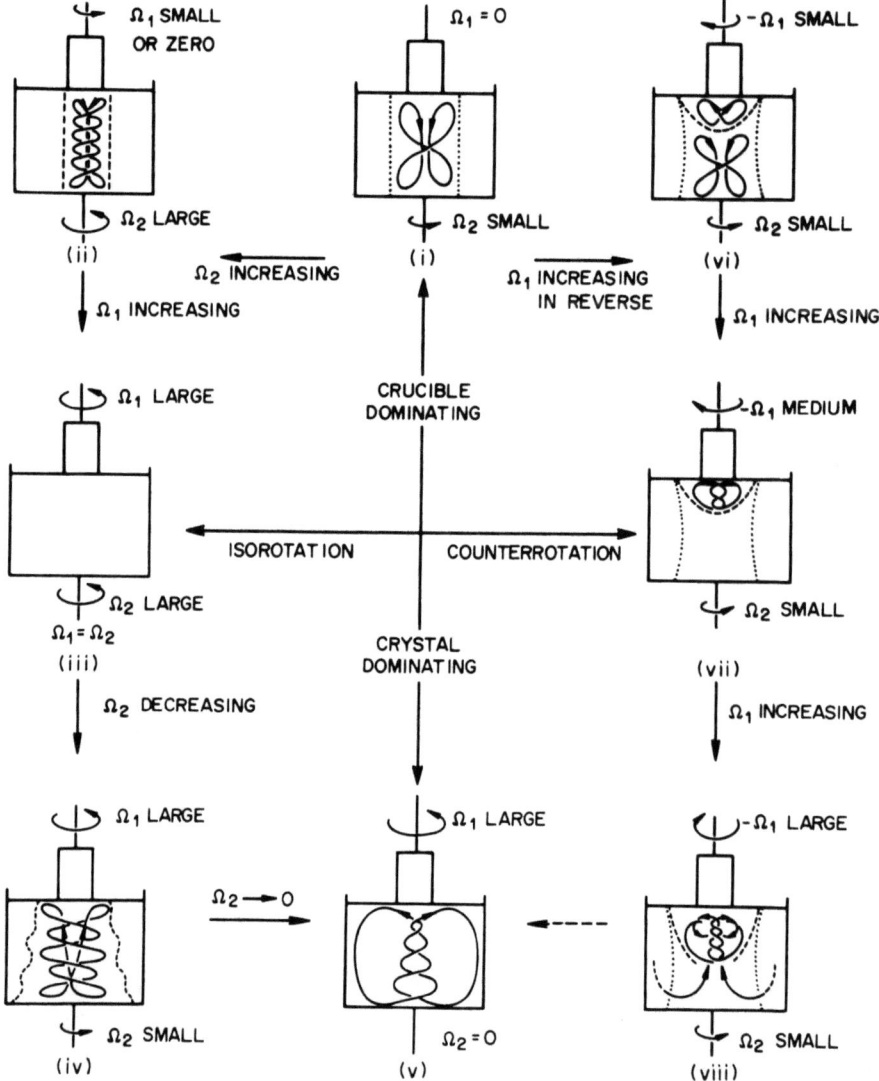

Fig. 12. Forced convection flow patterns in Czochralski melts resulting from various combinations of crystal (seed) rotation and crucible rotation. (From Ref. 32.)

spiral along a cylindrical surface with a diameter roughly equal to the crystal diameter and returns in a similar fashion just outside this surface. At the growth interface, the flows must reverse direction over a short distance quite close to its periphery, whereas the melt in the central portion of the interface remains relatively undisturbed.

Consequently, the diffusion boundary layer must be expected to be thicker at the center of the interface than at the periphery. Rotation of crystal and crucible in opposite directions produces a third region of the melt close to the growth interface where flows are controlled by crystal rotation alone. The quantitative correlation between these flows and the resulting segregation behavior at the growth interface is just beginning to be understood by using techniques to be described later. In very large, high-temperature melts, as used for industrial crystal growth, thermal convection flows may be completely dominant in the melt and neither crystal nor crucible rotation will induce flows that influence mass transport and solute segregation at the growth interface. These effects are currently the subject of extensive studies.

A variation of this process, known as the Kyropoulos method, involves a controlled reduction in power input to the melt with no withdrawal of the crystal. Thus the crystal grows into the melt as it follows the freezing point isotherm. Although the apparatus is simpler than that used for Czochralski growth, the crystal shape is not controlled and thermal shock results when the crystal is eventually separated from the melt.

Another variation of the Czochralski technique, which allows the growth of compounds with volatile components, involves the use of immiscible liquid encapsulants around the melt and a high inert gas ambient pressure. The most common encapsulant is carefully dried B_2O_3, which has been used for isolation of II–VI[33] and III–V[34] melts to minimize losses of volatile group VI and V components, respectively. The compounds most recently grown by this technique are listed by Mullin.[6] In general, liquid encapsulants should wet both the crystal melt and crucible and be less dense than the melt. A large temperature dependence of the viscosity is also helpful so that the crystal is coated as it is withdrawn from the melt. In practice there are problems of deployment of the encapsulant so as to avoid the entrainment of charge during the meltdown process, resulting in loss of visibility, and also slight reactivity between the encapsulant and both melt and crucible, resulting in contamination. Although inert gas overpressures may be used to avoid volatile losses, the encapsulation method does not permit full compositional control of the melt. A more complicated technique which does achieve this control is pressure balancing, where the dissociation pressure of the compound is dynamically balanced by a pressure of inert gas.[6] This balance can be achieved by using a liquid seal to isolate, sense, and control the pressure differential across the walls of the growth chamber.

2.2.2. Horizontal Growth

In this method the melt is contained in a boat which is placed within a horizontal furnace tube, which can be filled with an ambient gas or evacuated. A muffle furnace is placed around the tube and serves to melt the charge. Directional solidification is obtained by slowly withdrawing the boat from the furnace (or moving the furnace away from the boat).

At one time the solute distribution during horizontal growth was thought to be diffusion controlled in the sense that, if natural convection was present in the liquid, the diffusion boundary layer in the liquid δ was larger than the solute diffusion distance D/R (D is the liquid diffusion coefficient and R the growth rate). However, careful work by Weinberg,[35] using radioactive tracer techniques, showed by profiling the longitudinal distribution of solute that partial mixing was occurring in the liquid for most growth conditions. Abe et al.[36] similarly showed that in the horizontal growth of germanium the transverse distribution of solute was not uniform, which indicated a nonuniform thickness of the diffusion boundary layer. Carruthers and Winegard[22] used simulation experiments to show that the diffusion boundary layer existed as a result of fluid flow due to liquid density gradients arising from both composition and temperature variations. These experiments also showed that the influence of thermal convection on δ could be considerably reduced by minimizing vertical temperature gradients in the liquid.[19] This situation also applies to Czochralski growth.

Thermal convection in horizontal systems was shown to give rise to temperature fluctuations.[37,38] Such fluctuations had been observed to be the source of growth rate fluctuations in Ge-Si alloys by Goss et al.[39] More extensive work on temperature and growth rate fluctuations was reported by Mueller and Wilhelm.[40] However, the nature of these temperature fluctuations was not really understood until the pioneering work of Hurle,[41] which showed that at low-temperature gradients, the temperature oscillated in a simple sinusoidal mode. Hurle et al.[42] showed by spectral analysis techniques that as the temperature gradient increased, a superposition of modes occurred which gave the appearance of a complex temperature fluctuation like that caused by turbulent flow. This suggested that the fluctuations were a result of overstability. These results have subsequently been verified by careful work showing sinusoidal temperature oscillations in confined vertical cells for liquid gallium (Harp and Hurle[43]) and for liquid sodium nitrate (Carruthers and Pavilonis[44]). This work showed that temperature fluctuations could be reduced by reducing the adverse vertical temperature gradients, even though thermal convection will always exist for finite, non-vertical temperature gradients. Another method of reducing temperature fluctuations in liquid metals is by the application of a static magnetic field, and this was shown independently by Hurle[45] and Utech and Flemings[46] to be effective for metal and semiconductor crystal growth.

Forced convection in horizontal crystal growth has been provided by electromagnetic stirring.[47,48] Measurements of longitudinal normal freezing profiles in metal alloys indicated that a δ of several millimeters exists due to thermal convection in horizontal melts and that stirring with a rotating magnetic field reduced δ.[47] However, it is expected that Taylor–Proudman effects will exist in such rotating melts, so that the transverse macrosegregation may not be improved.

2.2.3. Vertical Growth

In this method the completely molten charge is solidified in a vertically upward direction. Motion of the crucible and heater relative to each other is known as Bridgman growth, while stationary systems from which heat is extracted through the bottom are referred to as Stockbarger growth. Consequently, the vertical temperature gradient is vertically stabilizing and very little thermal convection is expected. However, the small but finite radial temperature gradients which are always present in these systems will induce slow but important melt circulation.[29,49]

In this method the crystal–melt interface is completely surrounded by a container, so that contamination effects are maximized. Also, the differential thermal contraction of the grown crystal and the container gives rise to severe dislocation generation in metal, semiconductor, and ionic crystals and to cracking in oxide crystals. However, soft mold techniques can be used for metal and semiconductor systems[50] and, for oxide crystals, very thin-walled containers[51] to minimize such strains.

2.3. Zone Melting Techniques

Zone melting techniques differ basically from normal freezing methods in that only a small portion of the charge is molten at any time. There is an obvious reduction in melt contamination as well as a reduction in the heater power required. More importantly, the longitudinal solute distribution is subject to a much greater degree of control by zone melting methods (see Figure 5). For a comprehensive review of compositional distributions the reader should consult Pfann's book.[2] As for normal freezing methods, the most uniform longitudinal distribution is obtained when there is no convection in the zone. However, in practice, very pronounced thermal convection exists because of the temperature gradients required to maintain the zone.[52] In addition to increasing the variation in solute concentration along the growth direction, large transverse variations were seen due to the nonuniform diffusion boundary layer at the freezing interface. Forced convection by electromagnetic stirring will improve the transverse solute profiles but worsen the longitudinal variation.[2]

2.3.1. Horizontal Zone Melting

In horizontal methods a container can be used or electromagnetic levitation of the molten zone can be employed if the melt is conducting.[53] In horizontal zones the crystal–melt interface shapes from thermal convection are such that the distance between them is small at the bottom of the zone and large at the top. This effect was eliminated by Pfann et al.[54] by slowly rotating the horizontal container about its axis. This allowed a very small zone length to be used with a concomitant increase in zone refining efficiency. For crystal growth purposes, it is expected that such a rotation should dramatically reduce the transverse macrosegregation of solute.

2.3.2. Vertical Zone Melting

The most commonly used technique is the floating zone process, where a molten zone is maintained between a vertically mounted seed crystal and a charge rod by surface tension forces. This technique is the most widely used for materials where container reactivity is a serious problem. When the two rods are of equal diameter D_z, the maximum stable zone length l_{max} is given by[55]

$$l_{max}/D_Z = \pi(1 + N_{Bo}^g)^{-1/2} \qquad (20)$$

where the Bond number N_{Bo}^g is the dimensionless ratio of gravitational to surface tension forces

$$N_{Bo}^g = (\Delta\rho)gD_Z^2/\sigma \qquad (21)$$

where $\Delta\rho$ is the density difference between the zone and its surrounding medium, g is the acceleration due to gravity, and σ is the surface tension. Thus at zero gravity ($N_{Bo}^g = 0$), the maximum length is equal to the circumference, in accordance with the Rayleigh stability theory,[55] while at earth gravity, where $N_{Bo}^g \gg 1$,

$$l_{max} = \pi[\sigma/(\Delta\rho)g]^{1/2} \qquad (22)$$

This relation is similar to one derived by Heywang[56] by numerical integration of the Laplace equation, except for the constant, which was 2.84 instead of π. Consequently, under zero gravity the zone length is independent of surface tension, while on earth, it is independent of diameter. Rotation of the zone, which is commonly used to provide radial thermal symmetry, will reduce the maximum stable length,[57]

$$l_{max}/D = \pi(1 + N_{Bo}^g)^{-1/2}(1 + N_{Bo}^\omega)^{-1/2} \qquad (23)$$

where the rotational Bond number is

$$N_{Bo}^\omega = (\Delta\rho)R^3\omega^2/\sigma \qquad (24)$$

The reduction in surface stability due to rotation expressed in Eq. (23) has been verified both in simulated zero gravity[55] and in real zero gravity.[58]

The internal convective flows associated with the differential rotation of a liquid zone have been studied by Carruthers and Grasso[55] and some results are shown schematically in Figure 13. As for crucible rotation in Czochralski growth, the flows are radially nonuniform, so that a varying diffusion boundary layer thickness occurs across the freezing interface. Similarly, thermal convection flows, as shown in Figure 14,[59] and surface tension gradient flows[60] with radial nonuniformity exist in molten floating zones. The resulting excessive transverse macrosegregation associated with all these flows is a major limitation in the use of floating zone techniques to grow crystals of controlled composition.

The restrictions of surface stability on the length of vertical floating zones can be alleviated by increasing the diameter of the

Chapter 7

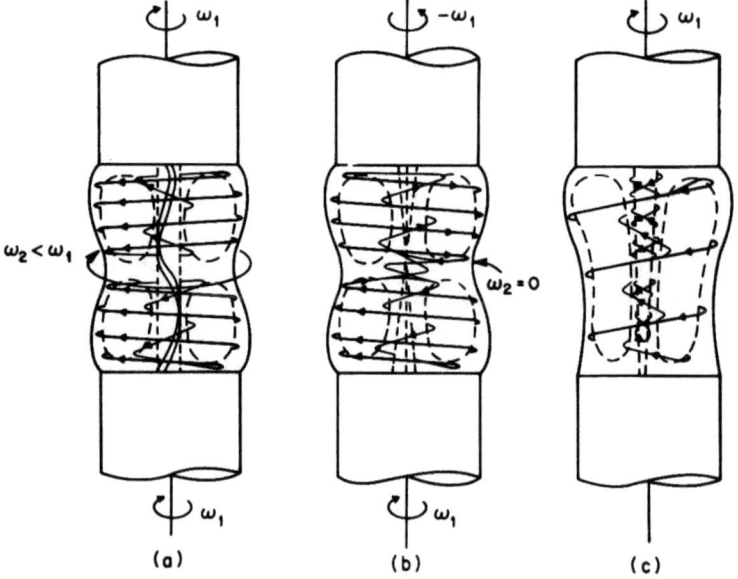

Fig. 13. Schematic illustration of flow patterns in liquid zones with rotating end members of equal diameter. (a) Equal isorotation; (b) equal counterrotation; (c) single rotation. (From Ref. 55.)

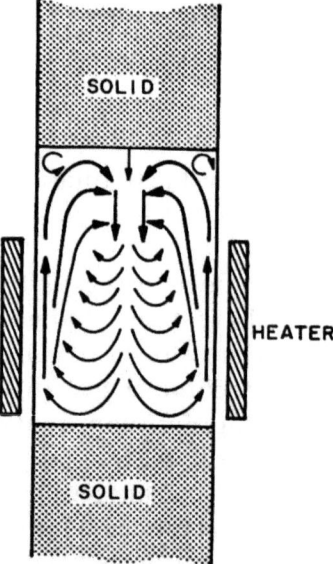

Fig. 14. Diagram of the observed thermal convection flows in a vertical zone of molten naphthalene. (From Wilcox; see Ref. 29.)

charge rod. This is the pedestal technique used to grow low-oxygen, dislocation-free silicon crystals. Methods such as this allow more control over the axial temperature gradients and hence flatter crystal–melt interface shapes as well as more uniform thermal convection flows and associated diffusion boundary layer thicknesses.[4,61,62] Another method useful for molten oxide floating zones is the heater-in-zone method first reported by Gasson.[63] The horizontal metallic strip is heated conductively and liquid flow occurs through holes. Here, as in the pedestal method, favorable axial heat flow is provided from the heating source so that flatter crystal–melt interfaces and more uniform thermal convection flows are generated. However, contamination may be a problem here.

Perhaps one of the oldest vertical zone melting techniques is the Verneuil method, where a film of liquid is formed by an oxyhydrogen flame on the upper surface of a vertically growing crystal. Growth occurs by the controlled introduction of fine, dry powder of the material through the flame, where it melts. Freezing occurs as the crystal is slowly lowered away from the flame. The art of the method is to precisely balance the rate of powder feed and the rate of lowering so as to maintain a constant growth rate and diameter.

2.4. Nonconservative Growth Techniques

All of the melt growth techniques described in the last two sections are conservative methods because the *total* concentrations of the components remain unchanged during the growth process. Very little attention in practice has been paid to the possibilities of adjusting the melt composition during growth so as to offset the normal freezing distribution and grow large single crystals of constant composition. Such techniques are particularly important for the growth of alloy or solid-solution crystals for which large composition shifts occur during growth due to the separation of liquidus and solidus curves.

In the practice of nonconservative growth methods, it is desirable that excessive melt supersaturations be avoided and that the crystal–melt interface be oriented so that it will not intercept spontaneously formed nuclei moving in the melt by Stokes flow. Thus, for example, when the higher melting component of a binary system is less dense than the lower melting component, the crystal–melt interface should be located beneath the source material. In addition, the ratio G_L°/R, where G_L° is the melt temperature gradient at the growth interface and R is the growth rate, must be large enough to avoid constitutional supercooling. In most systems the required G_L°/R value is highest in the middle of the compositional range, where the solidus and liquidus curves are the farthest apart.

Although some of the techniques to be described next can be considered as solution growth methods, they are included here because the growth is performed under isothermal conditions so as to maintain constant composition.

2.4.1. Vapor–Liquid–Solid (VLS) Transport

In this class of techniques solute is added to the liquid from the vapor phase and thence by convection and diffusion to the growth interface. Gas doping of floating zone silicon crystals was one of the first attempts to control composition by this technique,[64] while partial pressure control during the Czochralski growth of Si crystals has also been considered.[65] In a somewhat different set of experiments, the VLS growth of silicon whiskers was accomplished by depositing silicon from silicon tetrachloride onto eutectic droplets of Au–Si.[66] More recently, large crystals of $Pb_{1-x}Sn_xTe$ have been formed, by Mateika.[67] In this method, a sessile melt drop on the growth interface is continuously fed by condensation from the vapor phase so as to maintain x constant. Similarly, large crystals of GaP can be grown from a gallium solution fed by phosphorus vapor.[68]

The reverse process of evaporation from the melt has also been successfully used to control compositions. Bradshaw and Mlavsky[69] studied the influence of evaporation from floating zones of silicon on the longitudinal composition profile. In more heavily doped semiconductor melts, Trumbore and Porbansky[70] developed the isothermal solvent evaporation technique to grow heavily As-doped Ge crystals for tunnel diodes. The application of isothermal flux evaporation for the growth of oxide crystals of uniform composition was proposed by Roy,[71] and further experimental information can be found elsewhere.[1]

2.4.2. Liquid–Liquid–Solid Transport

In this class of nonconservative growth techniques the required compositional additions to the melt are provided through a second liquid which is either immiscible with the melt or separated from it by means of a membrane or capillary tube. The floating crucible method was developed by Leverton[72] and Goorisen[64] to provide a condition similar to zone melting for the Czochralski growth of uniformly doped Ge crystals. In this technique crystal growth occurs from a small volume of liquid contained in a crucible floating on the surface of the larger melt volume and connected with it only through a capillary tube. After an initial transient, the small liquid volume attains a constant composition if the capillary allows molten charge through to feed the crystal growth but does not permit back diffusion of solute in the reverse direction. Subsequent details have been discussed by Blackwell.[73] The application of the floating crucible technique to produce stoichiometric compounds when the dystectic (maximum melting) and stoichiometric compositions differ, as well as solid solution crystals of constant composition, has been discussed by Airapetyants and Shmelev.[67] Unfortunately, very little use of this technique for these materials has been reported to date.

No work has been reported where immiscible liquids serve as diffusion membranes between the crystal growth melt and some other liquid source material in a fashion analogous to the permeable membrane of electrotransport cells. However, this is an area of great potential. There are many possible immiscible liquid combinations for other metal[75] and inorganic[76] growth systems.

2.4.3. Solid–Liquid–Solid Transport

In this method the melt composition is continuously adjusted by either melting or dissolving a charge solid while the crystal is growing. The first use of such a technique was the feed-in, pull-out variation in the Czochralski growth of germanium crystals.[77,78] Here a polycrystalline Ge ingot with the same composition as the crystal being pulled is fed into the melt so that the melt volume stays constant during growth. In this way crystals with a uniform longitudinal solute distribution have been pulled. For more concentrated solid solutions, alloy melts are used with a solid source material maintained at a higher temperature than the seed so that a steady transfer of the material to be crystallized takes place from source to seed down the temperature gradient. Trumbore et al.[79] have used this technique to grow heavily doped Ge from Al and Ga melts. There are many other examples in the literature of the use of this technique to grow constant-composition solid-solution crystals isothermally. One of the more successful applications has been the growth of large crystals of $KTa_xNb_{1-x}O_3$ (KTN) by Gentile and Andres.[80] These workers used solid source material in an isothermal melt to prepare large crystals of constant composition. Previous workers had used cooling methods during growth, which resulted in large and undesirable Ta/Nb variations.

Strictly speaking, the nonconservative growth techniques, such as those described for KTN and other materials in previous sections, are classified as solution or flux growth methods. However, all these techniques are isothermal for large crystals of constant composition to be prepared. Consequently they are included in the chapter on melt growth. The efficiency of growth (in terms of mass transfer) can be considerably improved by the use of molten zones. Such techniques are referred to as temperature gradient zone melting[2] and variations of them have been discussed by Hurle et al.[81]

3. Materials Applications and Requirements

Any consideration of crystal growth from melts must involve some consideration of the uses for which the crystals are intended since these applications, in turn, place requirements on the material that must be fulfilled by the growth process. It is convenient here to deal with specific classes of materials. Consequently some commonly melt-grown oxide, halide, semiconductor, and metallic crystals are listed in Table 1 together with the methods used, the application and concomitant properties which must be controlled, and the resulting requirements imposed on the growth process. A list of symbol abbreviations is given in Table 2. This list of materials grown from the melt is by no means complete but merely represents particular materials which have been studied in enough detail to allow the necessary correlations to be made. For other materials or more detail on growth methods the reader is referred to Refs. 82–84. For elucidation of the device functions and property requirements of these materials, the reader can consult Ref. 85 for optical devices, Ref. 86 for magnetic devices, Ref. 87 for elemental semiconductor devices, Refs. 88 and 89 for III–V semiconductor devices, Ref. 90 for narrow-bandgap semiconductor devices, Ref. 84 for other ternary compound semiconductor devices, and Refs. 91 and 92 for mechanical property studies of metal and metal alloy crystals and ceramic crystals, respectively.

Examination of Table 1 will also show that many of the melt growth methods used to date are not really adequate for the particular material; for example, some compounds with volatile components have been grown in sealed tubes rather than with full vapor pressure control, and most solid-solution systems are still grown by normal freezing, conservative methods for which longitudinal compositional variations are unavoidable. In many instances other considerations

TABLE 1
Tabulation of the Applications and Requirements of Some Commonly Grown Crystals*

Material	Melt growth method (to date)	Application	Property to be controlled	Requirements (to date)
Semiconductors				
Si, Ge	CZ, FZ, HZ, EFG, CZ-FC, CZ-PP, CZ-SE, FZ-VP	eht, et, epc, e(pa)c, sub(eht)	ρ, τ	DF, SC-MA, SC-MI, SSP, DM, LC(O_2, C)
GaP, InP	CZ-LE	epc, sub(epc)	ξ_p, ξ_e	LD, SC-MA, LC(Si, O_2), NSC, DM
GaAs, InAs	CZ-LE, FZ-VP	et, epc sub(et, epc), IR	ρ, τ, μ	LD, SC-MA, LC(Si, O_2, transition metals), NSC
$ZnSnAs_2$	HNF	epc, nlo	ξ_p, ξ_e, n	LD, NSC
$CdGeP_2$				
$ZnGeP_2$				
(Zn, Cd) (S, Se, Te)	VNF-VP, HZ-VP	epc	ξ_p	LD, SC, LC (alkalis), NSC
$CuInSe_2$	HNF, VNF, CZ-LE	epc, nlo	ξ_p, ξ_e	LD, NSC, SSP
$CuGaS_2$				
$AgGaSe_2$				
$AgGaS_2$	HNF-VP, VNF-VP, CZ-LE	epc, me, ppc	ξ_p, ξ_e, ρ, P_1	LD, SC-MA, SC-MI, NSC, SSC
(Pb, Sn) (S, Se, Te)	VNF	e(pn)c		NSC, SSP
$Pb_{1-x}Sn_xTe$	VNF-VP	mo	$\alpha, \mu_m\mu_e, \rho$	NSC, NSV, SSP
$(Bi, As, Sb)_2Te_3$				
MnTe				
Oxides				
Al_2O_3	CZ, FZ, V, EFG	sub(epc, eht), ppc, es	N_d P_1, N_d	LD, SC-MA, SC-MI, DM, SH
$Al_2O_3:Cr^{3+}$				

Crystal Growth from the Melt

Table 1 (continued)

Material	Melt growth method (to date)	Application	Property to be controlled	Requirements (to date)
$MgAl_2O_4$	CZ, FZ, V, EFG	sub(epc, eht),	N_d, a_0	LD, NSC
$RAlO_3$	CZ	ppc	P_1, N_{l_w}, N_d	LD, NSC, PT
$RGaO_3$				
$R_3Al_5O_{12}:Nd^{3+}$	CZ, FZ	ppc, sub(md, mo)	P_1, a_0	LD, SC-MA, SC-MI, NSC, DM
$R_3Ga_5O_{12}$				
$LiNbO_3$	CZ	eo, nlo, es, pea	n, P_d	LD, LC (transition metals), NSC, NSV
$LiTaO_3$	CZ	nlo, epc	n, P_d	LD, LC (transition metals, alkaline earth ions), NSC
$Ba_xSr_{1-x}Nb_2O_6$				
$Ba_2NaNb_5O_{15}$				
$Sr_2LiNb_5O_{15}$				
$R_2(MoO_4)_3$	CZ	ppc, es, so	n, P_1	LD, SC, NSC, PT
$PbMoO_4$	CZ	so	nc	
$CaMoO_4$				
$CaWO_4:Nd^{3+}$	CZ, FZ	ppc	P_1	LD, SC
$Bi_2(GeO_4)_3$	CZ	so	nc	—
TeO_2	CZ	so	nc	
$RFeO_3$	CZ, FZ, VNF	md	M_s, μ_m	NSC, NSV
$NiFe_2O_4$	CZ, FZ, VNF	md	M_s, μ_m, ρ	NSC, NSV
$ZnFe_2O_4$				
Oxides/halides				
$Ca_5(PO_4)_3F:Nd^{3+}$	CZ	ppc	P_1, N_d, α	LD, SC-MA, SC-MI, NSC, SSP
Alkali halides				
MF_2	VNF, KY	IR	α	
(M = Ca, Sr, Ba, Co, Mn, Cd, Mg, Zn)	VNF, HZ, CZ	ppc	P_1, N_d, α	LD, SC-MA, SC-MI, NSC, SSP, LC(H_2O, HF)

Material	Growth Method	Properties	Measurements	Characterization
ABX_3 (A = K, Rb, Cs) (B = Mn, Fe, Co, Ni) (X = F, Cl, Br)	VNF, HZ, CZ	mo	$\alpha, \mu_m, \mu_e, \rho$	NSC, NSV, SSP
$BaMF_4$ (M = Mg, Mn, Fe, Co, Ni)	HZ	eo, nlo, mo	α, n	NSC, NSV, SSP
Metals and Alloys				
FCC — Al, Cu, Au, Pb, Ag	HNF, VNF, CZ, HZ, VNF-SOM, VNF-SPM	dd, ds	$N_d, \rho_{RT}/\rho_{4°K}$	LD, SC, LC, SSC, SSP; SH, PT for some specimens
Ni	FZ	dd, ds	$N_d, \rho_{RT}/\rho_{4°K}$	LD, SC, LC, SSC, SSP; SH, PT for some specimens
BCC — Fe, Cr, V, Nb, Ta, Mo, W, NaAl, Ti(U)	FZ	dd, ds	$N_d, \rho_{RT}/\rho_{4°K}$	LD, SC, LC, SSC, SSP; SH, PT for some specimens
HCP — Mg, Zn, Cd	HNF, VNF-SOM, HZ	dd, ds	$N_d, \rho_{RT}/\rho_{4°K}$	LD, SC, LC, SSC, SSP; SH, PT for some specimens
Ru, Co, Zr	FZ	dd, ds	$N_d, \rho_{RT}/\rho_{4°K}$	LD, SC, LC, SSC, SSP; SH, PT for some specimens

* See Table 2 for explanation of abbreviations.

Chapter 7

TABLE 2

Identification of Abbreviated Terminology Used in Table 1

I. Melt growth method
   A. Normal Freezing
      - CZ   Czochralski
      - KY   Kyropoulos
      - HNF   horizontal normal freezing
      - VNF   vertical normal freezing (Bridgeman, Stockburger)
   B. Zone melting
      - HZ   horizontal zone
      - FZ   floating zone and pedestal
      - V   Verneuil flame fusion
      - EFG   edge-defined, film-fed
   C. Variations due to nonconservative compositional control
      - LE   liquid encapsulated melt
      - VP   vapor pressure control
      - FC   floating crucible
      - PP   push-pull (feed-in, pull-out)
      - SE   solvent evaporation
   D. Variations due to excessive deformation in metals
      - SOM   soft mold technique
      - SPM   split mold technique (interior shaped to desired form)

II. Application
   A. Conversion devices
      - epc   electron-to-photon, or photon-to-electron, conversion
      - ppc   photon-to-photon conversion
      - e(pn)c   electron-to-phonon conversion
      - e(pa)c   charged-particle-to-electron conversion
   B. Transfer and storage devices
      - eht   electron–hole transfer devices (bipolar)
      - et   electron transfer devices (unipolar)
      - md   magnetic domain and magnetic storage devices
      - pea   photon–electron absorption (holographic and photographic storage)
      - s   controlled stress/strain applications
   C. Interaction devices
      - so   stress-optical and acoustooptical devices
      - eo   electrooptical devices
      - mo   magnetooptical devices
      - es   piezoelectric and ultrasonic delay devices
      - me   magnetoresistance devices (Fermi surface mapping)
      - nlo   nonlinear optical devices
   D. Passive functions
      - IR   infrared windows
   E. Solid state studies
      - dd   studies of dislocation dynamics (critical resolved shear stress, dislocation mobility, dislocation nucleation, multiplication and interaction)
      - ds   dislocation–solute interaction studies (solid-solution hardening, precipitation hardening)

Table 2 (continued)

    F. Substrates for epitaxial growth
       sub( )   substrate (for applications shown in IIA–E)

III. Property to be controlled
    A. Electronic
       ρ resistivity
       τ lifetime
       μ_e mobility
       ξ_p photoluminescent efficiency
       ξ_e electroluminescent efficiency
       ξ_q spectral quantum efficiency
       $\rho_{RT}/\rho_{4°K}$ residual resistivity ratio
    B. Optical
       n refractive index
       Δn_s strain-induced birefringence
       P_d power level for self-induced optical damage
       α absorption coefficient
       P_1 minimum input (threshold) power for lasing
       $V_{\lambda/2}$ half-wave retardation voltage
    C. Magnetic
       M_s saturation magnetization
       ϕ_s saturated magnetic rotation per unit length
       μ_m magnetic moment
    D. Mechanical
       N_d dislocation density
       N_{tw} twin density
       a_0 lattice parameter
    E. Noncritical
       nc

IV. Requirements
    A. Structural defects
       DF   dislocation free
       LD   low dislocation density, no lineage or low-angle boundaries
       PT   controlled cooling through solid-state phase transformation
    B. Composition
       SC   controlled and uniform solute (dopant) concentration
       SC-MA solute composition controlled on macroscale
       SC-MI   solute composition controlled on microscale
       LC( )   low contamination levels (from specified impurities)
       NSC   control of nonstoichiometric ratio of compound components
       NSV   control of nonstoichiometric valence changes of components
       SSC   control of concentration of solid-solution components
       SSP   avoidance of solid-state precipitation or pre-precipitation complexing
    C. Geometry
       DM   diameter control required for large-diameter crystals
       SH   shape control

have dictated the use of particular growth methods, such as the availability of existing equipment and the expense of new equipment, an initial unfamiliarity of the crystal requirements necessary (or even its ultimate use), a lack of knowledge of melt reactivity problems (with containers or atmospheres), and a lack of adequate characterization (for example, the use of an optical microscope to reveal second-phase precipitates). Even so, many exploratory materials programs are based on the use of a specific technique to grow small crystals of many different materials with relative ease. For example, the use of the Czochralski technique to grow oxide crystals in the early 1960's produced a plethora of compounds for use in laser devices and applications, while the application of the floating zone method to grow and refine refractory metals by electron beam melting has provided a tremendous impetus for plastic deformation studies. Similar advances may be expected in the future by the use of techniques such as the pedestal and edge-defined, film-fed methods for oxide and metal crystal growth.

Further refinements in techniques or even the development of new growth methods must rely, however, on our understanding of the dynamics of these processes and ultimately on our ability to control them. This understanding is basically stimulated by the need to reproducibly produce crystals with particular properties for a given application as shown in Tables 1 and 2. The result has been a series of basic studies of crystal growth from melts, which will be described next.

4. Studies of Crystal Growth from the Melt

Independent studies of the phenomena described in the preceding sections of this chapter have been carried out in simple systems and also by theoretical analyses. In addition, *in situ* growth studies and more extensive use of crystal characterization techniques have yielded a deep insight into the more subtle aspects of crystal growth from melts.

4.1. Morphological Stability of Crystal–Melt Interfaces

The instability of a crystal–melt interface in the presence of constitutional supercooling in the melt was recognized some 50 years ago in metals casting by Genders[93] and formalized by Northcott.[94] The concept was rediscovered and put on a quantitative basis by Tiller *et al.*[95] However, this analysis ignored the kinetic processes of heat and solute transfer as well as the stabilizing role of solid–liquid surface tension in the system. Such phenomena were considered in a more rigorous perturbation analysis by Mullins and Sekerka.[96] Basically, the initial instability is a set of sinusoidal shape perturbations on the interface which quickly degenerate to form a cellular substructure perpendicular to the original interface with the low-melting constituent being preferentially located in the grooves (which become cell walls). A tutorial review by Sekerka can be found in Ref. 97. Despite the fundamental importance of morphological instability to crystal growth from melts, surprisingly few direct experimental measurements have been made. The differences between the simple constitutional supercooling criterion and the full perturbation analysis are small under normal growth conditions and within the scatter of both the data and known values of the constants.[98] In addition, the development of the instabilities is critically dependent on the presence of defects such as dislocation lineage or container walls at the interface.[99]

Further theoretical work has been performed on the stabilization of the growth interface by interface atomic kinetics (important for high-α materials), surface diffusion (important for very slow growth), and convection effects; this work has been discussed by Sekerka.[97] The role of convection is more complicated than the discussion in Section 2.1.1 because of changes in *both* temperature and concentration gradients at the interface as well as the possibility of associated time-dependent temperature, flow velocity, and growth rate fluctuations. Some of these complications can be found in the experimental studies by Singh *et al.*[100] for Czochralski-grown Ge crystals, using the interface demarcation technique to be discussed later. In this work, although temperature gradients were not measured, convection was expected to be stabilizing (for a distribution coefficient less than unity), whereas, in fact, the associated time-varying growth rate (including some remelting) resulted in destabilization.

A detailed survey of more concentrated solid-solution systems by Dismukes and Yim[101] has also indicated that presently available experimental data cannot distinguish between the simple criterion and more complicated perturbation analyses.

4.2. Melt Statics and Dynamics

Many of the general aspects of fluid statics and dynamics in the various melt growth processes have already been referred to in

Section 2. In the present section some of the more common, general aspects of convective processes will be discussed, together with some description of the surface stability of containerless melts for space processing applications.

4.2.1. Melt Statics

A full understanding of the behavior of free melt surfaces is becoming increasingly important for situations involving menisci such as used for diameter control in Czochralski growth and shape control in edge-defined, film-fed (EFG) growth, as well as for the response of melt zone and drop surfaces to vibrational and rotational forces.

4.2.1.1. Meniscus Behavior

The behavior of the meniscus shape during Czochralski growth on the earth has been discussed thoroughly by Geist and Grosse.[102] The stabilities of such menisci have been dealt with by Huh and Scriven[103] and Padday and Pitts.[104] Basically the situation in Czochralski growth is as shown in Figure 15, where a solid cylinder is immersed in the free liquid surface at a low contact angle. The preferred meniscus height h_2 is shown in Figure 15(b), since the freezing interface will tend to remain at constant diameter for small nonuniform displacements of the crystal pulling rate V_0 or freezing point isotherm. The stabilities of menisci at zero gravity for the free drop have not been investigated as yet but are going to depend quite sensitively on the configurations of the force fields used to constrain

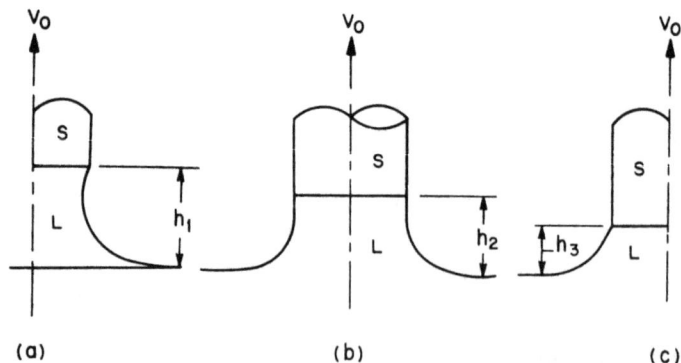

Fig. 15. Meniscus configurations important at the crystal–melt interface during Czochralski growth on the earth; (a) necking instability, (b) desired configuration, (c) crystal diameter control difficult. (From Ref. 58.)

the liquid drop position. The maximum height h of the crystal interface above the original drop surface is determined primarily by the radial departure R allowed at the bottom of the meniscus. If R is constrained to be close to the crystal radius R_c, then the height h cannot exceed $2\pi R$ without the meniscus becoming unstable. Otherwise longer liquid meniscus columns are allowed according to the approximate relation

$$h_{max} \approx 2\pi\{[R^2 - (R - R_c)]^2\}^{1/2} \qquad (25)$$

These considerations apply only if the liquid column base radius R is substantially smaller than the liquid drop radius. Otherwise the constant-volume condition of the drop or zone will enter into the stability picture as pressure changes in addition to those caused by

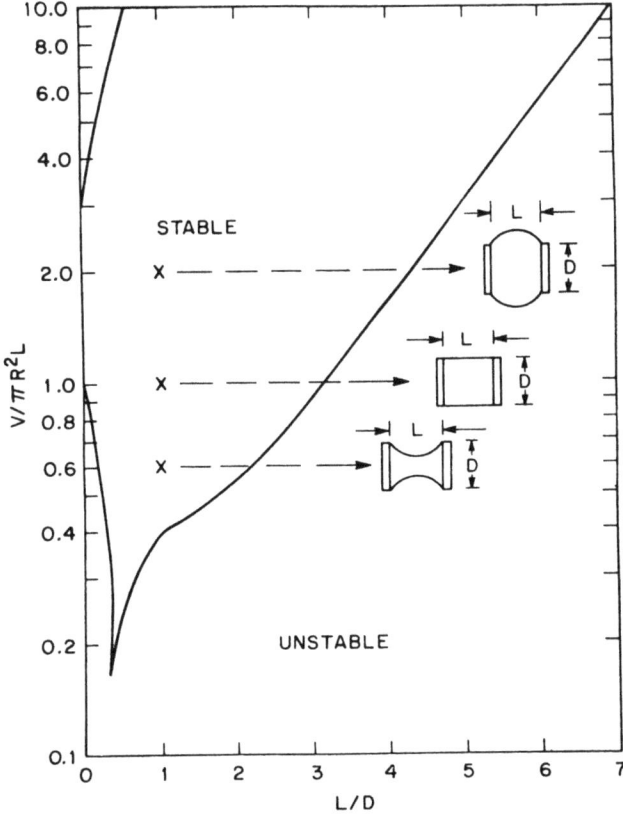

Fig. 16. Stability region of the surface of cylindrical liquid zones of various shapes contained between equal-diameter end disks (from Gillette and Dyson[105]).

Chapter 7

the liquid surface curvature. A treatment of the mathematical stability of liquid surfaces of revolution under such conditions has been given by Gillette and Dyson[105] for equal end radii under zero gravity. A summary of their calculated stability limits is shown in Figure 16 for liquid zones of volume V, length L, radius R, and diameter D. The limits for unequal end diameters and differential volumes have not been calculated or observed as yet.

4.2.1.2. Rotational Stability

The behavior of free liquid surfaces under rotational conditions is now examined. In crystal growth and solidification processes, rotation is usually required in order to provide radially symmetric temperature distributions since such symmetry may not be present in the heating source and perfect alignment of the crystal axis with the thermal center of symmetry is difficult to achieve. For purposes of maximum compositional homogeneity, however, *no* convection may be desired in the melt. Consequently, for growth in zero gravity, only uniform "solid-body" rotation of the liquid (equal isorotation) may

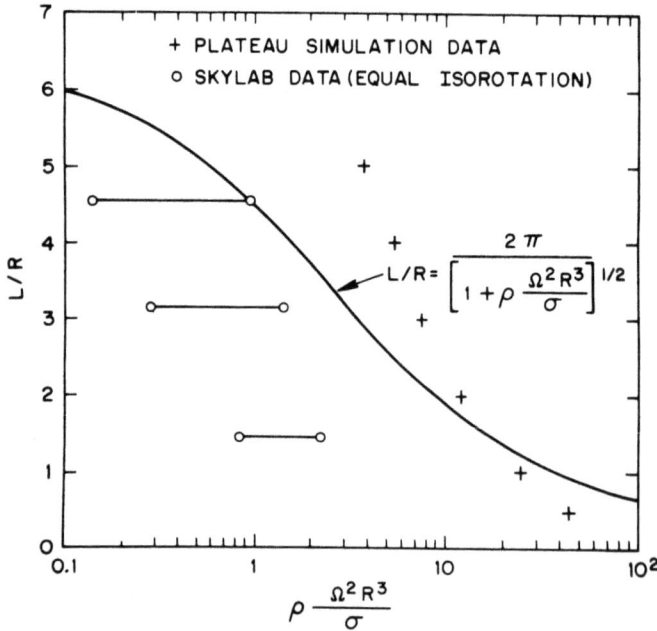

Fig. 17. Comparison of the rotational stability of cylindrical liquid zones in Plateau simulation and Skylab experiments with theory for axisymmetric modes (from Ref. 58).

be considered and the absolute rotation velocity ω must be small enough so that the centrifugal acceleration $\omega^2 r$ does not cause density-gradient-driven flows. The rotational stability criteria for cylindrical liquid columns has been tested experimentally in a Plateau simulation system by Carruthers and Grasso[55] and in real zero gravity in the Skylab manned space station.[58] These data are summarized in Figure 17 for equal isorotations and compared to Eq. (23). Discrepancies in the Plateau results occur because the actual zone rotation velocity is somewhat smaller than the end-member rotation velocity because of the viscous drag effects exerted by the viscous outer fluid containing the zone. The Skylab data also show a deviation from the theoretical curve, becoming unstable at *lower* rotation rates than expected. The reason for this behavior is that the failure mode in the Skylab experiments was nonaxisymmetric. The nonaxisymmetric failures arise from disturbances created by misalignment of the rotation axes and nonparallelism of the disks at the ends of the zone. However, viscosity plays a very important and unsuspected role in determining which types of disturbances are amplified. Additional theoretical work will be required to analyze the nonaxisymmetric failure modes for both rotating liquid spheres and cylinders.

4.2.1.3. Vibration Behavior

In crystal growth processes the time-dependent deformation of free liquid surfaces is undesirable because the meniscus shape at the freezing interface will change so as to cause crystal diameter perturbations and also because internal fluid flows will develop which influence the growth segregation behavior. Two liquid shapes are important, spherical drops and cylindrical zones. We are interested in the natural resonant oscillation frequencies, which depend primarily on surface tension, and the decay rates, which depend mostly on viscosity. Knowledge of this behavior will permit apparatus designers to avoid vibrations that cause resonant, large-amplitude oscillations in the liquid volumes used in space processing. Comparisons have been made[58] of the vibrational behavior of liquid drops in both a Plateau simulation model and in zero gravity with the theory of Lamb and of Miller and Scriven.[106] These preliminary results indicate satisfactory agreement with the behavior in real zero gravity but indicate some disagreement in the immiscible liquid drop model case.

The vibration behavior of cylindrical liquid zones was also studied extensively on Skylab IV.[58] Preliminary data for the oscillation frequency of longitudinal modes indicate that there is a serious discrepancy between theory[106] and experiment, especially for longer zone lengths, which may be attributable to effects of viscosity and fluid flow on the oscillation behavior of the liquid zone surface.

4.2.2. Melt Dynamics

Flows in melts may be divided into those occurring naturally as a result of the boundary conditions used in the growth process and those introduced by force fields to provide controlled diffusion boundary layers at the interface. An overview of these flows will be given together with some consideration of their influence on melt growth. One of the few reviews of this subject is the article by Hurle.[107]

4.2.2.1. Natural Convection

The two main types of natural convection which are important in crystal growth from melts are the density-gradient-driven flows in a gravitational field, and surface tension-gradient-driven flows along free melt surfaces. Both gradients may arise from either temperature or composition gradients in the system. The resulting flows may be either time-independent (steady) or time-dependent (unsteady), depending on the complex combination of boundary conditions present. The transition to unsteady flows is a matter of great concern in crystal growth from the melt since the onset of temperature and velocity fluctuations leads to undesirable microsegregation in the crystal. Jakeman and Hurle[108] have recently reviewed the origin of temperature oscillations and their influence on solidification processes and Carruthers[29] has reviewed the factors influencing thermal convection instabilities during crystal growth from liquids.

Thermal convection processes are particularly important in melt growth because the bounding surfaces are maintained at differing but constant (with time) temperatures. In order to understand the conditions at which time-dependent flows will occur it is necessary to know such facts as the following[29]:

(a) The degree of fluid confinement—melt shape, dimensions (and aspect ratio).
(b) The orientation of bounding surfaces from the vertical.
(c) The orientations of the heat flows from the vertical and relative to the container.

(d) The thermal diffusivity κ and kinematic viscosity, v of the melt.
(e) The conductivity of the melt relative to that of the container.
(f) The presence of concentration-induced density differences.
(g) The presence of free surfaces as opposed to solid bounding surfaces.
(h) The presence of surface tension-gradient-induced flows.
(i) The existence of internal radiative heat transfer in the melt.
(j) The existence of forced convection.
(k) The presence of other body force fields (than gravity) such as magnetic and electric fields and field gradients as well as centrifugal fields.

It appears that, despite all these complications, many crystal growth situations exhibit thermal convection flows which are unsteady but not turbulent in nature. To illustrate this fact, some data from the fluid dynamics literature are shown in Figure 18 for the bottom heating of completely confined fluids. Here the dimensionless Rayleigh number $N_{Ra}^{(1)}$ is a measure of the vertical temperature difference ΔT_v as follows:

$$N_{Ra}^{(1)} = g\alpha(\Delta T_v)H^3/\kappa v \qquad (26)$$

where g is the gravitational acceleration, α is the melt thermal expansion coefficient, H is the vertical melt height, κ is the thermal diffusivity, and v is the kinematic viscosity. The superscript c on N_{Ra}^{c1} in Figure 18 denotes the critical point of instability, which for this case corresponds to the onset of a steady convective flow in an initially quiescent melt. The solid curves have been theoretically computed for containers of square cross section (side length W), while the data points have been taken from containers of both square and circular cross sections (diameter W) with fluids ranging from air and water to liquid mercury (see Ref. 29 for details). The terms conducting and insulating refer to the wall thermal conductivities. It can be seen that there is good agreement between theory and experiment regardless of the cross-sectional shape of the cavity, although the form of the flows may certainly be different. The effect of increasing the height-to-width ratio of the container can thus be seen to stabilize the flow for bottom heating.

It is amazing that in all the work performed on actual crystal growth melts, the onset of unstable thermal convection flows occurs at values of the critical Rayleigh number very close to those for

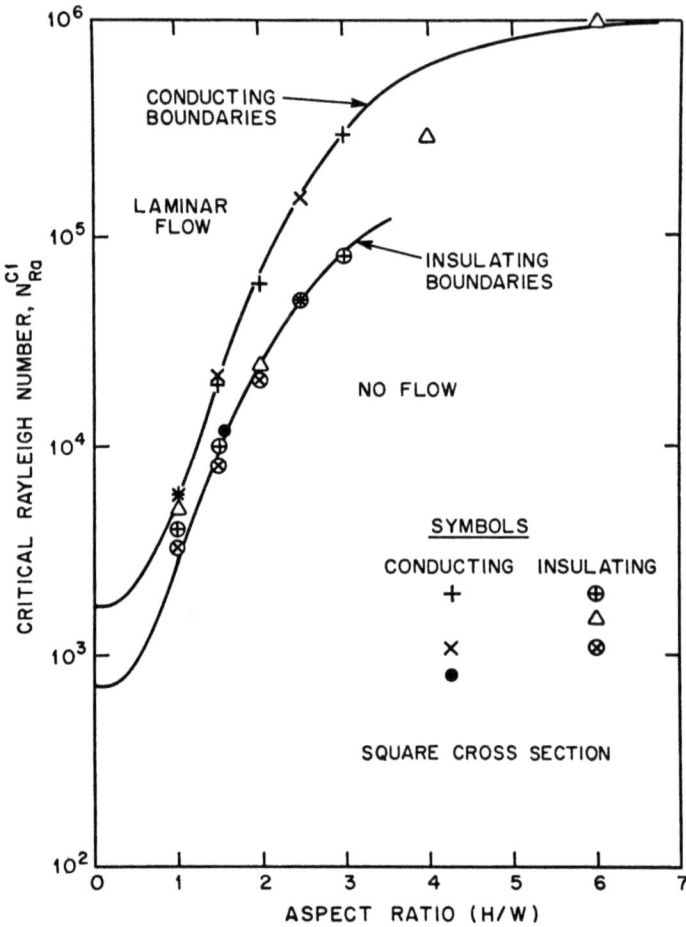

Fig. 18. Increase in critical Rayleigh number with aspect ratio for the bottom heating of confined fluids in containers with square cross sections and insulating or conducting horizontal boundaries. Solid curves are theoretical (see text). (From Ref. 29.)

which the fluid would be unstable if it were heated from below.[29] The presence of the circulating steady flows from the radial heat flows appears to stabilize the fluid more effectively if its Prandtl number $N_{Pr}\ (=v/\kappa)$ is higher. Thus, even though the vorticity of liquid metals driven by nonvertical heat flows is known to be higher than for oxide melts, apparently the reduction in stability from the thermal conductivity ratio effect has a greater influence. In addition, the occurrence of oscillatory instability is a direct consequence of the existence of an initial shear flow from nonvertical heat flows since such an unstable mode is not permitted for thermal convection from only vertical heat flows in confined regions by the principle of exchange of stabilities.[29]

Methods of controlling thermal convection in the Czochralski geometry have been used only recently. It has been pointed out that reducing heat losses at the upper free surface by afterheaters reduces temperature fluctuations. However, crystal growth under these conditions, such as, for example, the growth of striation-free barium sodium niobate by Zupp et al.,[109] is extremely difficult to control; i.e., the external surface of the crystal is faceted and any attempt to change the diameter results in completely freezing the melt surface or melting of the crystal. Another method, employing the use of baffles, was reported by Whiffin and Brice.[110] They showed that a horizontal baffle placed near the middle of the liquid volume was most effective in reducing temperature fluctuations. This result would be expected for axisymmetric flows but such baffles should exert far less effect on antisymmetric-type flows because no major streamlines would be intersected. Subsequently it was shown by Brice et al.[112,113] that horizontally positioning a baffle of diameter two-thirds that of the crucible a distance one-quarter of the way down from the upper surface provided optimum conditions for minimizing temperature oscillations in oxide melts. This use of a horizontal baffle also reduces the effective liquid height, a phenomenon shown by Cockayne et al.[115] to reduce temperature fluctuations. However, it is important to remember also the influence of the aspect ratio of the reduced liquid volume and the role of vertical heat losses at the upper free surface when designing baffles. Antisymmetric flows should not be influenced to as great an extent by baffles.

Another method of controlling thermal convection in the Czochralski geometry is to use localized cooling. Both cooling of the crystal by Brice et al.[111,112] and the crucible-base cooling method recently reported by Cockayne et al.[114] have been successfully applied. When a steady stream of gas impinges on the crystal periphery or on the bottom of the crucible at the center, the crystal diameter control is improved markedly. The change in temperature distribution is to move the position of maximum temperature upward, away from the crucible bottom (so as to reduce the effective H), to steepen the radial temperature gradients (so as to avoid facet formation on the external crystal surface), and to prevent slow meandering of the thermal center of symmetry (and its associated temperature excursions at the center of the melt surface). Although not discussed, it is likely that temperature fluctuations still exist in the melt since striations were used to determine the growth interface shapes.

In electrically conducting melts, thermal convection flows have been shown to be effectively damped by applying a static magnetic field of a few hundred gauss across the melt.[116,117] The growth of crystals in space eliminates all density-driven flows but not flows driven by surface tension gradients.

Natural convection flows due to compositionally induced density gradients is an important feature of the gravitational stability of diffusion boundary layers in melt growth. Since these occur simultaneously with temperature gradients, the interest is in studies of thermosolutal convection. A limited number of studies of thermosolutal convection in crystal growth configurations have confirmed the expected flow observations,[118-120] although most of these experiments were not conducted with isoconcentrate surfaces, so that the observations were transient. Sharp and Hellawell[119] have recently reported on the upward directional freezing of Al(Mg) alloys where the temperature gradient was stabilizing and the composition gradient was destabilizing. Examination of the (quenched) liquid composition profile ahead of the freezing interface showed the expected existence of compositional fluctuations, reminiscent of cellular convection, for several millimeters ahead of the growth interface.

Another source of natural convection in melts with free surfaces is a gradient in surface tension when a temperature gradient exists normal to the surface. Theoretical studies of such flows (sometimes called Marangoni convection) have been performed for horizontal melt surfaces,[121] liquid drops,[122] and liquid floating zones.[60] In all cases a substantial convective flow is predicted which is comparable to or even greater than that due to thermal convection. Although very few experiments have been performed in this area, it is a subject of great interest for future containerless space processing experiments.

4.2.2.2. Forced Convection

The primary role of forced convection in crystal growth systems is to offset uncontrolled natural convection of the type discussed previously in this section. However, all of the flow visualization studies, such as those shown in Figures 12 and 13, have not been performed in the simultaneous presence of natural convection. Consequently, the interaction of forced and natural convective flows must be inferred from the results produced on the solute segregation profiles in grown crystals and from independent heat transfer studies.

The directions of fluid flows due to crystal rotation and thermal convection are opposed in the Czochralski geometry. Consequently, for crystal rotation rates less than about 50 rpm, one expects thermal convection flows to be unaffected by rotation in the bulk liquid, with rotational and thermal convection flows being important adjacent to the crystal. The situation for rotating melts is radically different because of possible Taylor–Proudman constraints as well as because of centrifugal acceleration. The introduction of an overstable regime is possible in these cases with very large-amplitude, sinusoidal temperature oscillations. Carruthers[124] has correlated solute striation spacings in Czochralski-grown silicon crystals to the expected overstable frequencies at different rotation rates. Cockayne and Gates[125] have observed that the spoke pattern observed on the surface of oxide melts (Figure 11a) intensifies into well-defined spokes which rotate with the crucible rotation velocity, as shown in Figure 19. The associated temperature variations are very much larger for these rotating radial streams.

The radial solute distributions in Czochralski-grown Si crystals under various rotation conditions from small (150 g) melts has been reported in detail by Benson.[126] These data show that, for the degree of thermal convection existing in this particular melt, the use of crucible rotations produced a maximum variation in the macroscopic

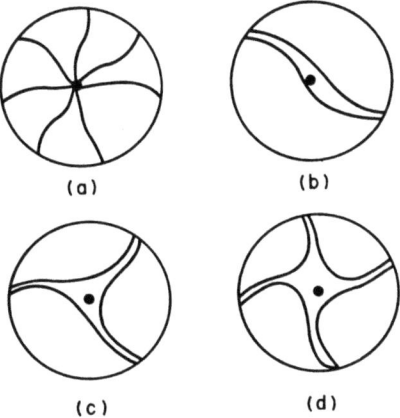

Fig. 19. Flow patterns observed by Cockayne and Gates[125] on the surface of oxide melts used in Czochralski growth; (a) stationary melt, (b)–(d) rotating melts.

radial solute variation at 25 rpm under a wide variety of conditions. This rotation rate probably corresponds to the stabilization of a Taylor–Proudman cell (see Figure 12) under this particular mode of thermal convection. However, a detailed understanding of the flows associated with combined thermal and forced convection has not yet been established.

4.3. Transient Segregation and Interface Demarcation

The existence of striations due to transient segregation phenomena during melt growth has been described in Section 2.1.2. However, the exact origin of these striations could not be easily determined without a detailed knowledge of the interface growth-rate behavior.

4.3.1. Growth Interface Demarcation Techniques

In recent years several techniques for effective growth interface demarcation have been developed. All these techniques are based on the incorporation of "controlled" striations into the growing crystal through the generation of transients which affect simultaneously all points of the growth interface. For useful interface demarcation the generated transients must lead to detectable segregation effects, which should be controllable in intensity and frequency, and whose introduction must in no way interfere with the overall growth and segregation process. Three different approaches are currently in use for interface demarcation: vibrational, thermal, and thermoelectric control striations.[28] Vibrational and thermal control striations can be used for any system, but perturb the interface segregation behavior excessively and cannot be used at high frequencies. Thermoelectric control striations are based on thermoelectric heating or cooling effects (Peltier effect) associated with current transport across the growth interface. The technique has thus far been most extensively used for semiconductor systems because of its versatility and high degree of controllability.[127] Its application in conventional Czochralski systems requires electrical insulation of the seed holder and the pedestal as well as commutator systems to provide for current flow during rotational pulling. The required current pulses may be obtained from an appropriate power supply controlled by a function generator. The versatility of this approach is indicated in Figure 20, in which the growth and segregation effects of extended cooling pulses (1 sec)

Crystal Growth from the Melt

Fig. 20. Etched off-facet section of a Te-doped indium antimonide single crystal grown without seed rotation and subjected to controlled current pulsing (see insert). The effects of extended current pulses from the crystal to the melt are investigated through superimposed current pulses of 15 msec duration, which result in effective interface demarcation. From the spacing of the demarcation lines it can be seen that the extended current pulses lead to abrupt, constant increases in the microscopic growth rate. Note also that with termination of current flow the microscopic growth rate reverts without any noticeable transient to its original low level. ×78. (From Ref. 128.)

across the growth interface were quantitatively investigated through superimposed current pulses of short duration[128] (15 msec). The photomicrograph reveals that the response of growth rate and segregation to current flow is virtually instantaneous and proportional to the current density. The resulting interface demarcation is sharp and completely controllable in terms of width and intensity (extent of segregation change). Although no quantitative analysis for current-induced segregation is available at this time, there are strong indications that the increase in segregation associated with current flow is in excess of that expected from the resulting increase in the microscopic growth rate.

Interface demarcation can be used for several purposes.[129] The demarcation lines can serve for the determination of the microscopic growth rate (rate striations); they can also be used for the identification of simultaneously grown areas (locators), for example, for the identification of exactly corresponding off-facet and facet growth regions; they can furthermore be used to establish precise time reference points in the growing crystal (time reference markers) for the identification of particular events during growth.

4.3.2. Nonrotational Striations—Origin and Characterization

It has become customary to classify impurity striations in Czochralski-type growth systems into two categories: (a) nonrotational striations and (b) rotational striations. While rotational striations (to be discussed in the next section) are characterized by a relatively simple structure and can be attributed to seed or crucible rotation, nonrotational striations are frequently complex and of varying origin.[130–138] The identification of the exact origin of the multiple of nonrotational striations has been complicated by our inability to establish unambiguous cause and effect relationships between phenomena in the melt and the corresponding growth and segregation behavior. From theoretical considerations it is clear that phenomena such as irregular crystal pulling, mechanical vibrations, and, particularly, fluctuations in melt temperature and flow will result in transient growth and segregation behavior. These phenomena, particularly temperature fluctuations, are not restricted to Czochralski systems but are also operative, for example, during top-seeded Bridgman growth and growth by the gradient freezing technique.

Nonrotational striations originating from thermal instabilities have been systematically studied in InSb and Ge grown with top seeding of a confined melt by programmed power reduction in a gradient furnace.[139] Growth and segregation under these conditions proceed in the presence of destabilizing thermal gradients, with the melt height decreasing under continuing solidification. Under these conditions the melt is expected to experience with continuing growth, according to thermohydrodynamic theory, successively turbulent convection, oscillatory instability, and finally, for small melt heights, thermal stability. In agreement with theory, irregular convection-controlled transients in growth and segregation behavior were observed for large melt heights, periodic transients caused by oscillatory thermal behavior occurred at intermediate heights, and homogeneous segregation (absence of transients) was encountered in the bottom section of a grown crystal.

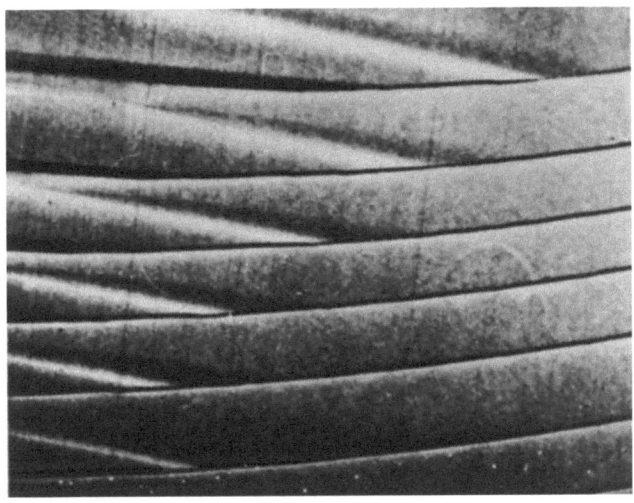

(a)

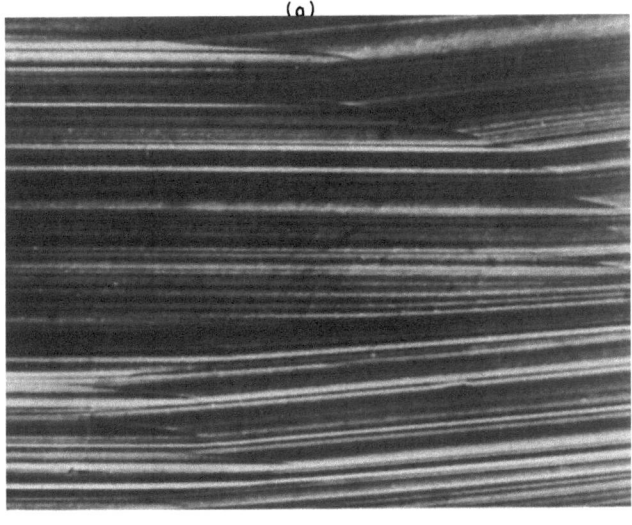

(b)

Fig. 21. (a) Segregation behavior in Te-doped ($5 \times 10^{17}/\text{cm}^3$) InSb pulled with seed rotation. Observe the elimination of random nonrotational striations and the appearance of periodic rotational striations. (The central portion of the crystal reveals a "coring" effect associated with {111} facet growth). ×8.5. (See Ref. 28.) (b) Segregation behavior in Te-doped ($10^{17}/\text{cm}^3$) InSb pulled without rotation in the ⟨111⟩ direction. The depicted longitudinal cut along the growth axis (etched with a differential $KMNO_4$ solution) reveals random segregation inhomogeneities (nonrotational striations) which are attributed to transient perturbations at the growth interface due to turbulent convection. ×28.

4.3.3. Rotational Striations

The effect of seed rotation on segregation during Czochralski pulling can readily be observed in Figure 21, which depicts differentially etched longitudinal sections of Te-doped InSb pulled (a) with rotation and (b) without seed rotation. The photomicrographs show that seed rotation is effective in suppressing adverse convective interference on segregation. They also reveal, however, that seed rotation in turn leads to more or less pronounced transient segregation inhomogeneities which exhibit the periodicity of rotation (rotational striations).

Studies of transient segregation effects associated with rotational pulling indicate that the complexity of striations increases with the melting point of the growth system, as seen in Figure 22, which presents a comparison of dopant inhomogeneities in indium antimonide (mp 525°C) and silicon (mp 1410°C). The observed complexities in high-melting-point materials are attributed to the inadequacy of seed rotation in suppressing increased turbulent convection effects associated with steeper destabilizing thermal gradients in the vicinity of the growth interface. A typical example of transient growth and segregation behavior associated with rotational pulling of Ga-doped Ge under reduced destabilizing gradients is shown in Figure 23, which depicts an etched longitudinal cut along the rotational axis of the crystal.[129] To permit a quantitative segregation analysis, control striations, clearly visible as distinct dark lines, were introduced during growth. The etching behavior of this crystal segment reveals pronounced rotational segregation transients with nonrotational transients superimposed on them. Since these convection-induced inhomogeneities are not associated with corresponding growth rate fluctuations (the spacings of control striations do not change with the appearance of nonrotational striations), it must be concluded that they result from transients in boundary layer conditions. Rotational striations are unambiguously identified as compositional maxima. Thus the transient segregation behavior appears to be in strong contrast to that in InSb, where rotational striations are identified as compositional minima. A segregation analysis on the microscale (comparison of growth rate and composition analysis) shows that compositional and growth rate minima coincide; however, growth rate and compositional maxima appear spatially displaced. In all instances the compositional maxima occur under conditions of decreasing growth rate. The extent of displacement of the maxima in

Crystal Growth from the Melt

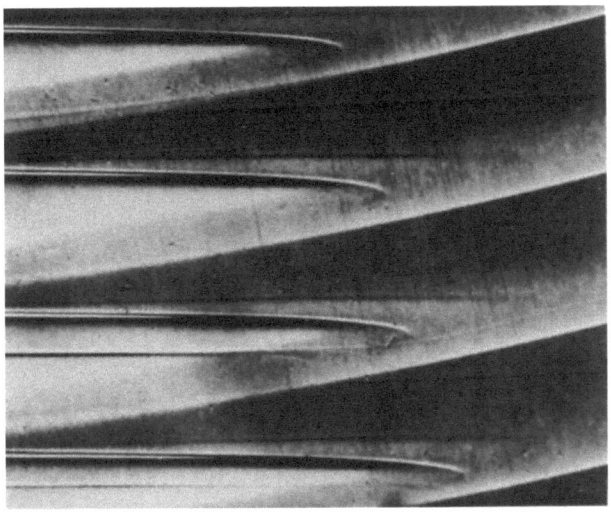
(a)

(b)

Fig. 22. (a) Rotational striations in Te-doped indium antimonide (mp 525°C). Note the high degree of periodicity. (From Ref. 31.) (b) Rotational striations in Sb-doped silicon (mp 1410°C). Note the lack of periodicity and the high density of nonrotational striations. (See Ref. 28).

Chapter 7

general decreases from the periphery to the interior of the crystals where coincidence does occur. Analysis of the data of Figure 23 showed that for a given rotation rate, dopant segregation associated with increasing growth rates during each rotational cycle could be treated as "steady state" segregation. However, during the decreasing growth rate portions, the Burton–Prim–Slichter theory could not be used, even qualitatively. Furthermore, striations were observed which could not be associated with growth rate variations and therefore must have been caused by transient perturbations of the diffusion boundary layer.[27] A further discrepancy from the analysis was observed for the large amplitude of the compositional striations due to the *n*-type dopant Sb as compared to the *p*-type dopant Ga, which

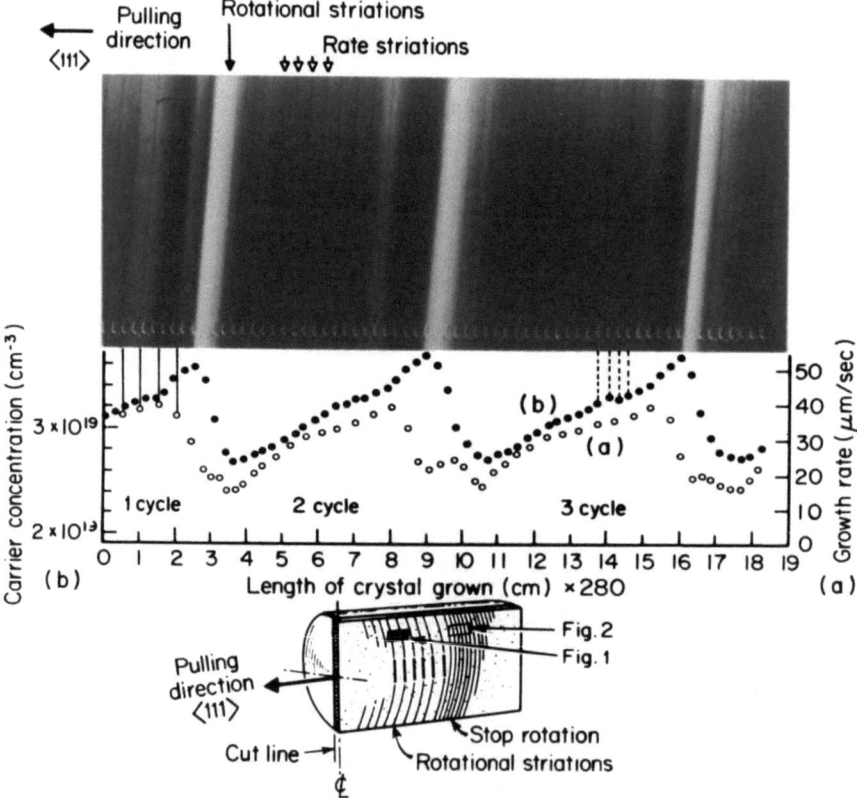

Fig. 23. Differentially etched segment of Ga-doped germanium single crystal pulled with seed rotation and quantitative growth and segregation analysis. Curve (a): microscopic growth rates as obtained from the spacing of the control striations introduced at 0.5-sec intervals; curve (b): carrier (dopant) distribution as obtained from spreading resistance measurements at spacings of 10 μm. ×98. (From Ref. 129.)

Crystal Growth from the Melt

Fig. 24. Compositional changes of the Gd/Ga ratio produced by transient growth rate behavior during the Czochralski growth of gadolinium gallium garnet with the seed rotation. The photomicrograph was taken by interference contrast microscopy and shows a longitudinal cut etched in hot orthophosphoric acid depicting regions of nonfaceted growth (left) and faceted growth (right). The fine striations are attributed to crystal rotation and the coarse set to periodic temperature fluctuations associated with thermal convection in the melt. Notice the long time periods of partial remelting in the off-facet region. ×195. (From Ref. 140.)

has a similar "equilibrium" distribution coefficient. This effect had been observed previously for *n*-type dopants in Si[15] and is clearly a manifestation of nonequilibrium segregation.

Rotational striations in systems with even higher melting points, such as Si, are complicated by the predominance of nonrotational striations from the more vigorous thermal convection in the melts (see Figure 22). In fact, in oxide and halide melts, where the thermal diffusivity is two orders of magnitude lower than that of Si, $N_{Ra}^{(1)}$ is correspondingly higher and thermal convection is always turbulent. Since oxide crystals are grown at rates which are slower than Si by an order of magnitude and the large-amplitude temperature fluctuations have periods which are of the order of minutes, then rotational striations are very closely spaced and the major striations are nonrotational

Chapter 7

in origin. A typical set of striations in a gadolinium gallium garnet ($Gd_{3+x}Ga_{5-x}O_{12}$) crystal is shown in Figure 24.[140] In this photomicrograph the growth direction is vertically downward with a noncentral facet visible in the right corner. The striations have been identified as nonstoichiometric variations in the Gd/Ga ratio[141] (see Ref. 142 for further details on the phase equilibria). The fine striations are rotational in origin, while the coarse set is due to temperature fluctuations large enough to probably have caused remelting. This system exhibits a pronounced tendency for the development of {110} facets on convex (into the melt) growth interfaces. Such facet regions (right top corner) exhibit pronounced departures of the Gd/Ga ratio from that of the surrounding region. The facetted type of compositional inhomogeneity is in most instances suppressed by the application of high rates of seed rotation, which result in a virtually flat growth interface. Measurements of the lattice parameter changes associated with the facets[143] indicate that the Gd/Ga variations are of the order of 0.04%.[142] Although not measured, it is probable that similar compositional changes are occurring across the growth striations. Similar behavior is expected even in Si melts when $N_{Ra}^{(1)}$ is increased for large melt sizes. A height increase from 1.5 to 6 in. will raise $N_{Ra}^{(1)}$ from $\sim 10^4$ to $\sim 10^7$ so that the melt thermal convection will go from a simple unsteady flow (Figure 18) to full turbulence. Under such conditions it must be expected that very little compositional control is possible from crystal rotation or any other form of forced convection.

4.3.4. Simultaneous Seed and Crucible Rotation

The simultaneous application of seed and crucible rotation during Czochralski pulling is based on the *a priori* assumption that crucible rotation will result in increased mixing of the melt and thus reduce the magnitude of thermal asymmetry which is responsible for the appearance of rotational striations. Recent experimental studies of segregation in crystals pulled with seed and crucible rotation reveal, however, major, as yet unexplained, complications. There is strong evidence that during counterrotation of seed and crucible, thermal asymmetry effects introduced through the crystal and through the melt become additive, thus leading to rotational segregation transients of increased complexity for Te-doped InSb. The application of synchronized simultaneous seed and crucible rotation in the same direction (isorotation) does result in rotational striations of decreased intensity; however, there is strong evidence that this improvement is accompanied by an increased radial segregation which more than offsets the beneficial effects.

The effects of crucible rotation on dopant segregation during Czochralski pulling have most recently been investigated in Ga-doped germanium. Preliminary results indicate that in spite of pronounced thermal asymmetry, the microscopic growth rate remains virtually constant, as seen from the constant spacing of the control striations. Crucible rotation does, however, lead to compositional segregation inhomogeneities which are most noticeable at the crystal periphery and decrease in intensity toward the central portions. The observed compositional heterogeneities never exceed about $\pm 2\%$, appear under virtually constant microscopic growth rate, and must be attributed to periodic transients in boundary layer conditions. Since the radial segregation effect present with seed rotation is virtually completely suppressed, this approach for achieving homogeneous single crystals deserves more intensive investigation. The effectiveness of crucible rotation appears, however, to be restricted to low rotational rates. At increased rates of seed rotation the growth interface exhibits a pronounced depression in the central portion of the crystal interface. It must be expected that such growth interface deformations, attributed to forced convection effects, will be associated with a substantially increased boundary layer thickness and thus should give rise to a pronounced segregation effect. However, quantitative data are not yet available.

4.4. Phase Equilibria and Nonstoichiometry

For crystal growth of compounds it has been found that the components frequently do not combine in the ratios dictated by Dalton's law. This behavior may be attributed to the fact that the sublattices involved allow for thermally generated defects (such as vacancies) with widely differing energies. Consequently, at high temperatures such lattices show a high solubility for one (or more) of the components. Furthermore, the dystectic composition, where liquidus and solidus curves meet, does not coincide with the stoichiometric composition in many systems, as shown schematically in Figure 25. While the nonstoichiometric composition range is quite narrow, for some compounds there are many for which it is of the order of a few mole percent, as was shown in the previous section for gadolinium gallium garnet. As a result, many compounds which melt

Chapter 7

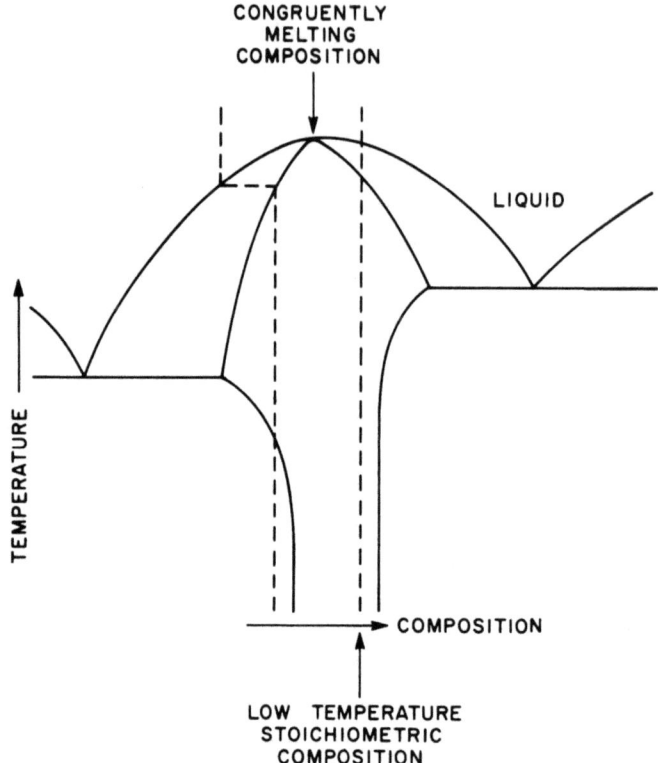

Fig. 25. Schematic phase diagram for a congruently melting, nonstoichiometric compound showing the displacement of the dystectic (maximum melting) composition from stoichiometry.

congruently (and hence can be grown from their own melts) will usually be nonstoichiometric unless the appropriate phase equilibria are known and corrective measures are taken. At absolute zero temperature, the solvus phase boundaries converge to the stoichiometric composition. If such crystals are grown at other than the dystectic composition, any variations in growth rate, diffusion boundary layer thickness, or melt composition will lead to compositional variations in the same manner as for solutes in dilute concentrations. The widespread nature of this problem can be deduced by referring to Table 1, where the need for control of nonstoichiometry is indicated for most compounds. Some cases for which extensive studies have been undertaken are indicated for binary compounds in Table 3. Compounds with transition metals which possess valence state changes requiring control of the ambient chalcogen or oxygen pressure are not included here.

TABLE 3
Nonstoichiometric Nature of Some Binary Congruently Melting Compounds

Compound	Maximum existence range	Dystectic composition	Ref.
$Ga_{0.5+x}P_{0.5-x}$	$+0.00025$ to -0.00005	$+0.00006$	144
$Ga_{0.5+x}As_{0.5-x}$	$+0.00023$ to -0.00018	$+0.00004$	212
$Pb_{0.5+x}S_{0.5-x}$	$+0.0005$ to -0.00005	$+0.0005$	82, p. 102
$Pb_{0.5+x}Se_{0.5-x}$	Not known	-0.00005	213
$Pb_{0.5+x}Te_{0.5-x}$	$+0.0005$ to -0.00013	-0.00012	145, 146
$Sn_{0.5+x}S_{0.5-x}$	0 to -0.00028	-0.00013	214
$Cd_{0.5+x}Te_{0.5-x}$	$+0.000002$ to -0.000008	-0.000005	215
$Mg_{1+x}Al_{2-2x}O_{4-2x}$	$+0.47$ to -0.42	Not known	216
$Gd_{3+x}Ga_{5-x}O_{12}$	$+0.30$ to -0.002	$+0.05$	142, 217
$Li_{1-5x}Nb_{1+x}O_3$	$+0.031$ to 0^-	$+0.0092$	152, 162
$Li_{1-5x}Ta_{1+x}O_3$	$+0.025$ to 0^-	$+0.0066$	151, 218

It can be seen immediately that oxides possess much wider ranges of compositional shifts than semiconductors. However, the values of these shifts for semiconductors are comparable to (or considerably in excess of) the doping levels which are desired, so that nonstoichiometric defects play an important role in determining properties such as resistivity type, resitivity, lifetime, and the ratio of nonradiative to radiative recombination rates.

The shift of the dystectic composition is generally in the direction of the component with the lower vapor pressure over the compound. When the vapor pressures are comparable, as for CdTe, the compositional shift is very small. The nature of the lattice defects associated with the nonstoichiometric compositions is a matter of great interest since at absolute zero temperature only the stoichiometric composition is thermodynamically stable and such defects will therefore give rise to properties which are highly temperature sensitive. Unfortunately, such studies are difficult and are often not performed on adequately characterized material. In addition, trace impurities can completely modify the nature of nonstoichiometric defects and their influence should be checked by independent doping experiments.

4.4.1. Measurement of Nonstoichiometric Phase Relations

In all phase relation studies pertaining to crystal growth, the measurement of liquidus curves as a function of composition is accomplished quite simply by the differential thermal analysis (DTA)

Chapter 7

of solid-state reacted and equilibrated specimens. However, the measurement of solidus curves cannot be performed by thermal arrest techniques and more sophisticated approaches are required to establish the tie lines; these will now be discussed. In addition, since the slope of the liquidus curve is zero at the maximum melting point, then, with the usual temperature measurement errors, DTA techniques lack sufficient sensitivity to accurately determine the dystectic composition.

4.4.1.1. Direct Chemical Analysis—GaP

Chemical analytical techniques normally possess a precision of a few mole percent. For major components, this makes deviations such as those shown in Table 3 inaccessible by normal analytical methods. Recently, however, coulometric titration techniques have been refined so that the precision is limited only by the purity and atomic weight value of the standard pure component. Application of the method to GaP by Jordan et al.[144] indicates that a precision of one part in 10^5 is possible, so that although the exact location of the dystectic composition in Table 3 is not known, it is clearly on the Ga-rich side of the stoichiometric composition. The experimental solidus data indicate that:

(a) All crystals prepared along the Ga-rich liquidus contain an excess of Ga.
(b) The crystals always show a normal freezing distribution of Ga in the growth direction.
(c) The solidus is retrograde, with a maximum departure from stoichiometry 65°C below the maximum melting point.
(d) Doping variations do not strongly influence the data.
(e) Equilibration of Ga-excess crystals in phosphorus induces a change toward stoichiometry.
(f) Crystals grown at the indicated dystectic composition become metastable at 950°C with respect to the precipitation of liquid Ga.

It was necessary to carefully remove all traces of second-phase Ga in the as-grown crystals. Since excessive constitutional supercooling was observed in melts with less than 30 at. % P, these specimens were crushed into a powder to allow Ga removal before performing the analysis.

Associated calculations showed that if neutral Ga and P vacancies are the predominant native defects, then their concentrations at the melting point (1465°C) are 8×10^{18} and 1.3×10^{19} cm^{-3}, respectively. It has become increasingly clear that Ga vacancies are an important part of the "killer centers" in GaP, which increase the nonradiative recombination rates and so lower the electroluminescent and photoluminescent efficiencies.

Application of the coulometric titration technique to other materials will depend upon whether efficient separation and complexing reactions are available for the ion of interest.

4.4.1.2. Crystal Equilibration—PbTe

In this technique initially as-grown narrow-bandgap semiconductor crystals are equilibrated in the solid state with fixed vapor pressures of the component species at various temperatures. Then the crystal is quenched and its resistivity measured. Assumptions about the nature of the nonstoichiometric defects allow the data to be converted to composition.

Brebrick et al.[145,146] have applied this method successfully to the determination of phase relations and solidi for PbTe. Small crystals of Bridgman- or Czochralski-grown PbTe were equilibrated through the vapor phase with pre-prepared ingots of Pb-Te with variable compositions so that solid–liquid equilibrium was established in the ingots at the temperature in question. The extrinsic carrier concentration was measured at room temperature for such an equilibrated and quenched specimen to provide a measure of the compositional stability limit at the particular equilibration temperature. Conversion to chemical concentration assumed that the difference between the atomic fraction of Te in PbTe and one-half is equal to the difference between the concentrations of holes and electrons divided by four times the concentration of sites on each sublattice.

Subsequent measurements by Chou et al.[147] on the retrograde nature of the solidi showed that for PbTe the dystectic composition can be located at $x = +0.00008$ rather than the value shown in Table 3. However, the equilibration method is not particularly suited to the measurement of the maximum melting composition since, in principle, the crystal should melt under such conditions.

Another serious problem with the use of only carrier concentration to determine chemical concentration is the necessity for assuming a specific defect model. Although straightforward for PbTe, Kroger[148] has indicated that for other, nonequimolecular compounds there may be complications in the assumption of simple defect models. In addition, the band structure must allow full ionization of either vacancies, interstitials, or antistructure substitutional atoms to allow use of resistivity methods. Concurrently, the ionized impurity level must be lower than the defect concentrations, to avoid interferences. It is desirable, therefore, to independently characterize post-equilibrated crystals by the use of precision hydrostatic density[149] and lattice parameter[150] measurements to precisions of 1 ppm. Thus, in principle, if there are no precipitates and the samples are uniform in composition, deviations from stoichiometry can be determined for specific models by comparing the hydrostatic and x-ray densities (see, for example, Ref. 151).

4.4.1.3. Comparative Methods—LiNbO$_3$

Somewhat simpler methods for determining nonstoichiometric phase relations can be used if properties can be measured for a material which depend only on its composition and not on its state of aggregation. One of the most common techniques here is to compare the lattice parameters of carefully equilibrated, solid-state reacted (ceramic) specimens to those of single crystals grown from known melt compositions. The lattice parameter varies continuously across the existence range (usually obeying Vegard's law) and is constant in the two-phase regions. In addition, the lattice parameters of the two types of specimens will be equal at the dystectic composition. However, for many compounds, the variation in lattice parameter across the existence range is quite small and measurements on polycrystalline specimens are complicated by line broadening due to particle size and strain effects.

Examples of other properties which can be measured on both polycrystalline and single-crystal specimens are the ferroelectric transition (Curie) temperature and ^{93}Nb broad-line NMR peak shapes in LiNbO$_3$.[152] The variation in Curie temperature T_C with solid composition is shown in Figure 26 for ceramic specimens (data points). Also included are data from crystals measured as a function of melt composition. These data were obtained by capacitance measurements at 1 kHz and the linewidths were much broader for ceramic specimens than for crystals. Nevertheless, there is a shift of 160°C in T_C over the existence range of 2 mole % Li$_2$O, so that the

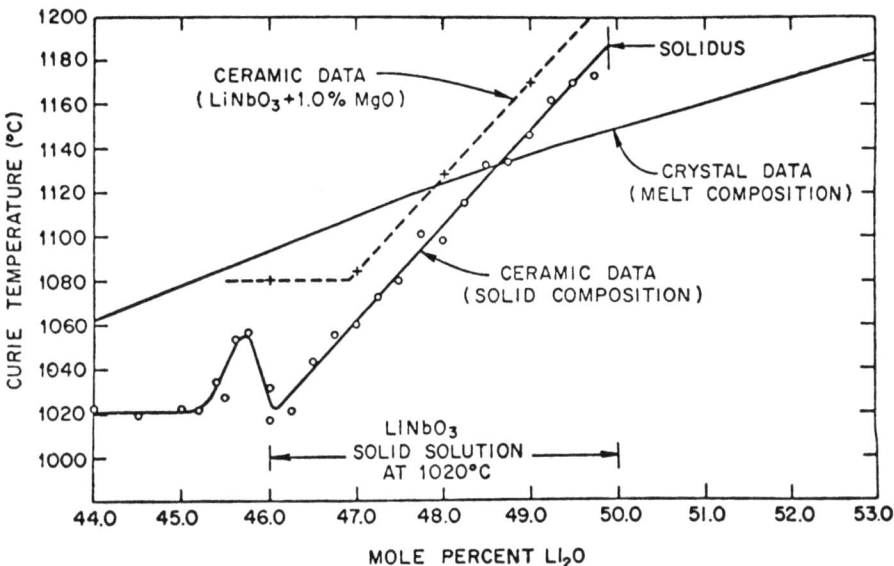

Fig. 26. Variation in ferroelectric transition (Curie) temperature with solid composition for ceramic (solid-state reacted) specimens and with melt composition for crystals. (From Ref. 152.)

technique is very sensitive except close to the solidus, where premelting anomalies occur. However, the ^{93}Nb low-field derivative NMR peak width is particularly sensitive to composition in the region close to 50% Li_2O, so that with both techniques, a full correlation between solid and melt compositions is possible. The results are shown in Figure 27 as the expanded phase diagram for $LiNbO_3$. The maximum melting or dystectic composition is seen to be 48.6 mole% Li_2O by the intersection of the ferroelectric transition curves in Figure 26. Crystals grown at the dystectic composition have been shown to be very uniform in composition by second harmonic generation mapping techniques.[153,154] In a similar fashion, the existence range and dystectic composition for $LiTaO_3$ have been identified.[151]

The existence range for $LiNbO_3$ is not shown below 1000°C in Figure 27 and crystals with the dystectic composition are stable in this temperature region. However, prolonged annealing at lower temperatures has indicated that this composition crosses the solvus line near 800–850°C and precipitation occurs.[155,156] Such behavior causes optical scattering and the crystal is no longer useful for optical applications. Re-solution is easily accomplished at 1000°C

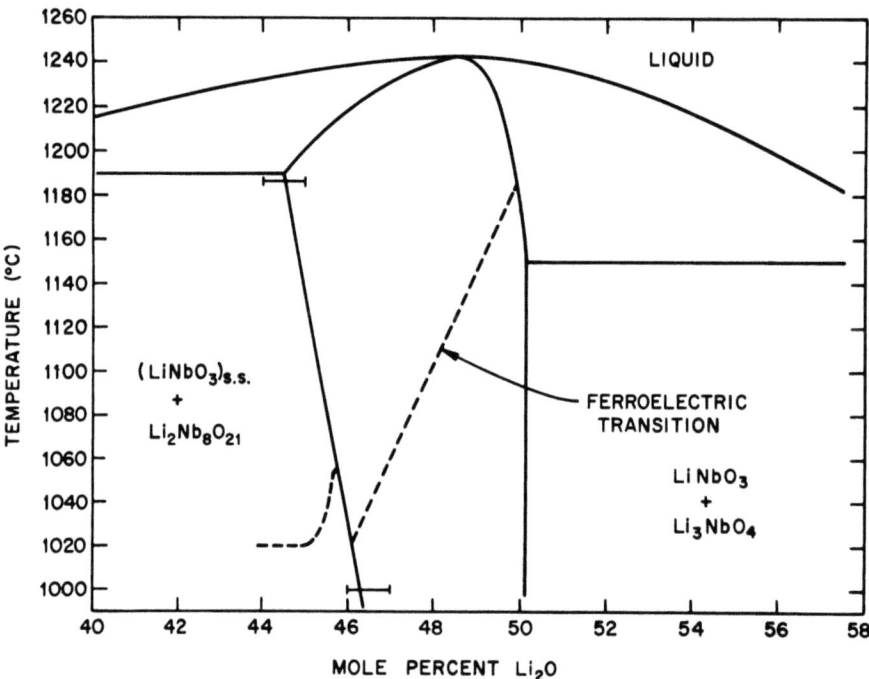

Fig. 27. High-temperature phase diagram of lithium niobate based on Curie temperature, ^{93}Nb NMR, and lattice parameter measurements. (From Ref. 152.)

and even slow cooling rates ($\sim 10°$C/min) can be used to avoid precipitation in the neighborhood of 850°C. At room temperature such a crystal exists in a state of metastable equilibrium as a supersaturated solid solution. Although not reported, such metastability may cause problems when crystals with the dystectic composition are used for second harmonic generation at high laser power intensities.

4.4.1.4. Methods for Ternary Compounds—$Ba_2NaNb_5O_{15}$

Measurements of existence regions, dystectic compositions, and phase tie lines become very complicated in compounds with more than two primary components. In ternary ferroelectric compounds the simple correlation in Figure 26 of T_C with composition becomes a three-dimensional surface which possesses a range of compositions for each temperature. Isothermal contours of such a surface are shown in Figure 28 for the ternary existence region of $Ba_2NaNb_5O_{15}$.[157] The boundaries of this region and general tie lines had been previously mapped by the identification of second-phase

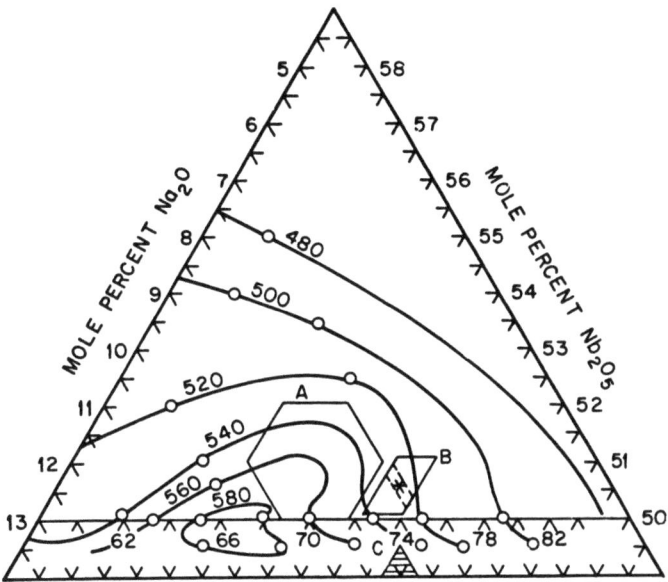

Fig. 28. Curie temperature measurements for ceramic specimens in the ternary solid-solution range of barium sodium niobate. Region A is the congruency estimate (from Ref. 157). Regions B and C are the congruency estimates from previous work. Horizontal axis shows coordinates of the pseudo-binary system $NaNbO_3$–$BaNb_2O_6$ in terms of mole percent $BaNb_2O_6$.

precipitation and thermal arrest techniques.[158] It is obvious that at least one other property should be mapped in three-component space for both ceramic and crystal compositions to isolate the dystectic composition. However, the use of high-resolution DTA liquidus temperature measurements and direct chemical analysis was used in this instance to narrow the dystectic composition to the region shown as A in Figure 28.

There have been no full phase relation investigations for ternary compounds that yield all the detailed information discussed above. In future work, if ternary and quarternary component compounds are to be available in controlled compositions, such work must be carried out.

4.4.2. Growth of Crystals at Nondystectic Compositions

The growth of crystals at dystectic compositions may produce crystals with the most uniform nonstoichiometric compositional homogeneity; however, such compositions may not be desirable because of second-phase precipitation effects at lower temperatures or because the associated physical properties may not be desirable. For example, in $LiNbO_3$ the dystectic composition yields crystals with a phase matching temperature that is so low that permanent laser damage can occur at moderate power levels, and higher Li_2O concentrations are required. In GaP the dystectic composition is not desirable because the nonradiative recombination rate is too high and Ga-rich crystals are required. Consequently, the problems associated with growth at nondystectic compositions must be considered in more detail.

It is necessary, in the growth of crystals at nondystectic compositions, to use melt compositions which differ widely from the crystal composition since there is usually a wide separation between solidus and liquidus curves. As a result, compositional inhomogeneities and structural defects arise as follows:

(a) Constitutional supercooling will be present for a wide range of growth conditions and result in interface shape destabilization. As a result, second-phase liquid inclusions with the adjacent eutectic composition will be formed in the cell boundaries. These inclusions are invariably trapped by the growing crystal and then undergo temperature gradient zone melting[2] back to the interface, leaving a string of second-phase solid particles in the crystal. Such substructure has been observed in both Li_2O-rich $LiNnO_3$* and in Ga-rich GaP.[160]

(b) Nonuniform segregation behavior will result from nondystectic growth unless the special nonconservative techniques outlined in Section 2.4 are used. Normal freezing variations have been observed in GaP by coulometric titration[144] and ^{69}Ga NMR[161] measurements and in $LiNbO_3$ by SHG mapping[153] and lattice parameter measurements.[162] In addition, growth interface facet-induced compositional variations and also striations have been observed by etching studies in GaP[160] and by SHG mapping,[153] visual examination,[163] and Curie temperature measurements[164] in $LiNbO_3$. It should be noted that any attempt to reduce or eliminate thermal convection-induced striations by lowering melt temperature gradients will increase the degree of constitutional supercooling.

(c) Solid-state precipitation will occur in principle for any nonstoichiometric composition at some temperature down to 0°K. Such

---

* See similar work in $Ba_2Nb_5O_{15}$ by Linares and Sigsway.[159]

precipitation effects have already been alluded to in the discussion of growth at dystectic compositions and will continue to exist for other nonstoichiometric compositions.

(d) Growth twinning occurs more readily in materials grown with increasingly nonstoichiometric compositions and have been observed in both GaP[160] and LiNbO$_3$.[165] Although the origins of such twins are not completely understood, the role of nonstoichiometric defects must be primarily responsible.

(e) Nonstoichiometric lattice defects have been identified in GaP by etching studies[166,167] and in LiNbO$_3$ by the angular variation of the ^{93}Nb NMR linewidth in a nonstoichiometric single crystal.[168] Although these studies are not completely capable of describing the defects, they indicate that some aggregation and complexing of the basic atomic defects is probably occurring during the growth process. Further work is needed to understand and control such phenomena.

4.5. Defect Studies

Although the literature is filled with studies of structural defects in crystals, there have been only a limited number which have related the nature and density of these defects to crystal growth processes. This fact is rather surprising since crystals of various metals were grown almost exclusively in the 1920's and 1930's for use in x-ray diffraction studies where substructure with low-angle grain boundaries (also known as lineage or mosaic substructure) was identified and found to be related to plastic deformation processes. The experimental confirmation of the existence of dislocations in the 1950's provided a new stimulus for crystals with lower dislocation densities since it was found that, for most materials, densities in excess of $\sim 10^5$ lines/cm^2 resulted in their coalescence into arrays which had previously been identified as lineage. The most significant work was that of Dash,[169] who showed that Si crystals could be grown free of dislocations during Czochralski growth by the use of controlled necking during the seeding process. Although Si is a relatively easy material to grow dislocation-free, extensive work has also yielded low dislocation densities (10^2–10^3 cm^{-2}) in Cu[170] and Al[171] crystals. Currently, dislocation-free Si crystals are routinely produced in 3-in. diameters by Czochralski growth and 2-in. diameters by the pedestal technique.

The elimination of dislocations in Si and Ge crystals has resulted in many studies of point-type defects in these materials. Dash[172] first indicated that the absence of dislocations removed the major source of vacancy sinks during cooling from the growth temperature. Evidence for the nucleation of vacancy clusters from this supersaturation was reported by Tweet[173] for Ge crystals and Patel et al.[174] and Plaskett[175] for Si crystals. In an extensive series of experiments, De Kock[176] has revealed that in fact there are two types of vacancy clusters in dislocation-free Si: small clusters close to the periphery of the crystal, and large clusters seen in the interior only. It is postulated that vacancy–oxygen association complexes serve as the nuclei for the precipitation of excess vacancies. On this basis De Kock has shown that the clusters can be eliminated by increasing the growth rate to above 3 mm/min and by doping the crystal with hydrogen during growth. Such cluster-free material is important for many silicon device applications involving the use of lightly doped (high resistivity) p-type crystals. It has been shown that the clusters serve as sinks for fast diffusing transition metal ions inadvertently introduced during processing and which subsequently lead to undesirable space-charge effects, including "soft" p–n junctions. In addition, such clusters collapse to form extrinsic stacking faults during high-temperature oxidation treatments. Subsequent precipitation of oxides then occurs at these sites, which leads to extensive dislocation generation in the presence of small thermal stress. Since the initial oxygen concentration is highest in the center of the as-grown Czochralski crystals, the generated dislocation density is highest here. Such an inhomogeneous distribution can lead to bowing or warping of wafers undergoing the oxidation.

Similar studies of the behavior of nonstoichiometric point defects of the type discussed in the last section have not been performed because these compounds have not yet been produced dislocation-free over large volumes.

5. Control of Crystal Growth from the Melt

The requirements placed upon the physical properties of crystals as discussed in Section 3 and the studies of melt growth in Section 4 have both led to significant advances in the control of crystal growth from melts. Although some of these techniques have been discussed earlier in this chapter, they will be mentioned here for completeness.

5.1. Defect Control

Control over structural defects in crystals is accomplished primarily by a reduction of the dislocation density during growth.

Crystal Growth from the Melt

This objective is achieved through the use of seeds with a low dislocation density and a lattice parameter equal to that of the material to be grown. Necking during seeding and diameter control are important to avoid dislocation generation through thermal shocks. In addition, the temperature distribution in the growing ingot must be carefully controlled, especially in the vicinity of the growth interface. In Si crystals the interface (and isotherm) shape must be concave into the liquid over the central portion of the interface and close to flat at the outer periphery to avoid the nucleation of dislocations.

Control over point defect densities in dislocation-free crystals is harder to control. However, phenomena such as vacancy clustering in lightly doped Si crystals appear to be eliminated at fast growth rates and by the use of hydrogen to avoid initial nucleation on oxygen impurities. Point defect densities in compounds are controllable through changes in melt stoichiometry and possibly through secondary charge compensation doping, although much work remains to be done here.

5.2. Compositional Control

The control of compositional uniformity during growth involves the use of nonconservative growth methods (Section 2.4) as well as control of the segregation processes. Primary emphasis must be placed on the control of mass transport processes at the growth interface and the elimination of nonequilibrium facet segregation effects. Facets are eliminated by maintaining the growth interface concave into the melt. Control of mass transport processes involves the elimination of unsteady natural convection effects and the reduction of steady bulk thermal convection. For a minimal normal freezing variation, either diffusion-controlled mass transport alone is desired at the growth interface or zone methods must be used. For minimal variation across the growth interface, uniform convection effects are required or, more simply, the methods used for minimal normal freezing variations can be used. Basically, therefore, control is required over the thermal convection processes normally present in the melt without using forced convection. These techniques have been discussed by Carruthers[29] and briefly are as follows.

(a) Temperature gradient control. This is achieved by reducing vertically destabilizing temperature gradients, by using centrally localized cooling of the melt, and by using heat pipes to reduce temperature gradients in the bulk melt but retain them at the growth interface.

(b) *Dimensional control.* The influence of container dimensions and shape on thermal convection is of particular interest in two areas: scaling-up problems, and the use of baffles in crystal growth processes. The dimensions enter into the driving force (through the Rayleigh number) to the third power if the temperature difference between bounding surfaces is considered, and to the fourth power if the liquid temperature gradient is considered. The distinction is thus made between the cases where (1) the boundaries are kept at the same temperature as the dimensions are changed, or (2) the temperature gradient in the liquid is kept contant. In both cases the driving force for thermal convection is increased so that a substantial enhancement of flow velocities and perturbations must be anticipated.

In scaling up growth processes under these conditions, the turbulent regime of thermal convection, together with secondary flows, will be more common than the simpler unstable forms of thermal convection. Such convection is undesirable because the growth interface will always respond to some portion of the temperature fluctuation spectrum. In addition, although bulk turbulent flows are more spatially isotropic then nonturbulent flows, there may be boundary layer separation problems which will result in highly non-uniform convective diffusion processes across growth interfaces. The aspect ratio and container geometry are not expected to be significant at high Rayleigh numbers in influencing thermal convection behavior.

The use of baffles in liquids in principle should reduce the Rayleigh number for thermal convection as well as inhibit the flow itself. However, careful consideration must be given to the placement of baffles since adverse vertical gradients and the maximum flow velocities are not uniformly distributed throughout the melts but are quite close to the bounding surfaces.

(c) *Body forces.* For electrically conducting liquids, small static magnetic fields aligned with field lines parallel to the major heat flow direction are very effective inhibitors of thermal convection flow. Uniform rotation around a vertical axis tends to stabilize liquids with adverse vertical gradients (temperature increasing downward). Dielectric liquids may be stabilized by static and alternating electric fields for temperature variations of the electrical conductivity and dielectric permittivity, respectively. However, most molten liquids possess enough electrical conductance to make internal Joule heating an additional problem. For such liquids a parallel alignment of static magnetic and electric fields may be a more feasible way of controlling thermal convection.

Another interesting possibility for ionic molten liquids is the use of gradients in magnetic field strength or electric field strength to interact with gradients in paramagnetic susceptibility and dielectric permittivity, respectively. If such fields oppose the interaction of gravity with density gradients, then, for modest values of magnetic or electric field strengths, the influence of gravity may be completely counteracted.

(d) Zero gravity. The removal of gravitational forces has been made possible in recent years on the manned orbiting laboratories of NASA. These include the three Skylab mssions of 1973 and the more elaborate laboratories being designed as a part of the Space Shuttle in the 1980's. In this environment, gravitational acceleration is actually offset by centrifugal acceleration so that this is really a special case of body forces. Residual levels of 10^{-4}–10^{-3} g are possible. However, before proper experiments can be designed, there are a number of immediately identifiable liquid handling problems which must be solved first. Some of these have been dealt with in Section 4.2.1. It will be imperative to avoid other sources of convective flow in such experiments, such as those due to shape changes, expansion or contraction of internal gas voids, solidification density changes, surface tension-gradient-induced flows, flows from force fields such as RF induction or centrifugal acceleration, and bulk flows from Taylor-Proudman effects (where one surface of a liquid rotates at a different angular velocity from some other portion of the surface).

5.3. Diameter Control

The adaptation of the Czochralski technique by Teal and Little[177] to the growth of Ge crystals required visual observation of the growing crystal diameter and manual power input adjustments to the heater to control the diameter. Subsequently, monitoring of melt temperature and power input allowed analog programming to continuously and automatically change the power input and was used to maintain constant diameter after the seeding and shouldering operations. The most sophisticated of these methods was that of Bachmann et al.[178] for the growth of metal single crystals. However, in the mid-1960's the development of automatic diameter control techniques in the silicon industry became stimulated by a number of factors:

- (a) The need to eliminate thermal shocks during growth so as to maintain dislocation-free configurations.
- (b) The increase in the diameter to 3 in. with the concomitant necessity for more refined control of the power input to the large melts (7–12 kg).
- (c) The waste associated with crystals when reduced to constant diameter by centerless grinding after growth.
- (d) The labor costs involved in the growth process.
- (e) The poor visibility associated with loss of encapsulant transparency in LEC growth.

Currently, there are three major automatic diameter control techniques: fixed-temperature sensing of the radiation emitted from the curved meniscus, scanning of the image projected onto a TV screen, and direct weighing of either the crystal or the melt. These will be discussed separately.

5.3.1. Fixed Radiation Sensing of the Meniscus

During Czochralski growth at high temperatures (over about 1000°C) the periphery of the growth interface appears to be surrounded by a bright ring of emitted radiation. This ring is caused by the focusing of reflected radiation from the curved liquid meniscus (although it is often erroneously associated with the generation of latent heat of fusion). This enhanced radiativity can be sensed easily by an optical thermopile (pyrometer) which views the melt surface at a small angle from the vertical.[179–181] The target area is about 2 mm in diameter and is located about 4 mm from the actual periphery of the growing crystal. The sensor output signal is amplified and used to control the crystal pull rate. The crucible lift rate is slaved to the crystal pull rate in order to maintain the melt surface at a constant vertical position in the heater.

For zero dislocation growth, upper and lower limits are placed on the pull rate and the heater power input is then adjusted to keep the diameter constant. In some systems, the crystal and crucible rotation rates and ambient gas flow rates may also be controlled from such fixed sensor outputs.

The fixed radiation sensing method has found its widest application in the growth of large-diameter silicon crystals. A similar technique had been conceived many years ago by Buhrendorf[182] to control the diameter of floating-zone silicon crystals. However, a simpler technique was devised by Buehler[183] based on the frequency shift of the RF tank circuit (used to heat the zone) produced by a change in melt volume.

5.3.2. Image Scanning

In this class of techniques, an image of the growing crystal and meniscus shape is projected onto a TV screen for electronic scanning by line blanking techniques. The image can be produced optically,[184-186] by infrared measurement,[187,188] or by x-ray radiography.[189,190]

In the optical image method, the diameter and meniscus are imaged from a horizontal mirror placed very close to the melt surface to optimize the viewing angle. This mirror surface tends to become coated by condensation of species evaporating from the melt. Nevertheless, the external viewing window can be shielded from such condensation by use of the mirror. The technique has been used for the Czochralski growth of Cu crystals but may not be useful at higher temperatures under blackbody conditions or if the emissivities of the melt and container wall are similar.

The infrared measurement method overcomes this problem by using a vidicon tube with a Si diode array to scan the IR portion of the spectrum. This technique can thus be used to record temperatures as well and in fact thermal convection patterns on the melt surface become visible for the niobate[187] and garnet[188] oxide melts from which growth has been controlled. In both optical imaging systems discussed so far, the diameter measurement resolution is about 0.1 mm and only the power input to the furnace is controlled.

The visibility problem associated with the viewing angle of the meniscus surface is complicated even further in LEC growth by the (sometimes) poor transmissivity of the B_2O_3 liquid encapsulant. Consequently, an ingenious x-ray radiographic method has been devised which utilizes the high absorption coefficient for x-rays of Ga in GaP crystals relative to its environment (SiO_2 crucible, graphite susceptor, B_2O_3 encapsulant, Al RF induction coil, and Be external windows). However, spatial resolution is lost in the fluorescent screen used to record the radiographic image because of the need to retain reasonable thicknesses for quantum conversion efficiency. Consequently, diameter control of only 1 mm is possible by using error signals from TV blanking to control the RF power input.[190]

5.3.3. Weighing

The visibility problems associated with the diameter control techniques of the last two sections can be eliminated completely by monitoring the weight of either the growing crystal[191,193,196] or the crucible and melt.[192,194,195]

In the crystal weighing scheme, provision must be made for crystal rotation since the load cell is incorporated into the pull rod itself. The electrical output from the load cell is compared to a predetermined ramp voltage and the difference signal is used to control the heater input power. Diameter control of $\pm 2\%$ has been achieved.[193] A control problem arises with materials, such as semiconductors, which decrease their density upon solidification, which can be corrected by additional signal processing.

In the crucible-plus-melt weighing scheme, a commercial digital weighing balance with a mechanical tare is used. The output signal can be processed digitally[195] or by analog methods[194] to provide diameter control of $\pm 2\%$ and less than $\pm 1\%$, respectively.

With weighing methods a complication occurs if the interface shape is changing during growth. This problem can be quite serious in oxide crystals grown under low thermal gradients, because the interface freezes into the melt as in the Kyropoulos technique, particularly during the initial seeding. Subsequently, during shouldering, the interface flattens out and the result appears as a reduction in the rate of weight change.[197]

One of the unexpected benefits from the diameter control of oxide crystals, just as for silicon crystals, has been the elimination of thermal shocks and associated excessive dislocation generation in the crystal. The result has been the elimination of low-angle boundaries and associated regions of misorientation in the crystals. These misorientations give rise to severe thermal stresses upon cooling if the thermal expansion coefficient is anisotropic and often result in cracking of the crystal at lower temperatures.[198] Reduction of dislocation densities by diameter control methods to below 10^4 cm^{-2} eliminates the cracking problem in oxide crystals with noncubic symmetry.

5.4. Shape Control

The first use of shape control in melt growth techniques was the use of split molds to grow metal single crystals in the shape of tensile specimens to avoid difficult and damaging machining operations subsequent to growth. In the Czochralski technique, the use of meniscus shaping was used to grow flat ribbons of semiconductor crystals with the intention of avoiding the slicing and polishing operations required for bulk crystals. Meniscus shaping was achieved through the use of induced RF fields,[199] twinned dendrites,[200,201] widely spaced pairs of dendrites (web Si),[202] dies,[203-205] and capillary–die combinations.[206-209]

These techniques all allow fast growth rates to be used since the dissipation of the latent heat of fusion is enhanced by the large surface area to volume ratio. However, there are major problems which have not been solved as yet with regard to surface nonplanarity, excessive dislocation generation, and solute segregation inhomogeneities. In addition, for methods using dies, the reactivity of the melt with the die can be a problem, so that dimensional stability, contamination, heterogeneous nucleation, twinning, and solid state precipitation may occur.

Nevertheless, the use of capillary–die combinations (known as the edge-defined, film-fed growth or EFG process) has been highly successful in the growth of complicated cross-sectional shapes of sapphire from molybdenum capillaries in inert atmospheres. The EFG process also resembles the floating crucible process in its ability to provide compositional control (see Section 2.4.2). The requirement for circular cross sections to grow with constant composition is[4]

$$R > (r_{cap}/r_{crys})^2 (D/L) \qquad (27)$$

where R is the crystal growth rate, r_{cap} and r_{crys} are the capillary and crystal radii, respectively, D is the liquid diffusion coefficient, and L is the capillary length. For $r_{cap}/r_{crys} = 1/2$, $D_L = 10^{-4}$ cm^2/sec, and $L = 0.2$ cm, growth rates greater than 0.45 cm/hr will be sufficient to maintain compositional uniformity. It thus appears that EFG possesses tremendous potential for the growth of controlled shapes and compositions. Future application of diameter control techniques, such as described in the previous section, to EFG may reduce surface irregularities as well as dislocation generation.

6. Future Directions

The future needs and opportunities in crystal growth have recently been reviewed by Laudise[210] from the viewpoint of

Chapter 7

materials requirements. The biggest impetus, common to semiconductor, magnetic, and optical materials, is the need to provide structural and compositional control over smaller dimensions (microminiaturization) and extend it over larger distances (large-scale integration). Therefore future emphasis must be placed on the controlled growth of larger crystals with no dislocations or point defect clusters and with compositional uniformity over the macroscale as well as down to the micrometer scale. These requirements dictate that studies of the type discussed in Section 4 become increasingly sophisticated and that new and old techniques, in synergistic combination with new and old materials, be pursued together with an increasing complexity of crystal growth control techniques. For instance, new methods, such as RF heating of materials with intermediate resistivities[211] and growth in zero gravity,[58] new materials, such as ternary and quaternary compounds, and more sophisticated control methods which could override transient growth rate fluctuations should all be pursued in parallel with studies of the detailed mechanistics of segregation processes.

References

1. R. A. Laudise, *The Growth of Single Crystals*, Prentice-Hall, Englewood Cliffs, New Jersey (1970).
2. W. G. Pfann, *Zone Melting*, 2nd ed., Wiley, New York (1966).
3. B. Chalmers, *Principles of Solidification*, Wiley, New York (1964).
4. J. C. Brice, *The Growth of Crystals from Liquids*, North-Holland/Elsevier, New York (1973).
5. D. T. J. Hurle, in *Crystal Growth; an Introduction* (P. Hartman, ed.), North-Holland/Elsevier, New York (1973), p. 210.
6. J. B. Mullin, in *Proc. Second Int. School on Crystal Growth*, North-Holland (1975).
7. B. Cockayne, in *Proc. Second Int. School on Crystal Growth*, North-Holland (1975).
8. K. A. Jackson, in *Treatise on Solid State Chemistry*, Vol. 5 (N. B. Hannay, ed.), Plenum, New York (1975), Chapter 5.
9. J. H. Wernick, in *Treatise on Solid State Chemistry*, Vol. 1 (N. B. Hannay, ed.), Plenum, New York (1974), Chapter 4.
10. R. N. Hall, *Phys. Rev.* **88**, 139 (1952).
11. R. N. Hall, *J. Phys. Chem.* **57**, 836 (1953).
12. J. A. M. Dickhoff, *Solid State Electron.* **1**, 202 (1960).
13. J. B. Mullin and K. F. Hulme, *J. Phys. Chem. Solids* **17**, 1 (1960).
14. A. Trainor and B. E. Bartlett, *Solid State Electronics* **2**, 106 (1961).
15. J. R. Carruthers and K. E. Benson, *Appl. Phys. Lett.* **3**, 100 (1963).
16. J. R. Carruthers, *Can. Met. Quart.* **5**, 55 (1966).
17. J. A. Burton, R. C. Prim, and W. P. Slichter, *J. Chem. Phys.* **21**, 1987 (1953).

18. W. G. Cochran, *Proc. Cambridge Phil. Soc.* **30**, 365 (1934).
19. J. R. Carruthers, *J. Cryst. Growth* **2**, 1 (1968).
20. K. E. Benson, *Electrochem. Technol.* **3**, 332 (1965).
21. J. R. Carruthers, *J. Electrochem. Soc.* **114**, 959 (1967).
22. J. R. Carruthers and W. C. Winegard, in *Crystal Growth* (H. S. Peiser, ed.), Pergamon, New York (1966), p. 645.
23. V. G. Levich, *Physicochemical Hydrodynamics*, Prentice-Hall, Englewood Cliffs, New Jersey (1962).
24. A. F. Witt, M. Lichtensteiger, and H. C. Gatos, *J. Electrochem. Soc.* **120**, 1119 (1973).
25. D. T. J. Hurle, E. Jakeman, and E. R. Pike, *J. Cryst. Growth* **34**, 633 (1968).
26. D. T. J. Hurle and E. Jakeman, *J. Cryst. Growth* **5**, 227 (1969).
27. A. F. Witt, M. Lichtensteiger, and H. C. Gatos, *J. Electrochem. Soc.* **121**, 787 (1974).
28. J. R. Carruthers and A. F. Witt, in *Proc. Second Int. School on Crystal Growth*, North-Holland (1975).
29. J. R. Carruthers, in *Preparation and Properties of Solid State Materials*, Vol. 2 (W. R. Wilcox, ed.), Marcel Dekker, New York (to be published).
30. J. C. Brice and P. A. C. Whiffin, *Brit. J. Appl. Phys.* **18**, 581 (1967).
31. K. Morizane, A. F. Witt, and H. C. Gatos, *J. Electrochem. Soc.* **114**, 738 (1967).
32. J. R. Carruthers and K. Nassau, *J. Appl. Phys.* **39**, 5205 (1968).
33. E. P. A. Metz, R. C. Miller, and R. Mazelsky, *J. Appl. Phys.* **33**, 2016 (1962).
34. J. B. Mullin, B. W. Straughan, and W. S. Brickell, *J. Phys. Chem. Solids* **26**, 782 (1965).
35. F. Weinberg, *Trans. Met. Sci. AIME* **227**, 231 (1963).
36. T. Abe, M. Karazawa, and M. Iida, *Jap. J. Appl. Phys.* **31**, 58 (1962).
37. A. J. Goss, private communication (1955).
38. G. S. Cole and W. C. Winegard, *J. Inst. Metals* **93**, 153 (1964/65).
39. A. J. Goss, K. E. Benson, and W. G. Pfann, *Acta Met.* **4**, 332 (1956).
40. A. Mueller and M. Wilhelm, *Z. Naturforsch.* **19a**, 254 (1964); **20a**, 1190 (1964).
41. D. T. J. Hurle, *Phil. Mag.* **13**, 305 (1966).
42. D. T. J. Hurle, J. Gillman, and E. L. Harp, *Phil. Mag.* **14**, 205 (1966).
43. E. J. Harp and D. T. J. Hurle, *Phil. Mag.* **17**, 1033 (1968).
44. J. R. Carruthers and J. Pavilonis, *J. Appl. Phys.* **39**, 5814 (1968).
45. D. T. J. Hurle, in *Crystal Growth* (H. S. Peiser, ed.), Pergamon (1967), p. 659.
46. H. P. Utech and M. C. Flemings, in *Crystal Growth* (H. S. Peiser, ed.), Pergamon (1967), p. 651.
47. W. C. Johnston and W. A. Tiller, *Trans. Met. Soc. AIME* **221**, 331 (1961).
48. W. C. Johnston and W. A. Tiller, *Trans. Met. Soc. AIME* **224**, 214 (1962).
49. J. D. Verhoeven, *Phys. Fluids* **12**, 1733 (1969).
50. T. S. Noggle, *Rev. Sci. Instr.* **24**, 184 (1953).
51. F. Schmid and D. Viechnicki, *J. Cryst. Growth* **11**, 345 (1971).
52. T. Abe, *J. Japan Inst. Metals* **21**, 611 (1957).
53. H. F. Sterling and R. W. Warren, *Nature* **192**, 745 (1961).
54. W. G. Pfann, C. E. Miller, and J. D. Hunt, *Rev. Sci. Instr.* **37**, 649 (1966).
55. J. R. Carruthers and M. Grasso, *J. Appl. Phys.* **43**, 436 (1972).
56. W. Heywang, *Z. Naturforsch.* **11a**, 238 (1956).
57. L. M. Hocking, *Mathematika* **7**, 1 (1960).

58. J. R. Carruthers, in *The Science of Liquid Drops and Bubbles* (D. J. Collins, ed.), Jet Propulsion Laboratory, Pasadena, California (1974).
59. W. R. Wilcox, in *Fractional Solidification* (M. Zief and W. R. Wilcox, eds.), Marcel Dekker, New York (1967), p. 157.
60. C. E. Chong and W. R. Wilcox, in *The Science of Liquid Drops and Bubbles* (D. J. Collins, ed.), Jet Propulsion Laboratory, Pasadena, California (1974).
61. B. N. Savel'ev, V. V. Drobrovensky, and V. S. Chudakov, *Soviet Phys.—Crystallog.* **15**, 473 (1970).
62. T. F. Ciszek, *J. Cryst. Growth* **12**, 281 (1972).
63. D. B. Gasson, *J. Sci. Instr.* **42**, 575 (1965).
64. J. Goorisen, *Philips Techn. Rev.* **21**, 185 (1960).
65. A. Trainor and B. E. Bartlett, *Proc. Phys. Soc.* **74**, 669 (1959).
66. R. S. Wagner and W. C. Ellis, *Trans. AIME* **233**, 1053 (1965).
67. D. Mateika, *J. Cryst. Growth* **9**, 249 (1971).
68. S. Akai, K. Aoyagi, Y. Hirata, and T. Suzuki, in *Crystal Growth 1974*, N. Kato, K. A. Jackson, and J. B. Mullin, eds.), North-Holland (1974).
69. S. E. Bradshaw and A. I. Mlavsky, *J. Electronics* **2**, 134 (1956).
70. F. A. Trumbore and E. M. Porbansky, *J. Appl. Phys.* **31**, 2068 (1960).
71. R. Roy, *Mat. Res. Bull.* **1**, 299 (1966).
72. W. F. Leverton, *J. Appl. Phys.* **29**, 1241 (1958).
73. G. R. Blackwell, *Solid State Electronics* **7**, 105 (1964).
74. S. V. Airapetyants and G. I. Shmelev, *Soviet Phys.—Solid State* **2**, 689 (1960).
75. B. W. Mott, *J. Mat. Sci.* **3**, 424 (1968).
76. I. N. Belyaev, *Russ. Chem. Rev.* **29**, 428 (1960).
77. H. Nelson, *Transistors I*, RCA Laboratories, Princeton, New Jersey (1956), p. 66.
78. M. Tannenbaum, in *Semiconductors* (N. B. Hannay, ed.), Rheinhold, New York (1959).
79. F. A. Trumbore, E. M. Porbansky, and A. A. Tartaglia, *J. Phys. Chem. Solids* **11**, 239 (1959).
80. A. L. Gentile and F. H. Andres, *Mat. Res. Bull.* **2**, 853 (1967).
81. D. T. J. Hurle, J. B. Mullin, and E. R. Pike, *J. Mat. Sci.* **2**, 46 (1967).
82. F. A. Kroger, *The Chemistry of Imperfect Crystals*, North-Holland, Amsterdam and Wiley, New York (1964), p. 67.
83. J. J. Gilman, ed., *The Art and Science of Growing Crystals*, Wiley, New York (1963).
84. J. L. Shay and J. H. Wernick, *Ternary Chalcopyrite Semiconductors—Growth, Electronic Properties, Applications*, Pergamon, New York (1974).
85. I. P. Kaminow and A. E. Siegman, eds., *Laser Devices and Applications*, IEEE Press, New York (1973).
86. J. H. Wernick, *Recent Developments in Magnetic Materials*, in *Ann. Rev. of Mat. Science*, Vol. 2 (R. A. Huggins, R. H. Bube, and R. W. Roberts, eds.), Annual Reviews, Palo Alto, California (1972), p. 607.
87. H. R. Huff and R. R. Burgess, eds., *Semiconductor Silicon 1973*, The Electrochemical Society, Princeton, New Jersey (1973).
88. M. B. Panish and I. Hayashi, Heterostructure junction lasers, in *Applied Solid State Science*, Vol. 4 (R. Wolfe, ed), Academic, New York (1974), p. 235.
89. *Gallium Arsenide and Related Compounds*, The Institute of Physics, London (1971).

90. T. C. Harman and I. Melngailis, Narrow gap semiconductors, in *Applied Solid State Science* (R. Wolfe, ed.), Academic, New York (1974), p. 1.
91. *Second International Conference on the Strength of Metals and Metal Alloys*, American Society for Metals (1970).
92. *Mechanical Behavior of Crystalline Solids*, NBS Monograph 59 (1963).
93. R. Genders, *J. Inst. Metals* **35**, 259 (1926).
94. L. Northcott, *J. Inst. Metals* **62**, 101 (1938).
95. W. A. Tiller, K. A. Jackson, J. W. Rutter, and B. Chalmers, *Acta Met.* **1**, 428 (1953).
96. W. W. Mullins and R. F. Sekerka, *J. Appl. Phys.* **35**, 444 (1964).
97. R. F. Sekerka, in *Crystal Growth* (W. Bardsley, D. T. J. Hurle, and J. B. Mullin, eds.), American Elsevier, New York (1973), p. 403.
98. K. G. Davis and P. Fryzuk, *J. Cryst. Growth* **8**, 57 (1971).
99. L. R. Morris and W. C. Winegard, *J. Cryst. Growth* **5**, 361 (1969).
100. R. Singh, A. F. Witt, and H. C. Gatos, *J. Electrochem. Soc.* **121**, 380 (1974).
101. J. P. Dismukes and W. M. Yim, *J. Cryst. Growth* **22**, 287 (1974).
102. D. Giest and P. Grosse, *Z. Angew. Phys.* **14**, 105 (1962).
103. C. Huh and L. Scriven, *J. Coll. Interface Sci.* **30**, 323 (1969).
104. J. F. Padday and A. R. Pitt, *Phil. Trans. Roy. Soc.* (*London*) **275**, 489 (1973).
105. R. D. Gillette and D. C. Dyson, *Chem. Eng. J.* **2**, 44 (1971).
106. C. A. Miller and L. E. Scriven, *J. Fluid Mech.* **32**, 417 (1968).
107. D. T. J. Hurle, *J. Cryst. Growth* **13/14**, 39 (1972).
108. E. Jakeman and D. T. J. Hurle, *Rev. Phys. in Technol* **3**, 3 (1972).
109. R. R. Zupp, J. W. Nielsen, and P. V. Vittorio, *J. Cryst. Growth* **5**, 269 (1969).
110. P. A. C. Whiffin and J. C. Brice, *J. Cryst. Growth* **10**, 91 (1971).
111. J. C. Brice, O. F. Hill, and P. A. C. Whiffin, *J. Cryst. Growth* **6**, 297 (1970).
112. J. C. Brice, O. F. Hill, P. A. C. Whiffin, and J. A. Wilkinson, *J. Cryst. Growth* **10**, 133 (1971).
113. J. C. Brice, *Acta Electronica* **16**, 291 (1973).
114. B. Cockayne, J. G. Plant, and A. W. Vere, *J. Cryst. Growth* **15**, 167 (1972).
115. B. Cockayne, M. Chesswas, J. G. Plant, and A. W. Vere, *J. Mat. Sci.* **4**, 565 (1969).
116. H. P. Utech and M. C. Flemings, in *Crystal Growth* (H. S. Peiser, ed.), Pergamon (1967), p. 651.
117. D. T. J. Hurle, in *Crystal Growth* (H. S. Peiser, ed.), Pergamon (1967), p. 659.
118. C. S. Cheng, D. A. Irvin, and B. G. Kyle, *AIChE J.* **13**, 739 (1967).
119. R. M. Sharp and A. Hellawell, *J. Cryst. Growth* **12**, 261 (1972).
120. J. R. Carruthers and W. C. Winegard, *Can. Met. Quart.* **6**, 233 (1967).
121. R. V. Birikh, *J. Appl. Mech. Tech. Phys.* **3**, 43 (1966).
122. A. L. Dragoo, in *Symposium on the Science of Liquid Drops and Bubbles* (D. J. Collins, ed.), Jet Propulsion Laboratory, Pasadena, California (1974).
123. A. Acrivos, *AIChE J.* **4**, 285 (1958).
124. J. R. Carruthers, *J. Electrochem. Soc.* **114**, 1077 (1967).
125. B. Cockayne and M. P. Gates, *J. Mat. Sci.* **2**, 118 (1967).
126. K. E. Benson, *Electrochem. Tech.* **3**, 332 (1965).
127. R. Singh, A. F. Witt, and H. C. Gatos, *J. Electrochem. Soc.* **115**, 112 (1968).
128. M. Lichtensteiger, A. F. Witt, and H. C. Gatos, *J. Electrochem. Soc.* **118**, 1013 (1971).

Chapter 7

129. A. F. Witt, M. Lichtensteiger, and H. C. Gatos, *J. Electrochem. Soc.* **120**, 1119 (1973).
130. A. Muller and M. Wilhelm, *Z. Naturforsch.* **20a**, 1190 (1965).
131. K. Morizane, A. F. Witt, and H. C. Gatos, *J. Electrochem. Soc.* **113**, 51 (1966).
132. A. B. Chase and W. R. Wilcox, *J. Am. Ceram. Soc.* **50**, 332 (1967).
133. R. L. Burns, *Mat. Res. Bull.* **2**, 273 (1967).
134. D. T. J. Hurle, *Phil. Mag.* **13**, 305 (1966).
135. P. I. Baranskii and E. A. Glushkov, *Inorg. Mat.* **5**, 1008 (1969).
136. N. Y. Shushlebina, Y. M. Shashkov, and B. A. Sakharov, *Inorg. Mat.* **5**, 712 (1969).
137. D. O. Morantz and K. K. Mathur, *J. Cryst. Growth* **16**, 147 (1972).
138. C. Brehon, J. Boniort, and P. Margotin, *J. Cryst. Growth* **18**, 191 (1973).
139. K. M. Kim, A. F. Witt, and H. C. Gatos, *J. Electrochem. Soc.* **119**, 1218 (1972).
140. D. C. Miller, *J. Electrochem. Soc.* **120**, 678 (1973).
141. R. F. Belt and J. P. Moss, *Mat. Res. Bull.* **8**, 1197 (1973).
142. J. R. Carruthers, M. Kokta, R. L. Barns, and M. Grasso, *J. Cryst. Growth* **19**, 204 (1973).
143. H. L. Glass, *Mat. Res. Bull.* **7**, 1087 (1972).
144. A. S. Jordan, A. R. Von Neida, R. Caruso, and C. K. Kim, *J. Electrochem. Soc.* **121**, 153 (1974).
145. R. F. Brebrick and R. S. Allgaier, *J. Chem. Phys.* **32**, 1826 (1960).
146. R. F. Brebrick and E. Gubner, *J. Chem. Phys.* **36**, 1283 (1962).
147. N. Chou, K. Komarek, and E. Miller, *Trans. Met. Soc. AIME* **245**, 1553 (1969).
148. F. A. Kroger, *Phys. Chem. Solids* **7**, 277 (1958).
149. H. A. Bowman and R. M. Schoonover, *J. Res. Nat. Bur. St.* **71C**, 179 (1967).
150. R. L. Barns, *Mat. Res. Bull.* **2**, 273 (1967).
151. R. L. Barns and J. R. Carruthers, *J. Appl. Cryst.* **3**, 395 (1970).
152. J. R. Carruthers, G. E. Peterson, M. Grasso, and P. M. Bridenbaugh, *J. Appl. Phys.* **42**, 1846 (1971).
153. F. R. Nash, G. D. Boyd, M. Sargent III, and P. M. Bridenbaugh, *J. Appl. Phys.* **41**, 2564 (1970).
154. R. L. Byer, J. F. Young, and R. S. Feigelson, *J. Appl. Phys.* **41**, 2320 (1970).
155. B. A. Scott and G. Burns, *J. Am. Ceram. Soc.* **55**, 225 (1972).
156. L. O. Svaasand, M. Eriskrud, G. Nakken, and A. P. Grande, *J. Cryst. Growth* **22**, 230 (1974).
157. J. R. Carruthers, S. L. Blank, M. Grasso, and W. A. Biolsi, *J. Cryst. Growth* **23**, 195 (1974).
158. J. R. Carruthers and M. Grasso, *Mat. Res. Bull.* **4**, 413 (1969).
159. R. C. Linares and R. L. Sigsway, *Mat. Res. Bull.* **3**, 825 (1968).
160. G. A. Rozgonyi, A. R. Von Neida, and D. C. Miller, *J. Electron. Mat.* **1**, 381 (1972).
161. G. E. Peterson, A. Carnevale, and H. W. Verleur, *Mat. Res. Bull.* **6**, 51 (1971).
162. P. Lerner, C. Legras, and J. P. Dumas, *J. Cryst. Growth* **3/4**, 231 (1968).
163. H. T. Parfitt and D. S. Robertson, *Brit. J. Appl. Phys.* **18**, 1709 (1967).
164. P. M. Bridenbaugh, *J. Cryst. Growth* **19**, 45 (1973).
165. C. A. Wallace, *J. Appl. Cryst.* **3**, 546 (1970).
166. T. Iizuka, *J. Electrochem. Soc.* **118**, 1190 (1971).

167. G. A. Rozgonyi and M. A. Afromowitz, *Appl. Phys. Lett.* **19**, 153 (1971).
168. G. E. Peterson and A. Carnevale, *J. Chem. Phys.* **56**, 4848 (1972).
169. W. C. Dash, *J. Appl. Phys.* **31**, 736, 2275 (1960).
170. F. W. Young and J. R. Savage, *J. Appl. Phys.* **35**, 1917 (1964).
171. S. Howe and C. Elbaum, *Phil. Mag.* **6**, 1227 (1961).
172. W. C. Dash, in *Growth and Perfection of Crystals* (R. H. Doremus, S. Roberts, and D. Turnbull, eds.), Wiley, New York (1958), p. 361.
173. A. G. Tweet, *J. Appl. Phys.* **29**, 1520 (1958); **30**, 2002 (1959).
174. J. R. Patel, R. S. Wagner, and S. Moss, in *Metallurgy of Elemental and Compound Semiconductors* (R. O. Grubel, ed.), Interscience, New York (1961), p. 465.
175. T. S. Plaskett, *Trans. Met. Soc. AIME* **233**, 809 (1965).
176. A. J. R. De Kock, *Philips Res. Repts. Suppl.* **1973**, No. 1.
177. G. K. Teal and J. B. Little, *Phys. Rev.* **78**, 647 (1950).
178. K. J. Bachmann, H. J. Kirsch, and K. J. Vetter, *J. Cryst. Growth* **7**, 290 (1970).
179. E. J. Patzner, R. G. Dessaur, and M. R. Poponiak, *Solid State Tech.* **10** (October), 25 (1967); U.S. Patent 3,493,770 (1970).
180. K. E. Domey, *Solid State Tech.* (October), 41 (1971).
181. R. E. Lorenzini, F. S. Neff, and D. J. Blair, *Solid State Tech.* (February), 33 (1974).
182. F. G. Buhrendorf, private communication.
183. E. Buehler, *Rev. Sci. Instr.* **28**, 453 (1957).
184. K. J. Gartner, K. F. Rittinghaus, A. Seeger, and W. Uelhoff, *J. Cryst. Growth* **13/14**, 619 (1972).
185. K. J. Gartner and W. Uelhoff, *Optik* **37**, 15 (1973).
186. K. J. Gartner and W. Uelhoff, *J. Cryst. Growth*, to be published.
187. D. F. O'Kane, T. W. Kwap, K. Gulitz, and A. L. Bednowitz, *J. Cryst. Growth* **13/14**, 624 (1972).
188. D. F. O'Kane, V. Sadagopan, E. A. Giess, and E. Mendel, *J. Electrochem. Soc.* **120**, 1272 (1973).
189. H. D. Pruett and S. Y. Lien, *J. Electrochem. Soc.* **121**, 822 (1974).
190. H. J. A. v Dijk and G. J. Scholl, *J. Cryst. Growth*, to be published.
191. J. Levinson, U.S. Patent No. 2,908,004, October 6, 1959.
192. T. Rummel, U.S. Patent 3,259,467, July 5, 1966.
193. W. Bardsley, G. W. Green, C. H. Holliday, and D. T. J. Hurle, *J. Cryst. Growth* **16**, 277 (1972).
194. T. R. Kyle and G. Zydzik, *Mat. Res. Bull.* **8**, 443 (1973).
195. A. E. Zinnes, B. E. Nevis, and C. D. Brandle, *J. Cryst. Growth* **19**, 187 (1973).
196. W. Bardsley, B. Cockayne, G. W. Green, M. Healey, D. T. J. Hurle, G. C. Joyce, J. M. Roslington, P. J. Tufton, and H. C. Webber, *J. Cryst. Growth*, **24/25**, 369 (1974).
197. G. Zydzik, *J. Cryst. Growth*, to be published.
198. C. D. Brandle and D. C. Miller, *J. Cryst. Growth*, **24/25**, 432 (1974).
199. G. K. Gaule and J. R. Pastore, in *Metallurgy of Elemental and Compound Semiconductors* (R. O. Grubel, ed.), Interscience, New York (1961), p. 201.
200. E. Billig and P. J. Holmes, *Acta Cryst.* **8**, 353 (1955).
201. H. F. John and J. W. Faust, in *Metallurgy of Elemental and Compound Semiconductors* (R. O. Grubel, ed.), Interscience, New York (1961), p. 127.

Chapter 7

202. S. N. Dermatis and H. F. John, *IEEE Trans. on Commun. and Electron.* **82**, 94 (1963).
203. E. von Gompertz, *Z. Phys.* **8**, 194 (1922).
204. A. V. Stepanov, *Bull. Acad. Sci. USSR* **33**, 1775 (1969).
205. J. J. Brissot and H. Reynaud, *Electrochem. Tech.* **1**, 304 (1963).
206. H. E. LaBelle and A. I. Mlavsky, *Mat. Res. Bull.* **6**, 571 (1971).
207. H. E. LaBelle, *Mat. Res. Bull.* **6**, 581 (1971).
208. B. Chalmers, H. E. LaBelle, and A. I. Mlavsky, *Mat. Res. Bull.* **6**, 681 (1971); and *J. Cryst. Growth* **13/14**, 84 (1972).
209. T. F. Ciszek, *Mat. Res. Bull.* **7**, 731 (1972).
210. R. A. Laudise, *J. Cryst. Growth*, to be published.
211. P. Guinet, *Rev. Sci. Instr.* **43**, 454 (1972).
212. V. V. Karataev, M. G. Mil'vidskii, V. B. Osvenskii, and O. G. Stolyarov, *Soviet Phys.—Cryst.* **18**, 521 (1974).
213. A. E. Goldberg and G. N. Mitchell, *J. Chem. Phys.* **22**, 220 (1954).
214. H. Rau, *J. Phys. Chem. Solids* **27**, 761 (1966).
215. W. Albers and C. Maas, *Philips Tech. Rev.* **30**, 142 (1969).
216. D. M. Roy, R. Roy, and E. F. Osborn, *J. Am. Ceram. Soc.* **36**, 149 (1953).
217. M. Allibert, C. Chatillon, J. Mareschal, and F. Lissalde, *J. Cryst. Growth*, to be published.
218. F. W. Ainger, *J. Cryst. Growth*, to be published.

8

Solution Growth

R. A. Laudise
Bell Laboratories
Murray Hill, New Jersey

1. Introduction

Crystal growth techniques can be classified into two broad groups—*monocomponent* and *polycomponent* methods.* In monocomponent methods only the chemical component forming the crystal is present in the growth system. In polycomponent methods an additional component has been added to the system. The usual reason for adding an additional component is to permit crystallization at a lower temperature than in a monocomponent system.† Among the reasons why crystallization at as low a temperature as possible is advantageous are: (1) to permit the direct growth of low-temperature polymorphs; (2) to avoid incongruent volatilization, incongruent melting, decomposition, or high vapor pressures at high temperature; (3) to minimize thermal gradients, strain caused by such gradients, and hence defects brought about by strain; (4) to reduce vacancy concentration; (5) to lower the solubility of impurities, to minimize reaction with containers, etc.; (6) to permit a dopant distribution impossible at a higher temperature (because of, for example, high dopant vapor pressure); and (7) for experimental convenience.

---

* Component is used in this discussion in the same sense it is used in the phase rule.
† By the strictest definition, all real systems are polycomponent since they contain accidental traces of impurities. However, practical systems may be considered monocomponent as long as the concentration of additional components (e.g., dopants, impurities) is low enough that diffusion processes are not of overriding importance. It is often useful to reserve the term polycomponent for systems where the additional component is deliberately added to aid in crystallization.

The disadvantages of polycomponent growth include the following. (1) The additional component will have a solubility in the crystal and will reduce its purity. If the component is electrically inactive or if it is a constituent of the crystal (e.g., Ga solvent for GaP), this may not be a severe problem, although growth using a common constituent of the crystal as solvent will usually result in a different stoichiometry than growth in the absence of the constituent. (2) The additional component will need to diffuse away from the growing crystal interface as growth proceeds and there will be a concentration gradient in front of the growing interface. Thus the rate-limiting steps in crystal growth from a polycomponent system will involve comparatively slow diffusion processes. Crystallization will need to proceed at a slower rate than in monocomponent growth. Problems arising from the concentration gradient, such as constitutional supercooling, banding, facet effects, cells, dendrites, and second component inclusions will often occur. Careful control of thermal conditions, mass flow, and the growth interface is required to overcome these problems. In short, one pays the price of speed of growth and purity for the privilege of low-temperature polycomponent growth. It is, however, a price which one often pays gladly in systems where higher temperatures are precluded.

Crystal growth techniques are also classified on the basis of the nature of the phase where the reaction forming the crystal takes place: (1) solid → crystal, (2) liquid → crystal, (3) gas → crystal. Thus, considering both monocomponent and polycomponent methods, there are six broad classes of growth.

Except for strain annealing, neither monocomponent nor polycomponent solid → crystal preparations have been widely used, principally because nucleation in a solid matrix is hard to control. While monocomponent gas-phase methods are limited by the fact that few materials have appropriate sublimation pressures, polycomponent gas-phase growth, essentially growth by gaseous reaction, is an important technique. However, because of low gas densities, it is usually difficult to achieve a rapid enough movement of matter toward the growing interface to permit the rapid growth of very large crystals. Thin films are easily grown and one of the main uses of gas-phase growth is in the preparation of epitaxial layers. Growth from the pure melt (monocomponent liquid → crystal) is especially advantageous in terms of speed, simplicity, and purity and is discussed elsewhere in this series. However, its principal disadvantage is that usually high temperatures are required.

Polycomponent liquid-phase growth, or solution growth, on the other hand, has the advantages of low-temperature growth together with those of comparatively rapid matter transport (in comparison with the gas phase). The role of the solvent can be viewed as an additional component to lower the melting point of the material being crystallized and growth methods may be characterized on the basis of how the supersaturation causing growth is provided:

1. Isothermal growth: (a) supersaturation by temperature difference (ΔT) between dissolving and growth zone; (b) supersaturation by solvent evaporation; (c) supersaturation by chemical reaction.
2. Nonisothermal methods: (a) slow cooling; (b) temperature gradient zone melting.

Since the distribution constants for impurities and dopants are temperature dependent, more uniform doping can often be obtained in isothermal methods. Growth by the temperature difference method has the advantage of producing crystals whose size is not limited by the amount of solute present in the solution. On the other hand, diffusion and solute transport problems are often more troublesome in temperature difference methods, while in slow cooling and evaporative methods diffusion limitations are less severe. Growth by reaction permits the growth of nearly insoluble materials but is hard to control.

From the practical viewpoint the choice of solvent dictates the experimental equipment and the sort of materials which can be grown. For this reason we have organized this chapter on the basis of the solvent used. Before discussing specific solution growth techniques, however, we will discuss some important physical chemical generalizations underlying all the solution methods. Solution growth has also been reviewed in the book by Laudise,[1] the compendium of Gilman,[2] and the monograph by Buckley[3] (also see the article by Roy and White[4]).

2. Underlying Principles
2.1. Phase Equilibria and Distribution Constants in Solution Growth

Figure 1 shows schematics of typical binary diagrams involving solvents and a crystalline material.* Depending upon whether the

---
* For the moment we ignore cases of compound formation with the solvent or incongruent saturation. However, our present discussion is valid with minor modifications even in these cases.

Chapter 8

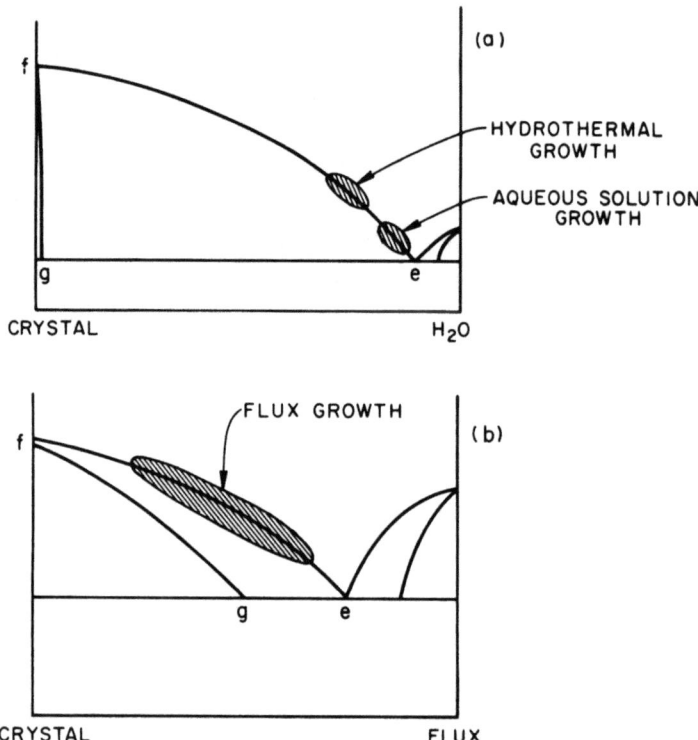

Fig. 1. Typical binary diagrams for solution growth.

solvent has a much lower melting point (Figure 1a; e.g., H_2O, Ga) than the solute or a more nearly comparable melting point (Figure 1b; e.g., fluxes) the diagrams are rather different. In both cases the role of the solvent is to depress the melting point of the material to be crystallized. In the case of the lower-melting-point solvent (Figure 1a); the depression is greater; the eutectic e occurs at a lower temperature and at a higher concentration of solvent than for the higher-melting-point solvent (Figure 1b). The solvent solubility in the crystal (f–g) is usually much higher for the higher-melting-point solvents. Typically in water the solubility of the solvent in the crystal will be of the order of ppm, while in fluxes the solubility ranges from tenths of a percent to a few percent. It should also be pointed out that somewhat higher temperatures (intermediate between ordinary H_2O systems and flux systems but at high pressures) are utilized in hydrothermal systems. The more conventional way of viewing solubility in a solvent is shown

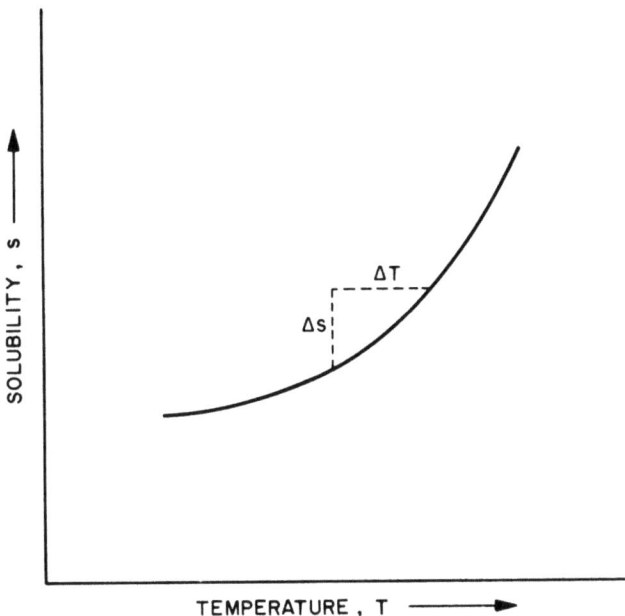

Fig. 2. Solubility–temperature curve.

in Figure 2, where the supersaturation ΔS produced by a given supercooling ΔT is shown. Usually the solubility obeys the van't Hoff equation, that is,

$$d(\ln s)/dT = \Delta H/RT^2 \qquad (1)$$

where s is the solubility and ΔH is the heat of solution. If the solubility is measured at constant pressure, the heat is the change in enthalpy ΔH. If, as is often the case in hydrothermal experiments, the solubility is measured at constant volume, then the heat is ΔE, the change in internal energy of the system.

The partition constant for an impurity grown from solution can be defined in a manner analogous to that used in monocomponent melt growth:

$$k_0 = C_{0,s}/C_{0,l} \qquad (2)$$

where k_0 is the partition (distribution constant) at equilibrium and $C_{0,s}$ and $C_{0,l}$ are the equilibrium concentrations of the impurity in the solid and liquid phases, respectively.

The effective distribution constant is defined as

$$k_{\text{eff}} = C_s/C_l \qquad (3)$$

where C_s and C_l are the actual concentrations of the impurity in the respective phases. In melt growth at very high rates $C_s \to C_l$ and $k_{\text{eff}} \to 1$. The analogous behavior in solution growth will be when the solid crystallizing from a given volume of solvent incorporates all the impurity available in that volume. Suppose that as a limiting case all the solute available crystallizes. If s is the solubility in moles/liter and C_l is the concentration of impurity in moles/liter, s moles crystallize and C_l moles of impurity will be incorporated so that

$$C_s = C_l \rho / sM \tag{4}$$

where M is the molecular weight of the solute and ρ is its density in g/liter. Thus

$$k_{\text{eff}} \to \rho / sM \tag{5}$$

as rate is increased. Solid-state diffusion at the temperatures of most solution growth is slower than in melt growth so that the impurity distribution frozen in at a growing crystal interface does not tend to equilibrate with the bulk of the crystal. Adsorption effects are also probably greater at the lower temperatures of solution growth.

Coupled substitution often occurs in solution growth in a manner causing modification of the simple expressions for distribution constant. When an atom of charge different from the atom it replaces enters a lattice, an additional atom of another impurity often enters the lattice at the same time so as to preserve electrical neutrality. The additional atom *charge-compensates* the original substituent. Similarly, an additional atom is often required for charge compensation when an atom enters the lattice interstitially.

When impurities are not available charge compensation will be by the formation of vacancies. In semiconductors charge compensation can occur by the formation of holes or the addition of electrons to the conduction band without the need of additional atoms entering the lattice. However, for the typical case of the growth of a large-band-gap material from a relatively impure solvent impurity, charge compensation by the entry of additional foreign atoms is the normal mode. This mechanism has been extensively studied in hydrothermal quartz growth and will be reviewed below where it will also be shown that k_{eff} depends on growth rate in the same manner as in melt growth, i.e., Burton–Slichter dependence.

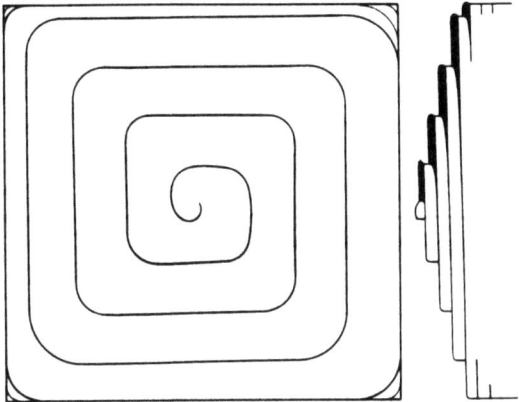

Fig. 3. Frank–Read dislocation.

2.2. Kinetics of Solution Growth

A detailed discussion of the kinetics of crystallization is beyond the scope of this chapter and is discussed elsewhere in this series. However, several unique features of the mechanism and kinetics of solution growth should be mentioned here. In the first place the best fit between theories of crystal growth and experimental results has been obtained under the comparatively experimentally uncomplicated conditions of monocomponent vapor–solid crystallization, where there is convincing experimental evidence that growth at low supercooling or supersaturation usually takes place on a spiral dislocation or Frank–Read source*. Figure 3 shows such a dislocation. On an atomically smooth surface once a layer is completed the initiation of a new layer would require a considerable supersaturation or supercooling. Thus it might be expected that growth would cease at low supersaturation or supercooling. Actually, of course, growth rate usually goes to zero only at zero supersaturation or supercooling. The spiral dislocation provides a surface which can can grow continuously at repeatable steps where no excess supersaturation or supercooling is required to initiate new layers. The experimental observation of spirals in real crystals, together with a detailed theory which predicted the growth rate supersaturation curve in I_2 vapor growth, led Frank and Cabrerra to suggest the Frank–Read source as mechanism whereby growth at low supersaturation on atomically smooth faces was possible. It is reasonable

---

* Attachment is usually formed at a point where the maximum number of bonds can be made. Thus a kink in a step is most favorable. See Laudise[1] for a more complete discussion of surface kinetics.

Chapter 8

to assume that the spiral dislocation plays a role in many solution growth mechanisms when growth is at low supersaturation on atomically smooth faces. However, it must be pointed out that much practical crystal growth is on atomically rough faces, where a spiral dislocation is not required. In addition, to achieve higher rates, the supersaturation is often made high enough that spiral dislocations would not be required.

The additional complexity of solution growth over, for instance, monocomponent vapor growth is brought about mainly because of the additional processes which the solvent introduces into the mechanism. Diffusion of the solute toward the growing interface and diffusion of the solvent away are often involved in the rate-limiting process. The solute species often is a complexate (e.g., in water a hydrate, in OH^- a hydroxide, etc.) so that desolvation or a reaction where the complexate is broken can often be involved in growth. In addition, interfacial steps such as chemisorption of the growing species and two-dimensional diffusion of the adsorbed species to the growing site can be important.

It is often convenient to treat solution growth as a consecutive process where diffusion and an interfacial reaction are both rate limiting.

If the equilibrium at the interface is so rapid as not to be involved in the rate-limiting step, so that bulk diffusion alone establishes the rate, Fick's first law expresses the rate of diffusion:

$$dm/dt = DA \, dc/dx \qquad (6)$$

where dm/dt is the mass added to a growing surface per unit time, dc/dx is the concentration gradient normal to the surface, D is the diffusion constant, and A is the area of the growing surface.

If the width of the diffusion layer is σ, c_B is the concentration in the bulk of the solution, and c_{su} is the concentration at the surface, Eq. (6) reduces to

$$dm/dt = (DA/\sigma)(c_B - c_{su}) \qquad (7)$$

If $\mathscr{R}$ is the linear rate of crystallization on the surface, where

$$dm/dt = \mathscr{R} A \rho \qquad (8)$$

with ρ the density of the crystal, Eq. (7) becomes

$$\mathscr{R} = (D/\rho\sigma)(c_B - c_{su}) \qquad (9)$$

If σ is a constant, it is convenient to replace D/σ with k_d, the "diffusion velocity constant," so that Eq. (9) becomes

$$\mathcal{R} = (k_d/\rho)(c_B - c_{su}) \tag{10}$$

Equation (10) is valid even if the rate of the interfacial reaction is not fast compared to diffusion. If the interfacial reaction is fast compared to diffusion, then $c_{su} = c_{eq}$, where c_{eq} is the equilibrium concentration.

If the interfacial reaction is overriding for a first-order heterogeneous reaction,

$$dm/dt = K_i A(c_{su} - c_{eq}) \tag{11}$$

where k_i is the "interfacial velocity constant." Thus by appropriate substitution,

$$\mathcal{R} = (k_i/\rho)(c_{su} - c_{eq}) \tag{12}$$

The value of k_i will generally depend on the crystallographic direction in which growth takes place. Equation (12) is valid even if the diffusion reaction is not rapid compared to the interfacial reaction. If the diffusion reaction is fast compared to the interfacial reaction, then $c_{su} = c_B$.

For consecutive reactions we may eliminate c_{su} between Eqs. (10) and (12), so that

$$\mathcal{R} = \frac{1}{\rho[(1/k_d) + (1/k_i)]}(c_B - c_{eq}) \tag{13}$$

which describes crystal growth when two consecutive reactions, diffusion and an interfacial step, are of comparable velocity. In Eq. (13) $\mathcal{R}$ will, of course, be directionally dependent and Eq. (13) will reduce to Eq. (10) for large values of k_i and to Eq. (12) for large values of k_d. If k_d and k_i are constants over the conditions of a given set of experiments, then

$$(1/k_d) + (1/k_i) = 1/\bar{k} \tag{14}$$

$$\mathcal{R} = (\bar{k}/\rho)(c_B - c_{eq}) \tag{15}$$

The values of k_d and k_i can be expected to be concentration and pressure independent over fairly broad ranges but will usually depend on temperature. The temperature dependence, at least to a first

Chapter 8

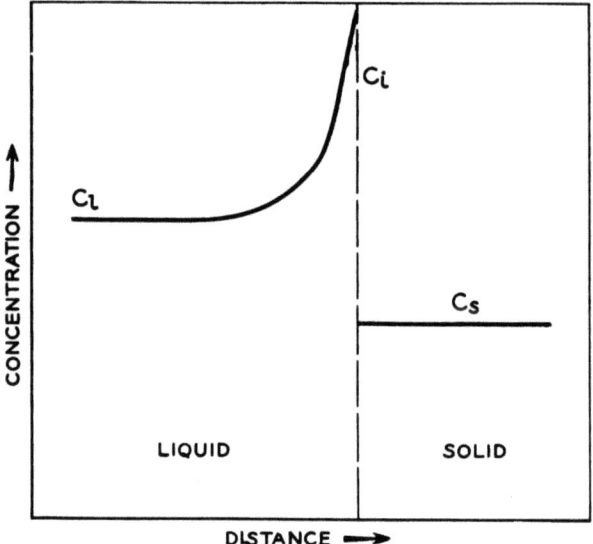

Fig. 4. Typical composition profile in front of a growing crystal.

approximation, will be, according to the Arrhenius equation,

$$k_d = \alpha_d \exp(-\Delta E_D^*/RT) \qquad (16)$$

$$k_i = \alpha_i \exp(-\Delta E_i^*/RT) \qquad (17)$$

where α_d and α_i are, to a first approximation, temperature-independent constants and ΔE_d^* and ΔE_i^* are the energies of activation of the diffusion and interfacial steps, respectively. Because ΔE_d^* is generally a few kilocalories per mole, while ΔE_i^* is considerably larger, k_i will be more strongly temperature dependent. In many cases $\bar{k}$ will obey the Arrhenius equation over a fairly wide range of conditions.

Constitutional supersaturation plays an important role in solution growth. Figure 4 shows the impurity concentration profile present in front of a growing crystal when the impurity tends to be rejected from the crystal (distribution constant $k_0 < 1$, typical for many accidental impurities, deliberate dopants, and most solvents). When the impurity lowers the melting or saturation temperature, as is typical for many impurities and most solvents, the saturation temperature profile in front of the growing crystal is like that of Figure 5. As can be seen, the saturation temperature of the solution decreases as the interface is approached. The dashed lines AB and $A'B$ are two

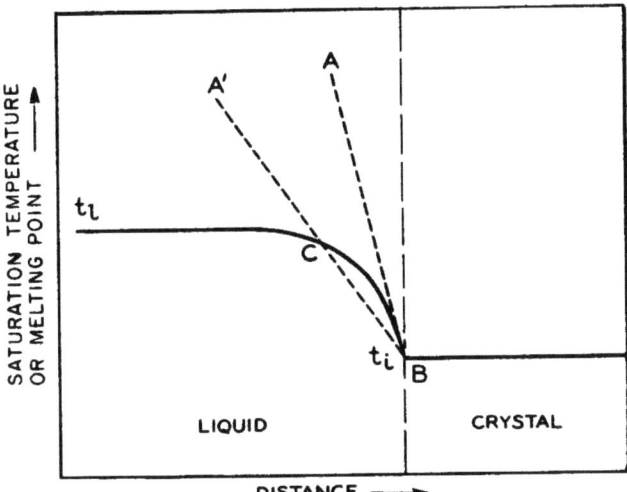

Fig. 5. Saturation temperature or melting point profile in front of a growing crystal.

different possible temperature gradients in the solution. For the larger gradient, AB, at all positions in front of the interface there is no supersaturation; for the smaller gradient, $A'B$, the region CB is supersaturated and crystallization will tend to take place in front of the solid interface. The region CB is referred to as constitutionally supersaturated (in melt growth the analogous situation is constitutional supercooling) and, as can be seen, large temperature gradients will lessen the tendency toward constitutional supersaturation.

3. Aqueous Solution Growth

Probably the oldest and certainly in many respects the simplest growth technique is crystallization from aqueous solution. The method has been reviewed in monographs by Holden and Thompson,[5a] Holden and Singer,[5b] and Petrov et al.[6] Its main restriction is that it is limited to water-soluble materials which are often hygroscopic, soft, and thermally not very stable and consequently not usually attractive for electronic device applications. However, the techniques we will discuss here are in many cases easily modifiable for use with nonaqueous solvents provided the nonaqueous solvent has a liquid range, vapor pressure, and corrosive nature similar to those of water.

Chapter 8

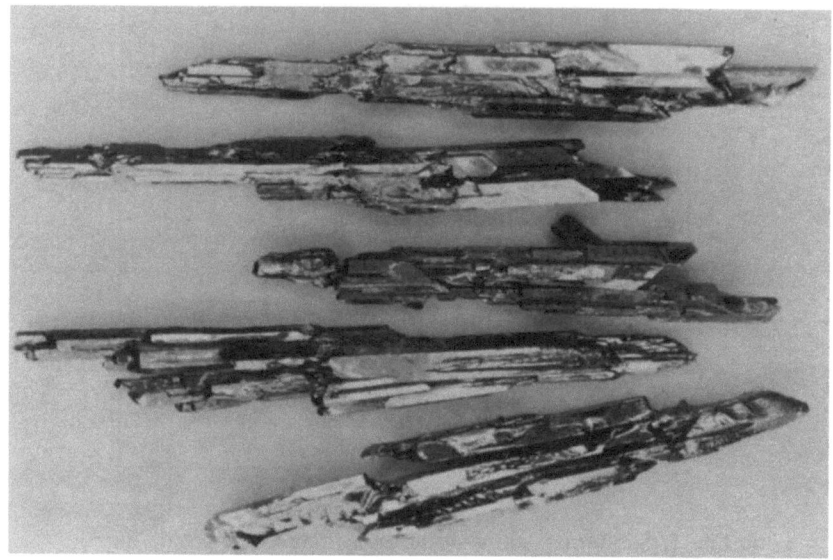

Fig. 6. Se crystals grown from Na$_2$S (length ~1 cm).

Materials which are water-insoluble can sometimes be brought into solution by the use of complexers (often called mineralizers in the geochemical and hydrothermal literature). This is illustrated by Se,[7] which is ordinarily insoluble in water but is easily complexed by aqueous sulfide solutions.[8] The reaction has been shown to be

$$2Se + S^{2-} \leftrightarrows Se_2S^{2-} \tag{18}$$

and is easily reversible, so that, for instance, Se is the stable phase in equilibrium with 0.25–1.75 M sulfide solutions between 25 and 250°C. The solubility has been measured and Figure 6 shows Se crystals grown from Na$_2$S. The technique for growth can be identical to those used in aqueous solution growth from pure water. Complexing techniques should be applicable wherever an appropriate complexer can be found. So far ambient complexing has been applied to only a few materials, including the following.

Material	Complexer	Ref.
CuCl	Cl$^-$ (in gels)	9, 10
HgS	S^{2-}	11–13
Se	S^{2-}	8

Solution Growth

Complexing is, of course, the basis of the use of (OH)⁻ and other "mineralizers" in hydrothermal growth.

3.1. Holden Crystallizer[4,5]

The slow cooling of a large volume of saturated solution in contact with seeds is probably the most generally used aqueous growth technique. Figure 7 shows a schematic of the apparatus developed by Holden. To begin, a saturated or nearly saturated solution is added to the tank. For ease in handling, it is usually convenient to transfer solutions at temperatures slightly above their saturation temperature. Suitably oriented seeds (prepared by evaporation of solutions on a watch glass or Petri dish or by growth in a mason jar; see below) are mounted on the seed holder. Small fragments of seeds can be forced into one end of a piece of plastic insulating tubing forced over the seed holder as shown in Figure 7. The support rod and seed frame should be made of material inert to the growth solution—lucite, Teflon, tungsten rod stock, glass, and noble metals or plating have all found use in various systems.

A test for saturation by observing schlieren lines near a small crystallite introduced near the surface is useful. With an appropriately adjusted light, index of refraction variations caused by concentration gradients are easily visible. If the current near the crystallite is descending, it is more dense than the surrounding fluid and the crystal is dissolving, indicating undersaturation. A rising current indicates supersaturation. The Hg seal prevents excessive

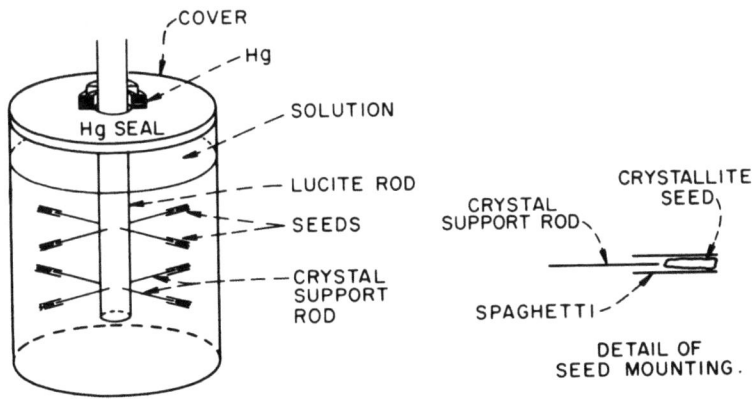

Fig. 7. Holden's crystallizer (after Laudise[1]).

Chapter 8

TABLE 1
Typical Crystals Grown from Aqueous Solution

Material	Method	Notes	Ref.
Potassium alum $KAl(SO_4)_2 \cdot 12H_2O$	Holden crystallizer	Saturated at 48°; cooled ~0.2°/day for several weeks	4
Sodium chlorate $NaClO_3$	Holden crystallizer	Saturated at 51°; cooled ~0.2°/day for several weeks	4
Nickel sulfate hydrates $NiSO_4 \cdot 6,7H_2O$	Holden crystallizer	Cooling from ~50 → 35° produces $NiSO_4 \cdot 6H_2O$; below 35° $NiSO_4 \cdot 7H_2O$ forms	4
Rock salt $NaCl$	Evaporation	Saturated brine containing 0.1% H_2SO_4 + 0.1% $Pb(NO_3)_2$ evaporated at 75% about 12 hr; without lead, cloudy crystals result	15
Rochelle salt (potassium sodium tartrate) $KNaC_4H_4O_6 \cdot 4H_2O$	Mason jar—cooling or evaporation	Make solution containing 130 g/100 cm^3 H_2O; heat until all dissolves; grow by evaporation or cooling	5
Sodium borate $NaBO_3$	Mason jar—cooling or evaporation	See Rochelle salt; 50 g/100 cm^3 for evaporation, 52 g/100 cm^3 for cooling	5
Sodium nitrate $NaNO_3$	Mason jar—cooling or evaporation	See Rochelle salt; 110 g/100 cm^3 for evaporation, 113 g/100 cm^3 for cooling	5
Nickel iodate hydrate $Ni(IO_3) \cdot 2H_2O$	Nassau technique	Five-liter solution using solvent boiling water, withdraw 25 cm^3/day from reflux from five-liter solution	16
Nickel iodate β-$NiIO_3$	Nassau technique	Five-liter solution solvent boiling HNO_3; withdraw about 25 cm^3/day	16
Ammonium dihydrogen phosphate—ADP $NH_4H_2PO_4$	Rotary crystallizer—temperature differential technique	(001) plates grow poorly until (101) "capping" faces formed; if solution made $pH \sim 5$, rate on faces increases	17–19
Ethylene diamine tartrate (EDT), $[NH_2CH_2CH_2NH_2]$ $[OC(CHOH)_2CO]$	Rotary crystallizer—temperature differential technique	Anhydrous form stable above 40.6°, hydrate below	17–19

TABLE 1—contd.

Material	Method	Notes	Ref.
Calcium tartrate $Ca(C_4H_4O_6)$	Gel growth	Silica gel media: tartaric acid + calcium chloride	22
Calcium tungstate $CaWO_4$	Gel growth	Silica gel media: sodium tungstate + calcium chloride	22
Cuprous chloride $CuCl$	Gel growth	Silica gel media: Dilution of CuCl complexer in HCl	23
Manganese iodate $Mn(IO_3)_2$	Gel growth	Silica gel media: 1 M $MnSO_4$ + 1 M HIO_3	20

evaporation; heaters include resistance immersion heaters, water baths, hot plates, IR lamps, and others. The commonly used controller is the mercury thermoregulator, although more sophisticated control would probably be advantageous. To decrease diffusion effects, it is common to slowly rotate the seed support at a few rpm. To avoid vortices, the rotation direction is usually reversed every few seconds. Supersaturation is provided by slow cooling, usually at from 0.1–1.5°/day. Motor-drive cams can be used to uniformly drop the set point of the controllers, but common practice is to reset the control point at a lower value manually each day. Thus such growth is usually neither isothermal nor at uniform rate, but nonetheless excellent crystals generally result.

Common problems are cloudy growth and flawed or dendritic growth. Impurity adsorption, which is easy at the low temperatures of solution growth, can have a great influence on growth rates, morphology, and perfection. Yamamoto[14] lists impurities effecting the growth of crystals from aqueous media. In general, small, highly charged ions with high polarizing power (so as to be easily adsorbed) will have the greatest effect on morphology and perfection. The use of Pb^{2+} as an additive to improve NaCl growth is a classic example of the effect[15] (see Table 1). Representative materials grown by the method are listed in Table 1.

3.2. Mason Jar Method

A simpler method for slow cooling is the mason jar technique illustrated in Figure 8.[1] Saturated or nearly saturated solution is

Chapter 8

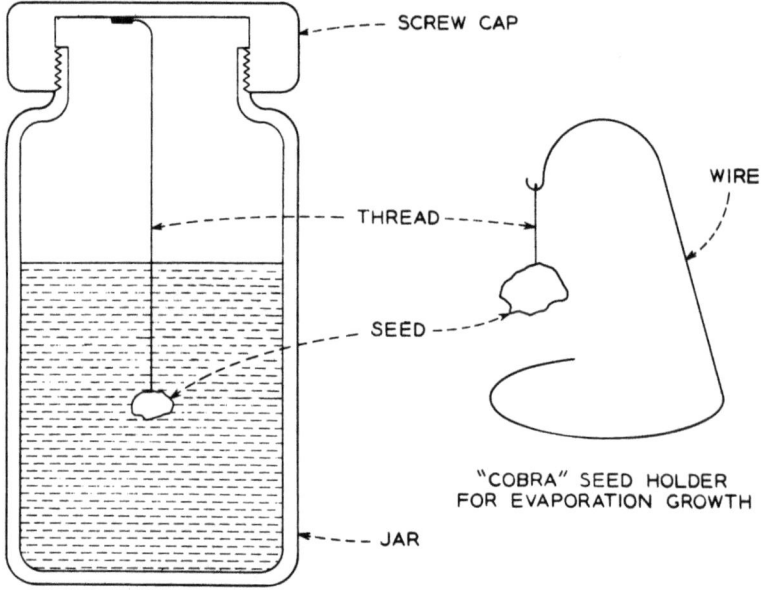

Fig. 8. Mason jar crystallizer (after Laudise[1]).

placed in the screw-cap jar together with a thread-mounted seed fragment. A solution slightly above saturation temperature is ordinarily to be preferred since it will dissolve small seed nuclei and etch the seed. The technique works best when the soubility is rather high and the cooling rate not too great. The solution is allowed to cool naturally to room temperature in a relatively constant environment. The technique is inexpensive and simple and many experiments can be run in parallel. The technique can be adapted for growth by evaporation by covering an open jar with cloth or gauze to allow evaporation but exclude dust nuclei. The "cobra" holder (Figure 8) for seeds is useful, since it does not project above the liquid and thus will not provide nucleation sites at the interface. A thin layer of petroleum jelly on the wall of the vessel just above the liquid–air interface is also helpful in reducing nucleation caused by the solution evaporating. Constant-temperature rooms (often a cellar or basement is sufficient), programmed cooling, etc., all are possible refinements of the technique. Controlled evaporation at higher temperature can be provided by evaporation through a capillary tube attached to a boiling erylenmeyer flask or, as in the method of Nassau,[16] by controlled outflow of the refluxed liquid of a reflux condensor attached to a boiling solution in which growth takes place. Some crystals grown by the method are summarized in Table 1.

3.3. Temperature Differential Method

Figure 9 shows a schematic of the temperature differential growth method. The solution is saturated in the higher-temperature chamber and moved (either by convection or with a pump) to the lower-temperature vessel where growth takes place. There are advantages in beginning with the growth chamber slightly undersaturated so that undesired nuclei are destroyed and the seed receives a modest etch. Appropriate stirring, including seed rotation, is also desirable. Use of an intermediate chamber or chambers between the dissolving and growth regions to ensure temperature equilibration is also advantageous. Ethylene diamine tartrate (EDT) and ammonium dihydrogen phosphate (ADP) were grown on a large scale in an apparatus based upon these principles.[17-19] Some conditions for ADP and EDT growth are given in Table 1.

3.4. Other Methods of Aqueous Growth

Growth by a chemical reaction in an aqueous medium does not usually produce very large or well-formed crystals. The reason for this is that it is extremely hard to avoid supersaturation sufficient to

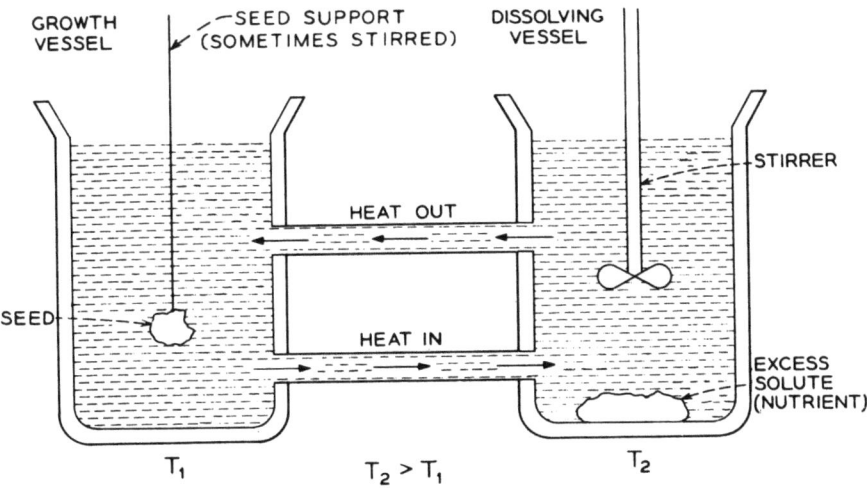

Fig. 9. Temperature differential method.

Chapter 8

initiate spontaneous nucleation. Growth by reaction is ordinarily applied to insoluble or nearly insoluble materials where growth by conventional aqueous methods is not applicable. For the case where ionic materials are involved we are thus dealing with materials where K_{SP}, the solubility product, is small. Thus, where AB is insoluble

$$AB_S \rightleftarrows A_l^+ + B_l^- \qquad (19)$$

$$K_{SP} = (A^+)(B^-) \qquad (20)$$

One might consider that growth by reaction would be feasible by mixing the soluble salts AC and DB. However, most mixing schemes will produce $(A^+)(B^-)$ products which locally much exceed K_{SP}, causing a rain of nuclei producing many small crystals.

One technique which effects mixing of AC and DB without large local supersaturations is gel growth. It is convenient to illustrate the technique with the growth of calcium tartrate in a silica gel contained in a U tube (Figure 10a). Soluble tartaric acid and calcium chloride solutions react to produce insoluble calcium tartrate:

$$H_2C_4O_6(l) + CaCl_2(l) \rightarrow CaC_4O_6(S) + 2HCl(l) \qquad (21)$$

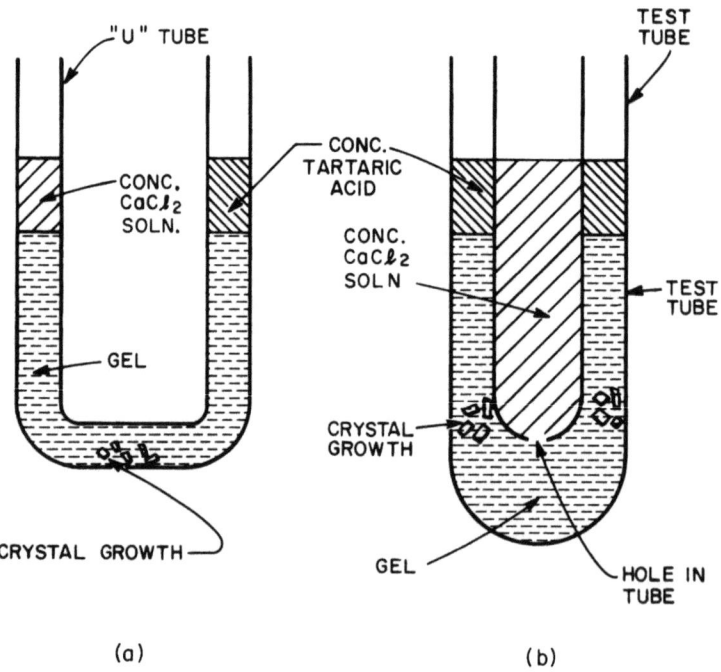

Fig. 10. Gel growth.

Tartaric acid is placed in one arm of the U tube and calcium chloride in the other. They diffuse toward one another through the silica gel. In a small local region the solubility will be exceeded gradually and a few nuclei will form which can grow to rather large (many mm) crystals. Other growth geometries where the gel contains one of the constituents are not as satisfactory because supersaturations where the soubility product exceeds K_{SP} occur over larger volumes of the system so that many more nuclei are formed, resulting in smaller crystals. Nassau and Shiever[20] have reduced the total apparatus size in gel growth by placing one test tube with a hole in it inside another. This method retains the advantages of the U tube geometry with respect to localizing the region of high supersaturation within the gel (Figure 10b). Some conditions for gel growth are given in Table 1.[22,23] Henisch has written an excellent monograph on the technique.[21]

Other solution growth techniques for growing low-solubility materials include growth by diffusion through a semipermeable membrane and electrochemical reaction. Neither of these has received wide exploitation in aqueous media, although electrochemical reaction has been used for the growth of tungsten bronze and related compounds in molten salt media and is discussed in Section 5.8. The electrochemical growth of metal single crystals would present many advantages and ought to be reasonably straightforward. Low current densities and single-crystal electrodes (seeds) could be advantageous. The mechanism of electrocrystallization and studies of the perfection and morphology of electrocrystallized copper have recently been reported.[24]

4. Hydrothermal Growth
4.1. General

When the solubility of a material in water is so low as to prevent its growth an attractive procedure is to raise the growth temperature. In order that the solvent density be high enough to give the solvent a reasonable solvent power, the pressure for reasonable growth temperatures will necessarily be rather high. In addition, it is common to add complexers (mineralizers) to the growth solution to further increase the solubility at comparatively modest pressure-temperature conditions. The use of high-temperature, high-pressure aqueous solvents to grow ordinarily insoluble materials is called hydrothermal crystallization. The technique was first applied by geochemists for the study of phase equilibria in mineral assemblages.

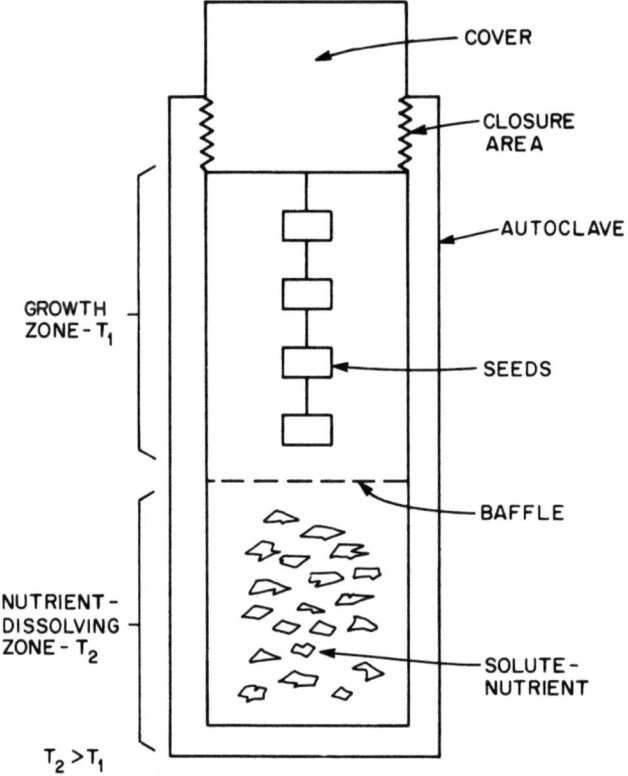

Fig. 11. Schematic of hydrothermal growth configuration.

Its first serious application to crystal growth was for the growth of single-crystal quartz. A number of reviews of the technique have been published,[25-28] so that historical and bibliographic details need not be given here.

The usual growth configuration is a cylindrical high-pressure autoclave of the sort shown in Figure 11. Solute is placed in the bottom of the autoclave in the dissolving or nutrient region. Suitably oriented single-crystal seeds are suspended in the growth region. The autoclave is filled to some predetermined factor, degree of fill, or percent fill with water or water plus mineralizer, closed, and placed in a furnace designed to heat the dissolving zone hotter than the growth zone. The dissolving and growth regions are often separated by a perforated metal disk or baffle which localizes the temperature gradient so that each zone is nearly isothermal. Solvent saturates in the dissolving zone, moves by convection to the cooler growth zone where it is supersaturated, and growth takes place.

Solution Growth

In a system of the above geometry crystals can usually be successfully grown when the following conditions are met:

1. The material must be the stable phase at the pressure–temperature conditions of growth.
2. The material should have a large enough solubility that high enough supersaturations for reasonable growth rates on seeds can be sustained without homo or heterogeneous (spontaneous) nucleation in the solution or on the vessel walls. The solubility should be high enough that reasonable rates of solution convection provide a high enough flow of solute to the crystal to sustain growth. For materials studied so far weight solubilities of 2–5% appear appropriate.
3. The area of the nutrient should exceed that of the seeds by a factor sufficient to exclude dissolving as a rate-limiting process. A dissolving area about five times growth area is usually sufficient.
4. The temperature coefficient of density at constant average density $(\partial \rho/\partial T)_{\bar{\rho}}$ should be large enough that with an appropriate temperature differential between dissolving and growth zones (ΔT) convective circulation will be sufficiently rapid so as to not be rate limiting.
5. The temperature coefficient of solubility $(\partial s/\partial T)_{\bar{\rho}}$ should be such that an appropriate ΔT will produce a supersaturation ΔS for the given system which will produce a rapid enough growth rate to be practical.
6. A vessel suitable for the pressure–temperature–corrosion conditions must be available.

The principal difficulties of hydrothermal growth are: (1) design of leak-proof vessels; (2) making predictions about phase stability, solvent power, growth rates, and perfection; and (3) the long times required to grow reasonable size crystals without easy observations of progress.

The main advantages of the method are: (1) its high solvent power at low temperatures; and (2) the comparative ease of scaling up once proper conditions are found.

4.2. Equipment

High-pressure equipment for hydrothermal growth is rather unique and causes the neophyte the greatest difficulty, so we will briefly discuss its design and use. Details of construction and schematics of autoclaves are given by Laudise.[1] Three sorts of apparatus are used for growth. The modified Bridgman closure autoclave is used for high-pressure growth in large systems and is especially useful for pressures above 500 atm. Commercial quartz growing autoclaves as large as 10 in. in diameter using the modified Bridgman design are regularly in use. The closure operates on Bridgman's principle of self-energization,[29] so that the pressure–temperature capabilities of the autoclave are set by the tensile properties of the construction material. For the growth of materials where the system is compatible with steel (mainly SiO_2 where low solubility acmite, $Na_2O \cdot Fe_2O_3 \cdot 4SiO_2$, forms and protectively coats the steel autoclave walls) the vessel may be used without a noble metal can. The direct lining, plating, or cladding of the modified Bridgman closure presents severe technical difficulties, so that when inert conditions are required a noble metal (Ag or Pt) can is used.[30] The can has a volume as close to that of the vessel as possible and the vessel and the can are filled to degrees of fill such that the fluid in the can (solvent + solute) and water placed between the can and the vessel will exert the same pressure at the growth temperature. Water is used outside the can to minimize corrosive attack on the vessel. Since the equations of state of most hydrothermal fluids are not known, it is usually necessary to estimate the appropriate degrees of fill. Thus the *small* volume of the space between the can and the vessel is advantageous. A slight pressure imbalance resulting in a small dimensional change of the can will cause a large change in this volume and a large pressure change in the water. Thus pressure balance may usually be achieved without dimensional changes in the can large enough to cause rupture of the tube.

For lower pressures the design of Morey[31] is most often used. Such vessels are comparatively easy to line with Pt, Ag, or Au. However, they are not self-energized. The operating pressure depends upon the torque used to close the vessel which must create a force on the plunger exceeding that exerted by the internal pressure. For 1-in.-diameter vessels ~600 atm seems a practical upper limit.

For phase equilibria, solubility, and small-scale crystal growth, the test tube–cold seal or Tuttle vessel[32] is generally used. The material to be studied is enclosed in a small Pt or Au capsule which is filled with the appropriate volume of solution and welded closed. Pressure is balanced by pumping water into the autoclave outside the capsule.

4.3. Phase Equilibria and Solubility

Preliminary to crystal growth it is important to establish that the desired phase is stable at the $P-T$ conditions contemplated. The usual vessel for such studies is the Tuttle–cold seal vessel. Powders or solutions of the appropriate components are placed in the capsule together with an appropriate volume of water or mineralizer solution. The capsule is welded shut and brought to the desired $P-T$ conditions. Pressure balance on the capsule is usually maintained by pumping water into the autoclave using an air-driven intensifier. At the conclusion of the experiment the vessel is quenched and the identity of the phases as determined by X-ray diffraction, petrographic microscopy, etc., is used to deduce the stability relationships at the pressure and temperature of the experiment. In solubility determinations weighed crystals are placed in the capsule and the weight loss is used to determine the solubility. It has been established[33] that provided only one fluid phase is present and other appropriate precautions are taken, solubilities of an accuracy sufficient for most hydrothermal investigations can easily be determined.

Three important phase equilibria problems can arise in hydrothermal work: (1) the formation of hydrated phases, (2) the formation of undesired polymorphs, and (3) incongruent saturation. Hydrated phases often occur with transition and post-transition metal oxides. For example, in the $H_2O-Al_2O_3$ system[34–36]

$$Al_2O_3 + H_2O \rightleftarrows AlOH \quad \text{(boehmite)} \quad (22)$$

$$\text{(boehmite)} \quad AlOOH \rightleftarrows AlOOH \quad \text{(diaspore)} \quad (23)$$

$$AlOOH + H_2O \rightleftarrows Al(OH)_3 \quad \text{(gibbsite)} \quad (24)$$

Equations (22) and (24) involve hydrate formation; Eq. (23) is a polymorphic transition. As would be expected, these transitions are pressure and temperature dependent as illustrated in Figure 12.[11–13] Although the boundaries of Figure 12 are not appreciably changed by the presence of moderate $(OH)^-$ concentrations, the possibility of this effect must be considered in other systems.

Congruent saturation is illustrated by the system $Fe_2O_3-Y_2O_3-H_2O$, where the garnet $Y_3Fe_5O_{12}$ is formed.[37] Garnet crystallizes from a solution where the ratio $Y_2O_3/Fe_2O_3 = 3/5$, the same as that in the solid; it saturates congruently. The line AB of Figure 13(a) (which is a schematic of part of the $Y_2O_3-Fe_2O_3$ system under hydrothermal conditions) drawn from the garnet composition on the

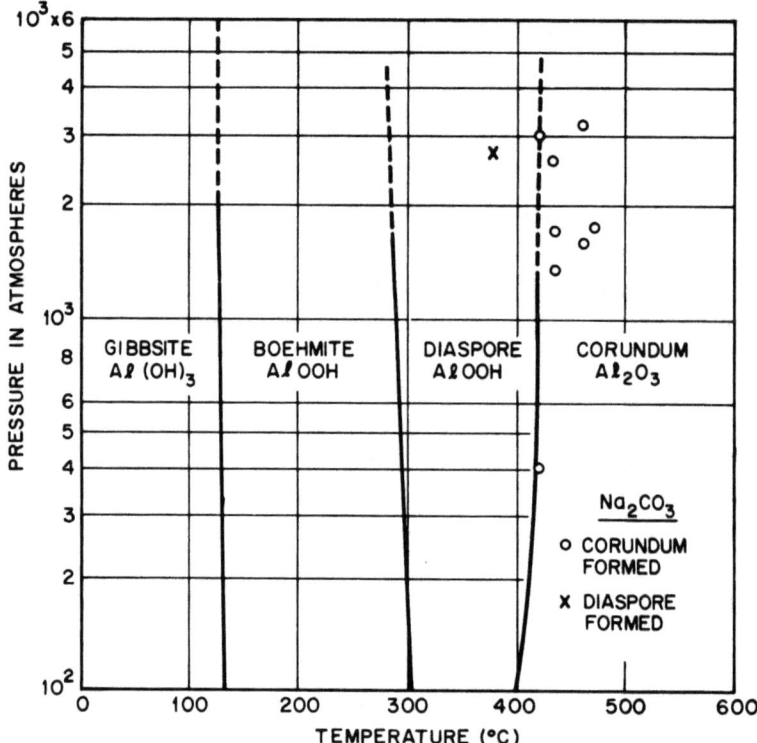

Fig. 12. $Al_2O_3-H_2O$ system. (——) Erwin and Osborne[34] (corrected); (---) extrapolated.

binary $Y_2O_3-Fe_2O_3$ joint to the H_2O apex in the ternary crosses the garnet solubility line. If the line does not cross the solubility line (compound AB of Figure 13b), the material is incongruently soluble. Special techniques, such as adjusting the composition of the solution by presaturation with a particular component, are required to crystallize an incongruently soluble material.

The solubility of a wide variety of materials has been studied under hydrothermal conditions. These results have been reviewed[28,38] and can be summarized as follows.

1. The log solubility is a linear function of the reciprocal of the absolute temperature, that is, solubility obeys the van't Hoff equation at constant fill or pressure and mineralizer concentration. The heats of solution of a variety of materials are summarized in Table 2. These heats can be rationalized[39,42] on the basis of bond energies and the hydration energies (Ref. 38, p. 116) of the species found in solution.

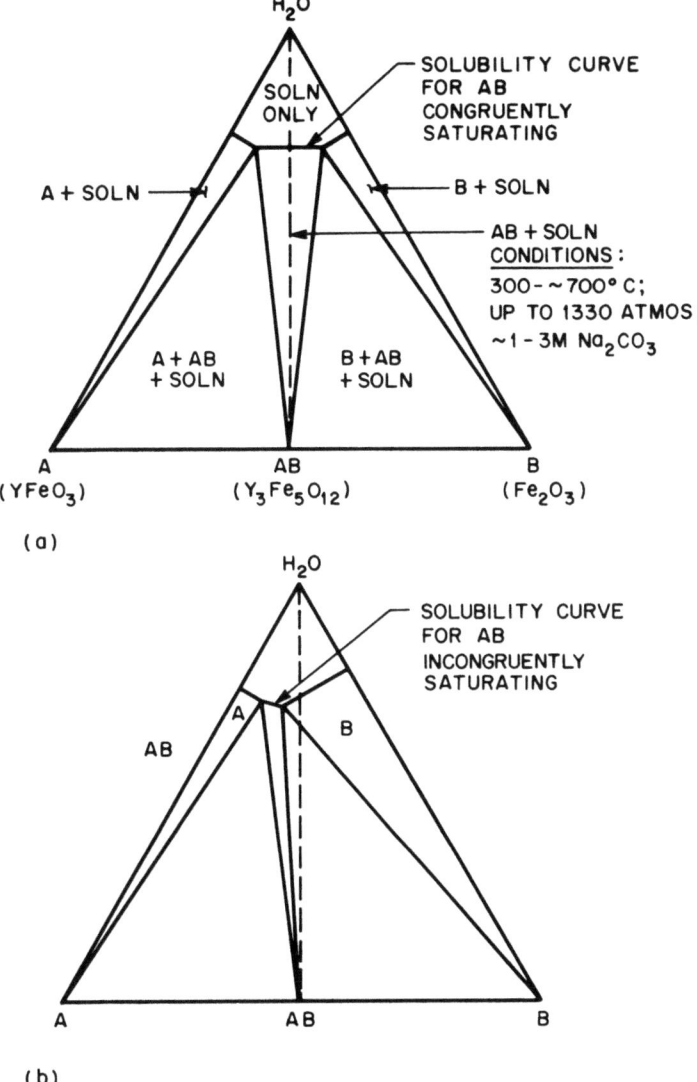

Fig. 13. A part of the Fe_2O_3–Y_2O_3–H_2O diagram.

2. The log solubility is a linear function of the percent fill at least in the case of quartz dissolving in NaOH solutions.[39] This relationship has been explained in terms of the reaction[39]

$$SiO_2 + 2H_2O \rightarrow Si(OH)_4 \quad (25)$$

and analogous reactions with the mineralizer. It is reasonable to expect a similar behavior for other solutes.

TABLE 2
Heats of Solution from van't Hoff Plots

Substance	Solution	Percent fill	Pressure, bar	Heat of solution, kcal/mole
Quartz[39]	Water	71.5	—	$\Delta E = 8.8$
Quartz	0.51 m NaOH	80	—	$\Delta E = 1.03$
Quartz	0.51 m NaOH	85	—	$\Delta E = 1.82$
Quartz	0.51 m NaOH	87	—	$\Delta E = 2.68$
Quartz	5 wt. % Na_2CO_3	80	—	$\Delta E = 3.42$
ZnO[40]	6.47 m KOH	—	550	$\Delta H = 1.1$
ZnO	9.07 m KOH	—	550	$\Delta H = 0.8$
ZnO	6.24 m NAOH	—	550	$\Delta H = 0.4$
Al_2O_3[33]	10 m NaOH	—	430–2800	$\Delta H = 0.2$
Al_2O_3	3.4 m Na_2CO_3	—	430–2800	$\Delta H = 1-5*$
ZnS (cubic)[41]	6.47–11.9 m KOH	—	550	$\Delta H = 7.2$
ZnS	6.24–38.4 m NaOH	—	550	$\Delta H = 7.2$
ZnS	8 m CsOH	—	550	$\Delta H = 7.2$
$KTa_{0.65}Nb_{0.35}O_3$[42]	14.7–15 m KOH	—	620–1720	$\Delta H = 18$
$LiGaO_2$[38,40]	0.5 m NaOH	50–70	—	$\Delta E = 4.5$
Se[38,58]	1.25 m Na_2S	—	1	$\Delta H = 3.88$

* Van't Hoff equation obeyed poorly.

3. The solubility s is a linear function of the $(OH)^-$ concentration in a great many cases.[38] In most instances the ratio $s/(OH)^-$ is constant and a small rational number over wide $(OH)^-$ concentrations, suggesting that a single species is mainly responsible for the observed solubility.[38] Figure 14[38] illustrates the solubility dependence of a number of solutes on $(OH)^-$ and Table 3 gives the species that $s/(OH)^-$ suggests in each case.[38]

4.4. Kinetics

Most of our information on the dependence of growth rate on external parameters is based upon studies of quartz growth.[26,43–44] Because of the commercial importance of hydrothermal quartz, this material has been extensively studied. Smaller-scale studies of the kinetics of growth of other materials tends to confirm the results of quartz investigations.[26] In the case of quartz a vast amount of data can be fitted to the simple rate equation[43]

$$\mathcal{R}_{(hkl)} = k_{hkl} \alpha \, \Delta S \tag{26}$$

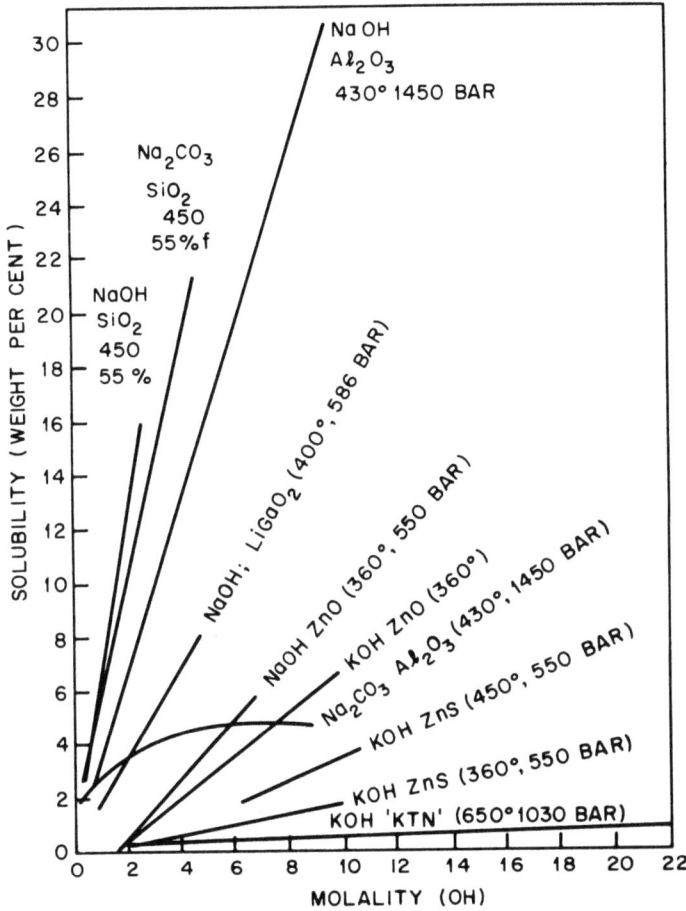

Fig. 14. Solubility as a function of (OH)$^-$ in several hydrothermal systems.

TABLE 3
Dissolving Reactions and Species Suggested by Data of Figure 14[38]

(1) $3SiO_2 + 2(OH)^- \rightarrow [3SiO_2 \cdot 2(OH)]^{2-} \rightarrow Si_3O_7^{2-} + H_2O$ (for NaOH)
(2) $SiO_2 + 2(OH)^- \rightarrow [SiO_2 \cdot 2(OH)]^{2-} \rightarrow SiO_3^{2-} + H_2O$ (for Na CO$_3$)
(3) $\frac{1}{2}Al_2O_3 + (OH)^- \rightarrow [\frac{1}{2}Al_2O_3 \cdot (OH)]^- \rightarrow AlO_2^- + \frac{1}{2}H_2O$
    (for NaOH; not calculable for Na$_2$CO$_3$)
(4) $ZnO + (OH)^- \rightarrow [ZnO(OH)]^- \rightarrow \frac{1}{2}Zn_2O_3^{2-} + \frac{1}{2}H_2O$
(5) $ZnS + OH^- \longrightarrow [ZnS(OH)^-]$
    or
    $ZnS + 2(OH)^- \rightarrow Zn_2O_3^{2-} + H_2O + HS^-$
(6) $KTa_{0.65}Nb_{0.35}O_3 + 2(OH)^- \rightarrow Ta_{0.65}Na_{0.35}O_4^{3-} + K^+ + H_2O$
(7) $LiGaO_2 + 2(OH)^- \rightarrow Li^+ + GaO_3^{3-} + H_2O$

where $\mathcal{R}_{(hkl)}$ is the growth rate in a particular crystallographic direction, $k_{(hkl)}$ is a velocity constant for that direction, ΔS is the supersaturation, and α is a dimensional conversion constant. The fact that a rate equation of simple form is obeyed, of course, does not necessarily imply a simple mechanism. Using this rate equation, the observed dependence upon experimental variables can in most cases be easily explained.

1. The growth rate depends very strongly on growth direction. A typical order for rates on the various seed surfaces is $(0001) > (11\bar{2}0) > (01\bar{1}1) > (10\bar{1}1) > (10\bar{1}0)$. Directional dependence arises mainly because k_{hkl} is directionally dependent. The order for growth rates can be approximately predicted on the basis of the recticular density of the atoms on the appropriate planes. However, such predictions, as would be expected, are very approximate. In general, nonequilibrium faces like (0001) are fastest growing and are often faceted.

2. The log of the rate depends linearly on the reciprocal of the absolute temperature because the velocity constants have an Arrhenius dependence on temperature. Typical energies of activation are 10–20 kcal/mole depending on seed orientation.

3. The rate depends approximately linearly on ΔT, the difference between dissolving and growth zone temperature, because the supersaturation ΔS is linear with ΔT for moderate ΔT values.

4. Rate depends approximately linearly on fill for fills below about 80% (at temperatures below about 400°C) and more strongly (approximately quadratically) on fill above these conditions. This dependence basically arises because of the solubility and ΔS dependence on fill.

5. Rate is independent of the surface area available for dissolving and the growing surface area provided the ratio of dissolving area to growth area is greater than about five. For lower values of this ratio dissolving becomes rate limiting.

6. Rate is independent of the area of the baffle left open (provided corrections for how baffle area changes perturb internal temperature are made), showing that convective transport is not rate limiting.

In general, 1–6 hold for NaOH solvents for 0.5–$\sim 1.5\ M$, temperatures of 300–$\sim 450°C$, fills of ~ 60–90%, and in vessels from laboratory size 1 in. diameter × 12 in. length to production size 10 in. diameter × 10 ft length. In production vessels when they are fully packed with quartz seeds, dissolving and circulation may be rate limiting near the end of runs when nutrient is nearly exhausted and the growth zone nearly full of quartz. The growth of "crevice flawed" dendritic morphology quartz under some conditions shows that constitutional supersaturation and diffusion problems can be important. The high ΔE values suggest, however, that the rate-limiting step is not pure diffusion but must also involve interfacial processes. Thus the growth reaction is probably consecutive with both diffusion and interfacial steps important.

4.5. Perfection of Hydrothermal Crystals

As mentioned above, under certain conditions [(0001) seeds and fast growth rates] faceted macroscopic imperfections occur in quartz brought about by constitutional supersaturation. Figure 15 shows such imperfections. The height of the so-called "hillocks" ("crevice flaws") gives some indication of the diffusion zone width, which can be of the order of a centimeter. The facet tips extend into regions of higher ΔS than their base because of the solute concentration gradient. Thus the cause of the imperfection is the same as that for constitutional supersaturation facets and dendrites in melt growth. Such growth can be avoided even at high growth rates by choosing conditions which speed the diffusion step relative to other steps in the growth. Unfortunately, the factors influencing the velocity of the various steps have not so far been quantitatively separable so that conditions for a nonflawed growth usually must be found empirically.

The most important physical property of quartz effecting its usefulness is acoustic Q.[45-46] Acoustic Q is associated with protons in the lattice which are thought to enter according to the reaction[47]

$$(M^{3+})_l + (H^+)_l \rightarrow (M^{3+} \cdot H^+)_s \qquad (27)$$

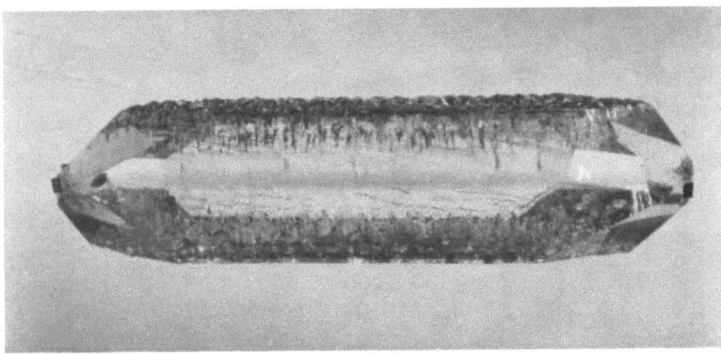

Fig. 15. Faceted faces on (0001).

Chapter 8

where M^{3+} is Al^{3+} or Fe^{3+} from the natural quartz particles ordinarily used as nutrient and occupies a Si^{4+} site, while H^+ probably actually enters the lattice as $(OH)^-$ from the mineralizer. However, for the purpose of charge compensation it can be considered H^+ and it is thought to occupy an interstitial position near M^{3+}. The partition coefficient K_0 describing Eq. (27) at equilibrium is

$$K_0 = (M^{3+} \cdot H^+)_s/(M^{3+})_l(H^+)_l \qquad (28)$$

where the subscripts s and l refer to the solid and liquid phases and concentrations are used to approximate activities. K_0 exhibits a retrograde dependence on temperature in NaOH and RbOH so that increases in crystallization temperature decrease $(H^+)_s$ and raise Q.[47,48] The value of K_0 also depends upon what other M^+ ions are present since these ions can also charge compensate for M^{3+}. Thus for Eq. (27) $K_0(Rb^+) > K_0(K^+) > K_0(Na^+) > K_0(Li^+)$, where

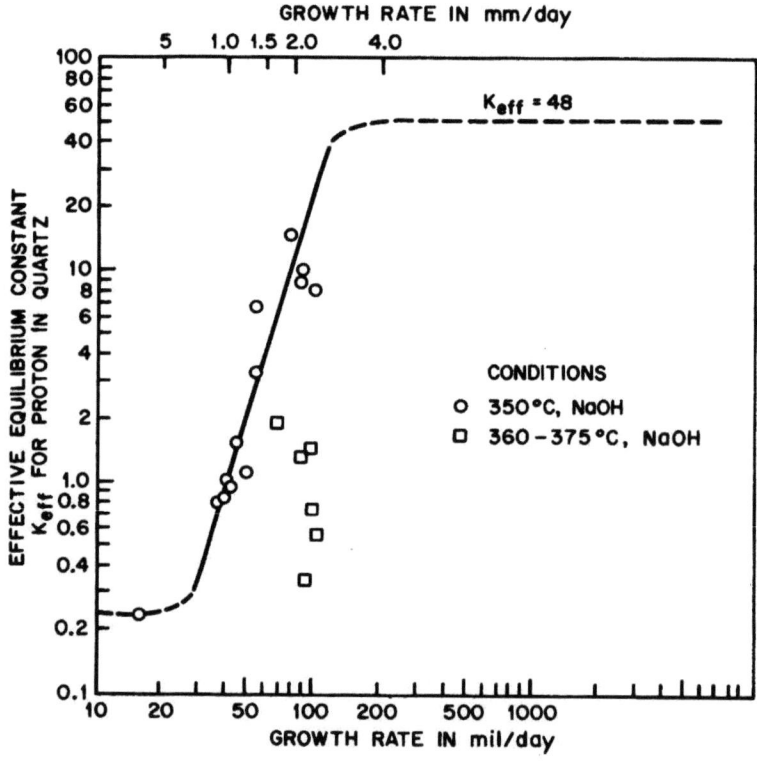

Fig. 16. Log K_{eff} versus log $\mathscr{R}$ for the partition of H^+ in hydrothermal quartz.

$K_0(\text{Rb}^+)$ is the K_0 for Eq. (15) in Rb(OH) mineralizer, $K_0(\text{K}^+)$ in KOH, etc. Smaller ions fit the interstitial positions better so that less proton is required to compensate a given concentration of M^{3+} when the mineralizer contains smaller ions. At rates much above 20 mils/day (0.5 mm/day) K becomes rate dependent in a manner similar to that described for the distribution coefficient by Burton and Slichter[50]

$$K_{\text{eff}} = K_0/[K_0 + (1 - K_0)e^{-\mathcal{R}\delta/D}] \qquad (29)$$

where δ is the width of the diffusion zone, $\mathcal{R}$ is growth rate, and D is the diffusion constant. The "effective partition constant" K_{eff} (also called the effective distribution constant or effective equilibrium constant) for Eq. (27) involves the actual concentrations at a rate substantially removed from equilibrium. Figure 16 shows the typical S-shaped dependence of log K_{eff} on log $\mathcal{R}$. A detailed understanding of the effects of rate, temperature, and charge-compensating ions has recently permitted the growth of quartz at rates above 100 mils/day (2.5 mm/day) with $Q > 10^6$.[47]

Studies of optical transmission,[49] etching, and dislocation density in hydrothermal quartz have been reviewed elsewhere (Ref. 1, pp. 287–289).

4.6. Other Hydrothermal Materials

Rooymans[28] gives a complete listing of the more than 50 materials where some hydrothermal attempts at crystal growth have been made up to about 1971. Table 4 gives conditions of growth and other data of interest concerning some representative materials.

5. Molten Salt Crystal Growth
5.1. General

Molten inorganic salts and oxides (fluxes) are powerful solvents for many inorganic materials. The technique has been reviewed by Laudise,[59] White,[60] Laurent,[61] and Nielsen[62] so that historical and bibliographic details will be ignored here. Pioneered by Remeika and Nielsen, exploited and extended by Van Uitert, White, and Timofeeva, the technique is probably the most widely used polycomponent method. The common molten salt solvents include KF, PbO, PbF_2, B_2O_3, and their mixtures. The usual technique is to dissolve sufficient amounts of the components to form the crystal at a

Chapter 8

TABLE 4
Representative Hydrothermally Grown Crystals

Material	Growth conditions	Comments	Ref.
ZnO	5 M KOH, 350°C, $\Delta T \sim 10°$, 85% fill. Rate ~ 0.25 mm/day on $(0001$–$000\bar{1})$ seeds; [rate $(0001) \neq (000\bar{1})$]	Addition of ~ 0.1 M Li$^+$ represses flawing on (0001)–$(000\bar{1})$ seeds; this can be interpreted as being due to rate reduction or because adsorption on (0001)–$(000\bar{1})$ reduces the surface free energy so that cusps rather than lobes occur in the Wulff plot at (0001) and $(000\bar{1})$; Li$^+$ acts as an acceptor so that resistivity can be controlled by adjusting (Li$^+$) in solution	41, 51, 52
ZnS (cubic)	Conditions similar to ZnO	Hydrolysis reactions like ZnS + OH$^-$ → ZnO + HS$^-$ do not take place even in 5 M (OH)$^-$; the cubic polymorph is grown free of stacking faults commonly seen in ZnS when growth is by techniques near the cubic–hexagonal transition temperature	53
Al$_2$O$_3$ (Corundum)	1 M K$_2$CO$_3$, 490°C, $\Delta T \sim 50°$, 80% fill, rate ~ 0.25 mm/day on (0001) seeds	Process above 400°C required to be in stability field of corundum; P–T boundary corundum–diaspore shown to be insensitive to (OH)$^-$; Cr doping (ruby) possible	33, 36
Y$_3$Fe$_5$O$_{12}$ and related garnets	20 M KOH, 350°C, $\Delta T \sim 10°$, 88% fill, rate ~ 0.1–0.2 mm/day	Quality comparable to best flux garnets; very large sizes possible; substituted uni-axial garnets and hetero-epitaxial garnet films can also be grown	54–57
LiGaO$_2$	3.5 M NaOH, 385°C, $\Delta T \sim 35°$, 70% fill, 2.25 mm/day	Careful solubility studies required to avoid retrograde solubility conditions; exceedingly high growth rates possible	40
Se	0.375 M Na$_2$S, 175°C, $\Delta T \sim 25$, 95% fill	Temperature cycling (see Section 5.5) improves size of spontaneously nucleated crystals; growth zone cycled so ΔT varies from ~ -20 to $+25°$.	58

Solution Growth

temperature slightly above the saturation temperature and slowly to cool the crucible. Crucibles are usually made of platinum and the growth is on nuclei which form spontaneously on the walls or at the liquid–air interface. When the cooling cycle is complete the flux is poured off, drained off, or leached away with an appropriate solvent (usually a mineral acid) and the grown crystals are removed. Seeded growth, stirring, temperature cycling, and epitaxial growth have all greatly extended the power of the technique.

5.2. Equipment

The principal unusual equipment requirement for molten salt growth is the ready availability of platinum ware. Crucibles of about 150 cm^3 volume are usually used for phase equilibria and preliminary crystal growth experiments. With good knowledge of the phase equilibria and growth kinetics, together with good control of cooling rates and of temperature gradients in the crucible, it is often possible to gow substantial (up to ~ 1 cm^3) crystals in such vessels. However, large crystals are usually grown in much bigger crucibles. For instance, crucibles up to several liters are regularly used in garnet growth. In using platinum ware it is especially important to avoid reducing conditions in the presence of bismuth and lead salts, since metallic bismuth and lead alloy with platinum. Platinum effectively limits flux growth experiments to about 1300°C, although in principle the use of rhodium or iridium could extend this range. Most furnaces for flux work are heated with silicon carbide resistance heaters which are suitable up to ~ 1400°C. "On–off" or proportional controllers provide $\pm 2°$ temperature control within this range, with a saturable core transformer used when better control is desired. Obviously cooling rates lower than the precision in temperature control are usually inappropriate. To produce unusual thermal patterns for seeding and other special experiments, ceramic cores wound with resistance wire are commonly used as heaters. Crucibles are often rotated to accelerate equilibrium and even out radial thermal inhomogeneities.

5.3. Phase Equilibria and Solubility

The choice of a solvent for flux growth is mainly a matter of experience and searching for parallels with previously successful systems. It is interesting to point out that probably 75% of the

Chapter 8

successful growth has been from PbO, PbF$_2$, KF, and B$_2$O$_3$ or combinations of these. The best guidelines for solvent choice are chemical:

1. Acidic fluxes are good solvents for basic crystals and vice versa provided reaction products with the flux are not stable.
2. Complex-formers are useful solvents provided complexates are not stable solid phases.
3. Chain-breaking cations are useful in lowering melt viscosity.

Laudise[59] has further discussed phase equilibrium considerations. The requirements of a good solvent include:

1. Solvent power such that the crystal has a solubility of 10–50 wt. % and is the stable solid phase.
2. The temperature coefficient of solubility should be such that growth by slow cooling is practical (~1 wt. %) or else flux evaporation or growth in a thermal gradient will be required.
3. Low solubility and low reactivity with Pt.
4. Low viscosity and low solubility of the flux in the grown crystal.

5.4. Crystal Growth by Slow Cooling

Yttrium iron garnet Y$_3$Fe$_5$O$_{12}$ (YIG) and its isomorphs have been the most studied flux grown material, so it is convenient to illustrate the technique using garnet. However, details of the flux growth of garnet have been reviewed elsewhere[59] so we will give only a broad outline here. YIG cannot be conveniently grown from a pure melt because it melts incongruently and because of the tendency toward reduction of iron at the melting point. Van Hook[63] investigated the ternary system Fe$_3$O$_4$–Fe$_2$O$_3$–YFeO$_3$ at several oxygen pressures. Figure 17 summarizes his results. The line of intersection of the 0.21-atm oxygen isobar and the liquidus surface determines the composition of melts in equilibrium with various crystalline phases in air. It should be pointed out that, although the traces of each of the line segments $A'A$, AB, and BB' in the Fe$_3$O$_4$–Fe$_2$O$_3$–YFeO$_3$ system is a straight line, the line $AA'BB'$ is not straight. The diagram lying along this isobar as determined by Van Hook is shown in Figure 18. It has the appearance of a binary diagram with a "eutectic" at A and a "peritectic" at B. The "eutectic" and "peritectic" points are actually the intersections of the air isobar with the invariant

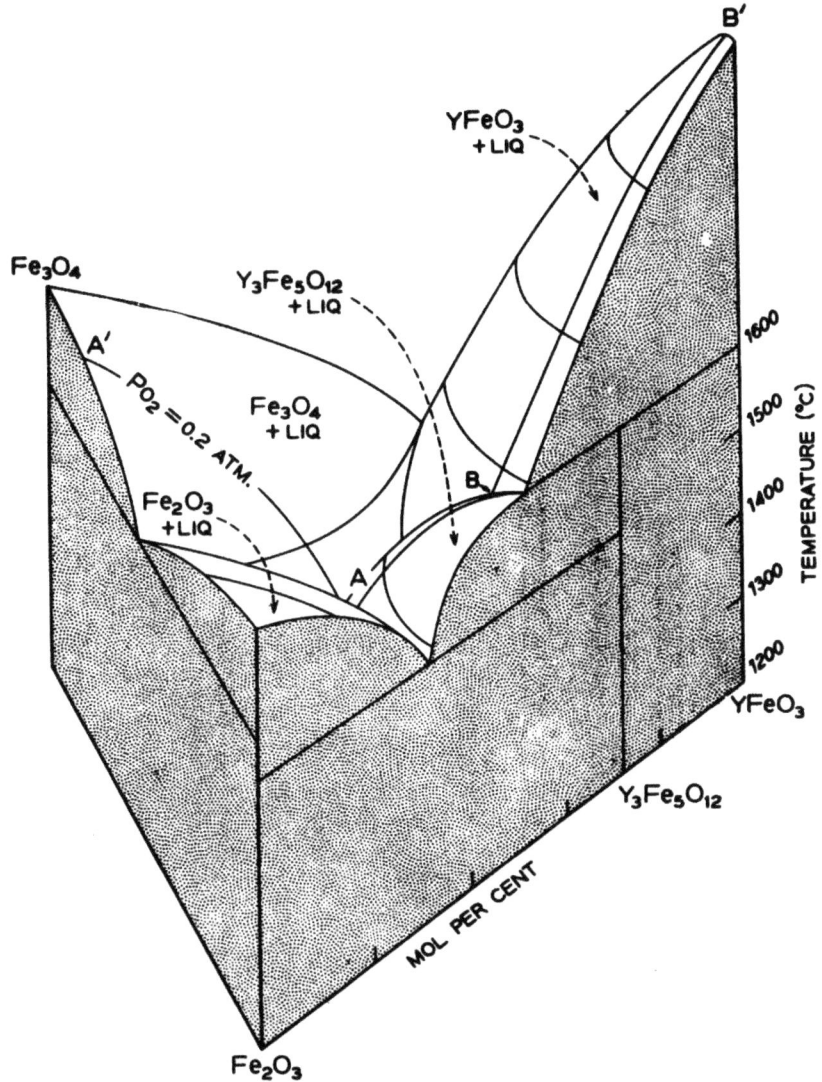

Fig. 17. Fe_3O_4–Fe_2O_3–$YFeO_3$ system (after Van Hook[63]).

boundary curves of the ternary system. Van Hook has determined the isobar for 1 atm of O_2 pressure and the "isobar" for the O_2 pressure in equilibrium with CO_2. This phase equilibrium work shows that garnet melts incongruently at 1555°C and is only stable in the melt composition range A–B of Figure 18. The compositions given for A and B in Figure 18 are starting compositions; the true melt compositions are undoubtedly oxygen deficient, as is the grown garnet.

441

Chapter 8

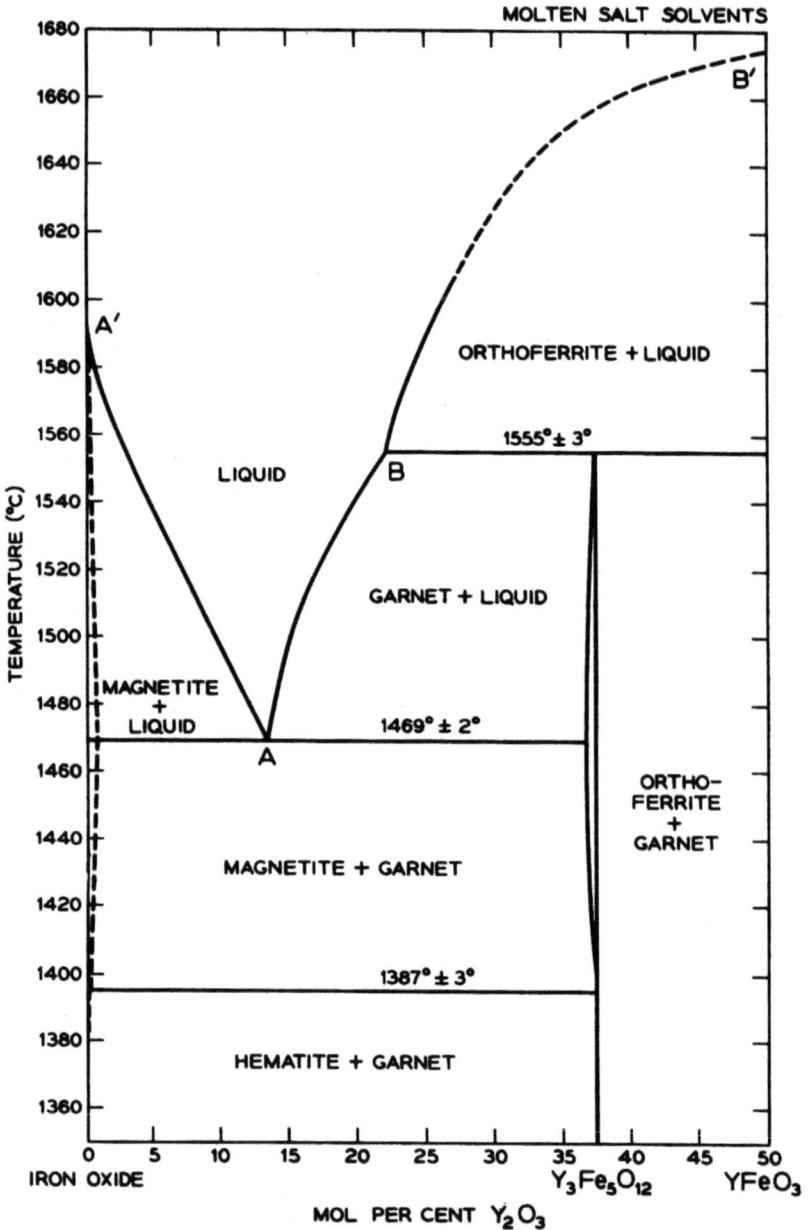

Fig. 18. Part of YIG phase diagram (after Van Hook[63]).

Consequently, the Fe^{2+} composition in grown garnet is probably a function of the temperature and O_2 pressure. Because of incongruent melting and reduction problems, it was recognized early that flux growth at lower temperatures was an appropriate garnet-growth method.

The Fe^{2+} ion is a particularly undesirable impurity in garnet because it effects the magnetic properties and the infrared transmission. Iron garnet has played a key role in our understanding of magnetic phenomena ever since it was discovered to be ferromagnetic. Single crystals found uses in microwave devices. Recently, substituted garnets have found high importance in magnetic memories. Thin sheets of garnets of appropriate composition have been found capable of sustaining very small bubble-shaped domains which can be used to store and manipulate information.[64] We will discuss the epitaxial preparation of garnet below.

In early experiments PbO was used as a molten salt solvent for garnet where it was found garnet was incongruently saturating. However, Nielsen[65] showed that when the ratio $Fe_2O_3/Y_2O_3 > 5/3$ (the oxide ratio in the solid) YIG can be crystallized between 1350 and 950°C. The best ratio appears to be $Fe_2O_3/Y_2O_3 \simeq 12.6$ and PbO volatilization must be suppressed by crimping Pt lids on the crucibles. Cooling rates generally used were from 0.5 to 5°/hr. HNO_3 is used to leach flux away from the grown crystals following quenching from ~900° to room temperature.

A variety of "improved" fluxes have been used for garnet. The flux 57% PbF_2–43% PbO was used by Nielsen.[66] Its advantages include increased YIG solubility, decreased viscosity, and avoidance of coprecipitating magnetoplumbite ($PbFe_{12}O_{19}$) which forms in PbO during the early stages of growth. Its disadvantages are mainly very high volatility and incongruent saturation for garnet. In BaO–$6B_2O_3$, YIG is congruently saturating,[67] but the biggest crystals have been grown by Grodkiewicz and co-workers[68] from PbO–PbF_2–B_2O_3. A typical recipe and mixture for large-scale garnet growth from a flux containing over 4 kg of PbF_2 has been given and CaO is reported to increase crystal size.[68] Garnets are grown from the mix by slow cooling at 0.5°/hr from 1300 to about 950°C. Below 950°C resolution is appreciable, so the crucible is tapped (i.e., flux is drained off by punching a hole in a thin platinum diaphragm welded into the crucible). Slow cooling *in situ* follows in order not to strain the grown garnets thermally. Rotation of the crucible in the furnace to accelerate dissolving during the equilibration period and even out thermal asymmetries during growth is helpful. A small, negative differential (~2.5°), arranged so that the bottom of the crucible is coolest, helps to localize growth and increase size. Typical crucibles

TABLE 5
Some Flux-Grown Crystals

Material	Method	Conditions	Ref.
Yttrium iron garnet (YIG), $Y_3Fe_5O_{12}$	Slow cooling	Flux PbO, equilibrate 1370°C; cool 1–5°/hr; can substitute for Y and Fe	65
Barium titanate, $BaTiO_3$	Evaporation + slow cooling	Flux KF, equilibrate at 1200°C; cool 20–40°/hr to ~850°C, pour off flux, leach with H_2O	69
Yttrium aluminum garnet YAG, $Y_3Fe_5O_{12}$	Slow cooling	Flux $PbO-PbF_2$, equilibrate at 1150°C; cool 4–5°/hr to 750°C	70
Sapphire or gallia Al_2O_3, Ga_2O_3	Slow cooling	Flux PbF_2, equilibrate 1200°C; cool to 900°C at 3°/hr	71
Rare earth orthoferrites $RFeO_3$	Slow cooling	Flux PbO, equilibrate at 1300°C; cool 30°/hr to ~850°C	72
Various oxides HfO_2, TiO_2, ThO_2, CeO_2, $YCrO_3$, Al_2O_3	Evaporation	Flux PbF_2 or $BiF_3 + B_2O_3$; evaporate at 1300°C	73
Lead zirconate $PbZnO_3$	Evaporation	Flux PbF_2 or $PbCl_2$; evaporate at 1200°C	74
Various sulfides $NaCrS_2$, $NaCrS_2$, CdS, ZnS, PbS, CuS, FeS_2, CoS_2, NiS_2, $KFeS_2$	Slow cooling	Flux, sodium polysulfide, "conventional flux methods"	75
Various oxides $BaTiO_3$, $Y_2Ti_2O_7$, $CaY_2Zn_2Ge_3O_{12}$ (germanate garnets)	Top seeding	See reference for details	88

are 8 in. high × 10 in. in diameter with a 0.050 in. wall thickness. Oxygen flowing through the furnace cools the drain tube, sweeps lead vapors out of the furnace, provides the negative temperature gradient, and suppresses Pb^{2+} reduction.

Similar results have been obtained with $Y_3Al_5O_{12}$: Nd (yttrium–aluminum garnet, YAG, doped with Nd).[68] Other representative materials grown by slow cooling are listed in Table 5.

5.5. Temperature Cycling and Accelerated Crucible Rotation

Two techniques which are applicable for increasing the size of slow-cooled, flux-grown crystals have recently been advanced: temperature cycling and accelerated crucible rotation.

Temperature cycling was first applied by Scholz and Kluckow[76] in vapor growth and has been used by Hintzman and Mueller-Vogt[77]

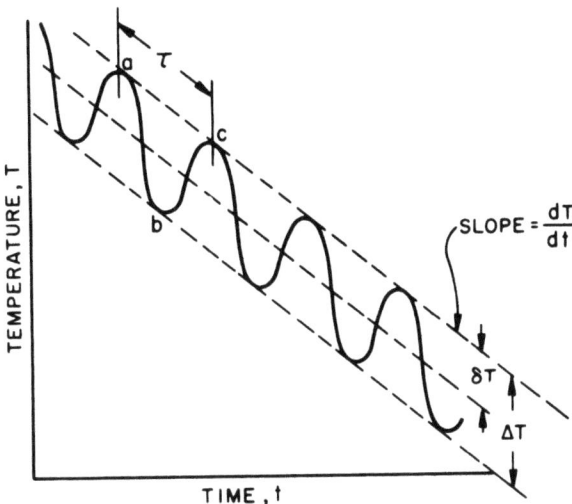

Fig. 19. Schematic of temperature cycling.

in flux growth and by Kolb and Laudise[58] in hydrothermal growth (see Section 4.6).

The method is simple in principle and can be generally used to increase the size of spontaneously nucleated crystals grown by any technique. In a typical embodiment in flux growth the temperature in the growth region is cycled as a function of time while simultaneously the average temperature in the growth region is lowered as shown in Figure 19. The average cooling rate dT/dt is usually of a magnitude that would be used to achieve good growth in a normal slow-cooling experiment. During the part of the cycle indicated by a–b the supersaturation is increasing and usually becomes large enough to induce spontaneous nucleation as well as growth on preexisting nuclei. However, during the heating part of the cycle b–c supersaturation decreases and the solution becomes undersaturated so that dissolution takes place. By the time the system reaches c many small crystallites and nuclei are completely dissolved and do not survive to grow during the next cooling cycle. The repetition of the process again and again tends to favor the perpetuation and growth of larger crystallites only. Thus temperature cycling enhances the yield of large crystals in any system where spontaneous nucleation is important. The main disadvantage is that cyclic growth on the surviving crystals can produce inhomogeneities in dopants, impurities, and stoichiometry causing inhomogeneities in properties. These inhomogeneities are often evidenced as "banding" or striations parallel to the growing interfaces. Hintzman and Mueller-Vogt[77] used the following conditions to grow crystals of $(A_{1-x}R_x)PO_4$, $(A_{1-x}R_x)AsO_4$, and $(A_{1-x}R_x)VO_4$ where A is Y or Sc and R is various rare earths: $dT/dt = 2°/hr$, $\Delta T = 40°$, $\tau = 8$ min (quantities defined in Figure 19).

Cooling usually began at about 1150°C and extended to about 850°C. The fluxes used were $Pb_2P_2O_7$, $Pb_2As_2O_7$, and V_2O_5 for the respective compounds. Crystals with a maximum dimension as large as 1.5 cm were obtained in favorable cases. It is surprising that the technique has not been applied more frequently in flux growth.

Stirring has always been recognized in crystal growth as an aid to accelerating solution when dissolution was taking place, reducing diffusion problems when growth was taking place, and homogenizing compositional and temperature inhomogeneities. Stirring is regularly used in Czochralski growth, and stirring in flux growth by means of crucible rotation has been used. Reversal of stirring direction in solution growth so as to prevent the formation of vortices is generally used in the Holden crystallizer. However, it remained for Scheel[78] to systematically apply reversing in stirring by crucible rotation (accelerated crucible rotation, ACRT) to the flux growth of crystals. Scheel[79] and Scheel and Schulz-DuBois[80] have analyzed in detail the flow patterns which result under various circumstances. Typical acceleration programs are from 0 to 100 rpm in 30 sec and then decelerate from 100 rpm to zero in 30 sec. $GdAlO_3$ was grown from $PbO-PbF_2-B_2O_3$ in crucibles of from 30 to 1000 cm³. Cooling was from 1300 to 900°C at 0.5°/hr. In every case much larger and more perfect crystals were obtained with ACRT. Tolksdorf and Welz[81] have studied the garnet system and obtained favorable results. The applications to other systems and the need for detailed investigations of the best stirring regime in each system are obvious. Indeed, controlled experiments with and without ACRT together with statistical analysis of crystal yield and size may be required to establish with certainty that crystal size improvements are not partly due to furnace design improvements. Indeed, the improvements in yield, size, and quality of crystals which are obtained by deliberately cooling one region of a system so as to induce nucleation there are such as to recommend this procedure wherever practical. For instance, perhaps the most important factor in improving the yield in both Van Uitert's large-flux-growth experiments and in Tolksdorf and Scheel's ACRT growth is that their furnaces are designed so as to produce cool

Solution Growth

regions in one part of the crucible. Van Uitert cools the bottom of the crucible by flowing oxygen into the bottom of the furnace, while Tolksdorf cools a sharp projection in the bottom of the crucible. Actually one problem with ACRT is that in evening out temperature inhomogeneities it may induce uniform nucleation throughout the system.

5.6. Flux Evaporation

Flux evaporation has the advantages of isothermal growth. If a material is stable over a narrow temperature range in a given flux so that slow cooling is impractical, it can sometimes be grown by evaporation provided the vapor pressure of the flux is high enough. If the material is incongruently volatile after appreciable volatilization, another phase may become stable and will begin to precipitate. If a dopant or impurity has a sharply temperature-dependent distribution constant, isothermal growth may be advantageous, but if the distribution constant is greatly dependent on flux composition, crystal component, stoichiometry, or dopant or impurity concentration, then evaporate growth may produce less homogeneous crystals.

It is quite probable that some of the growth, for instance, in KF and PbO fluxes, is caused by flux evaporation. Roy[82] has used flux evaporation of $(Na, K)_2O \cdot 3B_2O_3$, halide, and PbO fluxes to grow TiO_{2-x}, Cr_2O_3, $(Al, Cr)_2O_3$, VO_2, and In_2O_3 by isothermal flux evaporation at temperatures between 1500 and 1100°C.

Grodkiewicz and Nitti[73] have grown HfO_2, TiO_2, ThO_2, CeO_2, $YbCrO_3$, and Al_2O_3 by flux evaporation at 1300°C. Viting[83] has grown $MgFe_2O_4$ from PbF_2 by flux evaporation. Table 5 outlines several additional materials given by flux evaporation.

5.7. Seeding in Flux Growth

The advantages of seeding in flux growth are obvious and many workers have attempted seeding experiments. Clearly, whatever the configuration, deliberate seeding will present advantages. Two basic configurations for flux seeding have been used (Figure 20). Figure 20(a) is basically the same geometry as is used in hydrothermal growth, temperature differential (ΔT) geometry. Growth occurs because the solute-nutrient dissolves at a high temperature and the solvent-flux moves by convection to a cooler region where it is supersaturated. Baffles to localize the temperature difference, etc., can be used. Seeds

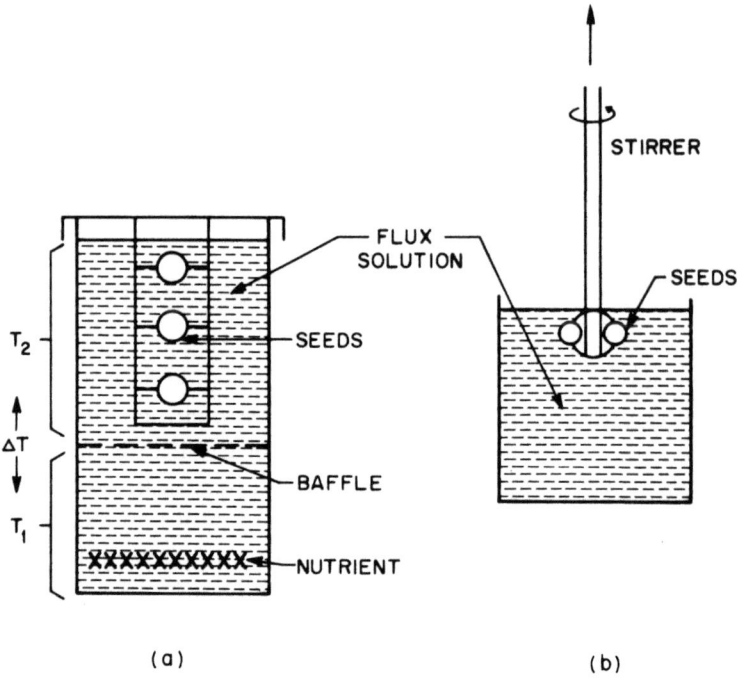

Fig. 20. Seeded flux growth configuration.

may be rotated and supersaturation can be produced without a ΔT and excess nutrient by slow cooling. Laudise et al.[84] used this geometry to grow YIG. An important requirement is that the nutrient be more dense than the flux or that chunks of the nutrient be held below the baffle. The advantages of the ΔT geometry are that many crystals can be grown simultaneously and that growth is not limited by the amount of material which can crystallize over a particular cooling interval. The main disadvantage is that since there is no thermal gradient near the growing interface, constitutional supercooling is hard to avoid. Constitutional supercooling is more likely in flux growth than in systems where water is the solvent, because of the higher viscosity of fluxes. The rapid convective stirring of hydrothermal growth allows growth in the ΔT geometry without severe constitutional supercooling. Laudise et al.[84] found that due to constitutional supercooling, flux inclusions, facets, and dendritic growth were problems in the ΔT geometry when they grew YIG.

To avoid these difficulties, a geometry more like that of Czochralski growth, where a big thermal gradient is present near the growing interface, is advantageous (Figure 20b). Growth with this geometry in flux growth is generally called top seeding and often the seed is rotated and withdrawn as in Czochralski growth. However, because the growth rate is diffusion limited, the pulling rate is much slower (~1 mm/hr or less) than in Czochralski growth from pure melts. Melts used in top seeding are often off-stoichiometric compositions containing the same components as the crystal. For example, $LiNbO_3$ congruently melts at 48.6 mole % Li_2O but is often Czochralski pulled at other compositions. Such an experiment is essentially growth by top seeding, although the "solution" and the crystal are almost identical (within a few percent) of one another. In composition $K(Ta, Nb)O_3$ as grown by Bonner et al.[85] and Whipps[86] is in the same category but the compositional difference between solution and crystal is somewhat greater. Indeed the first top-seeding experiment on record is probably that of Miller[87] with $KTaO_3$ where excess K_2CO_3 (i.e., K_2O) was the solvent. Perhaps the triumph of the top-seeding method is the exquisite crystals of $BaTiO_3$ grown by Linz and his colleagues.[88] Conventional fluxes including barium borate for $Y_2Ti_2O_7$ and $Y_3Fe_5O_{12}$ have also been used in top seeding experiments. Table 5 summarizes some additional materials grown by flux seeding.

5.8. Flux Growth by Electrolysis

The advantages of electrochemical growth both with respect to localizing the desired reaction on the seed-electrode, controlling the rate by limiting current, and selecting the desired reaction by voltage control have been mentioned above. The technique has been exploited, using molten salt media, especially by Perloff and Wold[89a] and Kunnman.[89b] Perloff and Wold[89a] found molten salt electrolysis especially useful because it allows the crystallization of ordinarily difficultly stable oxidation states. For instance, they grew crystals of MoO_2 by the electrolytic reduction of $Na_2MoO_3 + MoO_3$ at 675°C. Kunnman[89b] has reviewed the technique and discusses the growth of tungsten bronzes, borides, carbides, selenides, phosphides, sulfides; and arsenides. He particularly recommends the apparatus of Figure 21.[89b] The physical dimensions and conditions have been determined empirically and represent a number of compromises. Typical current density is 20 mA/cm² and a deposition rate equivalent to $1/68 \times 10^{-4}$ equiv/hr is possible. Most reactions take place by direct electrochemical decomposition of the flux, although sometimes a solute is

Chapter 8

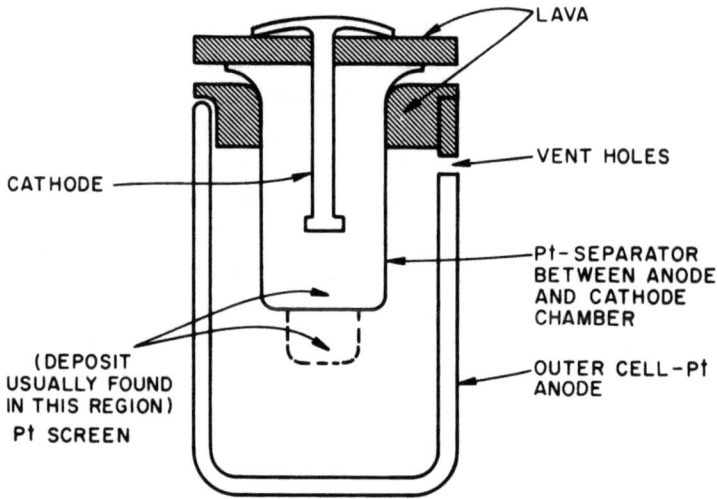

Fig. 21. Apparatus for molten salt electrolysis.[89b]

decomposed and occasionally electrochemical transport phenomena have been used.

6. Other Solvents

Clearly the solvent possibilities are endless. Organic solvents for organic materials present no difficulties and the techniques described for aqueous growth are usually directly applicable. Some "hydrothermal" experiments have been conducted with HF (for fluoride growth), NH_3 (for sulfides and selenium), and other solvents but so far the results have not been especially encouraging. Liquid metals have, however, found wide use. Diamond is grown from transition metal "catalysts" which are usually solvents for graphite.[90,91] It is interesting to note that diamond is stable at pressures as low as ~10,000 atm if the temperature is near room temperature. However, at low temperatures, the rate of transformation is impractically slow. To accelerate the kinetics, the temperature is increased and then the pressure must be increased to stay within the diamond stability field. Apparatus for achieving super pressure has been described in the literature and will not be reviewed here.[92,93] Until recently, no apparatus was available that was capable of reaching a high enough pressure to be within the diamond field at a temperature where the graphite → diamond conversion rate was observable. The first diamond synthesis, achieved by Bundy et al.,[90] was achieved in the so-called belt apparatus[93] and employed "catalyst" solvents that accelerated the conversion rate at a low enough temperature that the diamond field was accessible at "moderate" pressures. Chromium, manganese, cobalt, nickel, and palladium are among the successful catalyst-solvents used. Graphite is the usual source of carbon but other sources have been used. The solvent is molten and acts as a thin-film transport medium between the phases. It may also help to nucleate the diamond.

The supersaturation is provided because the dissolving phase (graphite) is metastable and thus has a higher solubility than the stable diamond phase. Thus the growth in addition to being from solution also involves a polymorphic transition. Large (carat size) diamond can now be grown by careful control of thermal conditions in long-term experiments.[94]

The most widely used electronic materials presently grown from metal solution are probably the III–V semiconductors, GaAs and especially GaP, where the solvent is usually Ga. The growth takes place at low enough temperatures that the volatility of As and P, which causes difficulties in melt growth, is low. The solvent is especially convenient because it does not introduce impurities into the system and it melts at low temperature.

In the case of GaP, one of the first and simplest solution growth methods was described by Wolff et al.[95,96] (see also Miller[97]) and Ga-grown GaP has been of great importance in electroluminescence studies because it produces pair spectra with many sharp lines and can form the basis of high-efficiency diodes.

Molten gallium containing 5–10 at. % phosphorus is sealed in an evacuated vitreous-silica ampoule and cooled slowly from temperatures between 1000 and 1275°C at rates of from 3 to 5°/hr down to temperatures 200–300° below the initial growth temperature. In some cases AlN, BN, and Al_2O_3 crucibles sealed inside the quartz ampoules are used to contain the melt. The system is cooled more rapidly once a temperature of from 900 to 700°C is reached. The resulting crystals are recovered by a combination of mechanical separation and leaching away of the Ga in concentrated HCl. Grown crystals are predominantly {111} platelets up to ~1 cm in maximum dimension (when growth is from 50-g melts).[98]

Chapter 8

7. Liquid-Phase Epitaxy (LPE)

In recent years high electroluminescent efficiencies in GaP have been obtained with Ga-grown crystals where semiconductor diodes are made by liquid-phase epitaxy. This method has been used to grow GaAs and GaP.[99–102] In the liquid-phase epitaxial growth of GaP a nearly saturated solution of GaP (Ga solvent) is brought into contact with a GaP seed (substrate) and the temperature is programmed to produce supersaturation and the growth of an appropriately doped, relatively thin layer on the seed. Figure 22 illustrates the

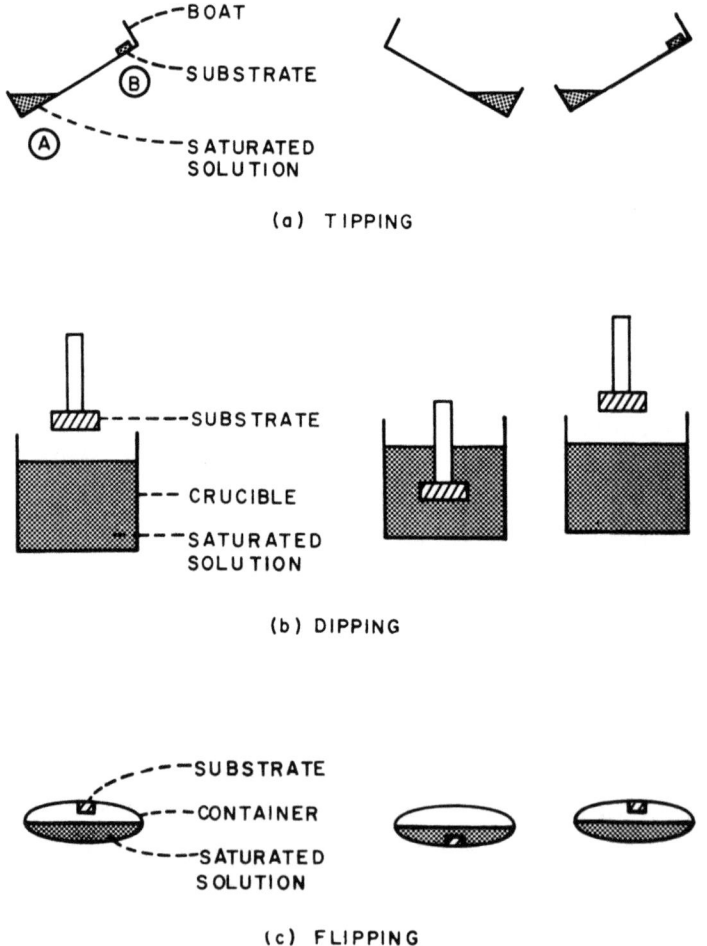

Fig. 22. Geometries for liquid-phase epitaxy.

geometries which have been used in liquid-phase epitaxy. In tipping (Figure 22a), the solution is contained in a boat tipped so as to keep it out of contact with the substrate-seed while equilibration takes place. Once the solution is saturated and brought to the proper temperature (generally a bit above the saturation temperature so as to prevent undesired nucleation) the boat is tipped so as to bring the solution into contact with the substrate and the temperature is programmed down.

In addition to GaAs and GaP, tipping has been used to grow epitaxial layers of Ge.[99] The usual solvents are

GaP–Ga (+Te → n-type material; +Zn → p)

GaAs–Sn (+Sn → n type; +Zn → p)

Ge–In   (+Ga → p type)

Common boat materials are graphite for Ge and BN for GaP.

For illustrative purposes, a GaAs run will be described. At the start of a run region A of Figure 22 is lower than region B and a Sn–GaAs mixture is placed at A, while a GaAs seed wafer is fastened at B. The graphite boat is fixed at the center of a constant-temperature zone in an appropriately designed furnace. The boat is heated to about 640°C and held for a period to dissolve GaAs in the Sn. Then the boat is tipped so that B is lower than A. It is probably important that when contact is made with the substrate the solution be slightly undersaturated. Seed dissolution removes surface damage and under some conditions helps to ensure that the growing interface starts off planar.* The furnace is cooled from ~640 to 500°C in about 30 min, the solution is decanted off, and the substrate with its doped epitaxial layer is recovered.

Appropriate heating cycles for other materials are as follows:[99]

GaAs from Ga: contact seed at 900°C, cool to 400°C in 30 min, decant.
Ge from In: contact seed at 500°C, cool to 400°C in 30 min.
Ge from Sn–Pb: contact seed at 400°C, cool to 300°C in 10 min.
GaP from Ga: contact seed at ~1100°C, cool to 700°C in 40 min.[100–102]

Linares[103] has used tipping to grow garnets where for $Y_3Fe_5O_{12}$ the PbO, PbF_2, and other fluxes were used in Pt boats. However, the

---

* The solvent should ideally be a polish etch for the substrate.

greatest success in garnet growth has been obtained using dipping (Figure 22b). This work has been stimulated by the need for thin magnetic garnet films for bubble domain memories. Levinstein et al.[104] have used $PbO-B_2O_3$ solvents which are stable to high supercooling. Films of complicated composition such as $Gd_{1.2}Y_{1.2}Yb_{0.6}Fe_{4.3}Ga_{0.7}O_{12}$ are deposited by dipping an appropriate substrate into a saturated melt contained in an appropriate furnace. It is important that the lattice parameters and coefficients of expansion of substrate and film match with about 1% or else the film will crack. Strain can have a great effect on properties. In magnetic bubble materials it can influence magnetocrystalline anisotropy and hence bubble size when magnetostriction is appreciable. In heteroepitaxial semiconductor diodes there is some indication the electroluminescent efficiency can be reduced by strain. Films are susceptible to greater strains without cracking in compression than in tension so that a smaller lattice parameter for the film than the substrate is generally advantageous. $Gd_3Ga_5O_{12}$ is a good match to many of the garnet compositions of interest for bubble studies. In garnet growth by dipping, the furnace is designed to provide a region above the crucible where the substrate can be equilibrated to reach the temperature of the solution. The growth temperature is usually 10–20° below the crystallization temperature. The substrate is dipped into the melt and held there during rotation at high speed. Growth rate has been observed to be linear with supercooling over most conditions. For good growth the substrate should be free or nearly free of dislocations, scratches, and saw damage. Special techniques of Czochralski melt growth and etching of $Gd_3Ga_5O_{12}$ are required to meet these criteria. Temperatures of growth much above about 1030°C are deleterious because substrate etching in the flux becomes severe.

The technique of flipping (Figure 22c) has been used occasionally for semiconductor growth. In this method the seed is introduced into the saturated solution by rotating or "flipping" the growth ampoule.

The need for miniaturization in semiconductors, magnetic bubbles, and integrated optics will be a great stimulus to future development of the powerful technique of liquid-phase epitaxy.

8. Temperature-Gradient Zone Melting (TGZM)

The process suggested by Pfann[105] of moving a narrow zone of solvent through a material by the use of a temperature gradient is called temperature-gradient zone melting. Consider the phase

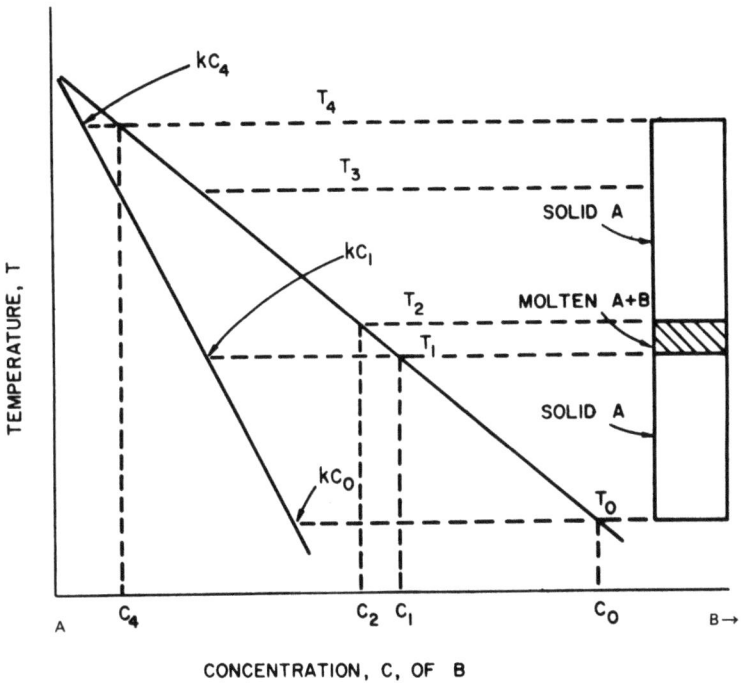

Fig. 23. Temperature-gradient zone melting (after Pfann[105]).

diagram AB of Figure 23,[105] where A, the solute, is the crystal to be grown, and B is the solvent. If we sandwich a thin layer of solid B between blocks of pure A and place the whole in a temperature gradient where the temperature of the layer is above the lowest melting point in the system, and the hottest temperature is below the melting point of A, the layer will dissolve some A and expand vertically in length. Now if $T_1 < T_2 < T_3 < T_4$, the zone will move toward T_4. When it is at the position T_1–T_2, the cool interface will dissolve enough A to correspond to the equilibrium liquidus concentration at C_1. Dissolving of A at the hot interface will continue until C_2 is reached. The concentration gradient will cause A to diffuse toward the cool interface, where the solution will be supersaturated (supercooled) and A, whose concentration of B is near kC_1, will freeze (where k is the distribution constant). Continuation of this solution–diffusion–freezing process causes the molten layer to move through the block.

Pfann has given a detailed analysis of the method. If the gradient is uniformly imposed across the whole ingot, the compositional change

Chapter 8

in the grown crystal if the zone starts at T_1 and moves to T_4 will be from kC_4 to kC_1. A zone can be held at an essentially constant temperature during its entire travel, with a resulting large improvement in homogeneity, by using a ring form heater and moving the heater along the ingot. This technique is sometimes called the traveling heater method and is actually a variant of zone leveling.[105] Among the single crystals grown by temperature-gradient zone melting are GaAs using Ga as solvent,[106] α-SiC using Cr as solvent,[107] Ge using Pb as solvent,[108] and GaP using Ga as solvent.[109,110]

Perhaps typical of the conditions and results obtained is the work of Plaskett et al.[110] on GaP. They moved a (Ga + GaP) zone through a GaP ingot by means of a traveling heater. Their procedure was a refinement of Broder and Wolff's[109] traveling-solvent method. A small seed of GaP was placed in the bottom of a pointed-tip BN cylindrical crucible. Above it a 1/4-in.-thick layer of Ga was placed, and above that a dense polycrystalline ingot of GaP. The crucible was placed in an evacuated, sealed quartz ampoule and inserted in a resistance furnace, which helped to control the slope of the profile. The furnace had a single-turn rf coil that provided a local sharp temperature gradient, produced the molten (Ga + GaP) zone, and could be moved. The zone was established at about 1160°C (which corresponds to a concentration of about 10 wt. % GaP in Ga). The rate of zone travel upward through the ingot was about 4 mm/day, which is of the same order of magnitude as other solution growth processes (such as aqueous solution, flux, and hydrothermal), where diffusion necessitates reduced rates. Crystals as long as 5 cm were grown by the method.

9. Vapor–Liquid–Solid Growth

Vapor–liquid–solid (VLS) growth, which was first described by Wagner and Ellis,[111] is illustrated by the schematic of Figure 24, which shows the method used in Si growth. Deposition from the vapor is by iodide disproportionation:

$$2SiI_2(g) \rightleftarrows SiI_4(g) + Si(soln) \qquad (30)$$

or by the hydrogen reduction of $SiCl_4$. The liquid phase is a solution of Si in Au. In the process of growth the vapor-phase reaction does not directly form solid Si, instead it forms Si in liquid solution. The solid Si is formed by growth from the liquid solution. Conditions for vapor deposition (flow rate, reactant rates, temperature, etc.) are

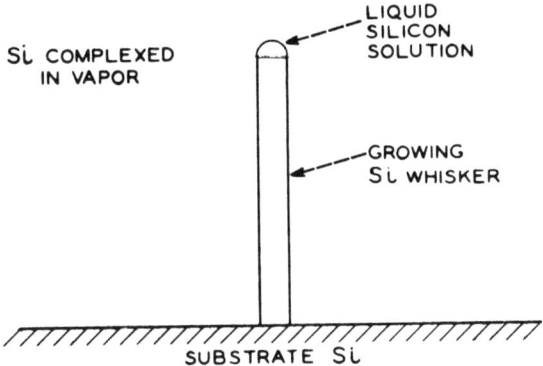

Fig. 24. Vapor–liquid–solid growth (after Laudise[1]).

generally similar to those used in vapor deposition with the same reactions. For instance, for $SiCl_4$ reduction the H_2 flow rate was typically 1000 cm^3/min and the $SiCl_4/H_2$ mole ratio was 0.02. Liquid drop temperature was between 875 and 1200°C, which corresponds to a Si solubility in gold of between 55 and 70 at. % Si.[112] The energy of activation for VLS growth was 7.5 kcal/mole as compared to about 31 kcal/mole for vapor growth on a Si substrate. The rate was about five times greater than for vapor–solid growth (about 10 μm/min) under similar conditions. Apparently a liquid surface is ideally atomically rough in comparison to a solid single-crystal surface, so that the vapor–liquid reaction proceeds much more easily than the analogous vapor–solid reaction. Deposition from the liquid (the liquid–solid reaction) is aided by the fact that a large thermal gradient at the growing interface is relatively easy to obtain and suppresses constitutional supercooling. The initial growth mechanism is somewhat obscure but growth is easy to initiate. Typically a small particle of gold is placed on a clean {111} surface of a single-crystal Si wafer and heated to 950°C, where it forms a small liquid Au–Si alloy droplet. Then the desired vapors are introduced and growth of the whisker begins. The alloy droplet rides atop the whisker and under certain conditions morphologies more complicated than a single whisker can be formed. A small concentration of Au remains behind in the grown whisker, and eventually, if the whisker is long enough, all the Au will be consumed.

Other materials besides Si, including Ge, GaAs, GaP, and (GaAs)P, have been grown by the technique,[113,114] but so far the

Chapter 8

morphologies obtained are basically whiskers and crystals large in three dimensions have not been grown. Wagner[115] has written an extensive review of the mechanism and of the materials grown.

References

1. R. A. Laudise, *The Growth of Single Crystals*, Prentice-Hall, New York (1970).
2. J. J. Gilman (ed.), *The Art and Science of Growing Crystals*, Wiley, New York (1963).
3. H. E. Buckley, *Crystal Growth*, Wiley, New York (1951).
4. R. Roy and W. B. White, *J. Cryst. Growth* **3/4**, 33 (1968).
5a. A. Holden and R. H. Thompson, *Growing Crystals with a Rotary Crystallizer*, Bell Telephone Laboratories, New York (1964).
5b. A. Holden and P. Singer, *Crystals and Crystal Growing*, Anchor—Doubleday, New York (1960).
6. T. G. Petrov, E. B. Treivus, and A. P. Kasatkin, *Growing Crystals from Solution*, Consultants Bureau, New York (1969).
7. E. D. Kolb, *The Physics of Selenium and Tellurium* (W. Charles Cooper, ed.), pp. 155ff, Pergamon, New York (1969).
8. E. D. Kolb and R. A. Laudise, *J. Cryst. Growth* **8**, 191–196 (1971).
9. H. K. Henisch, J. Dennis, and J. Hanoka, *J. Phys. Chem. Solids* **26**, 493 (1965).
10. A. F. Armington and J. J. O'Connor, *Mat. Res. Bull.* **2**(10), 907 (1967).
11. S. D. Scott and H. L. Barnes, *Mat. Res. Bull.* **4**, 897 (1969).
12. A. F. Armington and J. J. O'Connor, *J. Cryst. Growth* **6**, 278 (1970).
13. Y. Toudic and R. Aumont, *Compt. Rend.* **269**, 74 (1969).
14. T. Yamamoto, *Bull. Inst. Phys. Chem. Res.* (*Tokyo*) **17**, 1278 (1938).
15. W. E. Gibbs and W. Clayton, *Nature* **113**, 492 (1924).
16. K. Nassau, *J. Cryst. Growth*, **15**, 171 (1972).
17. A. C. Walker and G. T. Kohman, *Trans. Am. Inst. Elec. Eng.* **67**, 565 (1948).
18. A. C. Walker, *Bell Labs. Record* **25**, 357 (1947).
19. A. N. Holden, *Disc. Faraday Soc.* **5**, 312 (1949).
20. J. Shiever and K. Nassau, *Mat. Res. Bull.* **7**, 613 (1972).
21. H. K. Henisch, *Crystal Growth in Gels*, Penn. State Univ. Press, College Park, Pa. (1970).
22. H. K. Henisch, J. I. Hanoka, and J. Dennis, *J. Electrochem. Soc.* **112**, 627 (1965).
23. J. J. O'Connor, M. A. DiPietro, A. F. Armington, and B. Rubin, *Nature* **212**, 68 (1966).
24. I. Epelboin, F. Lenoir, and R. Wiart, p. 417; W. A. Schultze, p. 421; U. Bertocci, C. Bertocci, and B. C. Larson, p. 427; and E. Budevski, p. 93; in *Crystal Growth 1971* (R. A. Laudise, J. B. Mullin, and B. Mutaftschiev, eds.), North-Holland, Amsterdam (1972).
25. R. A. Laudise and J. W. Neilsen, in *Solid State Physics* (F. Seitz and D. Turnbull, eds.), Vol. XII, Academic, New York (1961).
26. R. A. Laudise, in *Progress in Inorganic Chemistry* (F. A. Cotton, ed.), Vol. III, p. 1, Wiley–Interscience, New York (1962).
27. R. Roy and O. F. Tuttle, *Phys. Chem. Earth* **1**, 138 (1956).

28. C. J. M. Rooymans, in *Preparative Methods in Solid State Chemistry* (P. Hagenmueller, ed.), pp. 71ff, Academic Press, New York (1972).
29. P. W. Bridgman, *The Physics of High Pressure*, Bell, London (1949).
30. E. D. Kolb, U.S. Pat. 3,271,114 (6 Sept. 1966).
31. G. W. Morey and P. Niggli, *J. Am. Chem. Soc.* **35**, 1086 (1913).
32. O. F. Tuttle, *Geol. Soc. Am. Bull.* **60**, 1727 (1949).
33. R. L. Barns, R. A. Laudise, and R. M. Shields, *J. Phys. Chem.* **67**, 835 (1963).
34. G. Erwin and E. F. Osborne, *J. Geol.* **59**, 385 (1951).
35. E. M. Levin, C. R. Robbins, and H. F. McMurdie (eds.), *Phase Diagrams for Ceramists*, Vol. I (1964), Vol. II (1969), pp. 1926–1928, 4021, American Ceramic Soc.
36. R. A. Laudise and A. A. Ballman, *J. Am. Chem. Soc.* **80**, 2655 (1958).
37. R. A. Laudise, J. H. Crocket, and A. A. Ballman, *J. Phys. Chem.* **65**, 359 (1961).
38. R. A. Laudise and E. D. Kolb, Endeavour **XXVIII**[105], 114–117 (1969).
39. R. A. Laudise and A. A. Ballman, *J. Phys. Chem.* **65**, 1396 (1961).
40. D. J. Marshall and R. A. Laudise, *J. Cryst. Growth* **1**, 88 (1967).
41. R. A. Laudise and E. D. Kolb, *Am. Miner.* **48**, 642 (1963).
42. D. J. Marshall and R. A. Laudise, *Crystal Growth* (Suppl. to J. Phys. Chem. Solids), p. 557, Pergamon, New York (1967).
43. R. A. Laudise, *J. Am. Chem. Soc.* **81**, 562 (1959).
44. R. A. Laudise and R. A. Sullivan, *Chem. Eng. Prog.* **55**, 55 (1959).
45. D. W. Rudd, E. E. Haughton, and W. J. Carrol, *Western Electric Engineer* **1966** (January), 22.
46. D. M. Dodd and D. B. Fraser, *J. Phys. Chem. Solids* **26**, 673 (1965).
47. E. D. Kolb, D. A. Pinnow, T. C. Rich, N. C. Lias, E. E. Grudenski, and R. A. Laudise, *Mat. Res. Bull.* **7**, 397 (1972).
48. R. A. Laudise, E. D. Kolb, N. C. Lias, and E. E. Grudenski, in *Proc. IVth All-Union Conf. on Crystal Growth, USSR, 1972*, to be published.
49. N. C. Lias, E. E. Grudenski, E. D. Kolb, and R. A. Laudise, *J. Cryst. Growth*, **18**, 1 (1973).
50. J. A. Burton and W. P. Slichter, in *Transistor Technology* (H. E. Bridgers, J. H. Scaff, and J. H. Shive, eds.), Vol. 1, Chapter 5, D. Van Nostrand, Princeton, N.J. (1958).
51. R. A. Laudise and A. A. Ballman, *J. Phys. Chem.* **64**, 688 (1960).
52. E. D. Kolb and R. A. Laudise, *J. Am. Ceram. Soc.* **48**, 342 (1965).
53. J. W. Nielsen and F. G. Foster, *Am. Miner.* **45**, 299 (1960).
54. E. D. Kolb, D. W. Wood, E. G. Spencer, and R. A. Laudise, *J. Appl. Phys.* **38**, 1027 (1967).
55. E. D. Kolb, D. L. Wood, and R. A. Laudise, *J. Appl. Phys.* **39**, 1362 (1968).
56. L. H. Shick, J. W. Nielsen, A. H. Bobeck, A. J. Kurtzig, P. C. Michaelis, and J. P. Reekstin, *Appl. Phys. Letters*, **18**, 89 (1971).
57. R. A. Laudise, *J. Cryst. Growth* **13/14**, 27 (1971).
58. E. D. Kolb and R. A. Laudise, *J. Cryst. Growth* **8**, 191 (1971).
59. R. A. Laudise, in *The Art and Science of Growing Crystals* (J. J. Gilman, ed.), pp. 252ff., Wiley, New York (1963).
60. E. A. D. White, in *Technique of Inorganic Chemistry* (H. B. Jonassen and A. Weissberger, eds.), Vol. IV, pp. 31ff, Interscience, New York (1965).
61. Y. Laurent, *Rev. Chem. Mineral* **6**, 1145 (1969).

62. J. W. Nielsen, in *Proc. IVth All-Union Conf. on Crystal Growth, USSR, 1972*, to be published.
63. H. J. Van Hook, *J. Am. Ceram. Soc.* **44**, 208 (1961).
64. A. H. Bobeck, *Bell System Tech. J.* **46**, 1901 (1967).
65. J. W. Nielsen and E. F. Dearborn, *J. Phys. Chem. Solids* **5**, 202 (1968).
66. J. W. Nielsen, *J. Appl. Phys.* **31**, 51S (1960).
67. R. C. Linares, *J. Am. Ceram. Soc.* **45**, 307 (1962).
68. W. H. Grodkiewicz, E. F. Dearborn, and L. G. Van Uitert, in *Crystal Growth* (H. S. Peiser, ed.), p. 441, Pergamon, New York (1967).
69. J. P. Remeika, *J. Am. Chem. Soc.* **76**, 940 (1954).
70. R. A. Lefever, J. W. Torpy, and A. B. Chase, *J. Appl. Phys.* **32**, 962 (1961).
71. J. Hart and E. A. D. White, unpublished work, reported in E. A. D. White, Ref. 64.
72. J. P. Remeika, *J. Am. Chem. Soc.* **78**, 4259 (1956).
73. W. H. Grodkiewicz and D. J. Nitti, *J. Am. Ceram. Soc.* **49**, 576 (1966).
74. F. Jona, G. Shirane, and R. Pepinsky, *Phys. Rev.* **97**, 1584 (1955).
75. H. J. Scheel, to be published, reported by Nielsen, Ref. 62.
76. H. Scholz and R. Kluckow, in *Crystal Growth* (H. S. Peiser, ed.), p. 475, Pergamon, New York (1967).
77. W. Hintzman and G. Mueller-Vogt, *J. Cryst. Growth* **5**, 274 (1969).
78. H. J. Scheel, *J. Cryst. Growth* **13/14**, 560 (1972).
79. H. J. Scheel, Oral presentation, ICCG-3, Marseille, July 1971.
80. H. J. Scheel and E. O. Schulz-DuBois, *J. Cryst. Growth* **8**, 304 (1972).
81. W. Tolksdorf and E. Welz, *J. Cryst. Growth* **13/14**, 566 (1972).
82. R. Roy, in *Crystal Growth* (H. S. Peiser, ed.), p. 505, Pergamon, New York (1967).
83. L. M. Viting, *Vestn. Mosk. Univ. Ser. II Khim.* **20**(4), 54 (1965).
84. R. A. Laudise, R. C. Linares, and E. F. Dearborn, *J. Appl. Phys. Suppl.* **33**, 1362 (1962).
85. W. A. Bonner, E. F. Dearborn, and L. G. Van Uitert, in *Crystal Growth* (H. S. Peiser, ed.), p. 437, Pergamon, Oxford (1967).
86. P. W. Whipps, *J. Cryst. Growth* **12**, 120 (1972).
87. C. E. Miller, *J. Appl. Phys.* **29**, 233 (1958).
88. V. Belruss, J. Kalnajs, and A. Linz, *Mat. Res. Bull.* **6**, 899 (1971).
89a. D. S. Perloff and A. Wold, in *Crystal Growth* (H. S. Peiser, ed.), p. 361, Pergamon, New York (1967).
89b. W. Kunnman, in *Preparation and Properties of Solid State Materials* (R. A. Lefever, ed.), pp. 1ff, Dekker, New York (1971)
90. F. P. Bundy, H. T. Hall, H. M. Strong, and R. H. Wentorf, *Nature* **176**, 51 (1955).
91. F. P. Bundy, in *Modern Very High Pressure Techniques* (R. H. Wentorf, Jr., ed.), Butterworth, London (1962).
92. R. E. Hanneman, H. M. Strong, and F. P. Bundy, *Science* **155**, 995 (1967).
93. H. T. Hall, *Rev. Sci. Instr.* **31**, 125 (1960).
94. H. M. Strong and R. H. Wentorf, Jr., *Naturwiss.* **59**[1], 1 (1972).
95. G. Wolff, P. H. Keck, and J. D. Broder, *Phys. Rev.* **94**, 753 (1954).
96. G. Wolff, R. A. Herbert, and J. D. Broder, *Semiconductors and Phosphors*, pp. 463ff, Wiley–Interscience, New York (1958).
97. J. F. Miller, in *Compound Semiconductors* (R. K. Willardson and H. L. Goering, eds.), Vol. I, pp. 200ff, Reinhold, New York (1962).

98. D. G. Thomas, M. Gershenzon, and F. A. Trumbore, *Phys. Rev.* **133**, A269 (1964).
99. H. Nelson, *RCA Rev.* **24**, 603 (1963).
100. M. R. Lorenz and M. Pilkuhn, *J. Appl. Phys.* **37**, 4094 (1966).
101. F. A. Trumbore, M. Kowalchik, and H. G. White, *J. Appl. Phys.* **38**, 1987 (1967).
102. R. A. Logan, H. G. White, and F. A. Trumbore, *Appl. Phys. Letters* **10**, 206 (1967).
103. R. C. Linares, *J. Cryst. Growth* **3/4**, 443 (1968).
104. H. J. Levinstein, S. Licht, R. W. Landorf, and S. L. Blank, *Appl. Phys. Letters* **19**, 486 (1971).
105. W. G. Pfann, *Trans. AIME* **203**, 961 (1955).
106. A. I. Mlavsky and M. Weinstein, *J. Appl. Phys.* **34**, 2885 (1963).
107. L. B. Griffiths and A. I. Mlavsky, *J. Electrochem. Soc.* **111**, 805 (1964).
108. W. G. Pfann, *Zone Melting*, 2nd ed., pp. 254ff, Wiley, New York (1966).
109. J. D. Broder and G. A. Wolff, *J. Electrochem. Soc.* **110**, 1150 (1963).
110. T. S. Plaskett, S. E. Blum, and L. M. Foster, *J. Electrochem. Soc.* **114**, 1303 (1967).
111. R. S. Wagner and W. C. Ellis, *Trans. AIME* **233**, 1053 (1965).
112. R. S. Wagner and C. J. Doherty, *J. Electrochem. Soc.* **113**, 1300 (1967).
113. R. L. Barns and W. C. Ellis, *J. Appl. Phys.* **36**, 2296 (1965).
114. C. M. Wolfe, C. J. Nuese, and N. Holonyak, *J. Appl. Phys.* **36**, 3790 (1965).
115. R. S. Wagner, in *Whisker Technology* (A. P. Levite, ed.), Wiley, New York (1969).

New Phases at High Pressure

J. C. Joubert and J. Chenavas

Laboratoire de Rayons X, CNRS
Grenoble, France

1. Introduction

In spite of the considerable amount of work by Bridgman and co-workers on both applied and fundamental research in high-pressure physics from 1915 to 1945, the use of high pressure for the synthesis of new materials was more or less neglected by chemists. Perhaps because of the relative complexity of high-pressure equipment, they paid more attention to the role of high temperatures for the preparation of new phases. In the period from 1945 to 1955 geologists began to use high-pressure, high-temperature conditions in order to synthesize minerals encountered in volcanic rock whose origin was the deeper portions of the earth's mantle (100–400 km). Only after 1955, the date of the announcement of the synthesis of diamond by the General Electric Co., did the advantages of using high-pressure conditions for the synthesis of new materials become evident to chemists. At that time many laboratories developed high-pressure facilities and chemists started to investigate pressure–temperature phase diagrams of the elements—more than forty years after the first synthesis of black phosphorus.[1,2] In several years a large number of new, denser phases of the elements were discovered. Some of the high-pressure phases had physical properties quite different from those of the ambient-pressure phases, e.g., high-pressure phases of silicon and germanium with the white tin structure were metallic and Bi II and III were both superconductors at low temperatures.

The investigation of two-component systems under pressure began in about 1960, with the discovery of dense metallic phases of the

Chapter 9

III–V and II–VI semiconducting compounds. Since 1965, due to the rapid developments of high-pressure technology, high-pressure, high-temperature conditions have been extensively used by chemists for the synthesis of new, denser phases of well-known binary or ternary compounds or for the synthesis of new compounds with a different stoichiometry.

The reason for using high pressure in addition to high temperature is clear: If one considers a two-component system, one can generally find two or three stable binary compounds with definite stoichiometry in the phase diagram at ambient pressure. When these compounds are submitted to high-pressure conditions, i.e., between 5 and 100 kbar, very few of them are stable. Many undergo a phase transition to a denser phase with identical stoichiometry; others disproportionate into the components and often into new binary phases. Frequently new compounds form from two-phase regions. Many of these new, dense phases can be quenched while under pressure and can thus be preserved as metastable phases at normal pressure and temperature. So it becomes evident that high-pressure equipment is a powerful tool for the synthesis of new materials. Occasionally single crystals can be prepared as is the case for synthetic diamonds and cubic boron nitride.

The purpose of this chapter is not to present an exhaustive review of all the dense phases which have been obtained up to now, but rather to give insight into materials synthesis where high-pressure conditions can be used with success. Typical examples of syntheses have been chosen from the literature and include the results of our recent investigations on high-pressure synthesis of mixed transition metal compounds.

2. High-Pressure Facilities Suitable for the Synthesis of New Materials

We first give a short description of a few devices which have proven to be convenient to use for the synthesis of new inorganic materials.

2.1. The 1–10 kbar Pressure Range

Among the high-pressure apparatus suitable for use in the 1–10 kbar pressure range (which we call the medium-pressure range) the simplest device is the conventional hydrothermal synthesis facility

New Phases at High Pressure

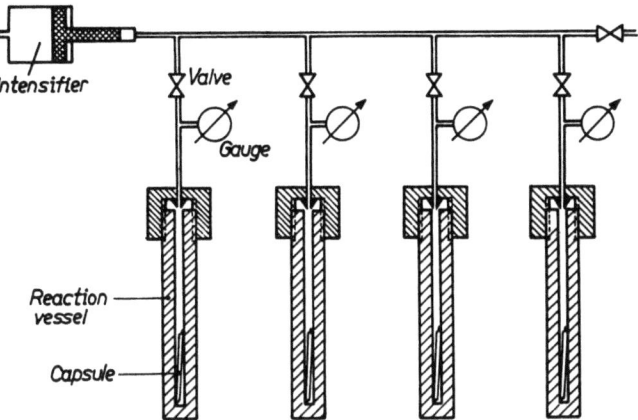

Fig. 1. The conventional hydrothermal synthesis apparatus. Pressure up to 5 kbar and temperatures to 850°C can be achieved in autoclaves having a volume of 10–50 cm^3.

represented in Figure 1. This type of device has been frequently used in the synthesis of oxide and fluoride compounds containing a high-valence-state cation, like Fe^{4+}, Ni^{3+}, Cr^{4+}, Mn^{4+}, and for the synthesis of nitrides. Two different types of runs can be performed with this apparatus. It can first be used as an open system, which means that the sample is in direct contact with the gas from the intensifier tank. Direct reactions with the oxygen, fluorine, nitrogen, etc., vapor phase can be conducted in an open capsule which does not allow reaction of the sample with the bomb wall. Gold is inert to oxygen as soon as the temperature reaches a few hundred degrees centigrade. This is not true for platinum (Figure 2), which readily oxidizes and thus is not a suitable sample container at high oxygen pressure.[3-6] Platinum oxide films form directly on solid slabs of metallic platinum and these films may in turn react with the specimen, introducing platinum impurities. The length of a typical run is 24 hr, although kinetics of individual reactions vary widely. Quenching can be accomplished by plunging the pressure vessel into a container of cold water.

Several new phases have been synthesized by this method, including RhO_2,[5] β-PtO_2,[5] $Na_2Ni^{4+}F_6$,[7] $Li_2Ni^{4+}F_6$,[7] etc. No doubt a large number of new compounds in which the transition metals are in an unusually high valence state will be obtained in the next few years even in this rather limited range of pressures. In the case of oxidizing atmospheres, in addition to the hazards associated

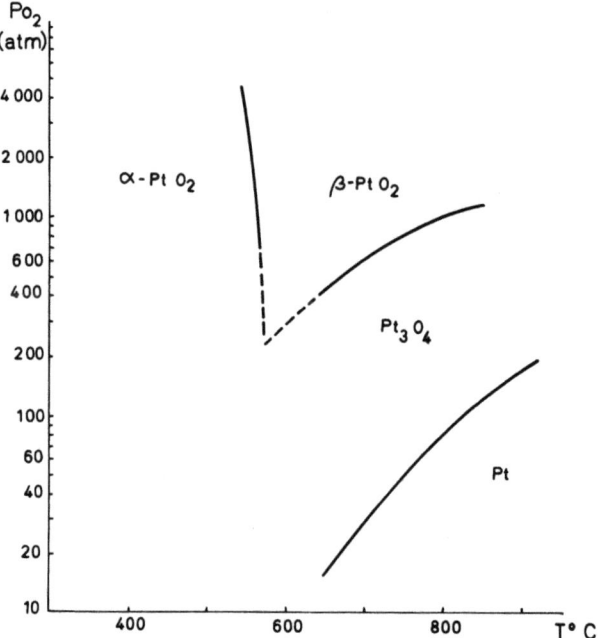

Fig. 2. *P–T* diagram of the Pt–O system (according to Refs. 5 and 6).

with the manipulation of a high-gas-pressure equipment one finds hazards due to combustion from the gas itself. All high-pressure fittings (gauges, valves) must be scrupulously free of grease and organic matter. High-iron alloys should be avoided. Stainless steel fitting and tubing will ignite at oxygen or fluorine pressures of 4 kbar at temperatures of several hundred degrees.[6,7] Stainless steel burns with explosive violence and hot oxygen should not be allowed to enter the connecting manifold. With new stellite vessels of small capacity (7 cm^3) acceptable operating conditions appear to be up to 700°C at 4 kbar, extending to 900°C at 200–300 bar, although the pressure vessels rapidly deteriorate at the higher temperatures.

In order to avoid the hazards associated with the accumulation of a rather large volume of oxidizing gas at high pressure and high temperature, one can use a closed system if some high-oxidation-state oxide or an internal oxidant can be encapsulated. The decomposition of the higher-valence oxide provides the oxygen which, because of the collapsible gold capsule, is in pressure equilibrium with the surrounding medium, namely supercritical water or argon. However, it is necessary to have a sufficient amount of oxygen released to fill all pore space in the capsule at the pressure of the system. If the quantity of gas is insufficient, there will be developed some unknown lower oxygen pressure in the pores, because the solid phase will continue to support the capsule walls. A second precaution is that the composition of the gas must be known. Unknown gaseous impurities, such as water, will contribute to the total pressure and result in an oxygen partial pressure that is lower than the pressure indicated by the hydrothermal gauge.[6]

This method has been used with success in the synthesis of rutile-type compounds RhO_2 and PtO_2 according to the reaction,[8]

$$\text{hexagonal ``PtO}_2\text{''} \xrightarrow[KClO_3, H_2O]{700°C, 3000 \text{ atm}} \text{orthorhombic } PtO_2 \text{ } (CaCl_2 \text{ type}), \text{ distorted rutile}$$

$$\text{unidentified structure ``RhO}_2\text{''} \xrightarrow[KClO_3, H_2O]{700°C, 3000 \text{ atm}} \text{tetragonal } RhO_2, \text{ rutile type}$$

At 700°C, $KClO_3$ decomposes and provides extra oxygen pressure within the limited volume of the gold capsule. A similar reaction has been used for the synthesis of almost stoichiometric cubic TbO_2[9]:

$$Tb_2O_3 \xrightarrow[700°C, 3 \text{ kbar}]{HClO_4, H_2O} TbO_2 \text{ cubic, fluorite type}$$

and for the synthesis of several rare earth pyrochlores containing tetravalent platinum and palladium[10]:

$$Ln_2O_3 + \text{hexagonal ``PtO}_2\text{''} \xrightarrow[700°C, 3 \text{ kbar}]{KClO_3, H_2O} Ln_2Pt_2^{4+}O_7$$

$$Ln_2O_3 + Pd^{2+}O \xrightarrow[700°C, 3 \text{ kbar}]{KClO_3, H_2O} Ln_2Pd_2^{4+}O_7$$

In order to avoid the easy reduction of mercury to metal, Sleight[11] synthesized the pyrochlores $Hg_2Nb_2O_7$ and $Hg_2Ta_2O_7$ at 700°C and 3 kbar in sealed capsules starting from HgO and Nb_2O_5 (or Ta_2O_5).

This technique would seem to have wide application in the preparation of higher-valence oxide materials since the presence of the water greatly increases the kinetics of many reactions. Preparation of single crystals should be possible in several cases. The method not only offers the advantage of being simple, but it can be used for the synthesis of rather large quantities of material, a few grams or more if necessary. This will not be the case for the methods having a higher-pressure working range.

When higher pressures and in particular higher temperatures are required one can use an internally heated pressure vessel. The vessel wall is kept cold with a circulating water cooling system, and consequently the pressure limits are those of the cold yield strength of the steel. The pressure is supplied by a two-stage gas intensifier and can reach the range of 10–15 kbar when inert gases (argon) are used. The internal furnace is a resistance winding on an inner alumina tube. With platinum windings working temperatures as high as 1200°C are possible.[6] A wide variety of compounds have been synthesized with this device, e.g., the $SrFeO_3$ and $BaFeO_3$ series[12] containing mostly Fe^{4+} cations, and the compounds $Ca_2Mn^{4+}O_4$, $Ca_3Mn_2^{4+}O_7$, $Ca_4Mn_3^{4+}O_{10}$.[13] The advantages of internally heated, large vertical vessels are the rather high range of attainable temperatures and the flexibility of the apparatus which should be convenient for use in crystal growth at rather high gas pressures by the flux method. In fact $CoFe_2O_4$ crystals free of ferrous iron have already been grown at high pressure[14]; it seems that insufficient attention has been paid to this versatile method and more use should be made of it in the near future.

2.2. The High-Pressure Range

Experimentation at pressures higher than 10 kbar requires the use of more sophisticated techniques. Many routine runs have been successfully performed in four types of apparatus.

2.2.1. The Piston–Cylinder Apparatus

The tungsten carbide chamber and piston assembly is represented in Figure 3. The sample is contained in a platinum capsule surrounded by pyrophyllite, a soft material which has been proven to be a good pressure transmitter. An internal microfurnace, made of molybdenum or graphite sleeves, is connected to a low-voltage, high-current power input through the tungsten carbide pistons. Temperatures of 1500–2000°C can be obtained and pressures up to 50 kbar are possible in a 0.1-cm^3-volume capsule using a 200-ton press.[15,16]

2.2.2. The Belt-Type Apparatus

Due to the geometry of the piston–cylinder apparatus, fractures are frequent at pressures higher than 40 kbar, and the belt-type apparatus is preferred (Figure 4).[17–19] It still offers a sufficiently large pressurized volume for the sample and internal resistance heating. The additional support given to the anvils by the surrounding

New Phases at High Pressure

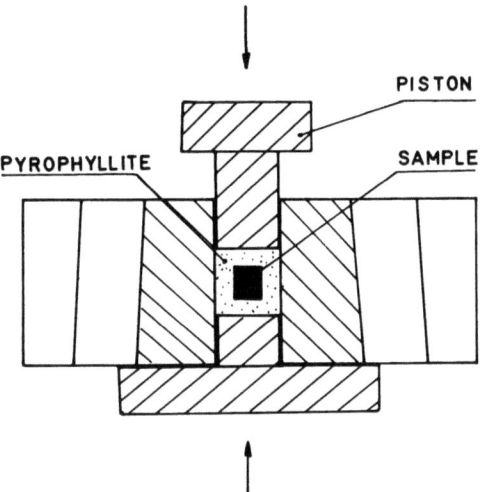

Fig. 3. The piston–cylinder apparatus. With graphite heaters temperatures of 1500–2000°C can be obtained at pressures up to 40 kbar with a usable volume of ~0.1 cm³.

hard steel ring allows pressures up to 150 kbar and temperatures up to 2000°C using a graphite heater.

Due to internal friction, a 400-ton press is needed in order to develop 100 kbar in a 0.1-cm³ capsule. Being simple to assemble and disassemble, this apparatus is particularly convenient for routine runs

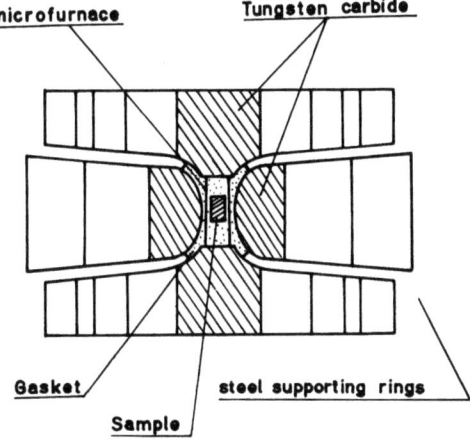

Fig. 4. The belt apparatus. With graphite heaters pressures up to 150 kbar and temperatures to 2000°C can be achieved.

Chapter 9

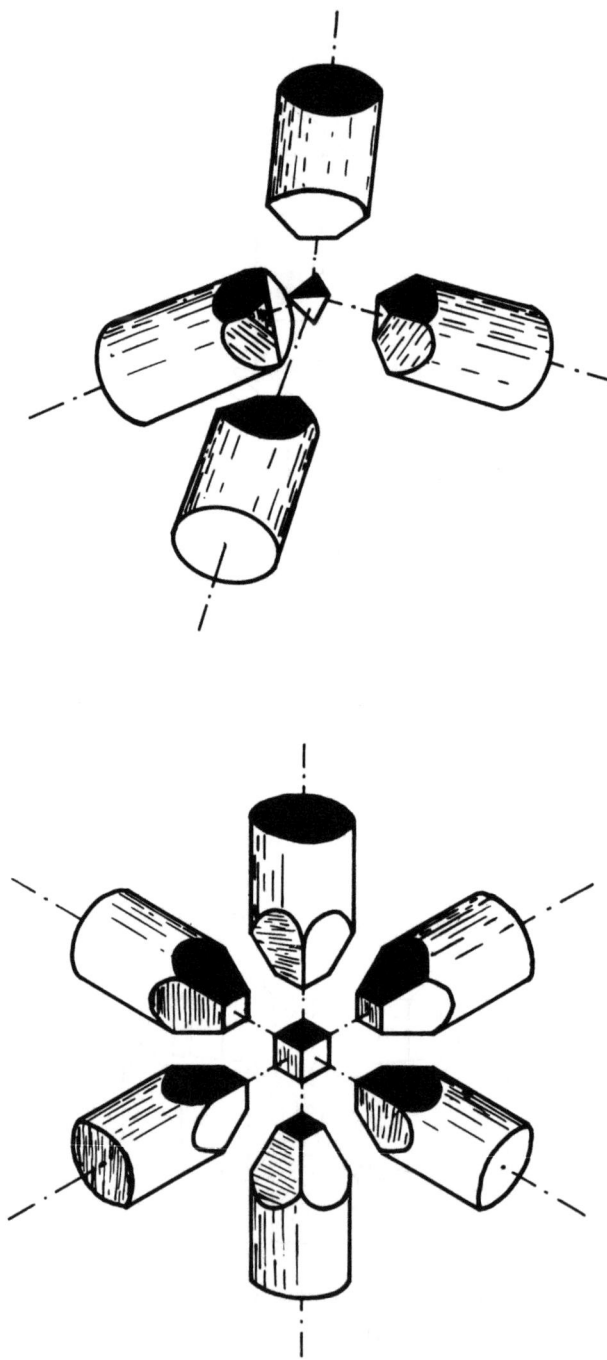

Fig. 5. Schematic drawing of the tetrahedral and the cubic anvil presses. Four or six independent pistons are pushed against each other by a hydraulic system.

in the 50–100 kbar pressure range; in addition, a rather homogeneous distribution of pressure can be achieved in the encapsulated sample.

2.2.3. The Tetrahedral Apparatus

A competitive although more sophisticated device is the tetrahedral anvil apparatus[20,21] represented in Figure 5 which can be operated in the same range of pressure and temperature as the belt system with comparable volume.

Two closely related facilities are the hexahedral press and the cubic press,[19,22] both having six anvils instead of four. Very few routine runs have been performed up to now with these types of apparatus, although the results obtained with the cubic press seem most encouraging.

2.2.4. The Bridgman Anvil Device

Static pressures ranging from 150 to 250 kbar can be achieved in the Bridgman anvils (Figure 6).[23] Two thin metal strips are used as microfurnace and sample container. The highest temperatures which can be reached are from 800 to 1000°C. This device is essentially

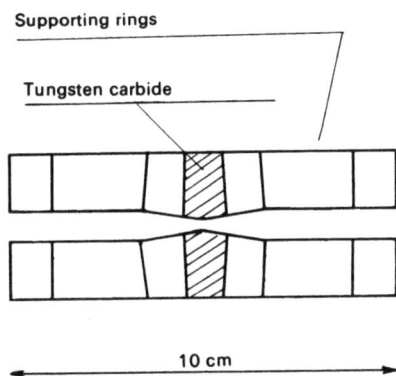

Fig. 6. The Bridgman anvils. Pressures ranging from 150 to 250 kbar can be obtained. Two thin metal strips are used as microfurnace and sample container. In this type of apparatus the sample is not in a sealed system. The highest temperature which can be reached is 800–1000°C.

used by geochemists, since milligram-size samples are inadequate for characterization of the products by methods other than powder X-ray diffraction.

2.2.5. Other Devices with Higher-Pressure Working Range

Static pressures up to 500 kbar can be achieved with the Drickamer anvils (Figure 7),[24] but such an apparatus is frequently inadequate for synthesis of inorganic materials due to the low-temperature limit (500°C) which can be obtained and the milligram size of the sample. Pressures in the range of 500–1000 kbar can only be achieved by shock-wave methods. This technique requires sophisticated facilities and thus is still not widely used. In addition, it seems that the difficulties in recovering the sample after the run as well as the hazards due to the manipulation of explosives have discouraged many chemists. The method still seems to hold some promise for the industrial synthesis of large amounts of sintered diamond and cubic boron nitride.

3. Empirical Rules Which Determine the Crystal Structure of Solids under Pressure

If we consider a model of identical spheres with a constant radius R, the fraction of total space occupied by such spheres is equal to 74% in the fcc and the hcp structures, but is only 68% in the bcc arrangement. The arrangement with tetrahedral coordination, found, e.g., in the structure of diamond, silicon, germanium, etc., is even less favorable with regard to space filling. Such considerations lead to the expectation that pressure will cause an increase in the coordination number. This has been observed in many elements, e.g., in graphite, which goes from three to four coordination (diamond), in silicon and germanium, which go from four to six coordination (white tin type), and iron, which goes from eight to twelve coordination (bcc → fcc).

Still, these simple considerations are only valid for a constant radius of the spheres and for equal compressibilities. In fact, it is easy to derive that an fcc and a bcc structure will have the same density if the radius of the spheres in the latter case is about 3% less than that in the fcc arrangement. Goldschmidt collected much experimental evidence that a decrease in coordination number is accompanied by a decrease in the interatomic distances.[25] In terms of the interatomic

New Phases at High Pressure

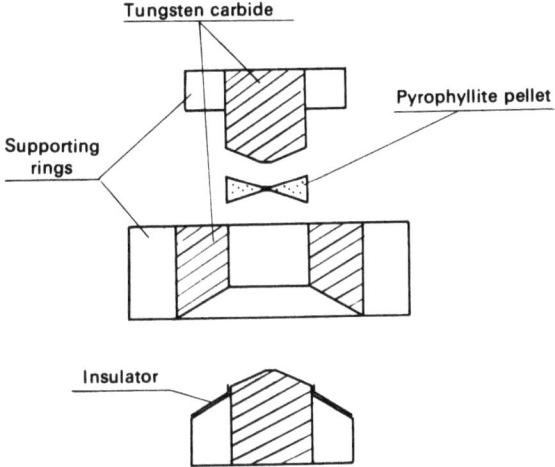

Fig. 7. The Drickamer anvils. Very small Bridgman-type anvils are supported by a pyrophyllite pellet and surrounding tungsten carbide rings.

distance in twelve coordination, he mentions a decrease of roughly 3% for eight coordination, 4% for six coordination, and 12% for four coordination. The effect of this dependence of radius on coordination is that the volume change of bcc to fcc transition is much smaller than would be expected from a hard-sphere model. Thus, in iron at the fcc to bcc transition temperature (916°C) the interatomic distance in the twelve-coordinated fcc structure is 2.578 Å, while in the eight-coordinated bcc, this distance is only 2.515 Å, a difference of about 2.5%. The expansion in volume on going from fcc to bcc is therefore only 1% instead of the 10% to be expected for rigid spheres.

A further consideration is that compressibility will depend on coordination number, being generally larger for smaller coordinations. Thus the fact that the compressibility in the bcc structure is greater than in the fcc structure may result in a difference in radius of even more than the 3% predicted by Goldschmidt. The density of the bcc phase will be greater and pressure-induced transitions from fcc to bcc will then be possible. As a matter of fact, many such transitions have been observed, particularly in transition elements.

In alloys and compounds, where two or more different elements are packed together in an ordered way, the situation is more complex, and one has to take into account the fact that the compressibilities of the various elements are different and will also depend on the type of chemical bonding. In general, compressibilities of large anions are thought to be greater than those of smaller cations. Such a situation leads, under pressure, to an increase of the radius ratio r_c/r_a and the stabilization of structures having increased coordination numbers. For instance, CdTe undergoes a pressure-induced phase transition from a four-coordinated zincblende structure type to a hexacoordinated rocksalt type[26]; HgTe and HgSe undergo a phase transition from the zincblende structure type to the hexacoordinated cinnabar structure type.[27] InSb goes from a four-coordinated zincblende structure to the hexacoordinated white tin structure type.[28] Coesite (SiO_2) with tetrahedral Si^{4+} undergoes a phase transition to a high-pressure rutile-type polymorph stishovite,[29] whereas MnF_2 with hexacoordinated rutile structure type goes to a 7–8-coordinated distorted fluorite type.[30] CaF_2 itself transforms into a nine-coordinated $PbCl_2$ structure type.[31] Shannon and Prewitt[32] listed a large number of high-pressure transitions accompanied by an increase in both cation and anion coordination and emphasized that the volume change of these transitions is generally considerably greater (10–40%) than those in which no coordination change exists (0–10%).

Although pressure-induced phase transformations of ionic compounds are often accompanied by an increase of the cation and anion coordinations, many transformations are known in which the cation and anion coordinations remain constant. This is, for instance, the case for the quartz to coesite transition, for the PbO_2 rutile to α-PbO_2 transition,[33] for the olivine to spinel transition (Fe_2SiO_4,[34,36] Co_2SiO_4,[35] Ni_2SiO_4[36,37]), etc. The only common trend among the transformations of ionic compounds is that the anion packing is more efficient in the high-pressure phase than in the low-pressure one.[38]

From the preceding considerations one can conclude that there is only one inviolable rule which all the pressure-induced phase transitions obey, namely the packing efficiency, and therefore the density of the high-pressure phase must be greater than that of the starting phase.[39] One could predict that the less efficiently packed a structure is to begin with, the greater will be the possibility of a pressure-induced phase change.

4. New Phases at High Pressure: Quenchable and Unquenchable Phases; Phase Diagrams under High Pressure

4.1. Quenchable and Unquenchable Phases

In order to increase the kinetics of a phase transformation of a compound, $(A_xB_y)_{LP} \to (A_xB_y)_{HP}$, one generally uses rather high temperatures, from about 700°C to 2000°C. If the sample can be cooled rapidly while being maintained under pressure, the dense phase can be "quenched" and recovered at normal conditions after the release of the pressure. Not all high-pressure phases can be "quenched" and recovered. In general before performing the experiment it is not possible to know if an expected dense phase will be preserved as a metastable phase under normal pressure and temperature conditions.

From empirical considerations based on the already extensive literature one can deduce some principles which help to predict and to understand the stability of the high-pressure phases.

First of all it seems that the more complex the chemical formula and the structure of the compound, the greater are the chances of recovering the dense form as a metastable phase. In such compounds one generally observes at the transition a reconstruction of the entire structure: Old bonds are broken and new bonds are established, while the relative positions of all the atoms are strongly modified. In order to return to their initial positions, the atoms have to jump over rather high potential barriers, and this generally requires considerable energy, No wonder then that some high-pressure phases have to be heated up to 1000°C before undergoing a phase transition to a less dense phase.

On the other hand, transitions in which atoms move relatively small distances without crossing will generally be difficult to quench, even in the cases where old bonds are broken and new bonds are formed with higher coordinations of the atoms. This is, for instance, the case for the six-coordinated metallic phases of germanium and silicon (white tin type)[40] which cannot be preserved when one releases the pressure. This is also the case for the monoclinic to tetragonal phase transition of ZrO_2.[41] In the monoclinic phase Zr atoms are heptacoordinated, while in the tetragonal phase with a distorted CaF_2 structure type they are octacoordinated. The density increase at the transition is about 7%. The transition can be induced by applying pressures higher than 40 kbar at 25°C. Still, since there exist no major changes in the relative distribution of the anions and cations, the tetragonal phase cannot be retained in a metastable form at ambient conditions.

Another example is the SnO I to SnO II transformation.[42] Under ordinary conditions stannous oxide has the PbO structure (tetragonal phase), each Sn cation having four close oxygen neighbors all located on the same side of the atom. At a pressure of 60 kbar a new phase can be observed of the wurtzite type with Sn cations located on regular tetrahedra. During the phase transformation the old bonds have been broken and new bonds have been established. Still, the process can occur by a rather limited displacement of the oxygen and the positions of the tin atoms undergo only minor changes. Again in that case the transformation is reversible and the high-pressure phase is unquenchable.

Even in the case of reconstructive transformations, the high-pressure phases may not be retained at normal pressure: This happens when the high-pressure phase can give a third phase without major rearrangements of the atoms which only slide against each other parallel to favorable crystallographic planes. For instance, rutile-type MnF_2 undergoes at 100 kbar a transformation to an eight-coordinated cubic fluorite type of structure; this structure transforms on the release of the pressure into the closely related α-PbO_2 type of structure, with octahedral coordination of the Mn atoms and an intermediate density between the starting phase and the high-pressure phase.[43-45] In such a case it is evident that if a new metastable phase has been synthesized using high-pressure techniques, a polycrystalline material will result because the single crystal will not be strong enough to resist the stress which appears at the phase transition.

4.2. Phase Diagrams under High Pressure

Phase equilibria in pure elements at high pressures have been studied rather well. The $P-T$ diagrams have been constructed for the majority of elements (see, for instance, Klement and Jayaraman[39,46]) and the crystal structures of polymorphic modifications formed at high pressure have been established in certain cases. Investigation of the effects of high pressures on phase equilibria in multicomponent systems (metallic, semiconductor, ionic compound) has only begun, even for binary systems. Most of the systems have been investigated in a rather limited range of compositions, generally the same as those giving definite compounds under normal conditions, in the search for high-pressure phases of these compounds.

It would appear that it is now time to investigate binary (and even ternary) phase diagrams not only with the goal of discovering high-pressure phases of compounds stable at normal pressures, but also of finding phases with different composition which require high-pressure conditions for their synthesis. In several cases which have been already carefully investigated both thermodynamic calculations and experimental data have shown that even comparatively low pressures can lead to drastic changes of the phase diagram.

Thus, for example, in the iron–carbon system[47] at a temperature of 1000°C a pressure of only 3–4 kbar shifts the equilibrium of the reaction

$$Fe_3C \leftrightarrow 3Fe + C$$

from the right to the left, leading to a considerable change in phase formation processes during annealing of both white and gray irons.

In addition, the use of high-pressure facilities allows chemists to investigate new phase diagrams in which one of the components, volatile in normal conditions, is condensed by the pressure and behaves as just another solid component. This is, for instance, the case of the Al_2O_3–H_2O phase diagram which leads to the compounds Al_5HO_8 and $Al(OH)_3$ and several phases with the AlOOH formula.[48,49] Another example is the phase diagram of boron–oxygen, which leads to the high-pressure synthesis of B_2O, the asymmetric anolog of graphite.[50] Many other examples can be found in the recent literature.

While physicists and especially thermodynamicists are concerned with problems such as the effect of pressure on the liquidus line and the phase transitions, solid-state chemists are more interested in the synthesis of new quenchable phases, with the object of using these phases as a source of new materials. In fact, the results published recently strongly support the idea that the systematic use of high-pressure conditions for the synthesis of new materials will yield a large number of new materials, some of them having interesting properties. From the large number of new phases which have been discovered even in the one-component systems, it becomes evident that many new phases will be discovered in most of the binary and ternary systems, probably several times more than the number which are known to exist under normal conditions. Perhaps if new compounds are particularly interesting, industrial concerns will spend money to develop high-pressure facilities large enough to ensure a wide-scale production, as in the case of diamond and cubic boron nitride.

4.3. Examples of High-Pressure Synthesis

We now give a few examples where high-pressure conditions have been successfully used for the synthesis of new materials. We first

Chapter 9

consider ionic compounds, oxides and fluorides; then we discuss a few examples of semiconducting compounds. chalcogenides, pnictides, and nitrides; and finally we discuss alloys in two-component systems.

4.3.1. Synthesis of Ionic Compounds under High-Pressure and High-Temperature Conditions

Ionic compounds are among the most extensively investigated compounds under high-pressure, high-temperature conditions; this is due essentially to the fact that the dense phases are easy to quench and generally have a rather high thermal stability.

4.3.1.1. Synthesis of Ionic Compounds Containing a Transition Metal Cation in an Unusually High Valence State or Containing a Volatile Component

When temperatures of several hundred degrees centigrade are required in order to yield well-crystallized products, oxygen or fluorine pressures of 4 or 5 kbar may not be sufficient to stabilize the high valence state of the cation in the expected material. For instance, one can see from the Cr_2O_3–CrO_2 and Mn_2O_3–MnO_2 phase diagrams represented in Figure 8[51–54] that synthesis of compounds containing Mn^{4+} or Cr^{4+} cations will require oxygen pressures of 15–20 kbar or even more as soon as the reaction temperature reaches 1000°C. In this case as well as in the synthesis of many compounds where at least one cation has a high valence state, any high-pressure, high-temperature apparatus of the piston–cylinder, belt, or tetrahedral type can be used. The decomposition reactions of oxygen or fluorine generators are used in much the same way as those in cold-seal hydrothermal vessels. The high-oxidation-state oxide is placed in the high-pressure apparatus and heated. It decomposes at some temperature to a lower-oxidation-state oxide with the release of oxygen. Since the sample volume is small and confined, the oxygen is in pressure equilibrium with the system and the measured pressure can be regarded as the oxygen pressure. This technique is subject to the same restrictions as other closed-system experiments: The apparatus must not leak; there must be no extraneous component in the vapor phase; the gas must not react with other parts of the system; and the gas in the pore volume must be in pressure equilibrium with the confining pressure. The last of these is a less serious problem because the

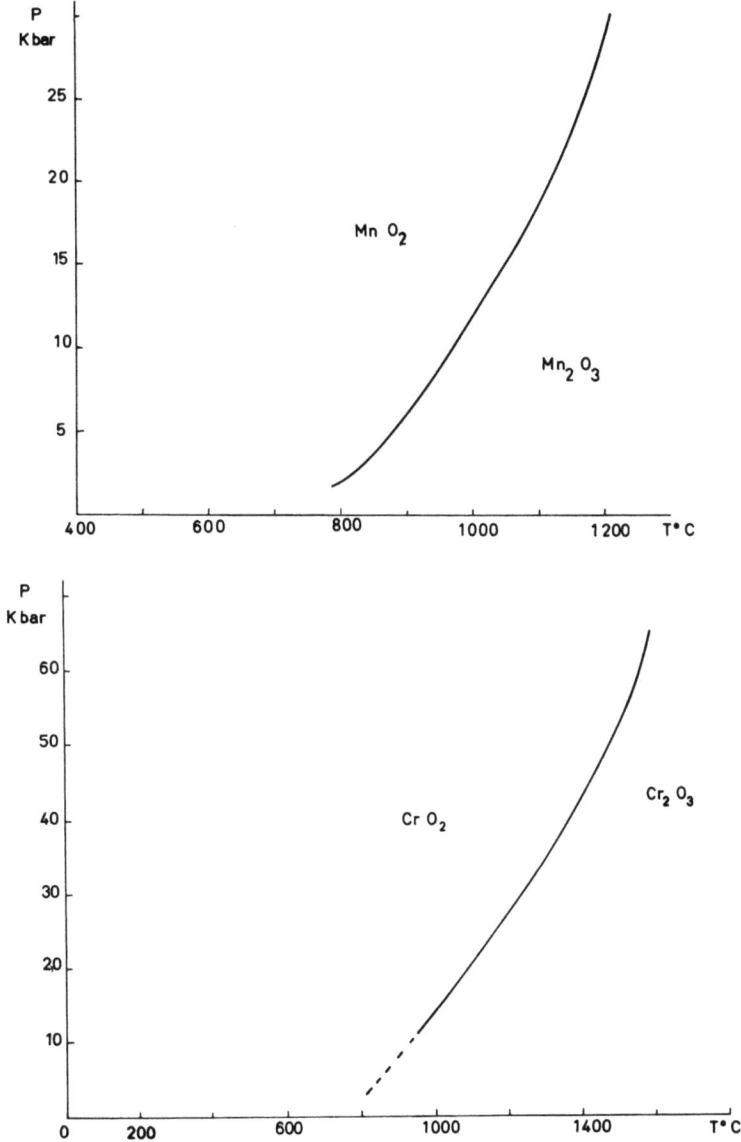

Fig. 8. The pressure–temperature diagrams for Mn_2O_3/MnO_2 and Cr_2O_3/CrO_2 systems.

high pressures generally exceed the crushing strength of the grains of the solid phases and pore volumes are very small.

Typical synthesis reactions are described below.

(a) Synthesis of Perovskite-Type Compounds Containing Fe^{4+}. In 1970 Kanamaru and co-workers[55] synthesized $CaFe^{4+}O_3$ at

Chapter 9

20 kbar and 1100°C in a piston–cylinder device using dry CrO_3 as an oxygen generator and $Ca_2Fe_2O_5$ as starting material. To prevent reaction of CrO_3 and $Ca_2Fe_2O_5$, a disk of ZrO_2 was inserted between them (Figure 9). After 1 hr a Mössbauer spectrum of the new perovskite compound revealed no trace of trivalent iron Fe^{3+}.

(b) Synthesis of Compounds Containing Trivalent or Tetravalent Nickel Cations. $Ln^{3+}Ni^{3+}O_3$ perovskite compounds (Ln^{3+} = lanthanide cation or Y^{3+}) have been synthesized recently by Demazeau et al.[56] in a belt apparatus at 1000°C and 65 kbar. The starting materials were nickel oxide, rare earth sesquioxide, and $KClO_3$ as the oxygen generator. Demazeau et al.[57] reported the synthesis of $LaCu^{3+}O_3$ using similar experimental conditions.

We have realized the synthesis of corundum $Ni^{3+}Cr^{3+}O_3$ at 1000°C and 80 kbar in a belt apparatus using nickel chromate $NiCr^{6+}O_4$ as starting material. The same compound was synthesized by Chamberland and Cloud,[58] who started from a mixture of CrO_2 and NiO at 65 kbar and 1000°C.[58] Previous attempts to synthesize this compound at moderate oxygen pressures were unsuccessful.[59] Magnetic measurements[60] indicate that the Ni^{3+} cations in this compound are in a rather unexpected high-spin state. $Li_2Ni^{4+}F_6$ and $Na_2Ni^{4+}F_6$ were prepared by Bougon[7] of Saclay Nuclear Center (France) starting from a mixture of NaF (or LiF) and NiF_2 with PtF_6 (or XeF_2) as fluorine generator at 600°C and a few kilobars total pressure.

(c) High-Pressure Synthesis of Co_2O_3. Two years ago we realized the first synthesis of cobalt sesquioxide Co_2O_3 as well-crystallized material with the corundum structure, performing the following solid-state exchange reaction under 80 kbar and 1000°C[61]:

$$2CoF_3 + 3Na_2O_2 \rightarrow Co_2O_3 + 6NaF + \tfrac{3}{2}O_2$$

In this new oxide of corundum type all Co^{3+} cations are in the low-spin state, as can be shown by magnetic measurements and X-ray analysis. When subjected for about 10 min to a temperature of 400°C under normal pressure conditions, the high-pressure phase undergoes a transition to a new corundum phase with all Co^{3+} cations in the high-spin state, according to the reaction

$$Co_2^{III}O_3 \text{ (low-spin)} \xrightarrow{400°C} Co_2^{3+}O_3 \text{ (high-spin)}$$

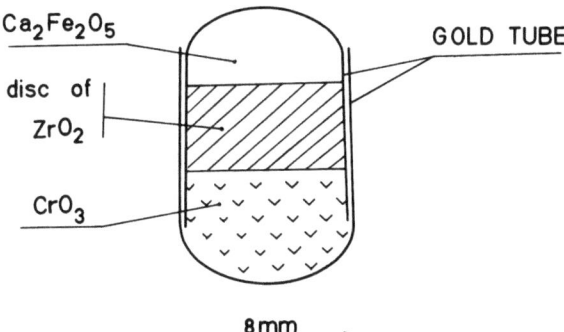

Fig. 9. Sample assembly for preparation of $CaFeO_3$.

The increase in volume is about 6.7% due to the change in radius of Co^{3+}. Exchange reactions similar to the preceding one should be possible for the synthesis of new compounds in more complicated systems.

(d) *Mixed Oxides Containing Pt^{4+}, Pd^{4+}, Rh^{4+}, Au^{3+} Cations.* $Tl_2^{3+}Pt_2^{4+}O_7$ with the pyrochlore structure was synthesized by Hoekstra and Siegel[62] in a tetrahedral anvil press. A mixture of PtO_2 and Tl_2O_3 was heated at 1000°C and 40 kbar for about 30 min. In spite of the fact that the new compound is thermally stable up to 750°C at atmospheric pressure, attempts to realize direct synthesis from Tl_2O_3 and PtO_2 at 520°C resulted only in the decomposition of PtO_2.

The Pt–O and Rh–O phase diagrams have been recently investigated by Muller and Roy,[4,5] who synthesized the compounds Pt_3O_4, β-PtO_2, RhO_2, and Au_2O_3 at intermediate oxygen pressures. The same authors synthesized the spinel phase $Mg_2Pd^{4+}O_4$[63] and the compound $Cd_2Pt^{4+}O_4$ with the Sr_2PbO_4 structure at 900°C and 200 atm oxygen pressure in a conventional cold-seal vessel.[64]

(e) *New Compounds Containing Cr^{4+} and Mn^{4+} Cations.* Recently several compounds containing tetravalent chromium were obtained at high pressure. Chamberland[65] synthesized several polymorphs of $BaCr^{4+}O_3$ in a tetrahedral press using BaO and CrO_2 as starting materials:

$$BaO + CrO_2 \xrightarrow{P > 3\ \text{kbar}} BaCr^{4+}O_3$$

Chapter 9

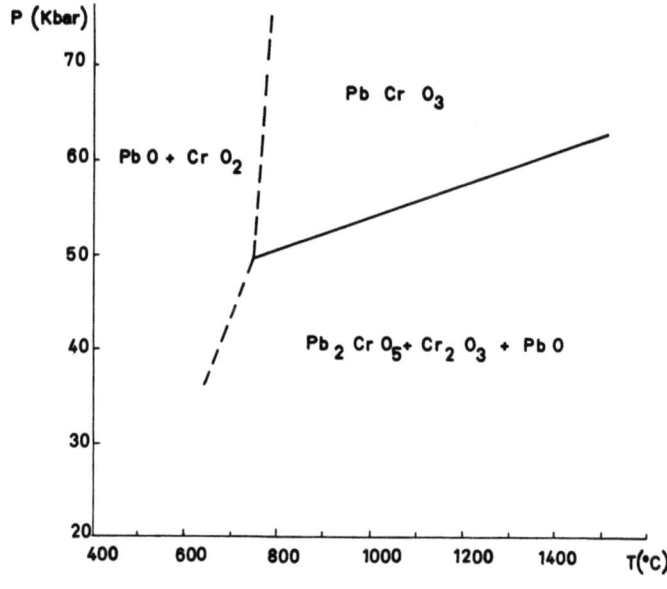

Fig. 10. P–T stability field of PbCrO$_3$.

Roth and De Vries[66,67] prepared PbCrO$_3$ of cubic perovskite type from CrO$_2$ and PbO at 50 kbar and 1150°C in a belt apparatus (Figure 10). Crystals of the same compound were obtained in a tetrahedral press at comparable pressure and temperature conditions by Chamberland and Moeller.[68] Goodenough et al.[69] and then Weiher et al.[70] prepared CaCr^{4+}O$_3$ with an orthorhombic distorted perovskite type of structure, starting from CaO and CrO$_2$ at 65 kbar and 700°C in a tetrahedral anvil press. Chamberland[71] synthesized SrCrO$_3$ at 65 kbar in a tetrahedral press by the action of SrO or Sr$_2$CrO$_4$ with CrO$_2$ at 800°C. The crystalline product has the cubic perovskite structure and was found to be a metallic conductor. Using the same starting materials, Kafalas and Longo[72] synthesized Sr$_3$Cr$_2^{4+}$O$_7$ in a belt by performing the reaction

$$3Sr_2Cr^{4+}O_4 + CrO_2 \xrightarrow[1000°C]{65 \text{ kbar}} 2Sr_3Cr_2^{4+}O_7$$

We would like to emphasize these results by pointing out that until a few years ago CrO$_2$ and Sr$_2$CrO$_4$ were the only two compounds known to contain Cr^{4+}. Now we already have ten or so.

Several BaMn^{4+}O$_3$ and SrMn^{4+}O$_3$ polytypes have been obtained by Chamberland et al.[73] at pressures ranging from 3 kbar (in hydrothermal vessels) to 65 kbar (in a tetrahedral press). Syntheses of several other Mn^{4+} compounds have been reported in the literature.

Compounds containing a volatile component generally start to decompose at temperatures of a few hundred degrees centigrade. For that reason single crystals of such phases are almost impossible to grow and quite often these compounds are only known as quasi-amorphous or at least poorly crystallized materials. Use of high pressure can considerably increase the temperature stability of the compounds, enabling one to prepare well-crystallized samples and sometimes even small single crystals.

For instance, oxyhydroxides of metals of the first transition metal series start to decompose at temperatures of 200–400°C. When prepared under normal pressures they are always poorly crystallized. Subjected to water pressures of 80 kbar, several of them start to decompose at temperatures above 1000°C, like the manganese oxyhydroxide MnOOH, the chromium oxyhydroxide CrOOH, and so on. We synthesized at Grenoble most of these compounds at 600°C and 80 kbar (Table 1).[74] The products were always well-crystallized materials, some of them containing crystals of 0.1 mm size, a few of them being suitable for determination of some physical properties.

Many other new compounds should be possible by this method, which we like to refer to as the "confined material under high gas

TABLE 1
Experimental Conditions for High-Pressure Synthesis of Some Oxyhydroxides

Starting material	Pressure, kbar	Temperature, °C	Product	Structure-type
Sc$_2$O$_3$	80	600	ScOOH	InOOH
V$_2$O$_3$	80	600	VOOH	InOOH
Cr$_2$O$_3$	80	600	CrOOH	InOOH
Mn$_2$O$_3$	80	600	MnOOH	Manganite
Fe$_2$O$_3$	80	500	FeOOH	InOOH
CoOOH	100	600	CoOOH	Diaspore
NiOOH	80	600	NiOOH	InOOH
Ga$_2$O$_3$	100	600	GaOOH	Diaspore
Al$_2$O$_3$	100	600	AlOOH	Diaspore
Rh$_2$O$_3$	80	600	RhOOH	InOOH
In$_2$O$_3$	80	600	InOOH	InOOH

pressure" method. Different gases can be used, i.e., oxygen, fluorine, chlorine, bromine, iodine, nitrogen, ammonia, CO_2, SH_2, SO_2, etc.

4.3.1.2. Synthesis of High-Pressure Polymorphs of Ionic Compounds Having a Stable Phase at Normal Pressures

(a) *Phase Transitions Accompanied by an Increase of Cation Coordination.* The search for high-pressure polymorphs has always been the most thoroughly investigated and the most fruitful field in high-pressure research on solids. We already mentioned that quite often the coordination of at least some of the cations increases under pressure. We can thus expect that in many cases the coordination of one kind of cation in the high-pressure phase will be the same as the coordination adopted at normal pressure by the heavier cation from the same group of the periodic system, the corresponding structures being isotypic.

Let us give a few examples of such new metastable phases prepared under high-pressure and high-temperature conditions. Particularly interesting is the synthesis of stishovite, the rutile-like high-pressure phase of SiO_2, containing octahedral Si^{4+}, first prepared by Stishov and Popova.[29] Stishov and Belov[75] investigated the structure and showed that the arrangement of the atoms is the same as in the compounds GeO_2, SnO_2, and PbO_2 at normal pressure. This new phase is stable at pressures above 100 kbar and temperatures above 900°C. After the announcement of the synthesis by Stishov, a mineral of similar composition and structure was discovered by Chao *et al.*[76] in the Arizona meteor crater; from all evidence, such samples were submitted to high-pressure and high-temperature conditions during the shock of meteorite impact. The increase of density between coesite—a dense phase of SiO_2 containing only tetrahedral $^{VI}Si^{4+}$—and stishovite—with octahedral $^{IV}Si^{4+}$—is from 2.95 to 4.28, namely about 45%! Pressure–temperature stability fields for the coesite–stishovite transition were determined by Akimoto and Syono,[77] who showed that the transition pressure increases linearly with the temperature, from about 80 kbar at 400°C to 110 kbar at 1000°C. More recently high-pressure phases of several silicates were shown to contain octahedral silicon. This is, for instance, the case for the synthetic potassium feldspar $KAlSi_3O_8$, which undergoes a polymorphic transition at 120 kbar and 900°C into the hollandite structure, with randomized Al^{3+} and Si^{4+} in octahedral

TABLE 2
Polymorphism of CaB_2O_4

		Symmetry	Stability,* kbar	$-\Delta V/V$, %
I	$^{VIII}Ca^{III}B_2^{III}O_2^{IV}O_2$	Orthorhombic	0–12	6.5
II	$^{VIII}Ca^{III}B^{IV}B^{III}O^{IV}O_3$	Orthorhombic	12–15	5.7
III	$^{VIII}Ca_2^{X}Ca^{III}B_2^{IV}B_4^{III}O^{IV}O_{10}^{V}O$	Orthorhombic	15–25	11.5
IV	$^{XII}Ca^{IV}B_2^{V}O_4$	Cubic	25–40	

* Approximate range of stability at 900°C.

positions[78]; the density increases from 2.55 to 3.84 g/cm^3. Many other silicates should undergo similar phase transitions to high-pressure polymorphs containing at least part of the Si^{4+} cations on octahedral sites.

Many of the synthetic or natural borates contain planar BO_3 groups with the B^{3+} cations located at the center of a regular triangle of oxygen. However, a number of examples are known in which boron occurs on tetrahedral sites. Thus one can predict that under increasing pressures new phases will appear containing less and less boron with planar triangular coordination. Such a behavior has already been found in many examples: For instance, hexagonal B_2O_3 undergoes at about 20 kbar and 500°C a transition to an orthorhombic phase with only tetrahedral boron and a density increase of about 27%.[79,80] Calcium metaborate CaB_2O_4 was shown by Marezio et al.[81] to undergo three allotropic transitions at different pressures between 0 and 40 kbar. Both boron and calcium coordination numbers increase with increasing pressure, as represented in Table 2.

$NiFe^{3+}BO_4$ with a warwickite-type structure undergoes at 40 kbar and 900°C a transition to a new phase with an olivine-type structure and tetracoordinated boron.[82] There is little doubt that many other dense phases of borates will be discovered in the near future. Tetravalent germanium adopts both tetrahedral and octahedral coordination in oxides. We can thus predict that high-pressure phases containing $^{VI}Ge^{4+}$ will be easier to obtain than in the case of isomorphous silicates. For instance, feldspar $KAlGe_3O_8$ transforms into a new phase possessing the hollandite structure[83] at about 30 kbar and 900°C as compared with 120 kbar and 900°C for isomorphous $KAlSi_3O_8$.[78] Germanium albite $Na^{IV}(AlGe_3)O_8$ and $Rb^{IV}(AlGe_3)O_8$ were also transformed to a hollandite form.[84]

Nepheline $^{IV}Na^{IV}AlGeO_4$ transforms at 120 kbar and 900°C into a new phase $^{VIII}Na^{VI}AlGeO_4$ with the calcium ferrite structure $^{VI}Fe_2{}^{VIII}CaO_4$). The density increase is about 38%.[85] $^{VI}Mn_2{}^{IV}GeO_4$ of the olivine group transforms at 120 kbar and 900°C into a new phase, $^{VI}Mn_2{}^{VI}GeO_4$ with the strontium plumbate $^{VI}Sr_2{}^{VI}PbO_4$ structure type. The density increase is about 17%.[86] $^{VI}Ca_2{}^{IV}GeO_4$, another compound of the olivine group, transforms at 110 kbar and 900°C into a new phase having the $^{IX}K_2{}^{VI}NiF_4$ structure, with a density increase of about 25%. In this new phase Ca and Ge cations have respectively nine and six oxygen neighbors.[87]

Rare earth pyrogermanates $^{VI}Ln_2{}^{IV}Ge_2O_7$, some of them having the thortveitite structure, crystallize into a pyrochlore type of structure when submitted to pressures of 65 kbar and temperatures of 1100°C, as shown by Shannon and Sleight,[88] who considerably extended the field of stability of the pyrochlore structure (Figure 11). According to the authors, it seems logical to assume that by the application of sufficient pressure one could make pyrochlores of the type $Ln_2Si_2O_7$, at least with Ln = Sc, In, and some rare earth having a small radius.

$^{VII}Ca^{IV}GeO_3$ with the pseudowollastonite structure transforms at 700°C and 70 kbar into a new phase with a distorted garnet structure, with part of the Ge^{4+} cations on octahedral sites: $^{VIII}Ca_3{}^{VI}(Ge, Ca)^{IV}Ge_3O_{12}$.[89] When the pressure is increased to 120 kbar and the temperature to 900°C, complete conversion to an orthorhombic distorted perovskite polymorph occurs, with only octahedral coordination of Ge^{4+} cations: $^{VIII}Ca^{VI}GeO_3$.[90] A similar behavior is observed for $CdGeO_3$.[90,91] $Sr^{IV}GeO_3$ and $Ba^{IV}GeO_3$ with the pseudowollastonite structure transformed at 50 kbar and 95 kbar, respectively, into cubic perovskite phases, with density increases of 46% for $^{XII}Sr^{VI}GeO_3$ and 40% for $BaGeO_3$.[92] $^{VI}Mn^{IV}GeO_3$ having an orthopyroxene structure transforms at 60 kbar and 700°C into an ilmenite structure $^{VI}Mn^{VI}GeO_3$,[93] while rhodonite $^{VI}Mn^{IV}SiO_3$ transforms at 120 kbar and 900°C into the garnet $^{VIII}Mn_3{}^{VI}(Mn, Si)^{IV}Si_3O_{12}$ phase.[90]

Some structures are known to be particularly stable under high-pressure conditions. This is the case for the perovskite, the strontium plumbate Sr_2PbO_4, and for the calcium ferrite $CaFe_2O_4$ structure types. We have already mentioned the transition from wollastonite to perovskite in the compounds $CdGeO_3$ and $CaGeO_3$. The ilmenite type $^{VI}Cd^{VI}TiO_3$ transforms at 300°C and a few kbar pressure into an orthorhombic distorted perovskite $^{VIII}Cd^{VI}TiO_3$.[94] Recently Syono et al.[95] found a similar phase transition in $Mn^{2+}V^{4+}O_3$ compound

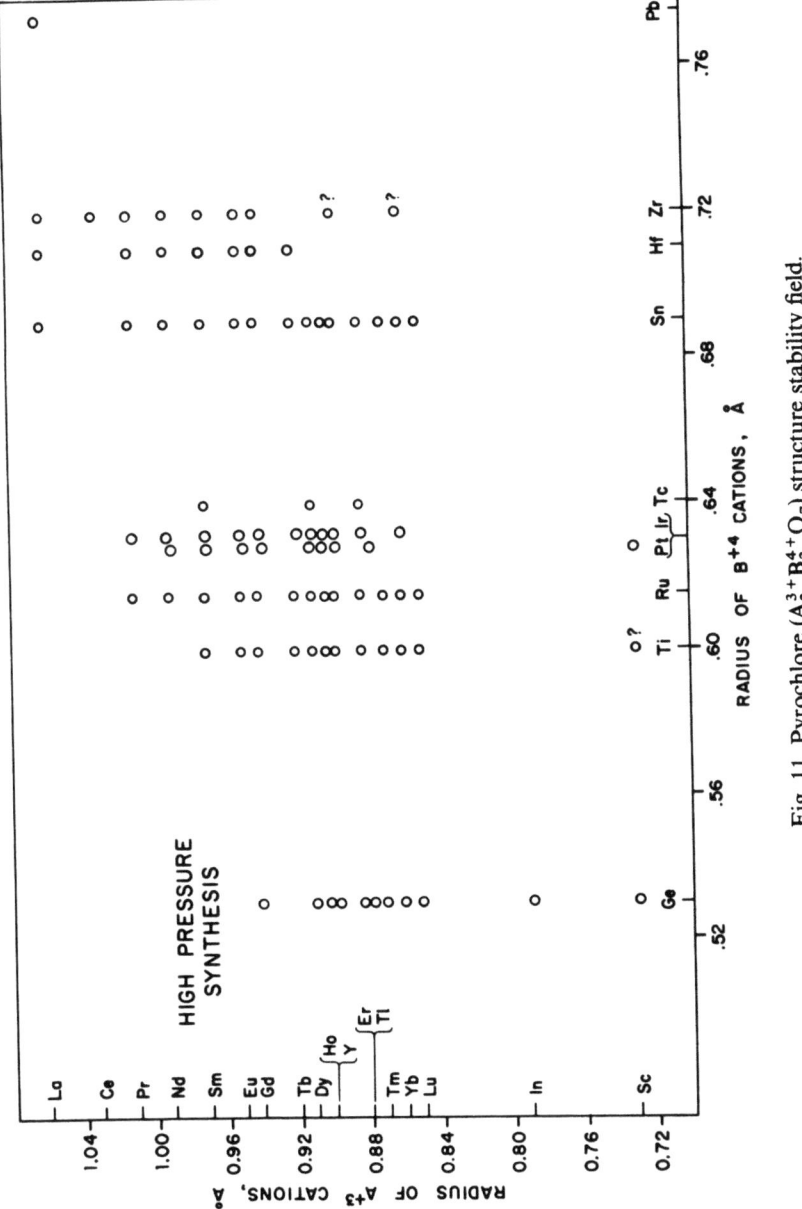

Fig. 11. Pyrochlore ($A_2^{3+}B_2^{4+}O_7$) structure stability field.

Chapter 9

$$^{VI}Mn^{VI}VO_3 \text{ (distorted ilmenite)} \xrightarrow[1000°C]{45 \text{ kbar}} {}^{VIII}Mn^{VI}VO_3 \text{ (GdFeO}_3 \text{ type)}$$

Waintal and Chenavas[96] have shown that several rare earth manganites $^{VII}Ln^{V}MnO_3$ of hexagonal symmetry also transform into a denser phase $^{VIII}Ln^{VI}MnO_3$ of distorted perovskite type.

Transformations from the spinel structure to the strontium plumbate type were shown to occur in Mn_2TiO_4, Mn_2SnO_4, Zn_2TiO_4, and Zn_2SnO_4.[97] Transformations into the calcium ferrite $CaFe_2O_4$ type were found for spinels $CdFe_2O_4$ and $CdCr_2O_4$,[97] while hausmanite Mn_3O_4 transforms into a $CaMn_2O_4$ type of structure[98]

$$^{VI}Mn_2{}^{IV}MnO_4 \text{ (distorted spinel)} \xrightarrow[\Delta V/V \sim 10\%]{900°C, 100 \text{ kbar}} {}^{VI}Mn_2{}^{VIII}MnO_4 \text{ (Mn}_2CaO_4 \text{ type)}$$

(b) *Phase Transitions with No Change of the Cation Coordination.* When the applied pressure is not too high and when the cations are small in comparison to the site they occupy, pressure phases can be obtained with no change in coordination.

This is the case for the quartz and the coesite phases of SiO_2, which can both be obtained from the less dense cristobalite structure:

$$^{IV}SiO_2 \text{ (cristobalite)} \xrightarrow{2 \text{ kbar}} {}^{IV}SiO_2 \text{ (quartz)} \xrightarrow{30 \text{ kbar}} {}^{IV}SiO_2 \text{ (coesite)}$$

In this case the increase in density, about 20%, is essentially due to the increase of the packing efficiency of the anions.

The olivine to spinel high-pressure induced phase transition shows a similar increase in efficiency of the oxygen packing, which goes from distorted hcp type to a rather perfect ccp type,[34-37]

$$^{VI}Fe_2{}^{IV}SiO_4 \text{ (olivine type)} \xrightarrow[\Delta V/V = 9\%]{} {}^{VI}Fe_2{}^{IV}SiO_4 \text{ (spinel type)}$$

Another example is the zircon to scheelite transition occurring in several rare earth arsenates and vanadates[98] as well as in zirconium silicate and zirconium germanate[98]

$$^{VIII}Ho^{IV}VO_4 \text{ (zircon ZrSiO}_4 \text{ type)} \xrightarrow{\Delta V/V = 9\%} {}^{VIII}Ho^{IV}VO_4 \text{ (scheelite CaWO}_4 \text{ type)}$$

$$^{VIII}Zr^{IV}SiO_4 \text{ (zircon type)} \xrightarrow{\Delta V/V = 8\%} {}^{VIII}Zr^{IV}SiO_4 \text{ (scheelite type)}$$

A rather general trend has been observed in several transformations of ionic compounds: When several consecutive phases of higher and higher densities appear with increasing pressure the space distribution of the cations of small size tend to be more and more symmetric. For instance, when highly charged cations are located on the octahedral sites of a quasi-close packing of anions, and when these octahedra share faces or at least several edges, the transitions induced

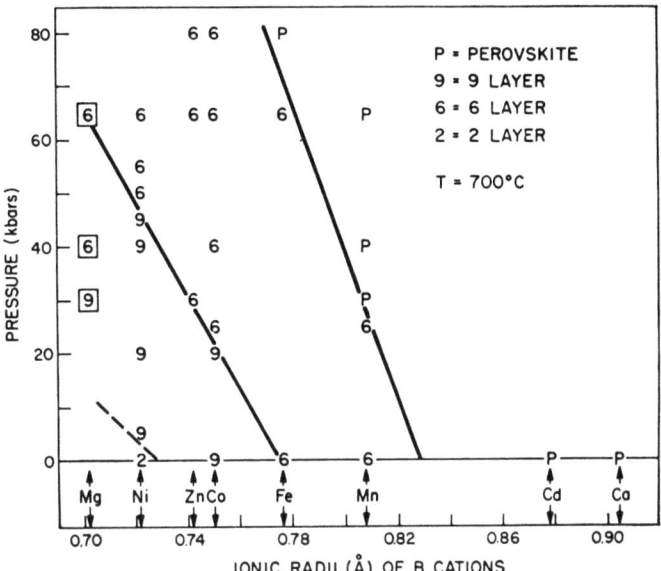

Fig. 12. Structural phase diagram of CsBF$_3$ compounds.

by pressure will tend to redistribute the small cations in order to decrease the electrostatic repulsion. Generally, the structure will be modified in order to prevent octahedra surrounding cations from sharing faces or from having too many edges in common. A good example is furnished by the compounds CsBF$_3$, where B = Mn, Fe, Co, Ni, Zn, Mg. They occur with four related structures which transform to one another with increasing pressure. These structures have different proportions of cubic and hexagonal stacking of close-packed CsF$_3$ layers and in all cases the proportion of cubic stacking increases with pressure. A structural phase diagram of CsBF$_3$ compounds is represented in Figure 12.[100] The two-, four-, six-, and nine-layer structures have a repeat distance of, respectively, two, four, six, and nine CsF$_3$ layers along the c axis. The cubic perovskite structure has a repeat distance of three CsF$_3$ layers along the same direction. Figure 13 shows the chains of octahedra through the successive CsF$_3$ layers.[100,101] Considering, for instance, the compound CsNiF$_3$, it occurs with the 2L structure under normal conditions (Figure 12). In the 2L structure octahedra share two opposite faces, building infinite chains in the c direction. When submitted to high-pressure conditions these chains are no longer stable and several polymorphic phases start to appear such as the 9L, 4L, and 6L types and others

Chapter 9

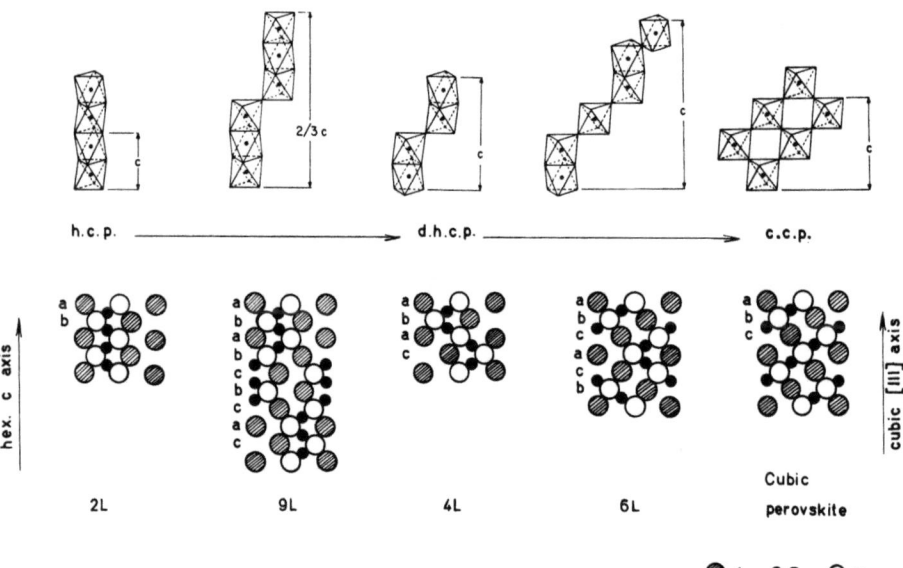

Fig. 13. Octahedra and cation–anion networks in $CsBF_3$ compounds.

more complicated, in which fewer and fewer octahedra share faces. Eventually under very high pressure, probably more than 150 kbar, a perovskite phase of $3L$ type will be stable in which Ni^{2+} octahedra will only share corners and no edges: Then the spatial distribution of the small octahedral cations will be very symmetric. Other compounds

TABLE 3
Allotropic Phases of $M^{3+}OOH$

Ionic radii[105,106] M^{3+}, Å	Formula	Boehmite type	$NaHF_2$ type	Diaspore type	InOOH type
0.525	CoOOH	—	×	×	—
0.53	AlOOH	×	—	×	—
0.60	NiOOH	—	×	—	×
0.615	CrOOH	×	×	—	×
0.62	GaOOH	—	—	×	—
0.64	VOOH	—	—	×	×
0.645	FeOOH	×	—	×	×
0.65	MnOOH	—	—	×	×*
0.67	RhOOH	—	—	—	×
0.73	ScOOH	×	—	×	×
0.79	InOOH	—	—	—	×

* Monoclinic distortion.

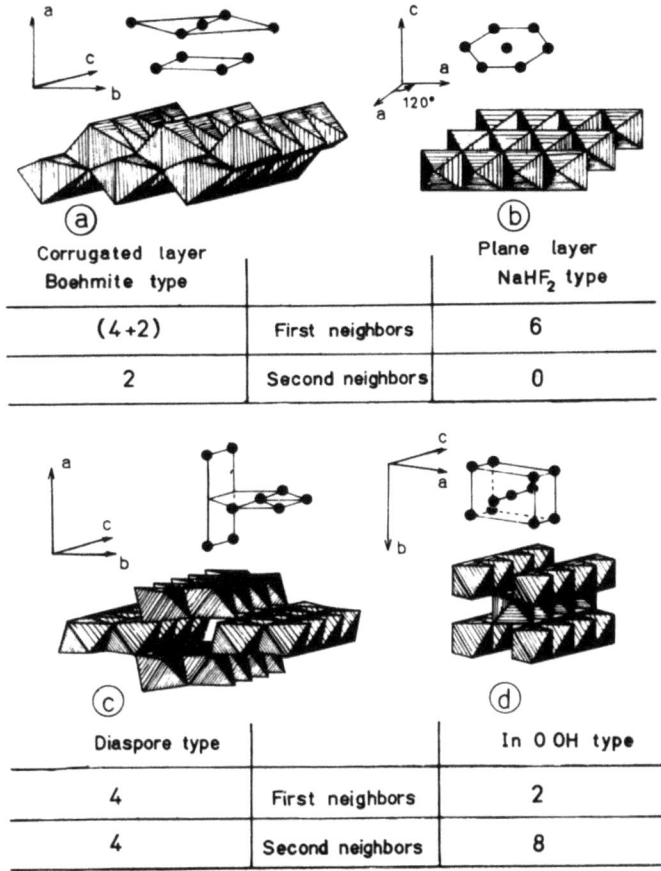

Fig. 14. Octahedra network in boehmite, diaspore, and InOOH structures.

are known to behave similarly, e.g., the $BaMnO_3$ and the $SrMnO_3$ polytypes,[73] the $Ba_{1-x}Sr_xRuO_3$ series,[102] $BaFeO_3$ and $SrFeO_3$,[12] $BaCrO_3$,[65] $RbFeCl_3$,[103] and $CsMnCl_3$ and $RbMnCl_3$ polytypes.[104]

We found similar behavior in the high-pressure phases of the $3d$ transition metal oxyhydroxides $M^{3+}OOH$.[74] Several of these compounds crystallize at normal pressure in the boehmite structure. Under high pressures they crystallize first in the diaspore-type structure and then in the InOOH type (Table 3), which is an orthorhombic distorsion of the rutile structure.[100,108] In Figure 14 one can find the main features of these three different types of structures: In the boehmite two-dimensional infinite puckered layers of octahedra are built by the combination of infinite one-dimensional octahedral chains sharing edges with the two adjacent chains. The puckered layers are bonded together by hydrogen bonds. Each trivalent cation has eight identical close neighbors located on the same side with oxygen only on the other side (and hydrogen bonds between the two layers). In the diaspore-type structure double chains of octahedra are linked by sharing only vertices. Each trivalent cation still has eight identical close neighbors but is located in a more symmetric way in space. Finally, in the distorted rutile type of InOOH single chains of octahedra are linked together by sharing vertices. Each trivalent cation has ten identical close neighbors located in a still more symmetric cation network.

4.3.1.3. High-Pressure Synthesis of Compounds Having No Stable Phase under Normal Pressure Conditions

Some of the compounds which are known to be stable under normal conditions transform into a denser phase at high pressure. Others disproportionate into a mixture of new phases if the average density of that mixture is higher than that of the starting phase. For instance, $LiFe_5O_8$ (spinel structure) decomposes into $LiFeO_2$ (disordered NaCl) and α-Fe_2O_3; the corresponding density increase is about 6%.[38] The spinels Mg_2TiO_4, Fe_2TiO_4, and Co_2TiO_4 disproportionate at pressures up to 40 kbar and temperatures 800–1500°C into mixtures of MO (rocksalt) and $MTiO_3$ (ilmenite type).[109] The spinels $MnAl_2O_4$, $FeAl_2O_4$, $NiAl_2O_4$, $CoAl_2O_4$, and $NiFe_2O_4$ were observed to disproportionate completely into their oxide components at 120 kbar.[97] $^{IV}Mn^{VI}V_2O_4$ transforms to ^{VI}MnO and $^{VI}V_2O_3$ and Li_2NiF_4 into NiF_2 (rutile) + LiF.[110] Synthetic garnets such as $^{VIII}Y_3{}^{VI}Fe_2{}^{IV}Fe_3O_{12}$ and isotypic compounds containing rare earth and aluminum or gallium were shown by Marezio et al.[111] to decompose into YMO_3 (M = Fe,Ga,Al) with the $GdFeO_3$ perovskite structure[112] and the sesquioxides M_2O_3 with the corundum structure. In Table 4 one can see that the transformation is always accompanied by an increase of the average cation coordination.

Some well-known compounds seem to have no dense phase stable under high pressure; conversely, many new metastable compounds have been found to have no stable phase under normal pressure. For instance, in the diagram Al_2O_3–B_2O_3, two compounds are stable under normal pressure, namely $9Al_2O_3$–$2B_2O_3$ and $2Al_2O_3$–B_2O_3.[113]

TABLE 4
High-Pressure Decomposition of Compounds Showing the Increase of Cation Coordination

$^{VI}(Fe_3Li)^{IV}Fe_2O_8$ (spinel) → $^{VI}Li^{VI}FeO_2$ (NaCl) + $2^{VI}Fe_2O_3$ (corundum)
$^{VI}(CoTi)^{IV}CoO_4$ (spinel) → $^{VI}Co^{VI}TiO_3$ (ilmenite) + ^{VI}CoO (NaCl)
$^{VI}V_2^{IV}MnO_4$ (spinel) → $^{VI}V_2O_3$ (corundum) + ^{VI}MnO (NaCl)
$^{VI}(NiFe)^{IV}FeO_4$ (spinel) → $^{VI}Fe_2O_3$ (corundum) + ^{VI}NiO (NaCl)
$^{VIII}Y_3^{VI}Fe_2^{IV}Fe_3O_{12}$ (garnet) → $^{VIII}Y^{VI}FeO_3$ (GdFeO$_3$, distorted perovskite) + $^{VI}Fe_2O_3$ (corundum)

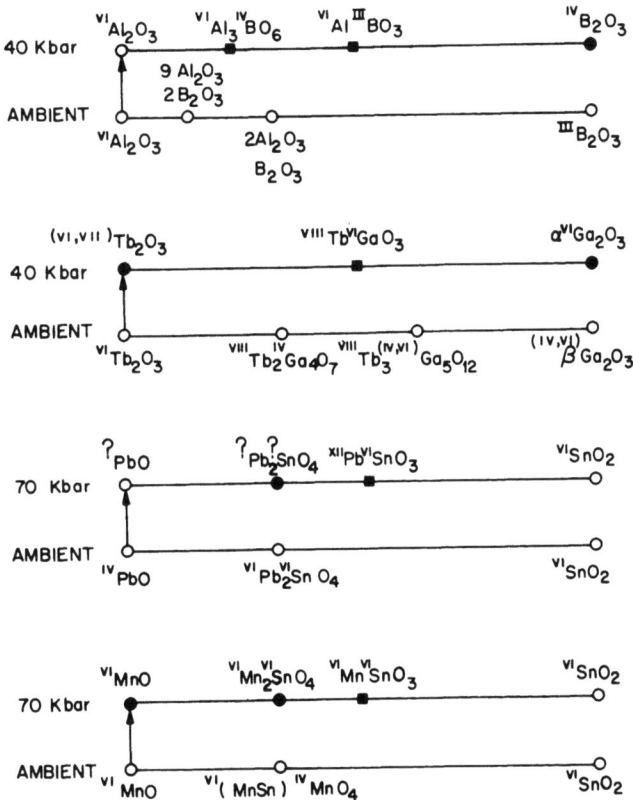

Fig. 15. Influence of pressure on phase diagrams. Full circles and squares represent high-pressure phases. Full and open circles are used when one can find two allotropic phases (at high and normal pressure, respectively) having the same composition. A full square is used when a new compound is found to be stable at high pressure without any corresponding phase at ambient pressure.

We reinvestigated the phase diagram at 40 kbar and found two new compounds, namely Al_3BO_6 and $AlBO_3$.[114] The latter compound has the calcite structure and was obtained as transparent regular rhombohedral crystals about 1 mm in size having strong birefringence. We found no sign of existence of the two compounds stable at normal pressure. Probably they are not stable at 40 kbar.

In the phase diagram Ln_2O_3–Ga_2O_3, where Ln^{3+} stands for a small rare earth ion, two compounds are stable at normal pressure: $Ln_4Ga_2O_9$ with tetrahedral Ga^{3+} and $Ln_3Ga_5O_{12}$ with the garnet structure. Under high pressure none of these compounds is stable and at least one new compound, $LnGaO_3$ with the perovskite structure, is known to be stable.[112] In the PbO–SnO_2 phase diagram at room pressure only the compound Pb_2SnO_4 appears as a stable phase.[115] Above 70 kbar and 400°C a new compound $PbSnO_3$ is formed which can be preserved as a metastable phase at normal pressure by quenching. This compound has a distorted perovskite structure.[116]

In the SnO_2–MnO phase diagram only the inverse spinel Mn_2SnO_4 is stable under normal pressure. When a mixture of composition $MnSnO_3$ is submitted to 60–70 kbar and 1000°C for a few minutes a new phase is formed with a corundum-type structure (disordered ilmenite).[117] In the SiO_2–FeO 1-atm phase diagram only the spinel Fe_2SiO_4 is known, while the $FeSiO_3$ pyroxene type is not known as a pure mineral and cannot be synthesized at atmospheric pressure. It is readily synthesized, however, at pressures from 18 to 45 kbar and temperatures from 1150 to 1400°C.[118] Figure 15 shows the influence of the pressure on several phase diagrams.

Many new compounds have been synthesized recently in systems where previously no intermediate phases were known: Generally the major effect of the pressure is to force cations to adopt a high coordination and to occupy sites in which they do not go naturally, because of their small size or their particular electronic configuration. Among those compounds that require high-pressure conditions for their synthesis are the perovskites $^{VIII}Bi^{VI}MnO_3$ and $^{VIII}Bi^{VI}CrO_3$,[119,120] the ilmenites $^{VI}Cu^{VI}VO_3$,[121] $^{VI}Co^{VI}VO_3$,[121] $^{VI}Mg^{VI}MnO_3$,[122] and $^{VI}Zn^{VI}MnO_3$,[122] the garnet $^{VIII}Co_3{}^{VI}Al_2{}^{IV}Si_3O_{12}$,[123] the perovskite $^{VIII}In^{VI}CrO_3$,[124] the compounds $^{VI}Cu^{VI}NbO_3$ and $^{VI}Cu^{VI}TaO_3$ of $LiNbO_3$ type,[125] the oxyfluorides VOF and $TiOF$ of rutile type,[126] the cubic perovskite $KTiO_2F$,[127] and stoichiometric FeO.[128]

4.3.2. High-Pressure Synthesis of New Semiconducting Phases

In contrast with the chemistry of oxides and halogenides which have been extensively investigated, the chemistry of chalcogenides, nitrides, pnictides, and other nonmetallic compounds has only been developed during the last ten years, and many basic systems have not yet been investigated. It is no wonder then that only a few syntheses of new materials have been performed under pressure, while so many binary and ternary compounds still remain to be discovered under normal conditions.

Nevertheless, there is no doubt that once the ambient-pressure phases have been investigated, high-pressure synthesis of new, dense phases will become of considerable interest, as is now the case with oxides. A few systems have been thoroughly investigated, like the ZnS–MnS[129] and the ZnS–FeS[130] systems, with the discovery of several new high-pressure phases. The compounds $^{IV}Ag^{IV}InTe_2$, $^{IV}Cu^{IV}InTe_2$, $^{IV}Ag^{IV}InSe_2$, and $^{IV}Ag^{IV}InS_2$ were shown to transform from the chalcopyrite structure to denser phases. In most of the cases the cation coordination increases from four to six.[131] Other chalcopyrite phases undergo phase transitions, like $^{IV}Cd^{IV}SnAs_2$, which transforms at 70 kbar and 700°C into a metastable disordered NaCl-type structure,[132] and $^{IV}Cd^{IV}GeAs_2$, which transforms into a metastable tetragonal phase, both compounds having superconducting properties at low temperatures.[133] A high-pressure polymorph of $FeCr_2S_4$ and some other compounds have been synthesized.[134] A few transition metal pnictides have been synthesized under pressure (65 kbar): CoP_2 of arsenopyrite type and $MnCoP_4$ of marcasite type.[135] The search for new nitrides has just begun: Of special interest is the synthesis of large amounts of sintered cubic boron nitride, a promising material for cutting tools.[136] For nitride compounds we mention the synthesis of $MgSiN_2$ at 30 kbar and 1200°C from a mixture of Mg_3N_2 and Si_3N_4[137] and the synthesis of the high-pressure phase of the compound Li_3BN_2.[138]

Several new metastable rare earth compounds have been synthesized by the high-pressure technique, e.g., the diantimonides $GdSb_2$, $TbSb_2$, $DySb_2$, $HoSb_2$, $ErSb_2$, $TmSb_2$, and YSb_2.[139,140] In contrast with the diantimonides of the light rare earths (La to Eu), these compounds could not be synthesized at normal pressure because of unsatisfactory size conditions.[141] Several new compounds or new phases were also obtained in the rare earth disulfides and

Chapter 9

diselenides subjected to high-pressure and high-temperature treatments.[139]

4.3.3. High-Pressure Synthesis of Metallic Compounds

Until recently not much attention has been paid to the synthesis of new intermetallic compounds using high pressures despite the fact that Bridgman obtained a new binary phase in the Bi–Sn system at high pressure as much as twenty years ago.[142,143] A few Laves phases of the $MgCu_2$ type were obtained recently, namely $LaCo_2$,[144] $PrFe_2$, $NdFe_2$, and $YbFe_2$.[145] But probably the most thorough investigation was made in systems in which the discovery of new superconducting compounds was anticipated. This is, for instance, the case in several carbide systems.

A whole series of new rare earth sesquicarbides was synthesized by Krupka and Bowman[146] in a belt apparatus at 1500°C and 15–40 kbar pressure. The bcc form of Y_2C_3 (Pu_2C_3 type) was obtained for the first time in this study.[147] This material was thermodynamically unstable with respect to ambient pressure as demonstrated by annealing experiments. However, reversion to its original crystal form had not occurred over a 12 month period at room temperature. A homogeneity range was shown to exist. Superconductivity was observed over the entire homogeneity range with rather high transition temperatures ranging from 6.0 to 11.5°K. This is the highest transition temperature measured for a compound composed of nonsuperconducting elements. Isotypic compounds Ln_2C_3 (Ln = La, Ce, Pr, Nd, Sm, Gd, Tb, Dy, and Ho) have been obtained as well as the compounds Er_2C_3, Tm_2C_3, and Yb_2C_3, which have a lower symmetry. In the system Y_2C_3–Th_2C_3, a bcc Pu_2C_3-type structure is stable over a wide ternary composition field. The new phase was superconducting over its entire homogeneity range with a very high transition temperature between 1.0 and 17.0°K for the nominal compositions $(Y_{0.60-0.75}Th_{0.40-0.25})C_{1.35-1.55}$. High-temperature ambient-pressure annealing destroyed both the bcc structure and superconductivity.[148] Among actinides, the synthesis of pure Th_2C_3 (Pu_2C_3 type) was accomplished under pressure.[149] It is unstable with respect to ambient pressure; decomposition to ThC + ThC_2 occurs almost completely within four months at room temperature. Other carbides have been synthesized under pressure, e.g., the dicarbides HoC_2, ErC_2, DyC_2, TbC_2, YC_2 or the compounds La_4C_3 and Sc_4C_3. While most of them can be obtained at ambient pressure, the kinetics of formation is greatly enhanced by the high-pressure method of synthesis. Of particular interest is the synthesis at high pressure (60 kbar) and at high temperature (1500°C) of the new rhenium monocarbide ReC with a γ'-MoC hexagonal type of structure.[150] At atmospheric pressure rhenium carbide decomposes at 1500°C into a rhenium–carbon interstitial solid solution.

Considerable work is being done in the USSR on the phase diagram Bi–Sn, in which a new superconducting phase was shown to be stable at pressures exceeding 7 kbar.[142,143,151,152] Other binary or ternary phase diagrams are presently under investigation, containing elements like Ti, Zn, Nb, Bi, Sn, Pb, In, Gd, Sb, etc.[153] A binary phase, Nb_3In, which can only be synthesized at high pressure, crystallizes in the β-W structure type and is known to be a superconductor.[154]

Several scientists have predicted that metallic hydrogen might be metastable, and others that it would be a room-temperature superconductor. This last speculation has certainly evoked interest in producing the substance. The transition pressure is predicted to be between 1.5 and 3 mbar. There are presently two active groups who are developing a program for measuring hydrogen's properties at high density. One group at Lawrence Radiation Laboratory at Livermore (USA) headed by R. S. Hawke is using dynamic high-pressure methods, while in the Soviet Union L. Vereschagin at the Institute of High Pressure Physics in Moscow is developing a 10-story-high static press designed to reach 2–3 mbar. Unlike the dynamic experiments, the static experiments would allow one to recover the metallic hydrogen if it is metastable. Whatever the case, to use metallic hydrogen on earth may be very tricky indeed. According to R. S. Hawke, the stored energy in metallic hydrogen is very high, 30–40 times greater than that of TNT, so that if metallic hydrogen is metastable, it will be extremely dangerous.

5. Crystal Growth under High Pressure
5.1. Crystal Growth in a Flux

The possibility of synthesizing single crystals of dense metastable phases was clearly demonstrated by the growth of centimeter-size diamonds from graphite in a belt-type apparatus.[155] Many "catalysts" have been tested, but the best results have been obtained in the ternary Ni–Fe–C system. The binary Fe–C system has been used in spite of the formation of the stable cementite Fe_3C phase under

Chapter 9

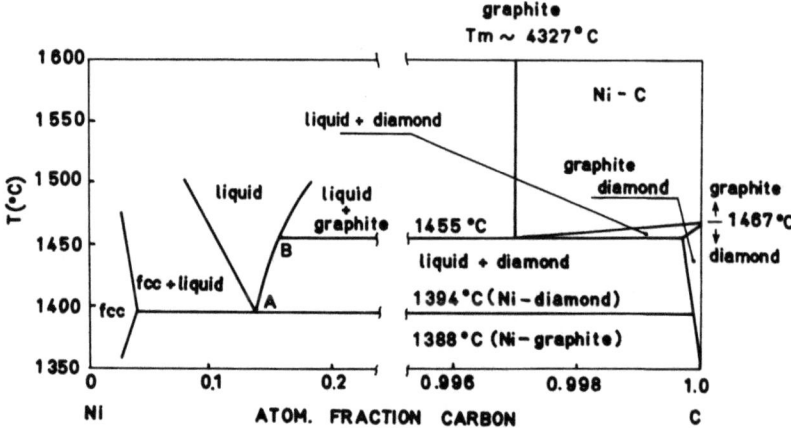

Fig. 16. Ni–C phase diagram at 54 kbar.

high-pressure conditions.[155] Figure 16 gives the Ni–C phase diagram at 54 kbar which was thoroughly investigated by Strong and Hanneman.[156,157] In the Ni–C system the temperature growth area is rather limited to the region from 1394 to 1455°C. It is easy to see from the position of the AB liquidus line, which is much closer to nickel than to carbon, that in this case the growth of diamonds occurs in a flux system with nickel playing the role of a solvent. Without using a flux, conditions required to obtain single crystals from the melt would be a pressure of 140 kbar at 3700°C.[158] Since the effective dimensions of the high-pressure cell are limited, it is necessary to decrease the relative amount of nickel versus the amount of carbon in order to obtain large crystals of diamond. So the quantity of solvent becomes too small to dissolve all the carbon present as should be done in the conventional flux method. To remedy this situation, the authors used the natural temperature gradient of the cell and the difference in solubility between the hotter central region and the two cooler regions, one near each end of the cell. The cylindrical cell is divided in two symmetric parts by an insulating ceramic disk. The carbon supply is made of a mixture of graphite and diamond being packed on each side (Figure 17).[159] Two disks of "catalyst" containing a few diamond seeds are packed in the two regions between the carbon supplies and the end ceramic disks (Figure 17). At high temperature, diamond and graphite float in the "catalyst" metal bath. The central nutrient mass remains in place by virtue of its rough surface against the container walls, but a few crystals of diamond that

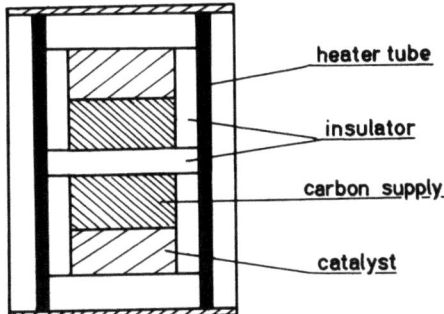

Fig. 17. High-pressure cell for growing diamond crystals.

are loosened by dissolution may float in the bath toward the ends of the cell. As the nutrient mass rapidly saturates the entire molten bath, the initial seeds and the new ones can develop on the cooler ends of the cell. The growth process is difficult to control, especially for long periods of time. Nevertheless, various kinds of diamonds have been grown with success by this method. Depending on the impurities added to the batch, blue, green, or transparent diamonds can be produced. While the process is still too expensive for the production of the large, perfect diamonds required for the jewelry industry, it has become quite competitive for the production of small, colored diamonds used in the abrasive industry.

The cubic high-pressure phase of boron nitride has also been extensively investigated[160,161] since it may prove to have commercial significance as an abrasive. Roughly the same pressure and temperature as required for diamond are necessary for cubic BN called "Borazon." However, the "catalyst" materials (actually the flux) are quite different, consisting of the alkali and alkaline earth metals and their nitrides. Cubic boron nitride can be grown as rather nice, clear crystals having the shape of tetrahedra or octahedra with 1–2 mm in cross section.[162–163] More recently some phase equilibria pertinent to the growth of cubic boron nitride have been carefully investigated by De Vries and Fleischer,[164] in particular the Li_3N–BN phase diagram at 50 kbar (Figure 18).

At this pressure cubic BN is stable over about 140°C for mole ratios of Li_3N:BN less than 47–53%. The solvent in that case is the intermediary compound Li_3BN_2, which melts at 1650°C at 50 kbar.

In addition to diamond and cubic boron nitride, a few more compounds have been synthesized as tiny single crystals by the flux

Chapter 9

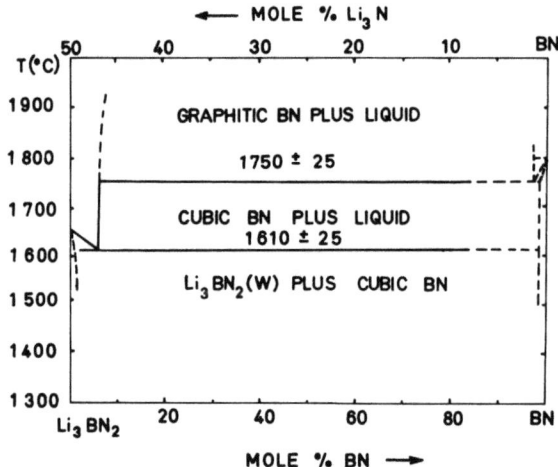

Fig. 18. Li_3N-BN phase diagram at 50 kbar.

method. For instance, the new compound CoP_2 was prepared using a germanium flux by reaction of $Co + Ge + 2P$ at temperatures between 800 and 1200°C at 65 kbar. The mixture was held at temperature 2 hr, cooled 2 hr at 100°/hr, then quenched. Crystals of CoP_2 formed at the ends of the pellet, whereas the Ge migrated to the center.[135]

One of the difficulties in growing crystals under high pressure by the flux method is to find the right solvent. Most of the conventional solvents are inadequate, either because they react with the material at high pressure or because their melting point increases sharply with the pressure, thus requiring high-temperature conditions. For instance, one can see in Figure 19 that the melting point of NaCl is about 1550°C at 65 kbar as against 800°C at room pressure.[165] If the pressure is not too high, one can still use a conventional flux: For instance, single crystals of the tetragonal high-pressure phase of $LiBO_2$ were grown from a flux consisting of LiCl at 15 kbar and a temperature of 950°C.[166] LiCl was chosen as a flux because it is liquid under the above conditions[167] and experiments indicated that it would not react with $LiBO_2$ if the ratio $LiCl/LiBO_2$ was kept within certain limits. Furthermore, the high solubility of LiCl in methyl alcohol made it attractive as a flux. When the pressure required to stabilize the denser phase is high, it is necessary to find new types of fluxes with relatively low melting points at high pressure. Dry NaOH has been proven to be an adequate flux for the growth of

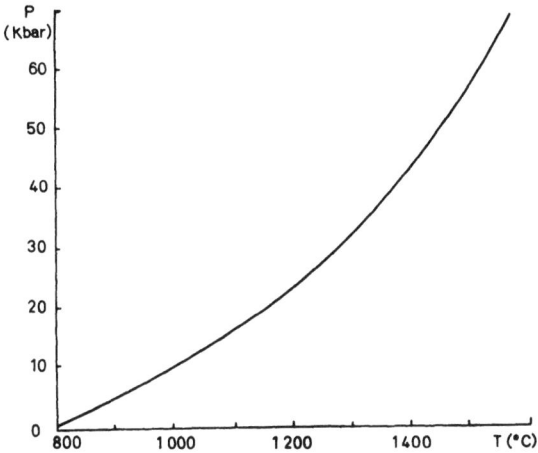

Fig. 19. NaCl melting point versus pressure.

transparent crystals of α-Ga_2O_3 at 44 kbar and 1000°C.[168] A mixture in a mole weight ratio of five β-Ga_2O_3 to NaOH was found optimum for the pressure–temperature conditions used.

5.2. Crystal Growth by Hydrothermal Synthesis* under Very High Pressure

The properties of water under high-pressure, high-temperature conditions are rather well known. For instance, measurements have shown that at supercritical temperatures, for example, at 500°C, the viscosity of the fluid does not vary much with density (and thus with pressure) and is lower than that of liquid water at room temperature by a factor of 10 or 20.[169,170]

Dense supercritical water is thus a medium of very high fluidity and consequently dissolved neutral or ionic particles have high diffusion coefficients and ion mobilities in this medium.[171] In addition, shock wave measurements to 1000°C and at more than 100 kbar demonstrate the increase of the ion product of water by 12 orders of magnitude or more at these conditions.[171] Under such conditions supercritical water becomes vigorously active and even when present as trace in the high-pressure cell it acts as a powerful

---

* See Chapter 8, this volume

TABLE 5
Crystals of New Dense Phases Obtained by Hydrothermal Synthesis under Extreme Conditions*

Formula	Structure of HP phase	Starting materials	Temperature, °C	Pressure, kbar	Size of crystals, mm
$AlBO_3$	Calcite	Al_2O_3, $B(OH)_3$	1000	40	0.5
$GaBO_3$	Calcite	Ga_2O_3 + $B(OH)_3$	1000	40	0.2
Al_3BO_6	Norbergite	Al_2O_3, $B(OH)_3$	1000	40	0.1–0.2
YBO_3	Pseudovaterite	Y_2O_3, $B(OH)_3$	1200	40	0.1
$LuBO_3$	Pseudovaterite	Lu_2O_3, $B(OH)_3$	1200	40	0.1
$Al_4NiB_2O_{10}$	Chondrotite	Al_2O_3, NiO, $B(OH)_3$	1200	40	0.1–0.2
$Ga_4NiB_2O_{10}$	Chondrotite	Ga_2O_3, NiO, $B(OH)_3$	1200	40	0.1–0.2
$Fe_4NiB_2O_{10}$	Chondrotite	Fe_2O_3, NiO, $B(OH)_3$	1200	40	0.1–0.2
$Co_3B_7O_{13}OH$	Boracite	CoO, $B(OH)_3$	1000	40	0.5–0.8
$Cd_3B_7O_{13}OH$	Boracite	CdO, $B(OH)_3$	1000	40	0.5–1.0
$FeNiBO_4$	Olivine	Fe_2O_3, NiO, $B(OH)_3$	1000	40	0.2–0.3
$AlMgBO_4$	Olivine	Al_2O_3, MgO, $B(OH)_3$	1000	40	0.2–0.3
$TeMn_2O_6$	Distorted trirutile	$Te(OH)_6$, Mn_2O_3	1000	80	0.1
Cr_2CdO_4	Fe_2CaO_4	$CrCdO_4$	1000	80	0.2
Mn_3O_4	Mn_2CaO_4	Mn_3O_4 (hausmannite)	1000	80	0.1

* Solvent is H_2O except in the case of Mn_3O_4, for which the solvent was NaOH. Time, 1 hr.

"catalyst" for the synthesis of single crystals. During the last few years we have developed a technique which allows us to perform hydrothermal synthesis of inorganic materials in a belt apparatus up to 100 kbar and more than 1500°C. Depending on the temperature and chemical conditions required, sealed capsules made of gold, platinum, or other metal are used.

In order to increase the kinetics of the crystal growth process, one can add several "catalysts" to the water solvent, like sodium hydroxide, lithium hydroxide, ammonium hydroxide, various acids, etc.

Table 5 contains a list of several high-pressure phases we have been able to identify and to obtain as single crystals. All the experiments have been performed in cylindrical platinum cells having a size of 6 mm in length by 4 mm in diameter. Probably if a bigger cell could be used, crystals of much larger dimensions could be obtained under the same conditions.

6. Conclusion

With the rapid progress of the technology in the field of high-pressure equipment, various kinds of apparatus have become commercially available for the synthesis of new materials under high-pressure and high-temperature conditions.

From a general survey of the new, dense metastable phases which have been prepared under extreme conditions one can predict that the field of high-pressure synthesis is going to be very fruitful. In spite of the extensive research on oxides in the last five years, many new metastable phases still remain to be discovered and thus this area is very promising. In the field of alloys as well as in that of semiconducting compounds the use of high-pressure conditions for the synthesis of new compounds has just begun. One can thus predict that in these two large domains even the systematic reinvestigation of the phase diagrams of only two-component systems at high pressure should lead to the discovery of many new phases. When the dense phases are quenchable small single crystals can frequently be prepared by the flux method or the hydrothermal method.

References

1. P. W. Bridgman, Two new modifications of phosphorus, *J. Am. Chem. Soc.* **36**, 1344–1363 (1914).

2. P. W. Bridgman, Further note on black phosphorus, *J. Am. Chem. Soc.* **38**, 609–612 (1916).
3. O. Muller and R. Roy, *Am. Ceram. Soc. Bull.* **46**, 881 (1967).
4. O. Muller and R. Roy, Preliminary study of new crystalline gold oxides, *J. Inorg. Nucl. Chem.* **31**, 2966 (1969).
5. O. Muller and R. Roy, Formation and stability of the platinum and rhodium oxides at high oxygen pressures and the structures of Pt_3O_4, βPtO_2, and RhO_2, *J. Less Common Metals* **16**, 129–146 (1968).
6. W. B. White, High oxygen pressure, in *Research Techniques for High Pressure and High Temperature* (G. Ulmer, ed.), Chapter 4, Springer Verlag (1971).
7. R. Bougon, Private communication.
8. R. D. Shannon, Synthesis and properties of two members of the rutile family RhO_2 and PtO_2; *Solid State Commun.* **6**, 139–143 (1968).
9. J. B. MacChesney, H. J. Williams, R. C. Sherwood, and J. F. Potter, Preparation and low temperature magnetic properties of terbium oxides, *J. Chem. Phys.* **44**(2), 596–601 (1966).
10. A. W. Sleight, New ternary oxides of tetravalent platinum and palladium with the pyrochlore structure, *Mat. Res. Bull.* **3**, 699–704 (1968).
11. A. W. Sleight, New ternary oxides of mercury with the pyrochlore structure, *Inorg. Chem.* **7**, 1704–1708 (1968).
12. J. B. MacChesney, J. F. Potter, R. C. Sherwood, and H. J. Williams, Oxygen stoichiometry in barium ferrates; its effect on magnetisation and resistivity, *J. Chem. Phys.* **43**, 3317–3322 (1965).
13. J. B. MacChesney, H. J. Williams, J. F. Potter, and R. C. Sherwood, Magnetic study of the manganate phases: $CaMnO_3$, $Ca_4Mn_3O_{10}$, $Ca_3Mn_2O_7$, Ca_2MnO_4, *Phys. Rev.* **164**(2), 779–785 (1967).
14. A. Feretti, D. G. Wickham, and A. Wold, Induction heated pressure vessel for growing oxide single crystals, *Rev. Sci. Instr.* **32**, 566–568 (1961).
15. P. W. Bridgman, The compression of 46 substances to 50,000 kg/cm^2, *Proc. Am. Acad. Arts Sci.* **74**, 21–51 (1940).
16. F. R. Boyd and J. L. England, Apparatus for phase-equilibrium measurements at pressures up to 50 kbar and temperatures up to 1750°C *J. Geophys. Res.* **65**, 741–748 (1960).
17. H. T. Hall, Ultra high-pressure, high-temperature apparatus: the belt, *Rev. Sci. Instr.* **31**, 125–131 (1960).
18. W. B. Daniels and M. T. Jones, Simple apparatus for the generation of pressure above 100,000 atmospheres simultaneously with temperatures above 3000°C, *Rev. Sci. Instr.* **32**, 885 (1961).
19. L. F. Vereshchagin, V. A. Golaktionov, A. A. Smerchan, and V. N. Slesarev, Apparatus for high pressure and high temperature with a conical piston, *Dokl. Akad. Nauk. SSSR* **132**(5), 1059–1061 (1960) [English transl.: *Soviet Phys.—Doklady* **5**, 602 (1960)].
20. H. T. Hall, Some high pressure, high temperature design considerations, equipment for use at 100,000 atm and 3000°C, *Rev. Sci. Instr.* **29**, 267–275 (1958).
21. E. C. Lloyd, U. O. Hutton, and D. P. Johnson, Compact multi-anvil wedge-type high pressure apparatus, *J. Res. Natl. Bur. Std.* **63**(C), 59–64 (1959).
22. L. F. Vereshagin, in *Progress in Very High-Pressure Research* (F. P. Bundy, W. R. Hibbard, and H. M. Strong, eds.), pp. 290–301, Wiley, New York (1961).

23. P. W. Bridgman, The resistance of 72 elements, alloys and compounds to 100,000 kg/cm², *Proc. Am. Acad. Arts Sci.* **81**, 165–251 (1952).
24. A. S. Balchan and H. G. Drickamer, High pressure electrical resistance cell and calibration points above 100 kilobars, *Rev. Sci. Instr.* **32**, 308 (1961).
25. V. M. Goldschmidt, *Skifter Norske Videnskaps Akad. Oslo, I: Mat. Naturv. K.* **1926**, 57–69.
26. I. Y. Borg and D. K. Smith, X-ray diffraction studies on CdTe at high pressure, *J. Phys. Chem. Solids* **28**, 49–53 (1967).
27. A. N. Mariano and E. P. Warekois, *Science* **142**, 672 (1963).
28. H. A. Gebbie, P. L. Smith, I. G. Austin, and J. H. King, Pressure dependence of resistivity of indium antimonide to 70,000 atmospheres, *Nature* **188**, 1095–1096 (1960).
29. S. M. Stishov and S. V. Popova, New dense polymorphic modification of silica, *Geokhemiya* **10**, 837–839 (1961) [English transl.: *Geochemistry* **10**, 923–926 (1961)].
30. D. P. Danderkar and J. C. Jamieson, Some high-pressure phases of RX_2 fluorides, *Trans. Am. Cryst. Assoc.* **5**, 19–27 (1969).
31. K. F. Seifert, Untersuchungen zur Druck-Kristallchemie der AX_2 Verbindungen, *Fortschr. der Mineral. Dtsch.* **45**(2), 214–280 (1967).
32. R. D. Shannon and C. T. Prewitt, Coordination and volume changes accompanying high-pressure phase transformations of oxides, *Mat. Res. Bull.* **4**, 57–62 (1969).
33. W. H. White, F. Dackille, and R. Roy, High-pressure, high-temperature polymorphism of the oxides of lead, *J. Am. Ceram. Soc.* **44**(4), 170–174 (1961).
34. A. E. Ringwood, The constitution of the mantle II. Further data on the olivine-spinel transition, *Geochim. Cosmochim. Acta* **15**, 18–29 (1958); *Am. Miner.* **44** 659–661 (1959).
35. A. E. Ringwood, Olivine-spinel transformation in cobalt orthosilicate, *Nature* **198**, 79–80 (1963).
36. A. E. Ringwood, Olivine-spinel transformation in Fe_2SiO_4 and Ni_2SiO_4, *Nature* **187**, 1019 (1960).
37. A. E. Ringwood, Prediction and confirmation of olivine-spinel transition in Ni_2SiO_4, *Geochem. Cosmochim. Acta* **26**, 457–469 (1962).
38. M. Marezio, Oxides at high-pressure, *Trans. Am. Cryst. Assoc.* **5**, 29–37 (1969).
39. W. Klement and A. Jayaraman, Phase relations and structures of solids at high pressures, in *Progress in Solid State Chemistry* Vol. 3, pp. 289–376, Pergamon (1966).
40. J. C. Jamieson, *Science* **139**, 762 (1963).
41. G. L. Kulcinski, High-pressure induced phase transition in ZrO_2, *J. Am. Ceram. Soc.* **51**(10), 582–584 (1968).
42. L. F. Vereshchagin, S. S. Kabalkina, and L. M. Lityagina, Effect of high pressure on the structure of tin oxide, *Soviet Phys.—Doklady* **10**, 622–624 (1966).
43. S. S. Kabalkina and S. V. Popova, Phase transitions in zinc and manganese fluorides at high pressures and temperatures, *Soviet Phys.—Doklady* **8**(12), 1141–1143 (1964).
44. L. M. Azzaria and F. Dachille, High-pressure polymorphism of manganous fluorides, *J. Phys. Chem.* **65**, 889–891 (1961).
45. L. F. Vereshchagin, S. S. Kabalkina, and A. A. Kotilevets, Phase transition in MnF_2 at high pressure, *Soviet Phys.—JETP* **22**, 1181–1184 (1966).
46. A. Jayaraman, Influence of pressure on phase transitions, in *Annual Review of Materials Science*, Vol. 2 (1972).

47. T. P. Ershova and E. T. Ponyatovskii, *Dokl. Akad. Nauk SSSR* **151**(6), 1364 (1963).
48. G. C. Kennedy, Phase relations in the system $Al_2O_3-H_2O$ at high temperatures and pressures, *Am. J. Science* **257**, 563–573 (1959).
49. G. Yamaguchi, H. Yanagida, and S. Ono, Condition of Tohdite $5Al_2O_3-H_2O$ formation, *J. Ceram. Assoc. Japan* **74**, 84–89 (1966).
50. H. T. Hall and L. A. Compton, Group IV analogs and high-pressure, high-temperature synthesis of B_2O, *Inorg. Chem.* **4**, 1213 (1965).
51. S. Somiya, S. Yamaoka, and S. Saito, Phase relation between CrO_2 and Cr_2O_3 by decomposition of CrO_3 under high oxygen pressure—Preliminary report, *Bull. Tokyo Inst. Technology* **66**, 81–84 (1965).
52. N. Kawai, A. Sawaska, S. Kikuchi, and N. Tamagawa, Reduction of CrO_3 into CrO_2 and Cr_2O_3 under very high pressure and high temperature, *Japan J. Appl. Phys.* **6**(12), 1397–1399 (1967).
53. R. C. De Vries, Stability of CrO_2 at high pressures and temperatures in the belt apparatus, *Mat. Res. Bull.* **2**, 999–1008 (1967).
54. O. Fukunaga, K. Takahaski, T. Fujita, and J. Yoshimoto, Phase equilibrium between MnO_2 and Mn_2O_3, *Mat. Res. Bull.* **4**(5), 315–322 (1969).
55. F. Kanamaru, H. Miyamoto, Y. Minomura, M. Koizumi, M. Shimad, and S. Kume, Synthesis of a new perovskite $CzFeO_3$, *Mat. Res. Bull.* **5**, 257–262 (1970).
56. G. Demazeau, A. Marbeuf, M. Pouchard, and P. Hagenmuller, Sur une série de composés oxygénés du nickel trivalent dérivés de la perovskite, *J. Solid State Chem.* **3**, 582–589 (1971).
57. G. Demazeau, C. Parent, M. Pouchard, and P. Hagenmuller, Sur deux nouvelles phases oxygénées du cuivre trivalent $LaCuO_3$ et $La_2Li_{0.50}Cu_{0.50}O_4$, *Mat. Res. Bull.* **7**, 913–920 (1972).
58. B. Chamberland and W. H. Cloud, Preparation and properties of $NiCrO_3$, *J. Appl. Phys.* **41**, 434–435 (1970).
59. O. Muller, R. Roy, and W. B. White, Phase equilibria in the systems $NiO-Cr_2O_3-O_2$, MgO, $Cr_2O_3-O_2$ and $CdO-Cr_2O_3-O_2$ at high oxygen pressures, *J. Am. Ceram. Soc.* **51**(12), 693–699 (1968).
60. J. Chenavas, J. J. Capponi, and J. C. Joubert, to be published.
61. J. Chenavas, J. C. Joubert, and M. Marezio, Low spin → high spin state transition in high pressure cobalt sesquioxide, *Solid State Commun.* **9**, 1057–1060 (1970).
62. H. R. Hoekstra and S. Siegel, Synthesis of thallium platinate at high pressure, *Inorg. Chem.* **7**(1), 141–145 (1968).
63. O. Muller and R. Roy, in *Advances in Chemistry*, Vol. 98, pp. 28–37, Am. Chem. Society (1971).
64. O. Muller and R. Roy, Synthesis and crystal structure of Cd_2PtO_4, *J. Less Common Metals* **20**, 161–163 (1970).
65. B. L. Chamberland, The preparation and crystallographic properties of $BaCrO_3$ polytypes, *Inorg. Chem.* **8**(2), 286–290 (1969).
66. W. L. Roth and R. C. De Vries, Crystal and magnetic structure of $PbCrO_3$, *J. Appl. Phys.* **38**(3), 951–952 (1967).
67. R. C. De Vries and W. L. Roth, High pressure synthesis of $PbCrO_3$, *J. Am. Ceram. Soc.* **51**(2), 72–75 (1968).
68. B. L. Chamberland and C. W. Moeller, A study on the $PbCrO_3$ perovskite, *J. Solid State Chem.* **5**, 39–41 (1972).

69. J. B. Goodenough, J. M. Longo, and J. A. Kafalas, Band antiferromagnetism and the new perovskite $CaCrO_3$, *Mat. Res. Bull.* **3**, 471–482 (1968).
70. J. F. Weiher, B. L. Chamberland, and J. L. Gillson, Magnetic and electrical transport properties of $CaCrO_3$, *J. Solid State Chem.* **3**, 529–532 (1971).
71. B. L. Chamberland, Preparation and properties of $SrCrO_3$, *Solid State Commun.* **5**, 663–666 (1967).
72. J. A. Kafalas and J. M. Longo, High pressure synthesis of $(ABX_3)(AX)_n$ compounds; *J. Solid State Chem.* **4**, 55–59 (1972).
73. B. L. Chamberland, A. W. Sleight, and J. F. Weiher, Preparation and characterization of $BaMnO_3$ and $SrMnO_3$ polytypes, *J. Solid State Chem.* **1**, 506–511 (1970).
74. J. Chenavas, J. J. Capponi, J. C. Joubert, and M. Marezio, Synthèse de nouvelles phases denses d'oxyhydroxydes $M^{3+}OOH$ des métaux de la première série de transition, en milieu hydrothermal à très haute pression, *J. Solid State Chem.* **6**, 1–15 (1973).
75. S. M. Stishov and N. V. Belov, Crystal structure of the new dense modification of silica, *Dokl. Akad. Nauk SSSR, Geochem.* **143**, 146–148 (1962).
76. E. C. T. Chao, J. J. Fahey, J. Littler, and D. J. Milton, Stishovite SiO_2 a very high pressure new mineral from meteor crater, Arizona, *J. Geophys. Res.* **67**(1), 419–421 (1962).
77. S. Akimoto and Y. Syono, Coesite–stishovite transition, *J. Geophys. Res.* **74**(6), 1653–1659 (1969).
78. A. E. Ringwood, A. Reid, and A. D. Wadsley, High pressure $KAlSi_3O_8$ and aluminosilicate with sixfold coordination, *Acta Cryst.* **23**, 1093–1095 (1967).
79. F. Dachille and R. Roy, A new high pressure form of B_2O_3 and inferences on cation coordination from infrared spectroscopy, *J. Am. Ceram. Soc.* **42**(2), 78–80 (1959).
80. C. T. Prewitt and R. D. Shannon, Crystal structure of a high-pressure form of B_2O_3, *Acta Cryst.* **B24**, 869–874 (1968).
81. M. Marezio, J. P. Remeika, and P. D. Dernier, The crystal structure of the high pressure CaB_2O_4 (III) and the crystal structure of the high pressure phase CaB_2O_4 (IV) and polymorphism in CaB_2O_4, *Acta Cryst.* **B25**, 955–970 (1969).
82. J. J. Capponi, J. Chenavas, and J. C. Joubert, Synthèse hydrothermale à très haute pression de deux borates de type olivine $AlMgBO_4$ et $FeNiBO_4$, *Mat. Res. Bull.* **8**, 275–282 (1973).
83. S. Kume, T. Matsumoto, and M. Koizumi, *J. Geophys. Res.* **71**, 4999 (1966).
84. A. E. Ringwood, A. F. Reid, and A. D. Wadsley, High pressure transformation of alkali aluminosilicates and aluminogermanates, *Earth Planet. Sci. Letters* **3**, 38–40 (1967).
85. A. F. Reid, A. D. Wadsley, and A. E. Ringwood, High pressure $NaAlGeO_4$, a calcium ferrite isotype and model structure for silicates at depth in the earth's mantle, *Acta Cryst.* **23**, 736–739 (1967).
86. A. D. Wadsley, A. F. Reid, and A. E. Ringwood, The high pressure form of $MnGeO_4$, a member of the olivine group, *Acta Cryst.* **B24**, 740–742 (1968).
87. A. F. Reid and A. E. Ringwood, The crystal chemistry of dense M_3O_4 polymorphs: high pressure Ca_2GeO_4 of K_2NiO_4 structure type, *J. Solid State Chem.* **1**, 557–565 (1970).
88. R. D. Shannon and A. W. Sleight, Synthesis of new high pressure pyrochlore phases, *Inorg. Chem.* **7**(8), 1649–1651 (1968).

89. A. E. Ringwood and M. Seabrook, High pressure phase transformations in germanate pyroxenes and related compounds, *J. Geophys. Res.* **8**(15), 4601–4609 (1963).
90. A. E. Ringwood and A. Major, Some high pressure transformations of geophysical significance, *Earth Planet. Sci. Letters* **2**, 106–110 (1967).
91. C. T. Prewitt and A. W. Sleight, Garnet-like structures of high pressure cadmium germanate, *Science* **163**, 386–389 (1969).
92. Y. Shimizu, Y. Syono, and S. Akimoto, High pressure transformations in $SrGeO_3$, $SrSiO_3$, $BaGeO_3$ and $BaSiO_3$, *High Temp. High Press.* **2**, 113–120 (1970).
93. A. Sawaoka, S. Miyahara, S. Akimoto, and H. Fujisawa, Magnetic properties of manganese–iron and manganese–cobalt germanate having ilmenite structure, *J. Phys. Soc. Japan* **19**, 1750–1751 (1964).
94. J. Liebertz and C. J. Rooymans, Die ilmenit/perowskit-Phasenumwandlung von $CdTiO_3$ unter hohem druck, *Z. phys. Chem. Neue Folge* **44**, 242–249 (1965).
95. Y. Syono, S. Akimoto, and Y. Endoh, High pressure synthesis of ilmenite and perovskite type $MnVO_3$ and their magnetic properties, *J. Phys. Chem. Solids* **32**, 243–249 (1971).
96. A. Waintal and J. Chenavas, Transformation sous haute pression de la forme hexagonale de $MnT'O_3$ (T' = Ho, Er, Tm, Yb, Lu) en une forme perovskite, *Mat. Res. Bull.* **2**, 819–822 (1967).
97. A. E. Ringwood and A. F. Reid, High pressure transformations of spinels, *Earth Planet. Sci. Letters* **5**, 245–250 (1969).
98. A. F. Reid and A. E. Ringwood, Newly observed high-pressure transformations in Mn_3O_4, $CaAl_2O_4$ and $ZrSiO_4$, *Earth Planet. Sci. Letters* **6**, 205–208 (1969).
99. V. E. Stubican and R. Roy, High pressure schellite-structure polymorphs of rare-earth vanadates and arsenates, *Z. Krist.* **119**, 90–97 (1963).
100. J. M. Longo and J. A. Kafalas, The effect of pressure and B-cation size on the crystal structure of $CsBF_3$ compounds (B = Mn, Fe, Co, Ni, Zn, Mg), *J. Solid State Chem.* **1**, 103–108 (1969).
101. Y. Syono, S. Akimoto, and K. Kohn, Structure relations of hexagonal perovskite-like compounds ABX_3 at high pressure, *J. Phys. Soc. Japan* **26**(4), 993–999 (1969).
102. J. M. Longo and J. A. Kafalas, Pressure-induced structural changes in the system $Ba_{1-x}Sr_xRuO_3$, *Mat. Res. Bull.* **3**, 687–692 (1968).
103. J. M. Longo, J. A. Kafalas, N. Menyuk, and K. Dwight, High pressure $RbFeCl_3$. A transparent ferrimagnet, *J. Appl. Phys.* **42**(4), 1561–1562 (1971).
104. J. M. Longo and J. A. Kafalas, Effect of pressure on the crystal structure of $CsMnCl_3$ and $RbMnCl_3$, *J. Solid State Chem.* **3**, 429–433 (1971).
105. R. D. Shannon and C. T. Prewitt, Effective ionic radii in oxides and fluorides, *Acta Cryst.* **B25**(5), 925–945 (1969).
106. R. D. Shannon and C. T. Prewitt, Revised values of effective ionic radii, *Acta Cryst.* **B26**, 1046–1047 (1970).
107. A. N. Christensen, R. Gronbaek, and S. E. Rasmussen, The crystal structure of InOOH, *Acta Chem. Scand.* **18**, 1261–1266 (1964).
108. M. S. Lehmann, F. K. Larsen, F. R. Poulsen, A. N. Christensen, and S. E. Rasmussen, Neutron and X-ray crystallographic studies on indium oxide hydroxide, *Acta Chem. Scand.* **24**, 1662 (1970).
109. S. Akimoto and Y. Syono, High pressure decomposition of some titanate spinels, *J. Chem. Phys.* **47**, 1813 (1967).

110. A. F. Reid, Private communication.
111. M. Marezio, J. P. Remeika, and A. Jayaraman, High pressure decomposition of synthetic garnets, *J. Chem. Phys.* **45**, 1821–1824 (1966).
112. M. Marezio, J. P. Remeika, and P. D. Dernier, High pressure synthesis of $YGaO_3$, $GdGaO_3$ and $YbGaO_3$, *Mat. Res. Bull.* **1**, 247–255 (1966).
113. K. H. Kim and F. A. Hummel, Private communication, quoted by E. M. Levin, C. R. Robbins, and H. F. McMurdie, *Phase Diagrams for Ceramists*, p. 121, The American Ceramic Society (1964).
114. J. J. Capponi, J. Chenavas, and J. C. Joubert, Nouveaux borates d'aluminium et de gallium obtenus par synthèse hydrothermale à haute pression, *Bull. Soc. fr. Minéral. Cristallogr.* **95**, 412–417 (1972).
115. G. G. Uvazov, E. I. Speranskaya, and Z. F. Gulyanitskaya, *Zh. Neorg. Khim.* **1**(6), 1414 (1956).
116. F. Sugawara, Y. Syono, and S. Akimoto, High pressure synthesis of a new perovskite $PbSnO_3$, *Mat. Res. Bull.* **3**, 529–532 (1968).
117. Y. Syono, H. Sawamoto, and S. Akimoto, Disordered ilmenite $MnSnO_3$ and its magnetic property, *Solid State Commun.* **7**, 713–716 (1969).
118. D. H. Lindsley, T. C. Davis, and I. D. MacGregor, Ferrosilite ($FeSiO_3$): synthesis at high pressures and high temperature, *Science* **144**(3614), 73–74 (1964).
119. F. Sugawara and S. Iida, New magnetic perovskites $BiMnO_3$ and $BiCrO_3$, *J. Phys. Soc. Japan* **20**, 1529 (1965).
120. Y. Y. Tomashpol'skii, E. V. Zubova, K. P. Burdina, and Y. N. Venevtsev, X-ray diffraction study of the ferroelectric and ferromagnetic materials $BiMnO_3$, $BiCrO_3$ and their solid solutions obtained at high pressures, *Izv. Akad. Nauk SSSR, Neorg. Mat.* **3**(11), 2132–2134 (1967).
121. B. L. Chamberland, The synthesis of new ilmenite-type derivates, $CuVO_3$ and $CoVO_3$, *J. Solid State Chem..* **1**, 138–142 (1970).
122. B. L. Chamberland, A. W. Sleight, and J. F. Weiher, Preparation and Characterization of $MgMnO_3$ and $ZnMnO_3$, *J. Solid State Chem.* **1**, 512–514 (1970).
123. T. Noda and M. Ushio, High pressure synthesis and stability field of cobalt garnet ($Co_3Al_2Si_3O_{12}$), *J. Ceram. Assoc. Japan* **75**(5), 125–135 (1967).
124. R. D. Shannon, Synthesis of some new perovskites containing indium and thallium, *Inorg. Chem.* **6**, 1474–1478 (1967).
125. A. W. Sleight and C. T. Prewitt, Preparation of $CuNbO_3$ and $CuTaO_3$ at high pressure, *Mat. Res. Bull.* **5**, 207–212 (1970).
126. B. L. Chamberland, A. W. Sleight, and W. Cloud, The preparation and properties of rutile-type transition metal oxyfluorides, *J. Solid State Chem.* **2**, 49–54 (1970).
127. B. L. Chamberland, A new oxyfluoride perovskite, $KTiO_2F$, *Mat. Res. Bull.* **6**, 311–316 (1971).
128. T. Katsura, B. Iwasaki, S. Kimura, and S. Akimoto, High-pressure synthesis of the stoichiometric compound FeO, *J. Chem. Phys.* **47**(11), 4559–4560 (1967).
129. A. Neuhaus and R. Steffen, Über das Zustands- und Mischbarkeitsverhalten des Systems ZnS–MnS im Druckbereich bis 140 kbar, *Z. Phys. Chem. Neue Folge* **73**, 188–214 (1970).
130. A. Neuhaus and L. Cemie, Struktur und Mischbarkeit im System ZnS–FeS im Druckbereich bis 60 kbar, *Naturwiss.* **57**(7), 354–355 (1970).

Chapter 9

131. K. J. Range, G. Engert, and A. Weiss, High-pressure transformations of ternary chalcogenides with chalcopyrite structure. I. Indium-containing compounds, *Solid State Commun.* **7**, 1749–1752 (1969).
132. H. Katzman, T. Donohue, W. F. Libby, H. L. Luo, and J. G. Huber, A high-pressure superconducting polymorph of cadmium tin diarsenide, *J. Phys. Chem. Solids* **30**, 1609–1611 (1969).
133. H. Katzman, T. Donohue, W. F. Libby, and H. L. Luo, A high-pressure superconducting polymorph of cadmium germanium diarsenide, *J. Phys. Chem. Solids* **30**, 2794–2795 (1969).
134. W. Albers and C. J. Rooymans, High-pressure polymorphism of spinel compound, *Solid State Comm.* **3**, 417–419 (1965).
135. P. C. Donohue, High-pressure syntheses and properties of CoP_2 and $MnCoP_4$, *Mat. Res. Bull.* **7**, 943–948 (1972).
136. M. Wakatsuki, K. Ichinose, and T. Aoki, Synthesis of polycrystalline cubic BN, *Mat. Res. Bull.* **7**, 999–1004 (1972).
137. E. D. Whitney and R. F. Giese, Preparation of a new ternary lithium silicon nitride, $LiSi_2N_3$, and the high pressure synthesis of magnesium silicon nitride, $MgSiN_2$, *Inorg. Chem.* **10**(5), 1090–1091 (1971).
138. R. C. de Vries and J. F. Flei, The system Li_3BN_2 at high pressures and temperatures, *Mat. Res. Bull.* **4**, 433 (1969).
139. H. T. Hall, High pressure syntheses involving rare earths, *Rev. Phys. Chem. Japan* **39**(2), 110–116 (1969).
140. Q. Johnson, The crystal structure of high pressure synthesized holmium diantimonide, *Inorg. Chem.* **10**(9), 2089–2090 (1971).
141. R. Wang and H. Steinfink, The crystal chemistry of selected AB_2 rare earth compounds with selenium, tellurium and antimony, *Inorg. Chem.* **6**, 1685 (1967).
142. P. W. Bridgman, The effect of pressure on the bismuth–tin system, *Bull. Soc. Chem. Belge* **62**(1–2), 26–33 (1963).
143. P. W. Bridgman, The effect of pressure on several properties of the alloys of bismuth–tin and of bismuth–cadmium, *Proc. Am. Acad. Arts Sci.* **82**(1–2), 101–156 (1953).
144. D. L. Robertson, J. F. Cannon, and H. T. Hall, High-pressure and high-temperature synthesis of $LaCo_2$, *Mat. Res. Bull.* **7**, 977–982 (1972).
145. J. F. Cannon, D. L. Robertson, and H. T. Hall, Synthesis of lanthanide–iron Laves phases at high pressures and temperatures, *Mat. Res. Bull.* **7**, 5–12 (1972).
146. M. C. Krupka and M. G. Bowman, New high-pressure phase transformations in rare earth and actinide carbide systems, in *Propriétés Physiques des Solides sous Pression*, pp. 409–414, Edition du CNRS, Paris (1970).
147. M. C. Krupka, A. L. Giorgi, N. H. Krikorian, and E. G. Szklarz, High-pressure synthesis and superconducting properties of yttrium sesquicarbide, *J. Less Common Metals* **17**, 91–98 (1969).
148. M. C. Krupka, A. L. Giorgi, N. H. Krikorian, and E. G. Szklarz, High-pressure synthesis of yttrium–thorium sesquicarbide: a new high-temperature superconductor, *J. Less Common Metals* **19**, 113–119 (1969).
149. M. C. Krupka, High-pressure synthesis of thorium sesquicarbide: a new actinide carbide, *J. Less Common Metals* **20**, 135–140 (1970).
150. S. V. Popova and L. G. Boiko, A new rhenium carbide formed by high pressure treatment, *High Temp. High Press.* **3**, 237–238 (1971).

151. E. G. Ponyatovskii, Phase transformation of the alloy 50 at.% Bi–50 at.% Sn under high constriction pressures, *Phys. Metals Metallogr.* **16**, 119–120 (1963).
152. E. G. Ponyatovskii, P.T.C. phase diagram of bismuth–tin alloys, *Dokl. USSR Chem. Sect.* **159**, 1374 (1964).
153. E. G. Ponyatovskii, Private communication, Grenoble (1972).
154. M. D. Banus, T. B. Reed, H. C. Gatos, M. C. Lavine, and J. A. Kafalas, Nb_3In: A β tungsten structure superconducting compound, *J. Phys. Chem. Solids* **23**, 971 (1962).
155. H. M. Strong, Further studies on diamond growth rates and physical properties of man-made diamond, in *Symp. on Rate Processes at High Pressure*, Chemical Institute of Canada, American Chemical Society, Toronto, Ontario, May 28, 1970.
156. H. M. Strong, Variation with pressure of the nickel–carbon eutectic, *Acta Met.* **12**, 1411–1419 (1964).
157. H. M. Strong and R. E. Hanneman, Crystallization of diamond and graphite, *J. Chem. Phys.* **46**, 3668–3676 (1967).
158. F. P. Bundy, Direct conversion of graphite to diamond in static pressure apparatus, *Science* **137**, 1057 (1962); *J. Chem. Phys.* **38**, 631–643 (1963).
159. R. H. Wentorf, Some studies of diamond growth rates, *J. Phys. Chem.* **75**(12), 1833–1837 (1971).
160. R. H. Wentorf, Cubic form of boron nitride, *J. Chem. Phys.* **26**, 956 (1957).
161. R. H. Wentorf, Synthesis of the cubic form of boron nitride, *J. Chem. Phys.* **34**, 809 (1961).
162. R. H. Wentorf, Preparation of simiconducting cubic boron nitride, *J. Chem. Phys.* **36**, 1990–1991 (1962).
163. F. P. Bundy and R. H. Wentorf, Direct transformation of hexagonal boron nitride, *J. Chem. Phys.* **38**, 1144 (1963).
164. R. C. de Vries and J. F. Fleischer, Phase equilibria pertinent to the growth of cubic boron nitride, *J. Crystal Growth* **13/14**, 88–92 (1972).
165. J. Akella, S. N. Vaidya, and G. C. Kennedy, Melting of sodium chloride at pressures to 65 kbar, *Phys. Rev.* **185**(3), 1135–1140 (1969).
166. M. Marezio and J. P. Remeika, Polymorphism of $LiMO_2$ compounds and high-pressure single crystal synthesis of $LiBO_2$, *J. Chem. Phys.* **44**(9), 3348–3353 (1966).
167. S. P. Clark, Effect of pressure on the melting points of eight alkali halides, *J. Chem. Phys.* **31**, 1526–1531 (1959).
168. J. P. Remeika and M. Marezio, Growth of $\alpha\text{-}Ga_2O_3$ single crystal at 44 kbar, *Appl. Phys. Letters* **8**(4), 87–88 (1966).
169. K. H. Dudziak and E. U. Franck, *Ber. Bunsenges. Phys. Chem.* **70**, 1120 (1966).
170. E. A. Binges and M. R. Gibson, Dynamic viscosity of compressed water to 10 kilobar and steam to 1500°C, *J. Mech. Engng. Sci.* **11**, 189–205 (1969).
171. E. U. Franck, Water and aqueous solutions at high pressures and temperatures, *Pure Appl. Chem.* **24**, 13–30 (1970).

10

Transitions in Viscous Liquids and Glasses

David Turnbull
Division of Engineering and Applied Physics
Harvard University

and

Brian G. Bagley
Bell Laboratories
Murry Hill, New Jersey

This chapter surveys transitions between the viscous liquid state and the glass, phase-separated, or crystalline states and some of the uses of these transitions in altering the structure and properties of solids. In addition to its intrinsic interest, the liquid ↔ glass transition is important for kinetically limiting the other two types of transition. It is these limiting effects which make possible the formation of quite unique structures in the transitions. We will begin by reviewing the liquid ↔ glass transition.

1. Liquid ↔ Glass Transition

The prime manifestation of the liquid → glass transition is a homogeneous and continuous hardening, as measured by shear viscosity, of a body with falling temperature[1] (Figure 1). This change may be accompanied by secondary effects,[2] such as abrupt decreases in heat capacity (Figure 2) and thermal expansion coefficient (Figure 3), due to the inability of the body to reach configurational equilibrium during the time of observation in the high-viscosity range. This secondary behavior reflects the experience that the time constants τ_c

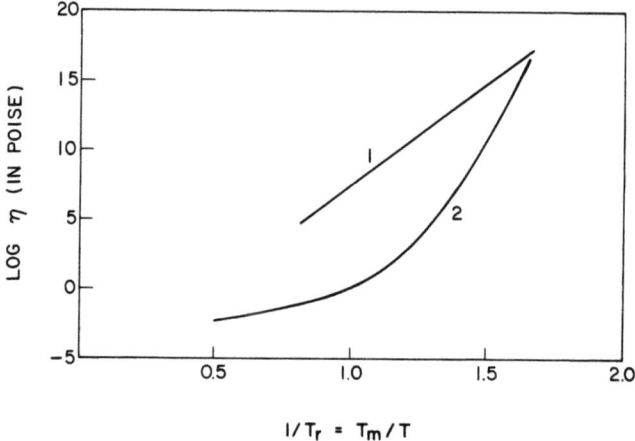

Fig. 1. Logarithm of the shear viscosity versus the reciprocal of the reduced temperature for glass-forming liquids. The data shown in curve 1 are for SiO_2, the viscosity of which (along with GeO_2) exhibits simple Arrhenius behavior [Eq. (1)] through the glass transition ($\eta = 10^{15}$ P). Curve 2 is the more generally observed viscosity behavior, in which the apparent activation energy is temperature dependent (after Turnbull[36]).

for molecular configurational changes in liquids scale roughly as the shear viscosity. According to this scaling relation, τ_c would be of the order of ½ hr and increasing rapidly with falling temperature at the viscosity $\eta \sim 10^{13}$ P where the thermal manifestation of the glass transition is often seen. It would be a day or so at the temperature, only slightly lower, where η passes the value of 10^{15} P usually taken to mark the transition from fluid to solidlike behavior. It is evident that at

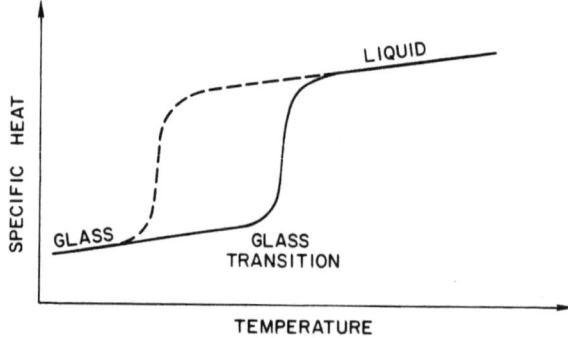

Fig. 2. The thermal manifestation of the glass transition observed upon cooling the liquid. Shown dashed are the results obtained on the same material cooled more slowly through T_g.

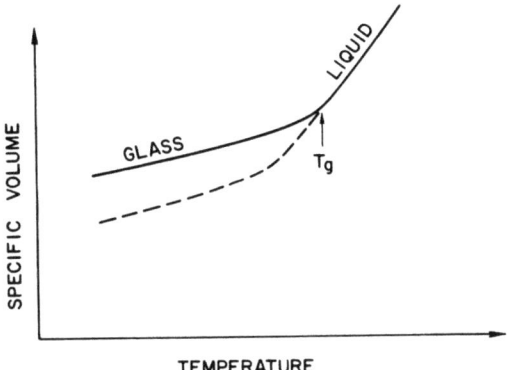

Fig. 3. The specific volume (the derivative of which with respect to temperature is the thermal expansion) of a liquid cooled to form a glass. The dashed curve is obtained when the same material is cooled more slowly.

temperatures a little lower still the time constant would become longer than the period which can be allotted, conveniently, to one experiment.

The temperature dependence of the viscosity of SiO_2[3] (Figure 1, curve 1) and GeO_2[4,5] from the fluid to the glass transition region seems adequately described by a single equation of the simple Arrhenius form:

$$\eta = \eta_0 \exp(\Delta H_\eta / RT) \qquad (1)$$

where ΔH_η is of the order of magnitude of a bond energy. According to simple rate theory, the preexponential factor can be expressed as[6,5]

$$\eta_0 = \alpha(kT/v)\tau_r \exp(-\Delta S_\eta / R) \qquad (1a)$$

where v is the molecular volume, τ_r is the time constant for the pertinent molecular oscillation, ΔS_η is the entropy of activation, and α is a constant which is usually within the order of magnitude of unity. Actually, the preexponential factors for GeO_2 and SiO_2 can be accounted for by this relation with plausible assignments of values to the entropies of activation. Thus we can suppose that these melts respond to shearing forces simply by the rupture and reformation of covalent bonds, with the former process rate limiting. The thermal manifestations of the glass transition in GeO_2 and SiO_2 are, at most, quite small and diffuse.[7–10]

In sharp contrast with the behavior of the tetrahedrally coordinated compounds, the viscosities of most glass-forming melts

Chapter 10

deviate sharply from simple Arrhenius behavior in the sense that the apparent activation energy for flow increases rapidly with falling temperature as the glass transition range is approached (Figure 1, curve 2). This behavior has been described by an equation, variously attributed to Vogel[11] or Fulcher,[12] of the form

$$\eta = A \exp[a/(T - T_0)] \tag{2}$$

We will call this the VF equation. However, the viscosity of any one substance cannot be fitted satisfactorily over the entire temperature range by this equation with a single set of the parameters. Rather, it seems that the behavior can be characterized best by distinguishing a higher temperature regime, where η ranges from 10^{-2} to 10^{4-6}, and a lower temperature regime, where η ranges from 10^{4-6} to 10^{15} P. The higher temperature behavior, which exhibits the greatest curvature in the $\log \eta$ vs. $1/T$ plot, can, in general, be fitted very well by the VF equation with one set of parameters.[13,14] It is generally supposed that the lower temperature behavior is satisfactorily fitted also by the VF equation but with a different set of parameters. Indeed, the well-known WLF universal equation[15] for the lower temperature behavior is of the VF form. However, due to the very limited temperature ranges spanned, it is really difficult to decide which temperature function best fits the data in this regime. Actually, Uhlmann and collaborators[14,16] have shown that, for a considerable number of substances, these data are better fitted by an equation of the Arrhenius than of the VF form.

Also, in contrast with the tetrahedrally coordinated compounds, the heat capacity changes which mark the liquid → glass transition of most glass-formers are sharp and quite large.[2] That these changes are, as we have noted, kinetic in origin is indicated by the decrease in the temperatures at which they occur with decreasing cooling rate (Figure 2). The possibility of a thermodynamic transition, at temperatures well below T_g, from the liquid to a hypothetical "ideal" glass state was first proposed and developed by Gibbs and DiMarzio.[17] It has been discussed further in several other papers and reviews.

We will survey briefly some of the prominent models which have been proposed to account for the molecular transport behavior of glass-forming melts. Often it is not clear whether a particular model is intended to apply to one or the other or both of the temperature regimes alluded to. In considering the models, it will be helpful to keep in view that the scaling relation, already referred to, between the time constants for diffusive transport and molecular reorientations and the shear viscosity seem to hold roughly, at least, over the range in which η changes from $\sim 10^{-2}$ to 10^{7-8} P. Further, at the highest temperatures the transport coefficients of simple liquids do not appear to deviate much from those predicted by the Enskog theory for dense gases, in which is is assumed that all molecular motions are translatory or diffusive in nature. Therefore any theory for molecular transport in glass-formers[18] must somehow account for a transition from this primarily diffusive regime to the oscillatory one which characterizes molecular motions in glasses and other solids.

We have noted that the transport behavior of most glass-forming melts cannot be interpreted in terms of a single-valued potential energy barrier. However, the potential barrier model of rate theory could be preserved formally and simply by supposing that the transport step is effected by the cooperative movement of n groups which must be activated simultaneously. The probability p that one group is activated can be expressed as

$$p = (1 - p)\exp(\Delta S'/R)\exp(-\Delta H'/RT) \qquad (3)$$

where $\Delta S'$ and $\Delta H'$ are, respectively, the entropy and enthalpy of activation. Then the probability that n groups are activated is

$$p^n = (1 - p)^n \exp(n\,\Delta S'/R)\exp(-n\,\Delta H'/RT) \qquad (4)$$

and the transport frequency can be taken to be $k_{\rm tr} \sim Kp^n$. At high temperatures, with increasing T, p would approach unity and the apparent activation enthalpy would tend toward zero, provided $\Delta S' > 0$. At low temperature, with T decreasing, p would approach zero while the apparent activation enthalpy would tend toward the constant value $n\,\Delta H'$. This general form of the model is, of course, inexplicit and it makes no obvious connection between the diffusive and oscillatory regimes.

The free volume model (FVM) does seem to provide a means for effecting this diffusive–oscillatory connection.[13,18] Indeed, this model is probably suited best for application to the high-temperature transport regime, though it is often used to fit low-temperature behaviour. It is based on the following central assumptions: (1) Transport is controlled entirely by intermolecular repulsive interactions. The attractive potentials are important only in their effect on the average density of the system. (2) Transport results from the occurrence of special configurations rather than from concentrations of kinetic energy. From these assumptions we see that the FVM actually is an adaptation of the mean field model (MFM) to transport phenomena. We will therefore review its application to an assembly of uniform hard and unattracting spheres to which the MFM should precisely apply.

We suppose that the self-diffusivity D in a hard-sphere system is simply related to the Enskog self-diffusivity D_E by

$$D = fD_E \qquad (5)$$

where f, the correlation factor, is the ratio of the actual diffusivity to that calculated by the random walk analysis; i.e., with all the scattering angles equally probable. The value of f will range from one to zero as the atomic motions change in nature from purely diffusive, in the dilute fluid, to purely oscillatory, in the ideal solid. The oscillatory motion is, of course, imposed by the increasing degree of backscattering from collisions as the system becomes more dense. The object of the model development then is to evaluate f.

In the simple FVM it is supposed that the probability of backscattering is determined by the local density. This density will fluctuate widely and it seems plausible that the motions will be mostly oscillatory, i.e., "solidlike," in the regions of highest density and diffusive (i.e., "gaslike") in the least dense regions. Bueche[19] developed a density fluctuation model based on this concept in the context of the harmonic motion approximation. Cohen and Turnbull[13,18] treated the problem strictly within the mean field framework. They assumed that a particle can move diffusively only when some critical density attenuation occurs in the cage delineated by its nearest neighbors, as defined, e.g., by a Voronoi construction. The total free volume of the system Nv_f, where N is the total number of particles, is considered to be partitioned randomly among the cages. Then, if the free volume Δv within a cage is less than a critical value v^*, the particle motion is oscillatory, but it may be diffusive when $\Delta v \geq v^*$. The diffusive displacement is taken to be proportional to Δv. On this basis Turnbull and Cohen[18] derived the following expression for the self-diffusion coefficient:

$$D = \tfrac{1}{3}\alpha \langle c \rangle [v^* + (v_f/\gamma)] \exp(-\gamma v^*/v) \qquad (6)$$

where $\langle c \rangle$ is the average speed of the particles, γ is a factor of order 1/2 to unity to correct for the overlap of free volume between the cages, and α is the proportionality factor relating the displacement $a(\Delta v)$ to Δv. The corresponding expression for the correlation factor is

$$f = [1 + (\gamma v^*/v_f)] \exp(-\gamma v^*/v_f) \qquad (7)$$

These expressions extrapolate to the proper limiting values as $v_f \to \infty$ or $v_f \to 0$ but they should be most applicable in the relatively high-density regime where $v^* \gg v_f/\gamma$; then

$$D \cong \tfrac{1}{3}\alpha v^* \langle c \rangle \exp(-\gamma v^*/v_f) \qquad (8)$$

and

$$f \cong (\gamma v^*/v_f) \exp(-\gamma v^*/v_f) \qquad (9)$$

A parallel formulation[18] of the model can be made in terms of the path lengths a between collisions. In this formulation the key assumption, analogous to that of a critical density fluctuation, is that the motion is diffusive if a exceeds a critical value a^* and is oscillatory if $a \leq a^*$. Then

$$D = \tfrac{1}{3}\langle c \rangle \lambda (1 + a^*/\lambda) \exp(-a^*/\lambda) \qquad (10)$$

and

$$f = (1 + a^*/\lambda) \exp(-a^*/\lambda) \qquad (11)$$

where λ is the mean free path of the particle at the given average density.

The equilibrium and transport behavior of three-dimensional hard-sphere systems have been investigated extensively by molecular dynamics and Monte Carlo techniques. These studies[20] revealed a first-order transition from a fluid with a specific volume of $1.49v_0$, where v_0 is the volume per particle in the ideally close-packed crystal, to a fcc crystal having a density of $1.35v_0$. The specific volume of the fluid, as well as of the crystal, at the crystallization point is very much higher than the value $v_{DRP} = 1.165v_0$ measured by Bernal and others of the structure (DRP) formed by the densest random packing of hard spheres. This structure is amorphous and apparently it does not collapse upon compression to a crystalline structure. Cohen and Turnbull[21] have identified it with the metastable "ideal" glass state of a monatomic system. This means that it should be internally stable with a self-diffusion coefficient of zero and $\eta = \infty$.

The self-diffusion coefficients of a three-dimensional hard-sphere fluid, as determined by molecular dynamics computations,[22] exhibit a component due to motions of vortices within the fluid as well as backscattering effects. However, the latter effects appear to become predominant at the higher densities. Indeed, at the crystallization point, D of the fluid is falling sharply, relative to the Enskog diffusivity, and apparently it would sink to negligibly small values at densities well below that of the DRP structure.

Chapter 10

In the specific volume range just above that at which thermodynamic crystallization occurs, the backscattering effects on D apparently are fitted satisfactorily[18] by an equation having the form of (6) with $v^* = 0.72 v_{\text{DRP}}$. In making this fit the free volume was taken to be

$$v_f = \bar{v} - v_{\text{DRP}} \qquad (12)$$

where $\bar{v}$ is the average volume per particle of the fluid.

C. H. Bennett (private communication) computed the diffusivity of a system in the DRP configuration and interacting by Lennard-Jones potentials. When the system was compressed to the density at which the total potential energy was zero he found $D \sim 0$. Weaire *et al.*[23] find that a DRP structure in which the particles interact pairwise by a fitted Morse potential reacts stably toward small shearing forces. Further support for the idea that D of the amorphous phase should go to zero at the DRP, rather than the ideal fcc, specific volume is provided by Le Fevre's[24] observation that the reciprocal compressibility factor ($1/z = kT/P\bar{v}$) of the hard-sphere fluid appears, upon extrapolation, to go to zero at the DRP density.

In applying the model to real molecular systems, specification of the free volume is a difficult problem. The difficulty arises mainly because the repulsive forces are not infinitely hard. In scaling the transport behavior of monatomic fluids to the results of molecular dynamic computations on the hard-sphere fluid, Dymond and Alder[25] and others have used an effective hard-core diameter R_{hc} of the molecule which is supposed to decrease with increasing temperature. If the intermolecular pair potential were known, R_{hc} might be taken, for example, as the intermolecular separation at which the potential energy on the repulsive branch is kT above the minimum. Alternatively, R_{hc} might be specified from experimental measurements of the temperature variation of the van der Waals parameter b of the dilute fluid. Applying these concepts to the FVM, it follows that the effective free volume at constant total volume must increase with increasing temperature, i.e., the transport behavior cannot be determined by the specific volume alone, and this variation must be taken into account in interpreting pressure effects.

The FVM leads simply to the VF form Eq. (2) for the isobaric temperature dependence of the transport coefficients. It fits the shear viscosities of simple molecular liquids in the high-temperature regime[14] ($\eta \sim 10^{-2}$–10^{4-6} P) with plausible values for the constants of the VF equation. We have noted that the viscosity in the low-temperature regime falls well below the extrapolation of the higher temperature equation, as though some short-circuiting mechanism were operative. This behavior may reflect that the longer range correlations in the DRP structure develop with more difficulty than those with shorter range.

The FVM is essentially an entropic interpretation of transport. A more generalized entropic interpretation for viscosity behavior in the glass transition was developed by Gibbs and co-workers. We have noted that Gibbs and DiMarzio[17] developed a statistical mechanical theory for a thermodynamic transition from the liquid to an "ideal" glass. An important outcome of this theory is that the configurational entropy S_c, practically taken to be the entropy difference between the amorphous and crystalline phases, should vanish in a second-order transition at some temperature T_0. However, this transition is not observed experimentally because of the intervention of the kinetic solidification, freezing the configuration, at T_g, which is expected to lie well above T_0.

Later, Adam and Gibbs[26] proposed a model based on the assumption that flow is limited by steps which can occur only in subsystems of the fluid in which the configurational entropy exceeds some critical value s_c^*. This assumption is the analog to that in the FVM development that transport is attendant upon a critical fluctuation in the local density ($\Delta v \geq v^*$). In the Adam and Gibbs development the size of the subsystem z^* required to encompass s_c^* increases with decrease in S_c, analogous to v_f in the FVM, of the whole system. The result of their development is a relation of the form

$$\eta = \eta_0 \exp(b/TS_c) \tag{13}$$

In a system to which the MFM applies, it appears that the FVM and configurational entropy model should reduce to essentially the same form, and an approximate analysis by Chen and Turnbull[27] indicates that this is so. Both models and their interrelation have been reviewed comprehensively and critically by Berry and Fox.[28]

Our treatment of the liquid ↔ glass transition has been quite selective and has not done justice to other interesting approaches to the problem, such as those by Angell,[29] Goldstein,[30] and Eyring[6,31] and their co-workers.

2. Phase Separation of Viscous Melts

We have noted that to reach the glass transition a melt must, in general, be cooled through a wide temperature interval in which it is metastable relative to some crystallized state of the system. In this

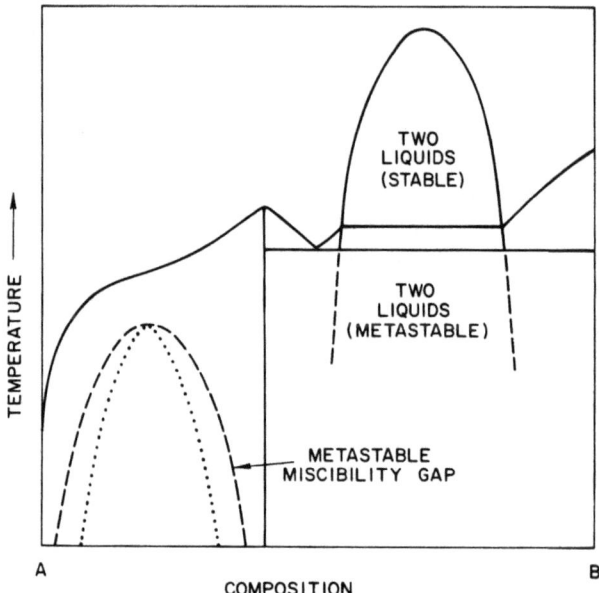

Fig. 4. Schematic phase diagram for a binary system exhibiting phase separation. A stable miscibility gap is illustrated, and regions of metastable immiscibility are delineated with dashed lines. Within the metastable miscibility gap, the dotted line denotes the spinodal.

metastable range, separation of the melt into two melts with different compositions may also become possible thermodynamically (Figure 4). In sharp contrast with crystallization, this transformation is easily initiated homogeneously at small departures from equilibrium. Consequently, it is extremely difficult to quench through the transformation range without the occurrence of separation.

In this type of transformation no topological discontinuity develops in the structure and the tension of the interface separating the two phases is relatively small.[32] Further, when cooled into the spinodal range, the homogeneous melt becomes unstable to composition fluctuations with wavelengths exceeding some critical value.[32,33] These critical wavelengths become small and as a result the separated phases may be very closely spaced (e.g., tens of angstroms apart). The average diffusion distance scales roughly as this spacing. If the time constant for diffusive motion scales as the shear viscosity, then at a spacing of 100 Å substantial phase separation would occur within a period of 1 sec at viscosities as high as 10^6 P.

Certain mechanical, magnetic, and electrical properties of solids are optimum when the internal morphology of the structure consists of an extremely fine interdispersion of one crystalline phase in another or of very small grains. One of the most effective methods of achieving these structures is by phase separation of viscous melts followed by crystallization. The inherently low liquid–liquid interfacial tensions favor, as we have noted, separation on a fine spatial scale, while the high viscosity of the melt retards coalescence and gravity segregation. Frequently, the formation of the fine-scale crystalline structures is facilitated by a high crystallization rate in one of the separated liquids. A classic example of an important structure produced by this phase separation and crystallization sequence is the fine-grained ceramic formed when certain silicate melts phase-separate and then devitrify internally.[34]

The thermodynamic likelihood of liquid–liquid separation at or just above T_g is quite high for systems containing two or more components. One reason for this is that the entropic stabilization of the disorder is relatively small at T_g. Further, most glass compositions will not correspond to those of the elements or stoichiometric compounds which are likely to be most favored energetically.[35] A corollary, pointed out by Morral[35] and Cahn, is that the motivation toward phase separation will tend to increase with the number of melt components.

3. Crystallization of Viscous Systems

We have noted that for most substances the glass temperature T_g lies well below the thermodynamic crystallization temperature T_m. Whether or not a melt can be undercooled to T_g will be determined by a set of conditions specified in the laboratory and by the kinetic constants of crystallization,[36] which are material characteristics.

The laboratory variables are the cooling rate $\dot{T}$, specimen volume v_l, and initial density of crystallization centers ρ_N; more particularly, the greater is $-\dot{T}$ and the lesser are v_l and ρ_N, the more likely is the formation of glass by melt quenching.[36] We note that these are the variables which determine the degree of finiteness of the system. In statistical theory, equilibrium is achieved if the number N of members of the statistical ensemble goes to infinity. Metastability results from limitations on the magnitude of N imposed by the physical conditions. Crystallization is initiated, even at considerable undercooling, by configurations so rare that they are often not realized in the times and specimen volumes accessible to experiment.

The kinetic constants in crystallization are the frequency of crystal nucleation per volume and the rate of crystal growth u. Simple nucleation theory leads to an expression of the following form for the frequency I of homogeneous nucleation of crystals in melts[36]:

$$I \cong (n_v/\tau_N) \exp[-b\alpha^3\beta/T_r(\Delta T_r)^2] \qquad (14)$$

where n_v is the number of molecules per volume; τ_N is the time constant for the jumping of a molecule across the nucleus–melt interface; b is a numerical factor determined by the shape of the nucleus; $\beta = \Delta S_m/R$, where ΔS_m is the entropy of melting; $T_r = T/T_m$; $\Delta T_r = (T_m - T)/T_m$; and $\alpha = (N\bar{V}^2)^{1/3}\sigma/\Delta H_m$, where N is Avogadro's number, $\bar{V}$ is volume per mole, ΔH_m is the molar heat of melting, and σ is the crystal–liquid interfacial tension. Essentially α is a measure of the fraction of a crystal monolayer per area which, in melting, would adsorb an amount of enthalpy equal in magnitude to σ. In this formulation the temperature dependence of α and β has been ignored and it is assumed that α is single-valued over the surface of the nucleus.

The actual number of crystallization centers is the sum of those formed homogeneously, $\int I v_l \, dt$, and those developed from heterogeneities, $\rho_N v_l$. Turnbull has specified[36] the conditions whereby the number of homogeneously nucleated centers would be negligible. For example, even if τ_N were no greater than 10^{-12} sec, homogeneous formation of a single nucleus would be rare, under most experimental conditions, if $\alpha\beta^{1/3} > 0.9$. With $\alpha\beta^{1/3}$ decreasing from 0.9, the range over T_r of melt metastability remains large but the magnitudes of cooling rate and limitation on v_l required to quench to a glass increases sharply (Figure 5).

Generally there are heterogeneities in contact with a melt which initiate crystallization at some undercooling in the fluid range. Then the tendency to form glass would be determined by the number of these heterogeneities, the undercoolings at which they initiate crystallization, and the crystal growth rate.

It is convenient to discuss the isobaric rate of growth of crystals into undercooled melts in the following terms[37,38]:

$$u = C(f_s/\tau_u)F(\Delta G) \qquad (15)$$

where $F(\Delta G)$ is a thermodynamic factor determined by the Gibbs free energy ΔG driving crystallization; f_s is the fraction of sites in the crystal surface on which growth may occur, and should depend strongly on ΔG and the crystallographic orientation of the crystal surface $\hat{\eta}$ normal to the growth direction; τ_u is the time constant for

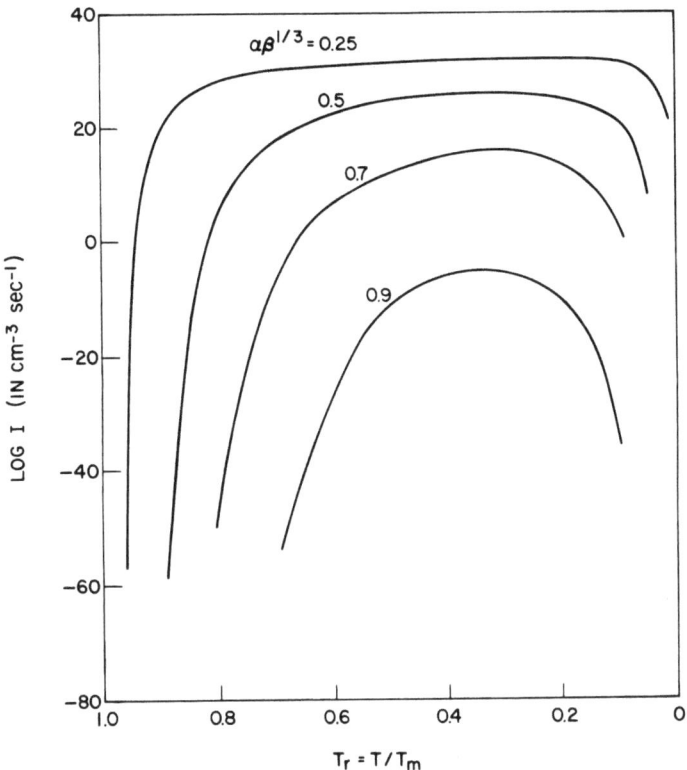

Fig. 5. Logarithm of the homogeneous nucleation frequency I versus undercooling for various values of $\alpha\beta^{1/3}$. This is a plot of Eq. (14) with n_v/τ_N taken as 10^{32} and b as $16\pi/3$ (after Turnbull[36]).

the attachment of a molecule from the melt to a crystal surface growth site, and should vary with both $\hat{\eta}$ and T; C is a dimensionality factor which may depend weakly on $\hat{\eta}$ and T.

In growth normal to low-index planes, f_s generally corresponds to the proportion of sites at the edges of steps on the crystal surface and it would be determined by the frequency of generation of these steps.[37] In the theory of surface step nucleation, this frequency is controlled, at a given undercooling, by σ_e, the melt–edge interfacial tension, and by the presence or absence of emergent screw dislocations in the surface. In any case, the spacing of step edges would be larger, corresponding to smaller f_s, the larger is σ_e. Jackson[39] noted that whether or not the growth appears to be limited by step generation correlates with the magnitude of the reduced entropy of melting β at T_m. More particularly, step generation seems to be infrequent when β is large compared with unity but it is copious at the smallest measurable undercooling for systems where β is of order unity or less.$^{(39,40)}$ To account for this behavior, Jackson$^{(39)}$ developed a statistical theory, analogous to that of Burton, Cabrera, and Frank$^{(41)}$ for the dilute fluid case, which predicted that at $\Delta T_r \to 0$ the equilibrium interface would be rough, i.e., $f \sim 1$, for $\beta \lesssim 1$, while at large β it would be essentially smooth, i.e., $f \sim 0$. However, this development was based on the assumption of a continuum model for the melt structure. An alternative point of view$^{(42)}$ is that σ_e, at small β, is so small that it poses no significant barrier to step nucleation except at undercoolings which may be too small to specify experimentally. There is currently much research in this area and recent developments are reviewed elsewhere in this Treatise.$^{(43)}$

Experience indicates that the time constant τ_u for crystal growth in molecular and covalently bound melts tends to scale roughly, as do the time constants for diffusive transport τ_D and molecular reorientation τ_c, as the shear viscosity of the melt. In covalently bound systems this correspondence may reflect that the attachment of a group to a growth site is limited, as are the other processes, by the breaking of covalent bonds in the melt. For molecular systems the correspondence suggests$^{(44)}$ that τ_u is determined by the reorientation of a molecule in the melt to an orientation suitable for attachment to the growth site.

In pure monatomic melts neither bond breaking nor molecular reorientation will limit crystal growth. Further, movement across the melt–crystal interface is not likely to be restricted by backscattering nearly as much as is diffusive transport within the melt. Consequently, τ_u for monatomic melts should be considerably smaller than the time constant τ_D for diffusive motion. Indeed, the lower limiting value of τ_u may be the average period of collision of atoms from the melt on the crystal surface. If so, the velocity of crystal growth, at given $F(\Delta G)$, may scale roughly as the Enskog, rather than the actual, transport coefficients.$^{(45)}$ Accordingly, it may be expressed as

$$u \leq \dot{n}_A v F(\Delta G) \qquad (16)$$

where v is the volume per atom and $\dot{n}_A$ is the collision frequency per area of atoms from the melt on the crystal surface. We shall estimate $\dot{n}_A$ by supposing that it is related to the ideal gas impingement frequency in the same way as is $\dot{n}_A$ of the three-dimensional hard-sphere fluid at its crystallization point; $F(\Delta G)$ will be equated to $\Delta S \Delta T_r / RT$.

Walker$^{(46)}$ measured the speed of growth of Ni and Co crystals into their pure melts at very large undercooling. These speeds, since they were still apparently increasing as $(\Delta T)^2$, must be considered to be lower limiting values at the specified undercoolings. At 175°C undercooling the measured speeds were 0.4–0.5×10^4 cm/sec, which is only a factor of 4–5 below the maximum, 2×10^4 cm/sec, calculated from Eq. (16).

Any redistribution of impurity which accompanies growth will occur at rates specified by the actual, rather than the Enskog, transport coefficients and the crystal growth rate, if limited by the redistribution, will scale accordingly. The increase in resistance to crystallization of metal glass-formers with impurity admixture may reflect, in part, the retardation of u by operation of the impurity effect. In monatomic systems there are no mechanisms evident to us whereby impurities in small amounts would catalyze crystal growth to a marked extent.

In covalently bound melts there is the possibility of catalysis of crystal growth by catalytic mechanisms akin to those which operate in chemical reactions between molecules. For example, impurities[47–49] or radiant energy which facilitate the breaking of covalent bonds may, under some conditions, catalyze growth. It has been established[48] that the rate of growth of Se crystals into undercooled Se melts is markedly increased by the addition of small amounts of halogen impurities. Apparently these effects reflect the lowered viscosity of the melt resulting from the reduced average length of Se chains following reaction with halogen. The rate of growth of cristobalite crystals into fused silica is increased sharply by the presence of small amounts of water.[47] It appears that the effect of water is to break "bridging"

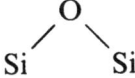

bonds to form Si–OH groupings which should be quite easily activated in the interface process. We suppose that water molecules are positively adsorbed at the cristobalite–melt interface and migrate with this interface as they act catalytically.

Irradiation with energies in excess of some threshold value may create "dangling" bonds in a covalent melt and thereby facilitate crystal growth. There is evidence, to be discussed in a later section, on the occurrence of this effect in some systems.

Isothermal application of pressure or of tension to the melt may change all of the factors [Eq. (15)] which determine the crystal growth rate.[50] There is no information on the change of f_s with P but, in principle, orders of magnitude variation seem possible in some systems. In view of the approximate scaling of τ_u with η and the simple models presented for the glass transition, we expect, at least for simple molecular systems, that τ_u should increase or decrease sharply with application of pressure or tension, respectively. Apparently the effects of tension on crystallization kinetics have not yet been explored. We think that they merit thorough investigation.

There are several well-controlled studies[50-52] which indicate that crystal growth rates in some covalently bound melts are, under certain conditions, sharply increased by application of high pressure. The increases were orders of magnitude greater[50,52] than can be accounted for by the increases in the thermodynamic factor $F(\Delta G)$ with pressure. These rather unexpected pressure responses have been exhibited by B_2O_3,[51,52] SiO_2,[50] and films of amorphous Ge.[53] Ulhmann et al.[50,52] have discussed several possible explanations for them. Where f_s is small at the ambient pressure, as it seems to be in B_2O_3, there is the possibility that it may increase sharply with pressure reflecting a corresponding decrease in σ_e. However, it is not evident why this should occur unless attended by some significant change in melt structure.

In the growth of cristobalite into fused silica it appears that the growth site fraction must be of order unity at the ambient pressure. Here the effect of pressure may be to accentuate the catalytic action of water, present as impurity, on the growth.[50] A thin and mobile water-rich film might form at the interface and transport silica from glass to cristobalite as in a hydrothermal process. A mechanism like this probably effects the room-temperature devitrification of GeO_2 in moist atmospheres at 1 atm pressure; under these conditions GeO_2 is much more soluble in H_2O than is SiO_2.

Amorphous states in which the interatomic cohesive forces are primarily centrosymmetric generally have smaller volumes, reflecting higher coordination numbers, than those in which the bonding is highly directional. Consequently, the thermodynamic resistance of the transformation from a covalently bound amorphous state to a nondirectionally coordinated one m_{nd}, in which the atomic mobility is likely to be much higher, will decrease with pressure application and vanish at some sufficiently high pressure. At such a pressure the m_{nd} might form at the crystal-melt interface and greatly facilitate its motion. The crystallization of a-Ge at room temperature under high pressure[53] may be effected by a mechanism of this type in which the mobile layer is constituted by the metallic liquid state of Ge. The operation of an analogous mechanism, attendant upon transition from covalent to more ionic type binding, provides an alternative explanation for the pressure accentuation of crystal growth in B_2O_3 and SiO_2.

4. Electrical Threshold Switch

The nondestructive, reversible, abrupt conductivity transition which occurs in some materials upon the application of a high electrical field is called electrical threshold switching. Switching has been observed in single crystals, polycrystals, powders, composites, amorphous solids, and liquids, and in such diverse materials as elemental and compound semiconductors, transition metal oxides (both crystalline and amorphous), amorphous chalcogenides, and organic polymers (for references see Ref. 54 or 55). However, we confine our discussion to switching in the nonoxide chalcogenide amorphous solids, the materials most extensively studied and best understood. The nonoxide chalcogenides contain, as a major constituent, one or more of the chalcogens S, Se, or Te. Other principal elements in these alloys are from columns IV (Si, Ge) and V (P, As, Sb) of the periodic table. Additional details on switching are available in several conference proceedings[56-59] and reviews.[54,55,60] The early history of the observations of switching (and memory) effects has been reviewed by Pearson[61] and Ovshinsky.[62]

A typical current–voltage trace for a semiconducting amorphous material between two metallic or carbon electrodes is illustrated in Figure 6. The high-resistance (off) state current is Ohmic at low voltages, but at high fields ($>10^4$ V/cm) it varies as[63,64] $I = I_0(T) \exp[V/V_0(T)]$, where I_0 and V_0 are experimentally determined parameters. This super-Ohmic conductivity has been experimentally demonstrated to be due to a nonthermal electronic process.[65] Conduction in the off state appears to be bulk- rather than contact-limited.[66]

At a threshold voltage V_T the material switches to the low-resistance (on) state along the circuit load line (Figure 6). There is a marked difference in the switching behavior of "thick" versus "thin" samples, with the delineation at a thickness of about 5 μm. For thin films the threshold field is independent of thickness at a value of $\sim 10^5$ V/cm, whereas for thick films it is thickness dependent (Figure 7).[67] In addition, thick samples exhibit a negative resistance region

Chapter 10

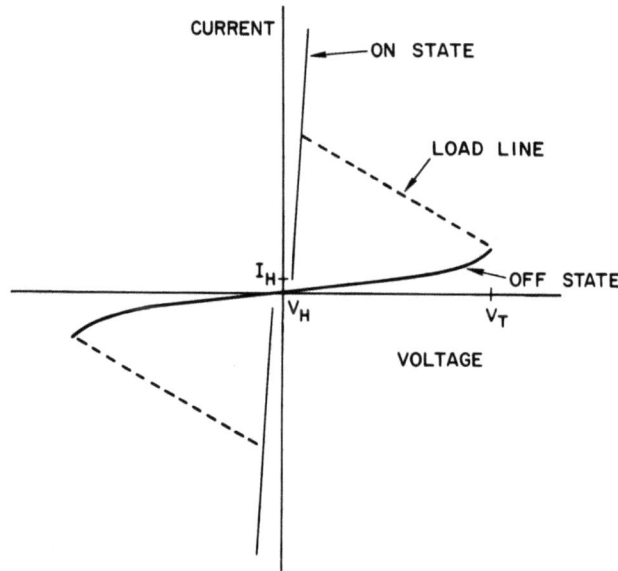

Fig. 6. The current–voltage characteristic of an amorphous semiconducting material which exhibits threshold switching.

in the off state not observed in thin films. The threshold voltage is temperature dependent, decreasing with increasing temperature.[63,68,69]

Upon the application of a voltage which exceeds V_T, there is a delay time τ_d, after which the material switches very rapidly ($\sim 1.5 \times 10^{-10}$ sec) into the on state. For thin films and an applied voltage which exceeds threshold by more than 20%, τ_d decreases exponentially with voltage,[66] approaching 10^{-9} sec at $V = 2V_T$. For bulk materials, the delay time is a more complicated function of applied voltage.[70]

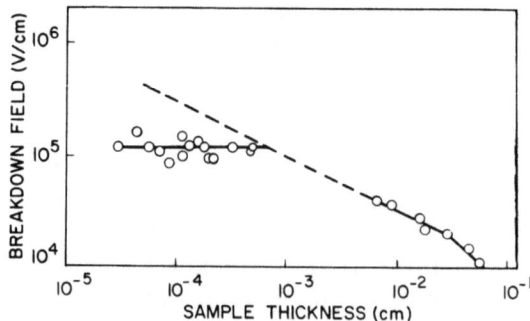

Fig. 7. The effect of thickness on switching (breakdown) field (after Kolomiets et al.[67]).

From their experiments on thin films, Haberland and co-workers[71,72] concluded that the onset of switching is controlled by the accumulation of charge during the delay time. The minimum stored charge required for switching was observed to be proportional to thickness and independent of temperature (273–333°K). However, Csillag and Jäger[73] concluded from their experiments that the onset of switching is controlled by the accumulation of energy during the delay time. Shanks[68] and Balberg[74] observed that if the pulse voltage polarity was reversed during the delay time, τ_d remained unchanged. This result is temperature sensitive, however, since Henisch and Pryor[75] observed a dependence of τ_d on polarity reversal in measurements made at lower temperatures; τ_d increased when the voltage polarity was reversed during the delay time.

In the low-impedance (on) state, the current collapses to a filament and the dynamic resistance is very small (1–10 ohm) and nearly temperature independent. The on-state holding voltage V_H is approximately 1 V, almost independent of temperature and film thickness (for thin films). The on state is not stable at zero bias and a holding current I_H is required to prevent switching to the off state. If the current is reduced below I_H, the device will switch off, requiring about 1 μsec to recover the original off-state characteristics. Pulsing the switch before this μsec recovery time has elapsed will result in switching at a voltage lower than V_H.[76] The switch will remain in the on state for very short-time (≥ 0.1 μsec) reductions in voltage. By probing with such short-time voltage pulses, Pryor and Henisch[76] have observed what they term a "blocked-on state" at voltages below V_H and resulting currents much smaller than I_H. This very high-resistance, blocked on state is continuous with the low-resistance, stable on state, and different, and disconnected, from the stable off state.

With metallic or carbon electrodes, the I–V curve is symmetric in the first and third quadrants. However, with doped crystalline Ge electrodes, asymmetries are produced in the I–V switching characteristics[77–80] which cannot be accounted for by assuming that the crystalline Ge acts as a rectifier in series with the switch.[77] From their experiments with doped crystalline Ge electrodes, Stiegler and Haberland[80] concluded that switching can be initiated only when electrons and holes are both injected into the switching material.

A number of mechanisms have been proposed to account for these observed switching characteristics. It is convenient to divide the mechanisms into three groups: thermal, in which the temperature dependence of the electrical conductivity is the most important factor;

electronic, in which the temperature increase of the material during switching is negligible; and a combination of these two.

The thermal mechanism (known as thermal breakdown, thermal runaway, or thermal avalanching) can occur in a material which has a strong dependence of electrical conductivity on temperature. The application of a field to such a material produces a positive feedback loop in which the resulting current produces Joule heat, which increases the temperature, which increases the conductivity, which increases the current, etc. An instability occurs when the heat generated exceeds the heat lost by conduction, whereby the current and the self-heated region collapse to a filament, establishing a new stationary state (if the current is limited by an external resistance, otherwise an irreversible breakdown through vaporization of the material will occur). The switching characteristics are obtained as solutions to the heat transport and charge conservation equations subject to the appropriate boundary conditions, device geometry, external circuitry, and materials properties (thermal and electrical conductivities and heat capacity), but to perform the calculations, simplifying assumptions are necessarily made. Such thermal avalanching calculations have been extensively done for various geometries and approximations since the early 1900's. Recently several authors have made calculations involving conditions and materials properties appropriate for switching devices and compared their results to observed switching characteristics.

In the "purely thermal" model calculations[81-94] the conductivity is taken as a function of temperature only. The results of these calculations are in very good agreement with experimental observations on switching in thick films. For example, the super-Ohmic conductivity is due simply to self-heating of the material, the delay time is regarded as the time required to heat to the instability, and the $I-V$ switching characteristics are quantitatively determined. However, for thin films the purely thermal theories are only qualitatively in agreement with observation and have several important discrepancies. For example, the theory predicts a threshold voltage which is higher than is observed, with a thickness dependence contrary to experiment (dashed line in Figure 2), and is unable to account for the blocked on state.[76]

More appropriate for thin films, with their higher switching voltages, are the electrothermal (or field-assisted thermal) theories.[95-105] In this approach, the conductivity is taken to have its experimentally observed field, as well as temperature dependence, and switching characteristics are then calculated in a manner similar to the purely thermal theories. The most comprehensive of these calculations[104,105] is in excellent agreement with all observed aspects of switching behavior. It is important to restate, however, that the agreement between the electrothermal theories and experiment depends crucially on the inclusion of the field dependence in the conductivity, and this field dependence is known to be nonthermal in nature.[65]

While good agreement obtains with electrothermal theories, the possibility exists that in thin films a purely electronic mechanism (such as can be observed in crystals) initiates switching and/or maintains the on state. Indeed, for very short-time (nsec) pulses[106] at high overvoltages and at low temperatures[107] electronic mechanisms become more likely. A number of such mechanisms have been suggested, including: electron avalanching;[108,109] switching initiated by strong field emission from electrodes with the on state maintained by tunneling of electrons and holes through Schottky barriers;[110] double injection of electrons and holes which results in space-charge overlap and a resulting highly conducting, neutral region maintained by the continued injection of electrons and holes;[111] trap filling by double injection reducing potential barriers with a resulting higher carrier mobility;[112] field-induced double injection through poor contacts;[113] a recombination instability resulting from double injections;[114] Schottky barriers created because electrodes cannot supply enough carriers to support the bulk high-field conductivity with the on state maintained by subsequent tunneling injection;[115] electron tunneling through an insulating layer (poor contacts);[116] a field-enhanced carrier mobility[117] (recent results indicate that the observed super-ohmic conductivity is due to a field-enhanced carrier mobility[118]); and finally, recombinative injection of minority electrons.[119,120] These mechanisms have had varying degrees of success in explaining experimental results (see discussions in Refs. 55, 60, and 120).

While it is clear that in thick films thermal effects dominate, in thin films field and temperature both play an important role, both in initiating the instability and in determining the characteristics of the resulting on state. The relative contributions of temperature and field to observed effects have been the source of a continuing controversy.

5. Memories

Whenever a marked change in properties accompanies a bistable phase transition, the transformation has device potential as a non-

Chapter 10

volatile memory for use in information storage. Heat, light, electron beams, or electric fields can be used to initiate, and enhance, the transformation. Because of the electrical and optical property changes which accompany the transformations in amorphous semiconductors, much attention has recently been focused on the use of these materials for electrical and optical memories and the balance of this chapter is devoted to these two classes of devices. In this section, our division into electrical and optical memories is not meant to imply that the phenomena and devices are necessarily mutually exclusive; for a marked change in optical properties generally accompanies a marked change in electrical properties and thus, for example, we could have devices which are written optically and read electrically or vice versa. As with the electrical threshold switch (Section 4), we limit our discussion in this section to the nonoxide chalcogenides, the materials upon which the most extensive research has been done.[54-60]

5.1. Electrical Memories

In an electrical memory, the stored information is written in, read out, and erased electrically.[61,62] The current–voltage characteristic of an electrical memory switch is depicted in Figure 8. The

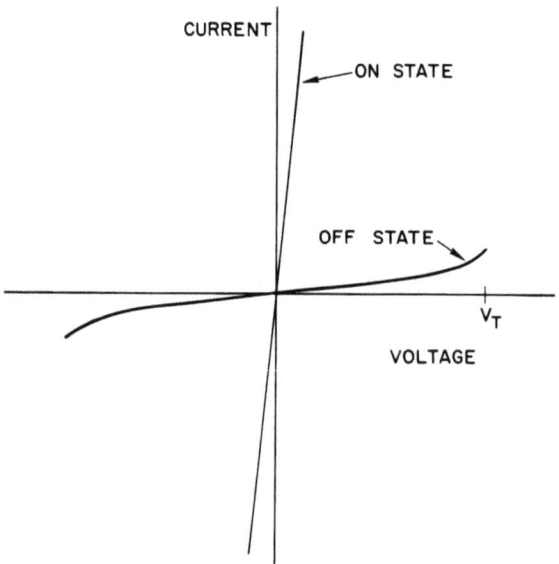

Fig. 8. The current–voltage characteristic of an amorphous semiconducting material which exhibits electrical memory switching.

material can exist in either of two conductivity states (bistability), both of which are stable at zero electrical bias (nonvolatility). The high-impedance (off) state has a resistivity of typically 10^4–10^{10} ohm-cm and the low-impedance (on) state has a resistivity of typically 10^{-1}–10^3 ohm-cm. To optimize reliability, materials are generally chosen such that off–on resistivity ratios are 10^5 or less.

The material is switched from the off to the on state with a voltage pulse which exceeds a threshold value V_T. Threshold switching into an on state (Section 4) is observed to precede the onset of memory. There is a critical time ($\sim 10^{-3}$ sec) in the threshold on state which must be exceeded to produce the memory on state.[121] If the voltage is reduced before this critical time, the material will return to the off state, as in the threshold switch. During this transient period (when the material has been threshold, but not memory, switched) large currents and therefore large amounts of Joule heat develop, but the electric field is small (Section 4). Thus thermal and high-current electromigration effects can play an important role, whereas high E-field effects contribute little.

The material is switched from the on to the off state by exceeding a critical current for a short time. To avoid accidental erasure, low current pulses are used to interrogate the device.

In contrast to the threshold switching mechanism, which is still controversial, the memory switching mechanism is well understood. Extensive research clearly ascribes the memory effect in the amorphous nonoxide chalcogenides to crystallization. Figure 9, taken from the data of Vengel and Kolomiets,[122] shows the conductivity for the glassy and polycrystalline phases for compositions in the As_2Te_3–As_2Se_3 pseudobinary. Note that in this system the crystallization of a glass can produce large conductivity increases. For the cooling conditions employed, compositions near As_2Te_3 could not be quenched to the glassy state. Similarly, for the particular heat treatment used, compositions near As_2Se_3 could not be crystallized. However, for compositions from ~ 45 to ~ 85 mole % As_2Te_3, the material could exist in either the crystalline or glassy state, and it is the bistability of this phase transformation, coupled with the associated large change in conductivity, which forms the basis for the memory effect. Using calorimetry and diffraction techniques, together with a concurrent determination of memory behavior in this system, Bagley and Bair[123] clearly demonstrated the association of crystallization and electrical memory; only in the composition region in which the glass-to-crystal transformation could be reversibly induced was the

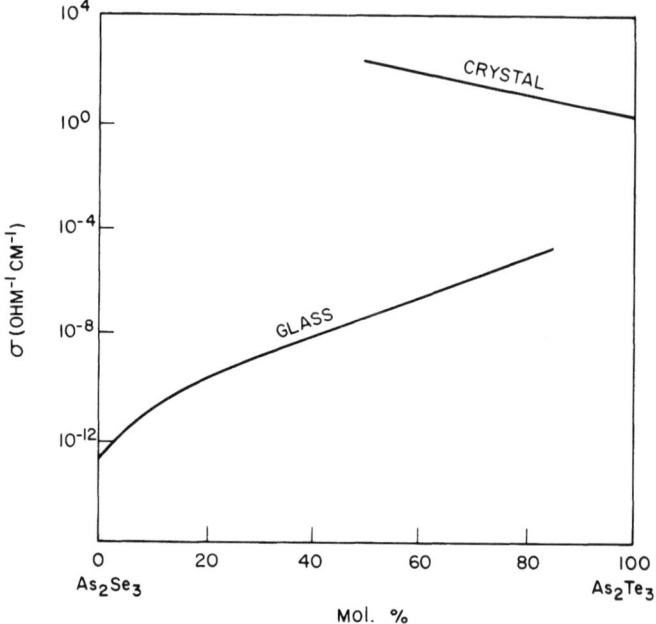

Fig. 9. The electrical conductivity σ of the glassy and polycrystalline phases for compositions in the As_2Se_3–As_2Te_3 pseudobinary (after Vengel and Kolomiets[122]).

bistable electrical memory effect observed, and the ease of inducing the memory-on state was directly related to observed fast crystallization kinetics. In addition, Bagley and Bair also demonstrated a reverse forming process whereby electrodes were applied to the polycrystalline form of a potentially glass-forming composition and electrically pulsed to form the glassy state.

A number of authors[124–127] have used calorimetric techniques on other systems to demonstrate the association of memory switching with crystallization and reversion to the glass. Others[128–139] have directly observed the crystalline memory filament. These extensive studies all agree as to the mechanism for electrical memory switching.

The overall memory switching process proceeds as follows: Exceeding the threshold voltage results in threshold switching, which generates a large amount of Joule heat in a current channel such that the material temperature in the channel is raised to a temperature between the glass transition and the melting point, i.e., to a temperature appropriate for crystallization (Section 3). The set time is the time required to crystallize a continuous filament path between the electrodes. If the voltage is reduced before the completed path is formed, the crystalline path is interrupted by a high-resistance glass barrier. At the conclusion of the set time, a continuous low-resistance crystalline path is formed and the switch is in the on state. The current pulse used to turn off the switch generates enough Joule heat in the crystalline filament to heat a section of the continous path to a temperature above the melting point. Turn-off is achieved by having a sharp trailing edge to the current pulse, resulting in a fast quench into the high-resistance, glassy state.

Chemical composition is an important parameter in determining switching behavior. Turn-off and turn-on reliability are intrinsically incompatible, since the first requires slow crystallization kinetics whereas the second requires fast kinetics. An appropriate composition always compromises between these two processes. As discussed in Section 3, crystallization is a nucleation and growth process. The stochastic nature of nucleation can be overcome by having crystals present (e.g., by having the reset current pulse not melt all the crystalline path), whereupon the switching kinetics is then only crystal-growth-limited.

Some compositions found to satisfy this compromise adequately are based on the eutectic composition $Ge_{15}Te_{85}$ with minor additions from groups V or VI of the periodic table, and these materials have been extensively studied.[140–150] In these materials, the conducting filament consists of dendritically grown Te crystallites ($\sim$ 200–400 Å in diameter) in, depending upon the thermal history, an amorphous or crystalline (GeTe) matrix. Cohen et al.[151] have noted that the memory-on state can be established with as little as 15–20 vol % of the conducting crystalline phase.

Feinleib et al.[152] have suggested that the crystallization which occurs in an electrical memory device under the application of an electric field is not a purely thermal effect, but that the growth rate is enhanced by the presence of a nonequilibrium electron–hole pair concentration (broken bonds) produced by the electric field, analogous to the photoenhanced crystallization discussed in Section 5.2. Such an enhancement has not yet been unambiguously verified.

Uttecht et al.[133] observed a marked electric field effect on the formation of the crystalline filament in an $As_{55}Te_{35}Ge_{10}$ glass. With the application of the field (exceeding the threshold) the crystal always nucleated at the positive voltage electrode and grew toward the negative. If the polarity was reversed during growth, the existing crystalline filament receded, reverting to the amorphous phase, and a new crystalline filament nucleated at the now positive electrode. The chemical composition of the crystalline filament, as determined by microprobe analysis, was observed to be $As_{39}Te_{57}Ge_4$; i.e., close to the composition As_2Te_3 and markedly different from the composition of the glass matrix. Extensive work in this ternary system by Tanaka et al.[153] has shown that the crystalline filament formation kinetics is markedly composition dependent and that electromigration effects play an important role in the observed field effects.

The role of phase separation in memory effects is not completely resolved. Pearson et al.[154] have demonstrated that changes in glass composition can result in large conductivity changes. Thus, in principle, phase separation into an amorphous-phase conducting filament, and its possible re-solution, could be a bistable memory device. However, such an effect has not yet been observed. As discussed in Section 2, however, phase separation can also play an important role as a precursor to a subsequent crystallization. Pearson[61,155] and Roy and Caslavska[156] have generally observed phase separation in threshold and memory switching glass compositions, thus indicating a potential effect. However, Bagley and Northover[157] (for the As–Se–Te system) and Moss and deNeufville[149] (for the Ge–Te system) did not observe phase separation as a precursor to crystallization in these memory-forming glasses. This indicates that phase separation is not a necessary condition, but it also does not rule out its possible effect in other compositions.

5.2. Optical Memories

Optical memories are capable, in principle, of packing densities of $\sim 10^8$ bits/cm^2, and thus have potential for large-capacity information storage. This large packing density is obtained because both the bit size and the spacing between bits can approximate the wavelength of light. Recently, transitions in amorphous materials have been considered for this application.

Brandes et al.[158] have described the formation of holograms in binary arsenic-sulfur glasses. The holograms were written with laser light ($\lambda = 4880$ Å), the energy of which exceeds the absorption edge (Figure 10). The image is formed directly upon irradiation and no further processing is required; the absence of a development step is an advantage but the inability to fix the image is a disadvantage. The holograms are formed in the sulfur-rich region of the As–S binary, with diffraction efficiencies as high as 18% observed. The efficiencies were observed to decrease with increasing arsenic content, approach-

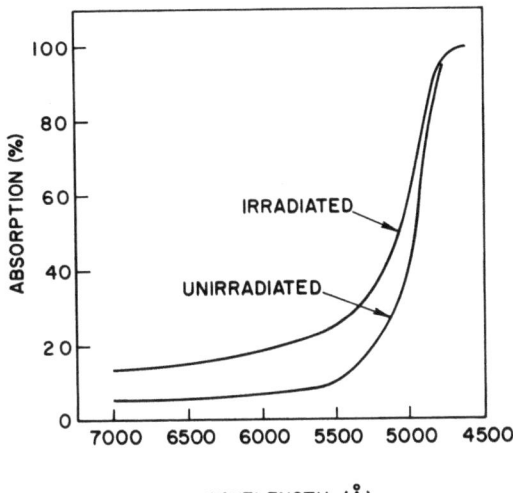

Fig. 10. The effect of irradiation ($\lambda = 4880$ Å) on the optical absorption of an arsenic–sulfur (40 wt % As) glass (after Brandes et al.[158]).

ing zero at the composition AsS_3. They note that the photoinduced absorption increase at 6328 Å was due to an increase in scattering with no apparent shift in the absorption edge (Figure 10). From their observations Brandes et al. concluded that the images formed are due to regions differing in refractive index (phase holograms).

Pearson and Bagley[159] examined a glass of composition $As_{0.07}S_{0.93}$ using calorimetry, chemical extraction, and electron spin resonance and observed that irradiation causes the precipitation of very small, almost pure, rhombic sulfur microcrystals in the glassy arsenic–sulfur matrix. Thus they concluded that the phase holograms recorded in this material are formed by the distribution and number of these microcrystals, and their differing refractive index from the glass matrix. This precipitation can also be effected thermally, but the kinetics is slower. The transformation in these dilute arsenic glasses can be viewed as an analog to the crystallization of a pure sulfur glass, in which the first stage is the precipitation of rhombic sulfur microcrystals in a polymeric sulfur glass matrix.

Very high-efficiency (up to 100%) phase holograms have been formed in vapor-deposited thin films of As_2S_3 by Keneman[160] and Ohmachi and Igo.[161] Diffraction gratings with spacings as small as 0.35 μm have been recorded. In the amorphous As_2S_3, light irradiation causes a shift in the absorption edge to longer wavelengths (Figure 11).

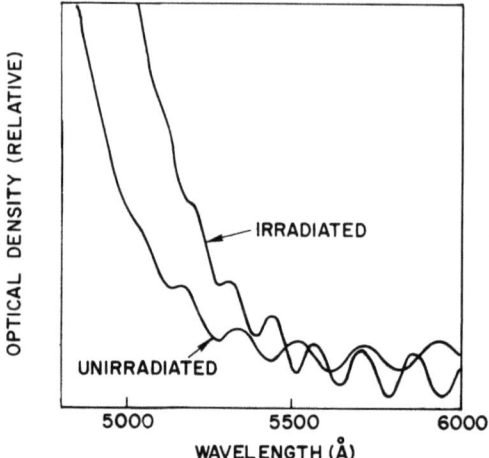

Fig. 11. The effect of irradiation (white light) on the optical absorption of an evaporated amorphous As_2S_3 thin film (after Keneman[160]).

In addition to As_2S_3, Ohmachi and Igo also studied other compositions within the As–S and As–S–Ge systems. In the As–S system, upon irradiation they observed that the maximum change in transmittance and refractive index occurred at the composition As_2S_3, and the effects decreased with increasing sulfur. This result is contrary to that observed by Brandes et al.[158] Ohmachi and Igo also observed that, upon irradiation, the absorption edge of an evaporated $As_{20}S_{60}Ge_{20}$ thin film shifted to shorter wavelengths, an opposite effect to that observed in the As_2S_3 thin film. In both films the change in refractive index Δn upon irradiation was positive, with values of $+0.056$ (As_2S_3) and $+0.010$ ($As_{20}S_{60}Ge_{20}$).

Igo and Toyoshima observed reversible changes in the optical properties of vapor-deposited thin films in the As–Se–Ge system.[162,163] The change in optical properties could not be accounted for by an amorphous-to-crystal transformation. High-power irradiation ($\lambda = 6328$ Å) of the thin films causes them to become more transparent due to a shift of the absorption edge to shorter wavelengths. Irradiation with light of a lower power (film heating is negligible) causes an absorption edge shift to longer wavelengths and thereby produces a darkening of the film. These optical changes were found to be reversible upon illumination and could not be duplicated with thermal treatments alone.

From their studies, Igo and co-workers[161,163] concluded that light irradiation produces two effects upon the optical properties. One, a thermal effect, shifts the absorption edge to shorter wavelengths; the other, a pure photoeffect, shifts the absorption edge to longer wavelengths (Figure 12). Which of these two effects is dominant during the irradiation of a film depends upon composition and light intensity. For the composition $As_{40}Se_{50}Ge_{10}$ the two effects were observed to be highly competitive, resulting in a marked effect of light intensity upon optical properties.

Asahara and Izumitani[164] have observed similar effects (photo-induced darkening, thermally recoverable) in thin films from the As–Se–Cu system. Their work also rules out the amorphous-to-crystal transformation as the source of the observed effects.

Berkes et al.[165] have studied photoeffects in vapor-deposited thin films of amorphous As_2Se_3 and As_2S_3. They observed that irradiation of these two materials with light whose energy is equal to or exceeds the band gap energy causes a darkening, in transmission,

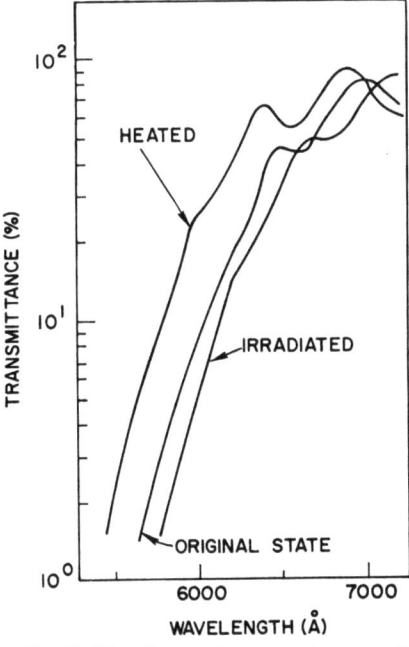

Fig. 12. The change in transmittance of an amorphous $As_{40}Se_{50}Ge_{10}$ thin film induced by either heating to 200°C ($T_g = 240°C$) or low-power irradiation (after Igo and Toyoshima[163]).

of the films. From their experiments they concluded that the light produces a photodecomposition as expressed by the relations

$$As_2Se_3 \xrightarrow{hv} xAs + As_{2-x}Se_3 \quad (0 < x < 2)$$

and

$$As_2S_3 \xrightarrow{hv} 2As + 3S.$$

They further suggest that, by diffraction, the free arsenic then nucleates and grows into particles, and that the observed darkening of the films upon exposure is due to the presence of the arsenic particles with their high apsorption in the visible. Berkes et al. did not detect the free arsenic directly, but its presence was inferred from observed As_2O_3 crystals on the film surface, the arsenic presumed to have rapidly oxidized. However, Igo and co-workers[161,163] did not observe As_2O_3 crystallites on their amorphous thin films after irradiation and photodarkening.

deNeufville et al.[166] offered a different explanation for the photoeffects in these thin films. From structural studies on evaporated thin films of amorphous As_2S_3 and As_2Se_3, they observed an irreversible change in local order upon annealing and/or illumination from that of a molecular solid (As_4S_6 or As_4Se_6 molecular configuration) present in the evaporated thin films to that of the cross-linked network characteristic of the bulk glasses. Concurrent with the change in structure is an observed irreversible decrease in the optical absorption edge energy, decreasing finally to that of the bulk glass. Another, reversible, effect was also observed in the vapor-deposited As_2Se_3 films; annealing of the thin film produced a small increase in the optical absorption edge energy, with the intial energy value being restored by illumination. This smaller effect is attributed to reversible changes in local order different from the irreversible effects noted above. The authors concluded that the ability to form holograms in these materials is due to these photoinduced structural changes. However, from similar studies on vapor-deposited thin films of amorphous $As_{40}Se_{50}Ge_{10}$, deNeufville et al.[167] concluded that the thermally reversible optical effects observed[163] in this material do not involve an obvious structural rearrangement, leading the authors to suggest that the optical changes are the result of a photoproduced nonequilibrium distribution of trapped charges.

If the optical properties of an amorphous material are sufficiently different from those of the crystal (in an appropriate energy region), then the amorphous $\rightleftarrows$ crystal transformation has application for

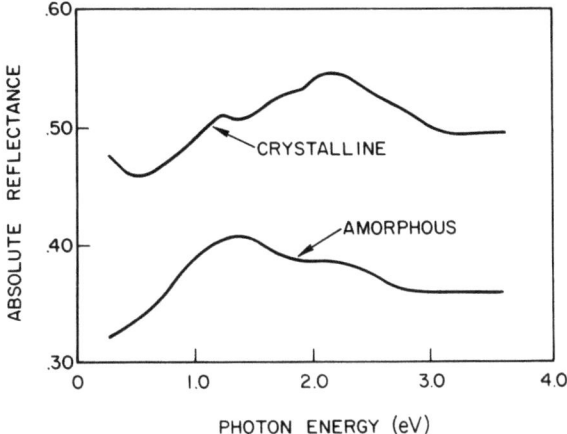

Fig. 13. The reflectivities of the amorphous and crystalline states of a Te-based alloy (after Feinleib and Ovshinsky[168]).

optical storage. Such is the case, for example, for the reflectivity of Te-based alloys,[168] where for an energy of 0.25–3.6 eV the reflectivity of the crystalline phase is markedly higher than that of the amorphous phase (Figure 13).

Utilizing a property change such as this, Feinleib and co-workers[152] have recently demonstrated the ability to reversibly write and erase descrete optical images in amorphous thin films of composition $Te_{81}Ge_{15}Sb_2S_2$. In this work, 3-μm spots were irradiated with light from a pulsed argon-ion laser ($\lambda = 5145$ Å). Light pulses of 1–16 μsec and corresponding intensities of 100–10 mW resulted in obvious optical property changes. Electron microscopic examination demonstrated that irradiation crystallized the initially amorphous film (writing sequence). A subsequent pulse could be used to heat the material above the melting point, which, together with rapid cooling, caused the film to revert to the amorphous state (erasure). They argued that the observed fast crystallization kinetics cannot be accounted for by thermal effects alone and that there must be a large photoenhancement of the crystallization rate. As with Dresner and Stringfellow,[169] Feinleib et al. attributed this enhancement to the fact that light with an energy exceeding the band gap creates electron–hole pairs, resulting in weakened or broken atomic bonds. These weakened bonds yield the higher atomic mobility necessary for crystallization. Hamada et al.[170] confirmed the fast crystallization rate of irradiated amorphous $Te_{81}Ge_{15}Sb_2S_2$ thin films. Using a

He–Cd laser ($\lambda = 4416$ Å) and a light intensity of 10 mW, they observed that crystallization of the irradiated spot is completed in 50 μsec. They also observed that erasure can be accomplished in 4 μsec at a light intensity of 16 mW.

A reverse mode optical switching suitable for an optical memory has been ascribed by Von Gutfield and Chaudhari.[171] For this reverse mode of writing and erasing, the thin film is initially crystalline. The writing pulse then makes the image area amorphous, and the erasure crystallizes it again. The authors suggest that an advantage for this mode of switching is that writing is not limited by crystallization kinetics, but simply by heating, and is therefore inherently more rapid. In their experiments on thin films of composition $Te_{88}Ge_7As_5$, they were able to form μm-sized spots with 2–5-μsec laser pulses ($\lambda \simeq 5800$ Å).

The role of light in the crystallization process is not completely understood. Paribok-Aleksandrovich[172] observed a photoenhancement of the surface crystallization rate of amorphous selenium when the incident photon energy exceeds 2.2 eV. He suggested that a photobreaking of the Se chain is responsible for the observed enhancement. Weiser et al.[173] determined the photo and thermally induced crystallization kinetics for an amorphous thin film of composition $Te_{77}Ge_{20}As_3$. They observed that a low-temperature thermally induced crystallization rate lies on an extrapolated curve for the laser-irradiated ($\lambda = 6471$ Å) photoinduced rate (observed activation energy 1.7 eV), and thus concluded that there is no evidence for photoenhanced crystallization in this material. Similarly, by direct observation of the growing crystallites, de Neufville[174] was unable to detect a photoenhancement of the crystallization of an amorphous film (composition $GeTe_2$) held at a temperature near T_g.

A thermal analysis by von Gutfield[175] led him to conclude that the absence of a crystallization halo surrounding a laser-produced amorphous region in a chalcogenide thin film (the usual observation) is not a definitive criterion for demonstrating photoenhancement, as was suggested by Feinleib et al.[152]

Crystallization is a nucleation and growth process and the rate can be increased by enhancing either or both of these factors. A photoenhancement of the nucleation frequency was observed by Ovshinsky and Klose.[176] They reported that, for a film of proprietary composition, the thermally induced surface nucleation frequency was 10^4 cm^{-2}, whereas an irradiated film (thermally developed) had a nucleation frequency of 6×10^5 cm^{-2}.

The effect of light on crystal growth has been investigated by Dresner and Stringfellow[169] and Kim and Turnbull.[177] Both of these studies were made on vapor-deposited amorphous selenium. Dresner and Stringfellow[169] reported a photoenhanced growth of surface crystallites. They studied the growth rate as a function of light intensity (mean photon energy ~ 3 eV) and temperature, and observed the growth rate to vary linearly with light intensity at low intensities and to vary as the square root of the intensity at intermediate intensities, finally saturating at high intensities (Figure 14). They concluded from their experiments that the growth enhancement is the result of electron–hole pairs being produced in the amorphous phase by the illumination. Through the application of an electric field during irradiation, they demonstrated that the growth rate is controlled by the flux of holes toward the interface. They noted that breaking a Se–Se bond is analogous to the creation of mobile holes and would enhance growth by aiding atomic mobility and reducing the interfacial energy barrier. However, Kim and Turnbull[177] observed that for high-purity Se crystallizing in a cylindrite morphology, the radial growth rate is not enhanced by illumination. Furthermore, they observed very fast thermal growth rates, as fast as the photoenhanced rates observed by Dresner and Stringfellow, indicating an interaction between photoeffects and growth-inhibiting impurities. Kim and Turnbull did observe a photoinduced thickening of the cylindrite

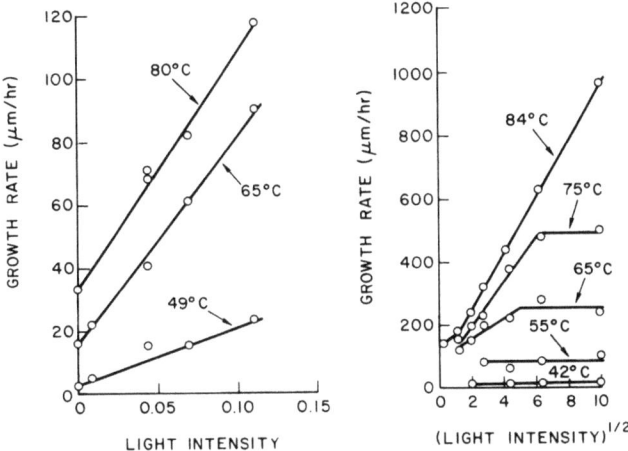

Fig. 14. The dependence of the selenium crystal growth rate on light intensity and temperature. The maximum light intensity (100) is 3×10^{18} quanta/cm^2 sec (after Dresner and Stringfellow[169]).

which did not occur thermally. More work is needed to completely resolve the several interrelated effects (e.g., photo, thermal, impurity) which play a role when a material is crystallized by illumination with light.

References

1. E. U. Condon, Physics of the glassy state. I. Constitution and structure, *Am. J. Phys.* **22**, 43–53 (1954).
2. W. Kauzman, The nature of the glassy state and the behavior of liquids at low temperatures, *Chem. Rev.* **43**, 219–256 (1948).
3. G. Hetherington, K. J. Jack, and J. C. Kennedy, The viscosity of vitreous SiO_2, *Phys. Chem. Glasses* **5**, 130–136 (1964).
4. C. R. Kurkjan and R. W. Douglas, The viscosity of glasses in the system Na_2O–GeO_2, *Phys. Chem. Glasses* **1**, 19–25 (1960).
5. J. P. De Neufville, C. H. Drummond III, and D. Turnbull, The effect of excess Ge on the viscosity of GeO_2, *Phys. Chem. Glasses* **11**, 186–191 (1970).
6. S. Glasstone, K. J. Laidler, and H. Eyring, *Theory of Rate Processes*, pp. 477–551, McGraw-Hill, New York (1941).
7. R. B. Sosman, *Properties of Fused SiO_2*, p. 313, Chem. Catalog Corp., New York (1927).
8. K. K. Kelley and A. U. Christensen, U.S. Bur. Mines, R.I. 5710 (1961).
9. R. Bruckner, Properties and structure of vitreous SiO_2, *J. Non-Crystalline Solids* **5**, 123–216 (1970).
10. D. E. Polk and D. Turnbull, Structure of amorphous semiconductors, *J. Non-Crystalline Solids* **8–10**, 19–35 (1972).
11. H. Vogel, Das Temperaturabhängigkeitsgesetz auf die Viscositat von Flussigkeiten, *Phys. Z.* **22**, 645–646 (1921).
12. G. S. Fulcher, Analysis of recent measurements of the viscosity of glasses, *J. Am. Ceramic Soc.* **6**, 339 (1925).
13. M. H. Cohen and D. Turnbull, Molecular transport in liquids and glasses, *J. Chem. Phys.* **31**, 1164–1169 (1959).
14. M. Cukierman, J. W. Lane, and D. R. Uhlmann, High temperature flow behavior of glass-forming liquids: A free volume interpretation, *J. Chem. Phys.* **59**, 3639–3644 (1973).
15. M. L. Williams, R. F. Landel, and J. D. Ferry, The temperature dependence of relaxation mechanisms in amorphous polymers and other glass-forming liquids, *J. Am. Chem. Soc.* **77**, 3701–3707 (1955).
16. W. T. Laughlin and D. R. Uhlmann, Viscous flow in simple organic liquids, *J. Phys. Chem.* **76**, 2317–2325 (1972).
17. J. H. Gibbs and E. A. DiMarzio, Nature of the glass transition and the glassy state, *J. Chem. Phys.* **28**, 373–383 (1958).
18. D. Turnbull and M. H. Cohen, On the free volume model of the liquid-glass transition, *J. Chem. Phys.* **52**, 3038–3041 (1970).
19. F. Bueche, Mobility of molecules in liquids near the glass temperature, *J. Chem. Phys.* **30**, 748–752 (1959).

20. B. J. Alder, W. G. Hoover, and D. A. Young, Studies in molecular dynamics. V. High density equation of state and entropy for hard discs and spheres, *J. Chem. Phys.* **49**, 3688–3696 (1968).
21. M. H. Cohen and D. Turnbull, Metastability of amorphous structures, *Nature* **203**, 964 (1964).
22. B. J. Alder, D. M. Gass, and T. E. Wainwright, Studies in molecular dynamics. VIII. The transport coefficients for a hard sphere fluid, *J. Chem. Phys.* **53**, 3813–3826 (1970).
23. D. Weaire, M. F. Ashby, J. Logan, and M. J. Weins, On the use of pair potentials to calculate the properties of amorphous metals, *Acta Met.* **19**, 779–788 (1971).
24. E. J. Le Fevre, Equation of state for hard-sphere fluid, *Nature, Phys. Sci.* **235**, 20 (1972).
25. J. H. Dymond and B. J. Alder, Van der Waals theory of transport in dense fluids, *J. Chem. Phys.* **45**, 2061–2068 (1966).
26. G. Adam and J. H. Gibbs, On the temperature dependence of Cooperative relaxation properties in glass-forming liquids, *J. Chem. Phys.* **43**, 139–146 (1965).
27. H. S. Chen and D. Turnbull, Evidence of a glass–liquid transition in a Au–Ge–Si Alloy, *J. Chem. Phys.* **48**, 2560 (1968).
28. G. C. Berry and T. G. Fox, The viscosity of polymers and their concentrated solutions, *Adv. Polymer Sci.* **5**, 261–357 (1968).
29. C. A. Angell and K. J. Rao, Configurational excitations in condensed matter and the bond–lattice model for the glass Transition, *J. Chem. Phys.* **57**, 470–481 (1972).
30. M. Goldstein, Viscous liquids and the glass transition: A potential energy barrier picture, *J. Chem. Phys.* **51**, 3728–3739 (1969).
31a. N. Hirai and H. Eyring, Bulk viscosity of polymeric systems, *J. Polymer Sci.* **37**, 51–70 (1959).
31b. N. Hirai, Liquid viscosity near the glass transition temperature, *Rep. Surface Sci.* **2**, 51 (1962).
32. J. W. Cahn, Spinodal decomposition, *Trans. Met. Soc. AIME* **242**, 166–180 (1968).
33. J. E. Hilliard, Spinodal decomposition, in *Phase Transformations*, pp. 497–560, Am. Soc. Metals, Metals Park, Ohio (1970).
34. R. D. Maurer, Crystallization of a titania-nucleated glass, in *Symposium on Nucleation and Crystallization in Glasses and Melts* (Reser, Smith, and Insley, eds.), pp. 5–9, Am. Ceramic Soc., Columbus, Ohio (1962).
35. J. E. Morral, Ph.D. Thesis, Dept. of Mat. Sci., M.I.T. (1968); J. E. Morral, Stability limits for ternary systems, *Acta Met.* **20**, 1061–1076 (1972).
36. D. Turnbull, Under what conditions can a glass be formed, *Contemporary Physics* **10**, 473–488 (1969).
37. W. B. Hillig and D. Turnbull, Theory of crystal growth in undercooled pure liquids, *J. Chem. Phys.* **24**, 914 (1956).
38. D. Turnbull and M. H. Cohen, Crystallization kinetics and glass formation, in *Modern Aspects of the Vitreous State* (J. D. MacKenzie, ed.), **1**, pp. 38–62, Butterworth's, London (1960).
39. K. A. Jackson, Nature of solid–liquid interfaces, in *Growth and Perfection of Crystals* (R. H. Doremus, B. W. Roberts, and D. Turnbull, eds.), pp. 319–325, Wiley, New York (1958).
40. K. A. Jackson, D. R. Uhlmann, and J. D. Hunt, On the nature of crystal growth from the melt, *J. Crystal Growth* **1**, 1–36 (1967).

41. W. K. Burton, N. Cabrera, and F. C. Frank, The growth of crystals and the equilibrium structure of their surfaces, *Phil. Trans. Roy. Soc.* (*London*) **243**, 299–358 (1951).
42. J. W. Cahn, Theory of crystal growth and interface motion in crystalline materials, *Acta Met.* **8**, 554–562 (1960).
43. K. A. Jackson, Theory of crystal growth, this volume, Chapter 5.
44. D. Turnbull, On the relation between crystallization rate and liquid structure, *J. Phys. Chem.* **66**, 609–613 (1962).
45. D. Turnbull, Amorphous solid formation and interstitial solution behavior in metallic systems, *Proceedings of Conference on Disordered Systems* (Strasbourg, 1973), *J. de Physique* **35**, colloque-4, C4.1–4.9 (1974).
46. J. L. Walker, reported by B. Chalmers, *Principles of Solidification*, pp. 114–116, Wiley, New York (1964).
47. N. G. Ainslie, C. R. Morelock, and D. Turnbull, Devitrification kinetics of fused silica, in *Symposium on Nucleation and Crystallization in Glasses and Melts* (M. K. Reser, G. Smith, and H. Insley, eds.), pp. 97–107, Am. Ceram. Soc., Columbus, Ohio (1962).
48. R. C. Keezer and M. W. Bailey, The structure of liquid Se from viscosity measurements, *Mat. Res. Bull.* **2**, 185–192 (1967).
49. P. J. Vergano and D. R. Uhlmann, Crystallization kinetics of GeO_2; the effects of stoichiometry on kinetics, *Phys. Chem. Glasses* **11**, 30–38 (1970).
50. D. R. Uhlmann, J. F. Hays, and D. Turnbull, The effect of high pressure on crystallization kinetics with special reference to fused silica, *Phys. and Chem. Glasses* **7**, 159–168 (1966).
51. J. D. MacKenzie and W. F. Claussen, Crystallization and phase relations of B_2O_3 at high pressures, *J. Am. Ceram. Soc.* **44**, 79–81 (1961).
52. D. R. Uhlmann, J. F. Hays, and D. Turnbull, The effect of high pressure on B_2O_3: crystallization, densification and the crystallization anomaly, *Phys. Chem. Glasses* **8**, 1–10 (1967).
53. O. Shimomura, S. Minomura, N. Sakai, K. Asaumi, K. Tamura, J. Fukushima, and H. Endo, Pressure-induced semiconductor–metal transitions in amorphous Si and Ge, *Phil. Mag.* **29**, 547–558 (1974).
54. The Ad Hoc Committee on the Fundamentals of Amorphous Semiconductors, *Fundamentals of Amorphous Semiconductors*, National Academy of Sciences–National Academy of Engineering Publication NMAB-284 (September 1971).
55. D. Adler, Amorphous semiconductors, *CRC Crit. Rev. Solid State Sci.* **2**, 317–465 (1971).
56. W. Doremus, ed., Proceedings of the symposium on semiconductor effects in Amorphous solids, New York, May 1969, *J. Non-Cryst. Solids* **2** (1970).
57. Sir Neville Mott, ed., Proceedings of the international conference on amorphous and liquid semiconductors, Cambridge, Sept. 1969, *J. Non-Cryst. Solids* **4** (1970).
58. M. H. Cohen and G. Lucovsky, eds., Proceedings of the fourth international conference on amorphous and liquid semiconductors, Ann Arbor, August 1971, *J. Non-Cryst. Solids* **8–10** (1972).
59. *Proceedings of the Fifth International Conference on Amorphous and Liquid Semiconductors*, Garmisch-Partenkirchen, September 1973 (J. Stuke and W. Brenig, eds.), Taylor and Francis, Ltd., London (1974).

60. H. Fritzsche, Switching and memory in amorphous semiconductors, in *Amorphous and Liquid Semiconductors* (J. Tauc, ed.), Chapter 6, pp. 313–359, Plenum Press, New York (1974).
61. A. D. Pearson, Memory and switching in semiconducting glasses, a review, *J. Non-Cryst. Solids* **2**, 1–15 (1970).
62. S. R. Ovshinsky, An introduction to ovonic research, *J. Non-Cryst. Solids* **2**, 99–106 (1970).
63. P. J. Walsh, R. Vogel, and E. J. Evans, Conduction and electrical switching in amorphous chalcogenide semiconductor films, *Phys. Rev.* **178**, 1274–1279 (1969).
64. E. A. Fagen and H. Fritzsche, Electrical conductivity of amorphous chalcogenide alloy films, *J. Non-Cryst. Solids* **2**, 170–179 (1970).
65. J. M. Robertson and A. E. Owen, Electronically-assisted thermal breakdown in chalcogenide glasses, *J. Non-Cryst. Solids* **8–10**, 439–444 (1972).
66. S. R. Ovshinsky, Reversible electrical switching phenomena in disordered structures, *Phys. Rev. Lett.* **21**, 1450–1453 (1968).
67. B. T. Kolomiets, E. A. Lebedev, and I. A. Taksami, Mechanism of the breakdown in films of glassy chalcogenide semiconductors, *Soviet Phys.—Semiconductors* **3**, 267–268 (1969).
68. R. R. Shanks, Ovonic threshold switching characteristics, *J. Non-Cryst. Solids* **2**, 504–514 (1970).
69. K. Tanaka, S. Iizima, M. Sugi, Y. Okada, and M. Kikuchi, Thermal effect on switching phenomenon in chalcogenide amorphous semiconductors, *Solid State Commun.* **8**, 387–389 (1970).
70. M. Sugi, M. Kikuchi, I. Iizima, and K. Tanaka, Switching characteristics of chalcogenide glass, *Solid State Commun.* **7**, 1805–1807 (1969).
71. D. R. Haberland, Ladungsbedingter Schaltmechanismus in Glashalbleitern, *Solid-State Electronics* **13**, 207–217 (1970).
72. D. R. Haberland and H. Stiegler, New experiments on the charge-controlled switching effect in amorphous semiconductors, *J. Non-Cryst. Solids* **8–10**, 408–414 (1972).
73. A. Csillag and H. Jäger, Energy-controlled switching process in the amorphous system Te–As–Ge–Si, *J. Non-Cryst. Solids* **2**, 133–140 (1970).
74. I. Balberg, Simple test for double injection initiation of switching, *Appl. Phys. Lett.* **16**, 491–493 (1970).
75. H. K. Henisch and R. W. Pryor, On the mechanism of ovonic threshold switching, *Solid-State Electronics* **14**, 765–774 (1971).
76. R. W. Pryor and H. K. Henisch, Nature of the on-state in chalcogenide glass threshold switches, *J. Non-Cryst. Solids* **7**, 181–191 (1972).
77. H. K. Henisch and G. J. Vendura, Jr., Characteristics of ovonic threshold switches with crystalline semiconductor electrodes, *Appl. Phys. Lett.* **19**, 363–365 (1971).
78. H. K. Henisch, R. W. Pryor, and G. J. Vendura, Jr., Characteristics and mechanism of threshold switching, *J. Non-Cryst. Solids* **8–10**, 415–421 (1972).
79. G. J. Vendura, Jr., and H. K. Henisch, Behavior of amorphous semiconductor films between asymmetric electrodes, *J. Non-Cryst. Solids* **11**, 105–112 (1972).
80. H. Stiegler and D. R. Haberland, The switching behavior of chalcogenide glass with semiconducting electrodes, *J. Non-Cryst. Solids* **11**, 147–152 (1972).
81. R. Holstrom, Switching and conduction behavior of amorphous semiconductor diodes, *Proc. IEEE* **57**, 1451–1453 (1969).

82. A. C. Warren, Switching mechanism in chalcogenide glasses, *Electronics Letters* **5**, 461–462 (1969).
83. P. Burton and R. W. Brander, Thermal breakdown and switching in chalcogenide glasses, *Int. J. Electronics* **27**, 517–525 (1969).
84. D. L. Thomas and A. C. Warren, Preswitching behavior of amorphous chalcogenide semiconductor films, *Electronic Letters* **6**, 62–64 (1970).
85. F. M. Collins, Switching by thermal avalanche in semiconducting glass films, *J. Non-Cryst. Solids* **2**, 496–503 (1970).
86. H. Fritzsche and S. R. Ovshinsky, Conduction and switching phenomena in covalent alloy semiconductors, *J. Non-Cryst. Solids* **4**, 464–479 (1970).
87. H. J. Stocker, C. A. Barlow, Jr., and D. F. Weirauch, Mechanism of threshold switching in semiconducting glasses, *J. Non-Cryst. Solids* **4**, 523–535 (1970).
88. A. C. Warren, Thermal switching in semiconducting glasses, *J. Non-Cryst. Solids* **4**, 613–616 (1970).
89. H. S. Chen and T. T. Wang, On the theory of switching phenomena in semiconducting glasses, *Phys. Stat. Sol.* **2**, 79–84 (1970).
90. D. D. Thornburg, Role of capacitive discharge energy in the switching of semiconducting glasses, *Phys. Rev. Lett.* **27**, 1208–1210 (1971).
91. D. L. Thomas and J. C. Male, Thermal breakdown in chalcogenide glasses, *J. Non-Cryst. Solids* **8–10**, 522–530 (1972).
92. N. Croitoru and C. Popescu, Approximations and boundary conditions in the theory of thermal instabilities, *J. Non-Cryst. Solids* **11**, 397–401 (1973).
93. J. C. Male and D. L. Thomas, Intrinsic instability and current channelling in thermally controlled two-terminal switching devices, *J. Non-Cryst. Solids* **13**, 409–422 (1973/74).
94. K. Shimakawa, Y. Inagaki, and T. Arizumi, Thermal switching in chalcogenide glass semiconductors, *Japan. J. Appl. Phys.* **12**, 1043–1046 (1973).
95. K. W. Böer and S. R. Ovshinsky, Electrothermal initiation of an electronic switching mechanism in semiconducting glasses, *J. Appl. Phys.* **41**, 2675–2681 (1970).
96. A. C. Warren and J. C. Male, Field-enhanced conductivity effects in thin chalcogenide-glass switches, *Electronics Letters* **6**, 567–569 (1970).
97. K. W. Böer, G. Döhler, and S. R. Ovshinsky, Time delay for reversible electric switching in semiconducting glasses, *J. Non-Cryst. Solids* **4**, 573–582 (1970).
98. W. W. Sheng and C. R. Westgate, On the preswitching phenomena in semiconducting glasses, *Solid State Commun.* **9**, 387–391 (1971).
99. T. Kaplan and D. Adler, Thermal effects in amorphous-semiconductor switching, *Appl. Phys. Lett.* **19**, 418–420 (1971).
100. K. W. Böer, Electro-thermal effects in ovonics, *Phys. Stat. Sol.* **4**, 571–596 (1971).
101. A. H. M. Shousha, Negative differential conductivity due to electrothermal instabilities in thin amorphous films, *J. Appl. Phys.* **42**, 5131–5136 (1971).
102. C. Popescu and N. Croitoru, The contribution of the lateral thermal instability to the switching phenomenon, *J. Non-Cryst. Solids* **8–10**, 531–537 (1972).
103. T. Kaplan and D. Adler, Electrothermal switching in amorphous semiconductors, *J. Non-Cryst. Solids* **8–10**, 538–543 (1972).
104. D. M. Kroll and M. H. Cohen, Theory of electrical instabilities of mixed electronic and thermal origin, *J. Non-Cryst. Solids* **8–10**, 544–551 (1972).
105. D. M. Kroll, Ph.D. thesis, Physics Department, The University of Chicago, 1973.

106. M. P. Shaw, S. H. Holmberg and S. A. Kostylev, Reversible switching in thin amorphous chalcogenide films—electronic effects, *Phys. Rev. Lett.* **31**, 542–545 (1973).
107. B. T. Kolomiets, E. A. Lebedev, and E. A. Smorgonskaya, Breakdown mechanism of chalcogenide glasses, *Soviet Phys.—Semiconductors* **6**, 1766–1767 (1973).
108. N. K. Hindley, Random phase model of amorphous semiconductors, I. Transport and optical properties. II. Hot electrons, *J. Non-Cryst. Solids* **5**, 17–40 (1970).
109. N. F. Mott, Conduction in non-crystalline systems. VII. Non-ohmic behavior and switching, *Phil. Mag.* **24**, 911–934 (1971).
110. N. F. Mott, Conduction and switching in non-crystalline materials, *Contemp. Phys.* **10**, 125–138 (1969).
111. H. K. Henisch, E. A. Fagen, and S. R. Ovshinsky, A qualitative theory of electrical switching processes in monostable amorphous structures, *J. Non-Cryst. Solids* **4**, 538–547 (1970).
112. W. Heywang and D. R. Haberland, Zur Frage des Schalteffekts in Amorphen Halbleitern, *Solid State Electronics* **13**, 1077–1079 (1970).
113. F. W. Schmidlin, Electrical switching device based on charge-controlled double injection, *Phys. Rev. B* **1**, 1583–1587 (1970).
114. I. Lucas, Interpretation of the switching effect in amorphous semiconductors as a recombination instability, *J. Non-Cryst. Solids* **6**, 136–144 (1971).
115. H. Fritzsche and S. R. Ovshinsky, Electrical conduction in amorphous semiconductors and the physics of the switching phenomena, *J. Non-Cryst. Solids* **2**, 393–405 (1970).
116. M. Iida and A. Hamada, Electrical switching in mobility-gap materials, *Japan. J. Appl. Phys.* **10**, 224–227 (1971).
117. B. G. Bagley, The field dependent mobility of localized electronic carriers, *Solid State Commun.* **8**, 345–348 (1970).
118. H. K. Henisch, W. R. Smith, and W. Wihl, Field-dependent photoresponse of threshold switching systems, in *Proceedings of the Fifth International Conference on Amorphous and Liquid Semiconductors*, Garmisch-Partenkirchen, September 1973 (J. Stuke and W. Brenig, eds.), Vol. 1, 567–570, Taylor and Francis Ltd., London (1974).
119. W. van Roosbroeck, Electronic basis of switching in amorphous semiconductor alloys, *Phys. Rev. Lett.* **28**, 1120–1123 (1972).
120. W. van Roosbroeck, Principles of electrical behavior of amorphous semiconductor alloys, *J. Non-Cryst. Solids* **12**, 232–262 (1973).
121. E. J. Evans, J. H. Helbers, and S. R. Ovshinsky, Reversible conductivity transformations in chalcogenide alloy films, *J. Non-Cryst. Solids* **2**, 334–346 (1970).
122. T. N. Vengel and B. T. Kolomiets, Vitreous semiconductors; some properties of materials in the As_2Se_3–As_2Te_3 system, *Soviet Phys.—Technical Physics* **2**, 2314–2319 (1957).
123. B. G. Bagley and H. E. Bair, Thermally induced transformations in glassy chalcogenides, *J. Non-Cryst. Solids* **2**, 155–160 (1970).
124. H. Fritzsche and S. R. Ovshinsky, Calorimetric and dilatometric studies on chalcogenide glasses, *J. Non-Cryst. Solids* **2**, 148–154 (1970).
125. S. V. Phillips, R. E. Booth, and P. W. McMillan, Structural changes related to electrical properties of bulk chalcogenide glasses, *J. Non-Cryst. Solids* **4**, 510–517 (1970).

126. R. Pinto, Threshold and memory switching in thin films of the chalcogenide systems Ge–As–Te and Ge–As–Se, *Thin Solid Films* **7**, 391–404 (1971).
127. J. A. Savage, Glass forming region and DTA survey in the Ge–As–Te memory switching glass system, *J. Mat. Sci.* **6**, 964–968 (1971).
128. D. L. Eaton, Electrical conduction anomaly of semiconducting glasses in the system As–Te–I, *J. Am. Ceramic Soc.* **47**, 554–558 (1964).
129. H. J. Stocker, Bulk and thin film switching and memory effects in semiconducting chalcogenide glasses, *Appl. Phys. Lett.* **15**, 55–57 (1969).
130. M. Kikuchi, S. Iizima, M. Sugi, and K. Tanaka, Lock-on phenomenon in amorphous semiconductors, *Japan Soc. Appl. Phys.* **39** (*Suppl.*), 203–210 (1970).
131. M. Kikuchi and S. Iizima, Memory exchange in amorphous semiconductors, *Appl. Phys. Lett.* **15**, 323–325 (1969).
132. K. Tanaka, S. Iizima, M. Sugi, and M. Kikuchi, Electrical nature of the lock-on filament in amorphous semiconductors, *Solid State Commun.* **8**, 75–78 (1970).
133. R. Uttecht, H. Stevenson, C. H. Sie, J. D. Griener, and K. S. Raghavan, Electric field-induced filament formation in As–Te–Ge glass, *J. Non-Cryst. Solids* **2**, 358–370 (1970).
134. C. H. Sie, Electron microprobe analysis and radiometric microscopy of electric field induced filament formation on the surface of As–Te–Ge glass, *J. Non-Cryst. Solids* **4**, 548–553 (1970).
135. D. R. Haberland and H. P. Kehrer, Mikroskopische Untersuchungen an Festkörperschaltern aus Halbleitendem Glas, *Solid-State Electronics* **13**, 451–455 (1970).
136. D. Armitage and C. H. Champness, Memory switching and crystallization in selenium, *Can. J. Phys.* **49**, 2718–2723 (1971).
137. R. Pinto and K. V. Ramanathan, Electric field induced memory switching in thin films of the chalcogenide system Ge–As–Se, *Appl. Phys. Lett.* **19**, 221–223 (1971).
138. D. Armitage and C. H. Champness, Switching in amorphous selenium, *J. Non-Cryst. Solids* **7**, 410–416 (1972).
139. C. H. Sie, M. P. Dugan, and S. C. Moss, Direct observations of filaments in the ovonic read-mostly memory, *J. Non-Cryst. Solids* **8–10**, 877–884 (1972).
140. R. H. Willens, Dendritic crystallization of an amorphous alloy. *J. Appl. Phys.* **33**, 3269–3270 (1962).
141. A. Bienenstock, F. Betts, and S. R. Ovshinsky, Structural studies of amorphous semiconductors, *J. Non-Cryst. Solids* **2**, 347–357 (1970).
142. T. Takamori, R. Roy, and G. J. McCarthy, Structure of memory-switching glasses, I. Crystallization temperature and its control in Ge–Te glasses, *Mat. Res. Bull.* **5**, 529–540 (1970).
143. D. Adler, J. M. Franz, C. R. Hewes, B. P. Kraemer, D. J. Sellmyer, and S. D. Senturia, Transport properties of a memory-type chalcogenide glass, *J. Non-Cryst. Solids* **4**, 330–337 (1970).
144. T. Takamori, R. Roy, and G. J. McCarthy, Observations of surface-nucleated crystallization in memory-switching glasses, *J. Appl. Phys.* **42**, 2577–2578 (1971).
145. S. R. Ovshinsky and H. Fritzsche, Reversible structural transformations in amorphous semiconductors for memory and logic, *Metallurgical Trans.* **2**, 641–645 (1971).

146. J. R. Bosnell and C. B. Thomas, Preswitching electrical properties, forming, and switching in amorphous chalcogenide alloy threshold and memory devices, *Solid-State Electronics* **15**, 1261–1271 (1972).
147. P. Chaudhari and S. R. Herd, Crystallization of certain chalcogenide glasses, *J. Non-Cryst. Solids* **8–10**, 56–63 (1972).
148. E. J. Evans, Atomic transport in liquid chalcogenide alloys, *J. Non-Cryst. Solids* **8–10**, 702–707 (1972).
149. S. C. Moss and J. P. deNeufville, Thermal crystallization of selected thin films of Te-based memory glasses, *Mat. Res. Bull.* **7**, 423–442 (1972).
150. J. A. Savage, Glass formation and D.S.C. data in the Ge–Te and As–Te memory glass systems, *J. Non-Cryst. Solids* **11**, 121–130 (1972).
151. M. H. Cohen, R. G. Neale, and A. Paskin, A model for an amorphous semiconductor memory device, *J. Non-Cryst. Solids* **8–10**, 885–891 (1972).
152. J. Feinleib, J. deNeufville, S. C. Moss, and S. R. Ovshinsky, Rapid reversible light-induced crystallization of amorphous semiconductors, *Appl. Phys. Lett.* **18**, 254–257 (1971).
153. K. Tanaka, Y. Okada, M. Sugi, S. Iizima, and M. Kikuchi, Kinetics of growth of conductive filament in As–Te–Ge glasses, *J. Non-Cryst. Solids* **12**, 100–114 (1973).
154. A. D. Pearson, W. R. Northover, J. F. Dewald, and W. F. Peck, Jr., Chemical physical and electrical properties of some unusual inorganic glasses, in *Advances in Glass Technology*, pp. 357–365, Plenum Press, New York (1962).
155. A. D. Pearson, The Hall effect—Seebeck effect sign anomaly in semiconducting glasses, *J. Electrochem. Soc.* **111**, 753–755 (1964).
156. R. Roy and V. Caslavska, Di-phasic structure of switching and memory device glasses, *Solid State Commun.* **7**, 1467–1473 (1969).
157. B. G. Bagley and W. R. Northover, Electron microscopic observations of thermally induced transformations in amorphous chalcogenide thin films, *J. Non-Cryst. Solids* **2**, 161–169 (1970).
158. R. G. Brandes, F. P. Laming, and A. D. Pearson, Optically formed dielectric gratings in thick films of arsenic-sulfur glass, *Appl. Opt.* **9**, 1712–1714 (1970).
159. A. D. Pearson and B. G. Bagley, The mechanism of hologram formation in arsenic-sulfur glass, *Mat. Res. Bull.* **6**, 1041–1046 (1971).
160. S. A. Keneman, Hologram storage in arsenic trisulfide thin films, *Appl. Phys. Lett.* **19**, 205–207 (1971).
161. Y. Ohmachi and T. Igo, Laser-induced refractive-index change in As–S–Ge glasses, *Appl. Phys. Lett.* **20**, 506–508 (1972).
162. T. Igo and Y. Toyoshima, Optically induced reversible change in amorphous semiconductors, *Japan. J. Appl. Phys.* **11**, 117–118 (1972).
163. T. Igo and Y. Toyoshima, A reversible optical change in the As–Se–Ge glass, *J. Non-Cryst. Solids* **11**, 304–308 (1973).
164. Y. Asahara and T. Izumitani, Light-induced memory effect in Cu–As–Se glasses, *Japan. J. Appl. Phys.* **11**, 1748 (1972).
165. J. S. Berkes, S. W. Ing, Jr., and W. J. Hillegas, Photodecomposition of amorphous As_2Se_3 and As_2S_3, *J. Appl. Phys.* **42**, 4908–4916 (1971).
166. J. P. deNeufville, S. C. Moss, and S. R. Ovshinsky, Photostructural transformations in amorphous As_2Se_3 and As_2S_3 films, *J. Non-Cryst. Solids* **13**, 191–223 (1973/74).

Chapter 10

167. J. P. deNeufville, R. Seguin, S. C. Moss, and S. R. Ovshinsky, Mechanism of reversible optical storage in evaporated amorphous AsSe and $Ge_{10}As_{40}Se_{50}$, in *Proceedings of the Fifth International Conference on Amorphous and Liquid Semiconductors*, Garmisch-Partenkirchen, September 1973 (J. Stuke and W. Brenig, eds.), Vol. 2, 737–743, Taylor and Francis Ltd., London (1974).
168. J. Feinleib and S. R. Ovshinsky, Reflectivity studies of the Te (Ge,As)-based amorphous semiconductor in the conducting and insulating states, *J. Non-Cryst. Solids* **4**, 564–572 (1970).
169. J. Dresner and G. B. Stringfellow, Electronic processes in the photo-crystallization of vitreous selenium, *J. Phys. Chem. Solids* **29**, 303–311 (1968).
170. A. Hamada, T. Kurosu, M. Saito, and M. Kikuchi, Transient phenomena of the light-induced memory in amorphous semiconductor films, *Appl. Phys. Lett.* **20**, 9–11 (1972).
171. R. J. von Gutfeld and P. Chaudhari, Laser writing and erasing on chalcogenide films, *J. Appl. Phys.* **43**, 4688–4693 (1972).
172. I. A. Paribok-Aleksandrovich, Photocrystallization of amorphous selenium, *Soviet Phys.—Solid State* **11**, 1631 (1970).
173. K. Weiser, R. J. Gambino, and J. A. Reinhold, Laser-beam writing on amorphous chalcogenide films: crystallization kinetics and analysis of amorphizing energy, *Appl. Phys. Lett.* **22**, 48–49 (1973).
174. J. P. deNeufville, Optical information storage, in *Proceedings of the Fifth International Conference on Amorphous and Liquid Semiconductors*, Garmisch-Partenkirchen, September 1973 (J. Stuke and W. Brenig, eds.), Vol. 2, 1351–1360, Taylor and Francis Ltd., London (1974).
175. R. J. von Gutfield, The extent of crystallization resulting from submicrosecond optical pulses on Te-based memory materials, *Appl. Phys. Lett.* **22**, 257–258 (1973).
176. S. R. Ovshinsky and P. H. Klose, Reversible high-speed high-resolution imaging in amorphous semiconductors, in *1971 SID International Symposium Digest of Technical Papers*, pp. 58–61, Lewis Winner, New York (1971).
177. K. S. Kim and D. Turnbull, Crystallization of amorphous selenium films, I. Morphology and kinetics, *J. Appl. Phys.* **44**, 5237–5244 (1973); II. Photo and impurity effects, *J. Appl. Phys.* **45**, 3447–3452 (1974).

11

Transitions in Solid Polymers

Shogo Saito
Materials Division, Electrotechnical Laboratory
Agency of Industrial Science and Technology, Japan

1. Introduction

The simplest molecules, for instance, rare gas molecules, crystallize under suitable conditions. These simplest substances can exist in the gas, liquid, or solid phase. Diatomic or polyatomic molecules also crystallize and can exist in any of the three phases as long as the molecules are not very large. The shape of the polyatomic molecules, however, is different from that of the rare gas molecules, which have spherical symmetry. The intermolecular interaction in the polyatomic molecules is anisotropic.

A polymer molecule has a great amount of motional freedom and various possible conformations. Since the intermolecular force in a polymer often exceeds the intramolecular force, the polymer does not exist in the gas phase but only in the liquid or solid phase. Though a polymer system is represented in terms of a chemical formula, generally the system is a mixture of various molecules of different molecular weights. Therefore the polymer system is "impure" with respect to simpler substances. Recent development of crystallization techniques has produced so-called polymer single crystals. If such a crystal is compared with one from a simpler molecule, however, it will be regarded as an imperfect one. Often polymer molecules do not crystallize but become amorphous and glassy at low temperatures.

Polymers can be divided into two categories, those that are wholly amorphous and those that are partly crystalline. The partly crystalline polymers may of course be amorphous under certain

Chapter 11

conditions. Existence of an amorphous region is a distinctive feature of polymers in comparison with simpler substances. Polymer molecules in an amorphous region exhibit various types of molecular motion, depending upon such conditions as temperature, pressure, etc. Customarily, a transition temperature can be defined as the temperature below which a molecular motion stops (strictly speaking, rate of the molecular motion becomes slower than the time scale of the measurement). Thus one can observe several transition temperatures for the amorphous region in polymers. In a crystalline polymer the crystals transform into a liquid state at the melting point and sometimes a polymorphic transition is exhibited. Certain types of molecular motion often occur in an imperfect crystalline region.

The primary purpose of this chapter is to discuss the mechanisms of transitions and to summerize relationships between structure and transitions for polymers.

2. Multiple Transition Behavior in Solid Polymers
2.1. Multiple Transitions

When we measure certain physical properties of a polymer as a function of temperature, we can observe regions of temperature in

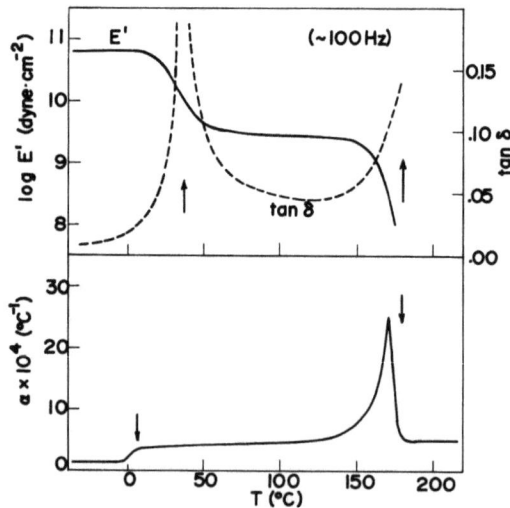

Fig. 1. Temperature dependence of thermal expansion coefficient, Young's modulus and mechanical loss for poly(3,3-bischloromethyl-oxacyclobutane), illustrating two transitions (after Sandiford[1]).

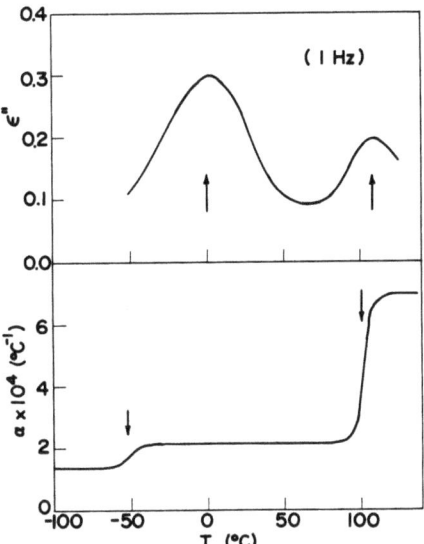

Fig. 2. Temperature dependence of thermal expansion coefficient and dielectric loss for poly(methyl methacrylate), illustrating two transitions.

which they change abruptly. Figure 1 shows the temperature dependence of thermal expansion coefficient α, Young's modulus E, and mechanical loss tangent tan δ for poly(3,3-bischloromethyloxacyclobutane).[1] The discontinuity of α at around 7°C locates the glass transition. The increase of α corresponds to the onset of crystalline melting and the discontinuity at around 180°C indicates that the melting process is complete. The mechanical properties also reveal these two transitions: The shift of the glass transition region toward higher temperature is due to the difference in the time scale of the experiment, which will be discussed later. Figure 2 shows the temperature dependence of α and dielectric loss factor ε'' in poly(methyl methacrylate). The transition at higher temperature is the glass transition and that at lower temperature is attributed to the side-chain movement in the glassy polymer.

The following types of transition have been investigated in solid polymers:

1. Melting transition (crystalline phase → amorphous phase).
2. Polymorphic transition (crystalline phase 1 → crystalline phase 2).

Chapter 11

These types of transformation are thermodynamic phase transitions. Experiments designed to yield results of thermodynamic significance must involve the measurement of thermodynamic quantities sensitive to small changes in the degree of crystallinity or the crystalline structure. In many cases the mechanical or the electrical measurements are not sufficiently sensitive. Specific volume or specific heat measurements have been widely utilized in studying these types of transition. Generally the melting transition occurs not at a critical point but in a range of temperature, as shown in Figure 1. The temperature at which the last traces of crystallinity disappear is well defined, and, most significantly, the abrupt termination of the melting process is indicated. This temperature is defined as the melting temperature T_m.

The following types of transition are rate processes. As the temperature is lowered certain degrees of internal mobility are lost and a molecular motion is apparently frozen in. In many instances experiments designed to detect the chain mobility are relaxation measurements: dynamic mechanical measurement for complex modulus, dielectric measurement for complex dielectric constant, and nuclear magnetic resonance measurement for the absorption linewidth. Therefore the term "relaxation" may be adequate for "transition."

3. Glass transition: relaxation concerning micro-Brownian motion of polymer segments in the amorphous region.
4. Side-chain transition: relaxation concerning rotational movement of long side chains in a polymer.
5. Local mode transition: relaxation concerning torsional oscillation of frozen main chains in the glassy state.
6. Crystalline transition: relaxation concerning torsional oscillation of main chains in the crystalline region.
7. Other transitions: relaxation concerning molecular motions in an intermediate phase or in the crystalline boundary region, and due to the defects in the crystalline region, and relaxation concerning rotational motions of methyl groups, phenyl groups, and chain ends.

2.2. Relaxation Phenomena

As described in the preceding section, the existence of several apparent transitions which are explainable as a relaxation process is

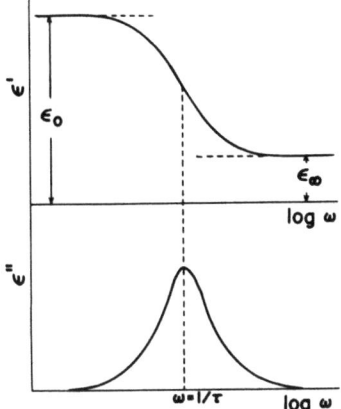

Fig. 3. Schematic representation of the dielectric relaxation.

a special characteristic of solid polymers. The physical meaning of the relaxation phenomena will be described in this section.

Dielectric relaxation in the simplest case is described by the following equations[2]:

$$\varepsilon^* = \varepsilon' - i\varepsilon'' \qquad (1)$$

$$\varepsilon' = \varepsilon_\infty + [(\varepsilon_0 - \varepsilon_\infty)/(1 + \omega^2\tau^2)] \qquad (2)$$

$$\varepsilon'' = (\varepsilon_0 - \varepsilon_\infty)\omega\tau/(1 + \omega^2\tau^2) \qquad (3)$$

where ε^* is the complex dielectric constant, ε' is the dielectric constant, ε'' is the dielectric loss factor, ω is the angular frequency, and τ is the relaxation time; $\varepsilon''/\varepsilon' = \tan\delta$ is called the dissipation factor or dielectric loss tangent. Figure 3 is a schematic representation of the dielectric relaxation. Mechanical relaxation is represented by relations similar to those for dielectric relaxation[3]:

$$G^* = G' + iG'' \qquad (4)$$

$$G' = G_0 + [(G_\infty - G_0)\omega^2\tau^2/(1 + \omega^2\tau^2)] \qquad (5)$$

$$G'' = [(G_\infty - G_0)\omega\tau/(1 + \omega^2\tau^2)] \qquad (6)$$

where G^* is the complex modulus, G' is the storage modulus, G'' is the loss modulus, and $G''/G' = \tan\delta$ is called the internal friction or mechanical loss tangent. The mechanical relaxation curve given by these equations is shown in Figure 4.

Chapter 11

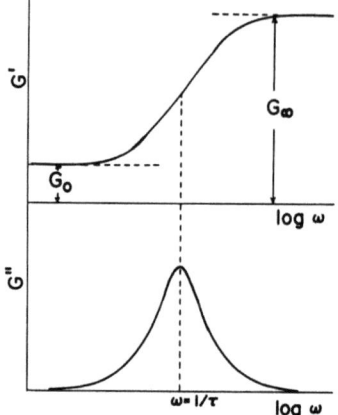

Fig. 4. Schematic representation of the mechanical relaxation.

In the dielectric and the mechanical cases the relaxation time is determined experimentally as the reciprocal of the frequency at which the loss component becomes maximum. Molecular movement to attain new equilibrium conditions under an external excitation (for instance, electric field or mechanical stress) requires a certain period of time. This is the reason for the existence of the relaxation phenomena in polymers. At frequencies much lower than $1/\tau$ the molecular motion follows the variation of the external field, while it does not for a field of much higher frequencies than $1/\tau$. Therefore τ is considered to be a measure of the rate of the molecular movement.

Of course, τ depends upon temperature and decreases remarkably with increasing temperature. When one measures ε^* or G^* at a fixed frequency as a function of temperature, therefore, one can observe a rapid change of ε' or G' accompanied by a loss maximum in a certain range of temperature. The loss maximum appears for $\omega\tau = 1$.

Usually, the temperature dependence of τ is given by the following equation:

$$\tau = Ae^{\Delta H/RT} \tag{7}$$

where A is constant, ΔH is apparent activation energy for the molecular motion, and R is the gas constant. If the temperature at which the loss maximum occurs (see Figures 1 and 2) is designated as T_{max}, we can write

$$T_{max} = \Delta H/(B - \ln \omega), \quad B = R - \ln A \tag{8}$$

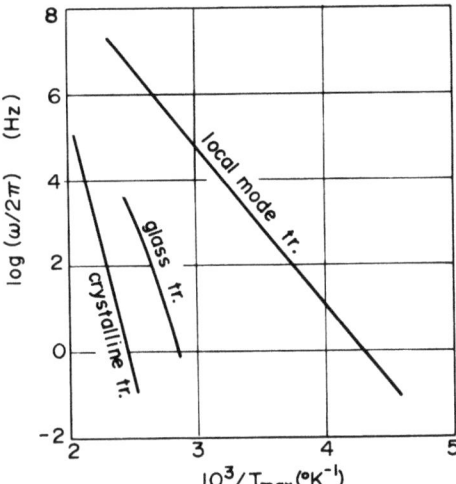

Fig. 5. Transition map for polychlorotrifluoroethylene, showing the apparent transition temperature as a function of measuring frequency.

Therefore T_{max} depends upon the frequency of the dynamic experiment and shifts toward higher temperature with increasing ω. An example of the location of the loss maximum as a function of ω is shown in Figure 5. Usually, this type of diagram is called a "transition map" and T_{max} is regarded as a transition temperature.

3. Melting of Crystalline Polymers
3.1. Melting Temperature for Homopolymers

The distinct differences in thermodynamic properties between the crystalline state and the liquid state are very similar to those which occur during the melting of a crystal of a simpler substance. The melting process can be described as a first-order phase transition and should be infinitely sharp; the critical temperature at which melting occurs is defined as the melting temperature. This temperature is governed by the relation

$$T_m = \Delta H_u / \Delta S_u \quad (9)$$

where ΔH_u and ΔS_u are the enthalpy and entropy of fusion, respectively.

Chapter 11

Crystallites in partially crystallized polymers include considerable numbers of defects and not very large in size. The latter factor requires consideration of excess contributions to the free energy caused by the surface or by junctions between the crystalline and the amorphous phases. Accordingly, a broadening of the melting range is often observed in commercial polymers.

Since the crystal growth is regarded as a rate process, the thickness l of the polymer crystal depends upon the crystallization temperature. The grown crystal does not always have the maximum stability in the thermodynamic sense: If the crystallite length l does not attain its equilibrium value, then the optimum stability of the system is reduced. The melting temperature of the crystallite of length l is given by the following equation[4]:

$$\frac{1}{T_m(l)} - \frac{1}{T_m^0} = -\frac{R}{\Delta H_u} \frac{\ln D}{l} \qquad (10)$$

where T_m^0 is the melting temperature of an infinitely large crystal, $\ln D = 2\sigma_e/RT_m(l)$, with σ_e the excess free energy per mole of unit at the crystallite ends, and ΔH_u is the heat of fusion. Many experimental results obey Eq. (10). Figure 6 shows $T_m(l)$ for the crystal of Nylon 6.

When a thin crystallite is annealed at a temperature lower than but close to T_m^0 it changes into a thick crystallite by a recrystallization (thickening phenomenon). If the crystallite of thickness l is heated very slowly, therefore, it does not melt at $T_m(l)$ but at a temperature higher than $T_m(l)$ due to the thickening. Accordingly, the slower the rate of heating is, the higher is the melting point of the polymer.

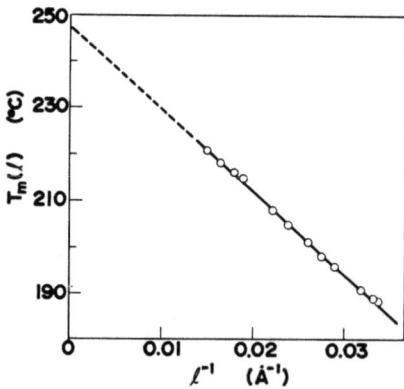

Fig. 6. Melting temperature of Nylon 6 as a function of thickness of crystalline lamella (after Arakawa et al.[5]).

As mentioned above, the melting point depression of polymer crystallites is mainly attributed to the surface (excess) free energy of the small crystallite. Another factor for the depression of the melting temperature is the various types of defects in the crystallites; the role of the chain ends in the crystallites will be discussed in the following section.

The variation of the melting temperature with hydrostatic pressure P obeys the Clausius–Clapeyron equation

$$dT_m^0/dP = T_m^0 \Delta V_u/\Delta H_u \tag{11}$$

where ΔV_u is the latent volume change that occurs on melting. Since the volume increases on melting in common polymers, dT_m^0/dP is usually positive.[6,7]

Determination of T_m^0 of polymer crystals is not very easy. We will discuss here several approaches to this problem. The first method is shown graphically in Figure 6: T_m^0 can be determined by extrapolation of the linear relation between $T_m(l)$ and $1/l$ to $l \to \infty$. The second method is a modification of the first one: Since crystal thickness l is a function of the crystallization temperature T_c,[8] i.e.,

$$l = \frac{kT_c}{b\sigma_s} + \frac{2\sigma_e T_m^0}{\Delta h(T_m^0 - T_c)} \tag{12}$$

the relation between $T_m(l)$ and T_c is used for estimation of T_m^0 instead of the relation between $T_m(l)$ and $1/l$. Plots of $T_m(l)$ for polychlorotrifluoroethylene against T_c are shown in Figure 7.[9] Intersection of

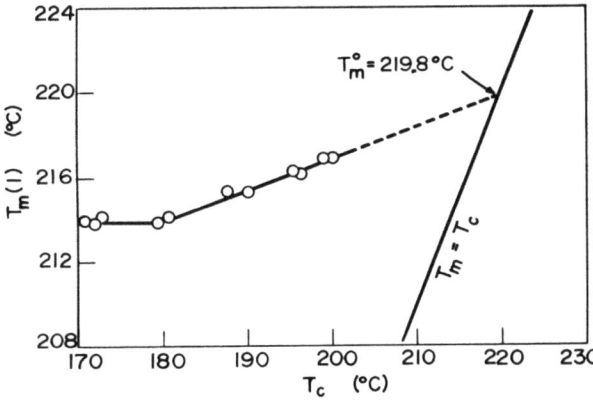

Fig. 7. Relation between melting temperature and crystallization temperature of polychlorotrifluoroethylene (after Hoffman and Weeks[9]).

Chapter 11

the straight line which represents the relation between $T_m(l)$ and T_c with the line $T_m(l) = T_c$ gives T_m^0.

For several polymers extended-chain crystals have been prepared by crystallization from the melt under high hydrostatic pressures. The thickness of the extended-chain crystal is much greater than that of ordinary folded-chain crystals. Therefore the melting temperature of the former is considered to be approximately equal to T_m^0. For instance, the extended-chain crystal of polyethylene melts at 414.6°K.

3.2. Effect of Molecular Weight on Melting Temperature

The effect of polymer chain length on the melting process depends upon whether the chain ends participate in the crystallization. If the chain ends are excluded from the crystallite and can be treated as a foreign ingredient, it is possible to derive simple relationships between the melting temperature and the degree of polymerization. Flory[10] has studied this problem for the polymer possessing a most probable molecular weight distribution

$$w_x = x(1-p)^2 p^{x-1} \qquad (13)$$

where w_x is the weight fraction of species composed of x units and p is a parameter representing the probability of the continuation of the chain from one unit to the next. If the chain ends whose distribution obeys Eq. (13) are excluded from the crystallite, the following relation is derived from the thermodynamic argument:

$$\frac{1}{T_m(\bar{x}_n)} - \frac{1}{T_m(\infty)} = \frac{R}{\Delta H_u} \frac{2}{\bar{x}_n} \qquad (14)$$

where $\bar{x}_n$ is the number-average degree of polymerization and $2/\bar{x}_n$ represents the mole fraction of noncrystallizing units. The experimental result for polydecamethylene adipate is shown in Figure 8.[11] It is found that Eq. (14) is valid for this result. According to Eq. (14), as $\bar{x}_n$ is increased the influence of $\bar{x}_n$ on T_m is expected to be diminished. The value of ΔH_u depends on the structure of polymer and varies from 1 to 10 kcal/mole. Therefore the asymptotic value of T_m should be reached for $\bar{x}_n$ of several hundred.

Chain ends of n-paraffin were found to establish well-defined crystallographic planes. Broadhurst[12] had discussed the molecular weight dependence of the melting temperature in terms of the heat and entropy of fusion: on the assumption that these thermodynamic quantities are both linear function of number of carbon atoms n in

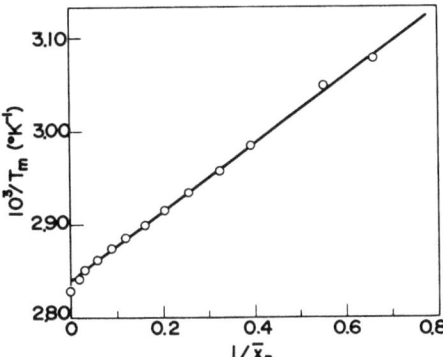

Fig. 8. Relation between melting temperature and degree of polymerization for polydecamethylene adipates (after Evans et al.[11]).

the molecular chain, the following equation has been derived:

$$T_m(n) = T_m(\infty)[(n + a)/(n + b)] \tag{15}$$

The experimental results obey Eq. (15), as shown in Figure 9.[12] It should be noted that $T_m(\infty)$ corresponds to T_m^0 for high-molecular-weight polyethylene.

3.3. Dependence of Melting Temperature on Chemical Structure

Starting with the basic thermodynamic equation for melting, one can point out two dominant factors which affect the melting temperature: intermolecular forces and chain stiffness.

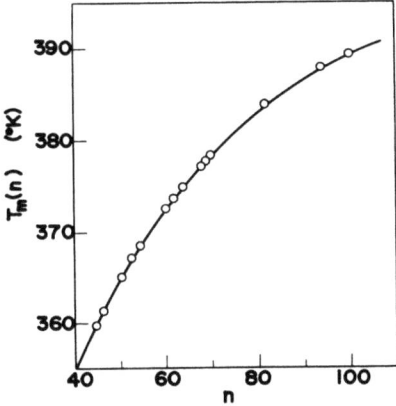

Fig. 9. Melting temperature of n-paraffin as a function of number of carbon atoms (after Broadhurst[12]).

Chapter 11

According to Dole's treatment,[13] the heat of fusion ΔH_u is described by

$$\Delta H_u = \Delta H_0 + \Delta H_h \tag{16}$$

where ΔH_h is that part of ΔH_u arising from hole formation and ΔH_0 is constant. Remembering that the cohesive energy is that energy required to remove a molecule to a position far from its neighbors, ΔH_h is represented in terms of cohesive energy density (CED). Then we can write

$$\Delta H_h = \Delta H_0 + k_1(\text{CED}) \tag{17}$$

Attempts to correlate T_m with CED have been unsuccessful. For instance, polydimethyl siloxane possesses T_m of $-80°C$ and $(\text{CED})^{1/2}$ of 7.3, while polytetrafluoroethylene has T_m of 327°C and $(\text{CED})^{1/2}$ of 6.2. From a survey of T_m for a variety of polymers and the thermodynamic quantities governing melting, it is evident that ΔS_u is of prime importance in establishing the value of T_m.

The quantity ΔS_u is a measure of the entropy difference for a chain repeating unit between the crystalline and liquid states. Though crystals can be disordered so that the entropy of the crystalline state varies to some extent, the most important factor affecting ΔS_u is increase of the configurational entropy of the system upon melting. Depending upon the polymer, the molecules can range from random coil to elongated rodlike forms, from flexible chains to stiff ones. The chain stiffness is a complex function of ε, the energy difference between the trans and gauche configurations. Following Boyer,[14] ΔS_u is represented in terms of ε by

$$\Delta S_u = \Delta S_0 - k_2 \varepsilon / kT \tag{18}$$

The difference of T_m between polydimethyl siloxane and polytetrafluoroethylene cited above can be explained by the difference in ε: polydimethyl siloxane possesses $\varepsilon = 0$ cal/mole and polytetrafluoroethylene has $\varepsilon = 4300$ cal/mole. The thermodynamic quantities governing the melting process are listed in Table 1.

It gives insight into the melting process in polymers to observe the variation of T_m by a systematic change of the chemical structure in homologous series. The aliphatic polyesters melt at a lower temperature than polyethylene, in analogy with the melting temperature of the monomeric chain esters as compared with the corresponding hydrocarbons. A general increase in T_m is observed as the proportion of methylene units $(-CH_2-)_n$ in the chain increases,[15] as

TABLE 1
Thermodynamic Quantities Governing the Melting Process

Polymer	T_m, °C	ΔH_u, cal/mole	ΔS_u, cal/deg mole
Polyethylene	137.5	960	2.34
Polypropylene	186	2100	4.6
Polystyrene	239	2000	3.9
Poly(ethylene oxide)	66	1980	5.35
Poly(oxymethylene)	180	1590	3.5
Poly(ethylene sebacate)	76	8000	23
Poly(hexamethylene adipamide)	267	10300	19.1
Polytetrafluoroethylene	327	1460	2.9

shown in Figure 10. The aliphatic polyamides melt at much higher temperatures than polyethylene, as shown in Figure 10. In contrast to the polyesters, there is a general decrease of T_m as the proportion of methylene units in the chain increases. This is explainable by the hydrogen bonding capacity of the amide groups. This is also the case with the aliphatic polyurethanes.

If we observe the variation of T_m with the number of methylene units in a repeating unit in detail, a periodic variation in T_m is found for polyesters, polyamides, and polyurethanes: as shown in Figure 11, the polymers that contain an odd number of CH_2 sequences melt at a lower temperature than those containing an even number of

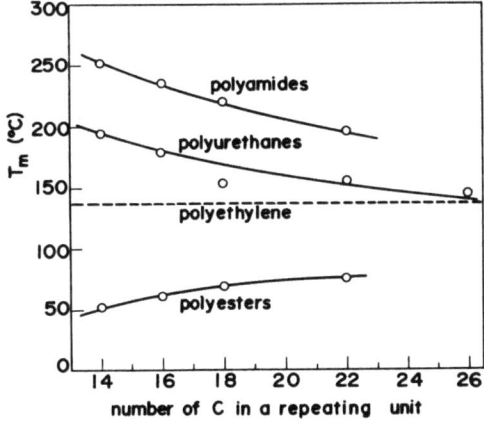

Fig. 10. Variation of melting temperature with systematic change of the chemical structure of a polymer (after Bunn[15]).

Chapter 11

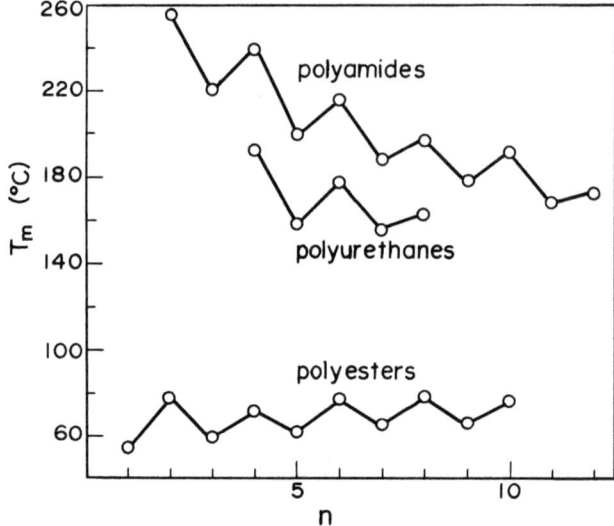

Fig. 11. Variation of melting temperature with systematic change of the chemical structure of a polymer (after Bunn[15]).

sequences.[15,16] This difference between odd and even polymers can be explained by the difference in ΔH_u due to the packing manner of plane zigzag chains in the crystallites.

It should be pointed out that in reality ΔS_u contains a contribution of the volume change on melting in addition to the influence of the configurational entropy. Then we can write

$$\Delta S_u = (\Delta S_u)_V + \Delta S_V \qquad (19)$$

At T_m^0 the change in entropy caused by the volume change ΔS_V can be expressed as

$$\Delta S_V = \Delta V_u (\partial P/\partial T)_V - (\alpha/\beta)\Delta V_u \qquad (20)$$

where α is the thermal expansion coefficient at constant pressure and β is the compressibility at constant temperature.[17] For the quantitative analysis of the effect of molecular configuration on the melting process, the entropy of fusion at constant volume $(\Delta S_u)_V$ should be used.

3.4. Melting Temperature for Copolymers

The problems involved in the melting process of copolymers are influenced by too many factors to be uniquely formulated: The composition of a copolymer, the length of a crystalline sequence, the distribution of the lengths, the stereochemical relation between the different repeating units, and the crystallization conditions are important factors.

A simple theory of the equilibrium crystallization and melting has been developed by Flory.[18] He has treated a model copolymer which contains only one type of crystallizable unit, designated as the A unit. In the molten state these units occur in a specified distribution of sequence lengths. The noncrystallizable comonomeric units are designated as B units. A quantitative formulation is developed by relating to the melt composition the probability P_ζ that a given A unit in the amorphous polymer is located within a sequence of at least ζ A units. For the necessary and sufficient condition for crystallization P_ζ in the molten copolymer should be larger than $P_\zeta{}^*$, that is, the probability when crystallization occurs and thermodynamic equilibrium is maintained. The limiting temperature for crystallization or the equilibrium melting temperature of the copolymer $T^0_{m,AB}$ is given by the following equation:

$$\frac{1}{T^0_{m,AB}} - \frac{1}{T^0_{m,A}} = -\frac{R}{\Delta H_u} \ln p \qquad (21)$$

where $T^0_{m,A}$ is the equilibrium melting temperature of A homopolymer, ΔH_u is the heat of fusion per mole of the crystallizable units, and p is the probability of an A unit being succeeded by another A unit.

For a random-type copolymer, p can be identified with the mole fraction of A units in the copolymer, X_A. Then Eq. (21) can be written

$$\frac{1}{T^0_{m,AB}} - \frac{1}{T^0_{m,A}} = -\frac{R}{\Delta H_u} \ln X_A \qquad (22)$$

For a block-type copolymer in which the crystallizable units occur in very long sequences p greatly exceeds X_A. Therefore the depression of the melting temperature is not as great as for random copolymers. For an alternating-type copolymer p is less than X_A and a much greater depression of the melting temperature is observed.

Figure 12[17] shows the variation of T_m of copolyesters and copolyamides with the composition. For these systems in which A and B units can crystallize independently of one another the units composing a major portion of the molecular chain can crystallize and then p is considered to be identical with the mole fraction of the crystallizable units. The melting temperature depends only upon

Chapter 11

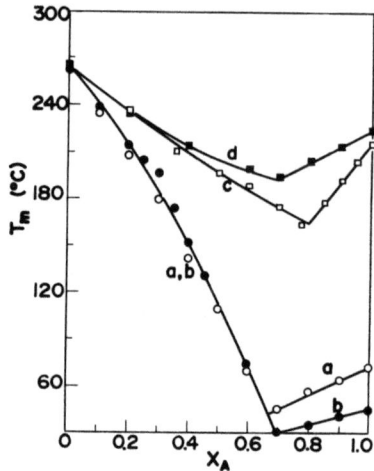

Fig. 12. Variation of melting temperature of copolyesters and copolyamides with composition: (a) polyethylene terephthalate/sebacate; (b) polyethylene terephthalate/adipate; (c) polyhexamethylene adipamide/caproamide; (d) polyhexamethylene adipamide/sebecamide (after Mandelkern[17]).

composition and is independent of the chemical nature of the minor coingredient that is introduced, as indicated for the copolymers of polyethylene terephthalate and those of polyhexamethylene adipamide. This result is in accordance with the random distribution of sequences in the copolymer. At high concentrations of the added ingredient Eq. (21) is not valid, since the ingredient can itself undergo crystallization.

For a block-type copolymer the melting temperature–composition relation is expected to be different from that for the random copolymers. Figure 13[19] shows the melting temperature of block copolymers of polyethylene adipate and polyethylene sebacate as a function of composition. The melting temperature of the block copolymers remains constant with added coingredient. For these copolymers the melting of only the major component is observed and a step-shaped melting temperature–composition curve thus results.

In Flory's theory it is assumed that the comonomer units added do not participate in crystallization. In some copolymers the added ingredients can mter into the crystalline lattice. The melting tempera-

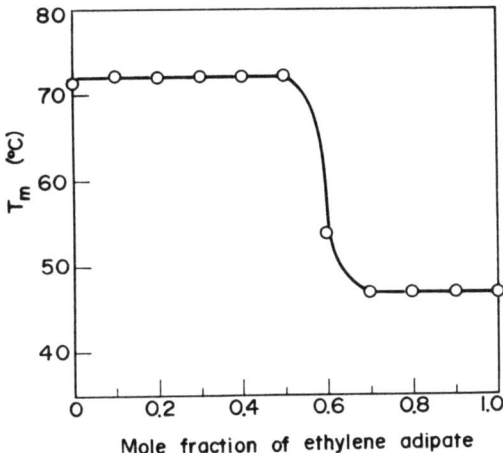

Fig. 13. Melting temperatures of block copolymers of polyethylene adipate and polyethylene adipate and polyethylene sebacate (after Coffey and Meyrick[19]).

ture–composition relations for such systems are different from those for random or block copolymers. The random introduction of isomorphous units causes cocrystallization and a continuous variation of the melting temperature which assumes values intermediate between those of the pure individual homopolymer. Figure 14[20] shows the melting temperature–composition relation for copolymers

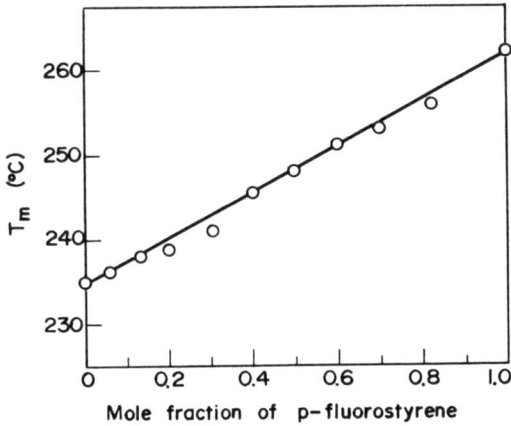

Fig. 14. Melting temperature of styrene and p-fluorostyrene copolymers as a function of mole fraction of p-fluorostyrene (after Natta[20]).

Chapter 11

of styrene and *p*-fluorostyrene. Eby[21,22] has discussed theoretically the melting of a cocrystallizing copolymer from the point of view that the minor comonomer units assume the role of defects in the crystallites. The relation derived is

$$\frac{1}{T^0_{m,AB}} - \frac{1}{T^0_{m,A}} = \frac{a}{T^0_{m,A}} X_B(1 + X_B) + \frac{b}{T_m^0 l} \tag{23}$$

where l is the thickness of crystalline lamella and a and b are related to excess free energy at the defects and at the folding part, respectively.

4. Glass Transition in Solid Polymers

4.1. Thermodynamic Phase Transition or Relaxation Phenomena

When a molten polymer is cooled down to temperatures far below the melting temperature, occasionally it does not crystallize and transforms into a glasslike solid which is regarded as a frozen liquid. This state of aggregation is found to be essentially amorphous by X-ray analysis and is called the glassy or vitreous state. The transformation of the supercooled liquid into the glassy state is called the glass transition. If a polymer crystallizes on cooling, the remaining amorphous region in it undergoes the glass transition at a certain temperature. Thus the glass transition is the transition that is observed in all polymers except in single crystals.

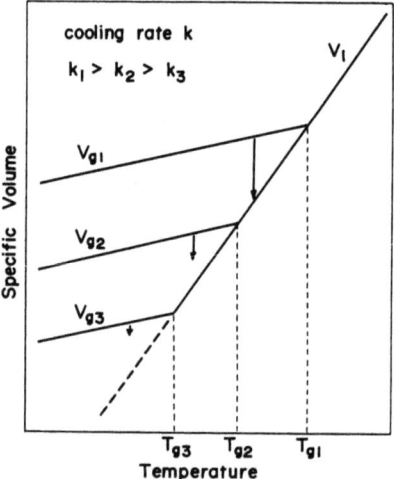

Fig. 15. Schematic representation of cooling rate dependence of glass transition temperature.

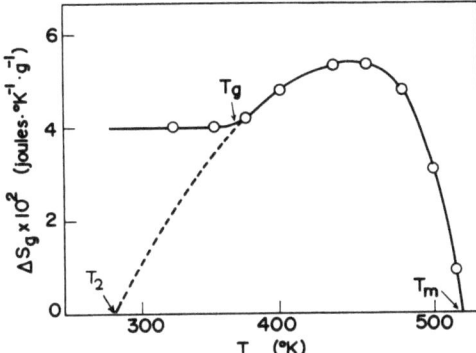

Fig. 16. Excess entropy of amorphous polystyrene relative to crystalline polystyrene as a function of temperature (after Karasz et al.[24]).

It has been found experimentally that the thermodynamic and other properties of polymers (or glass-forming substances) change abruptly at the glass transition point: Thermal expansion coefficient, compressibility, and specific heat change discontinuously. Since these thermodynamic quantities correspond to the second derivatives of Gibbs free energy, the transition was called a second-order transition for a time. A very unfavorable evidence for the thermodynamic transition is that a pronounced relaxation effect is observed in the physical experiments in the vicinity of the glass transition temperature T_g. An example of this effect is the cooling rate dependence of the T_g observed experimentally, as shown in Figure 15. The slower the cooling rate is, the lower is the observed transition temperature. An alternative possible origin of the glass transition has been proposed: The transition is essentially an apparent transition caused by the relaxational effect associated with attainment of the equilibrium condition of the system. This opinion is prevalent at present and may be called the relaxation theory.

In fact, the characteristic relaxation times for molecular motions and viscosity are so large near T_g that the value obtained for T_g may depend on the time scale of the experiment by which it is measured. However, difficulties arise in a purely kinetic view of the glass transition. Kauzmann[23] has shown that if the thermodynamic behavior observed experimentally in a material above its T_g is extrapolated through and below T_g to obtain the supposed equilibrium behavior at these lower temperatures, absurd results, such as negative entropies, are obtained. Figure 16[24] shows the excess entropy of supercooled amorphous polystyrene $\Delta S_g(T)$ relative to that of the crystalline polystyrene as a function of temperature. It is found that the extrapolated values for $\Delta S_g(T)$ to low temperatures become negative. Gibbs and DiMarzio[25] have shown theoretically that the configurational entropy of a linear polymer system becomes zero at a temperature higher than 0°K and the thermodynamic second-order transition occurs at this temperature. This second-order transition temperature is denoted by T_2 in Figure 16. Existence of T_2 has not been confirmed experimentally because at temperatures below T_g the physical and thermodynamic properties of the equilibrium amorphous state cannot be determined due to extremely large molecular relaxation times.

4.2. Glass Transition and Molecular Relaxation Time

Williams et al.[26] have proposed that the temperature dependence of relaxation time τ characterizing the rate of micro-Brownian motion of amorphous chains is described by the following (WLF) equation:

$$\log a_T = \log \frac{\tau(T)}{\tau(T_g)} = -\frac{C_1(T - T_g)}{C_2 + (T - T_g)} \qquad (24)$$

where C_1 and C_2 are constants. From the point of view that τ or the bulk viscosity in an amorphous polymer is governed by the free volume in it, the constants C_1 and C_2 are expressed as

$$C_1 = B/2.303 f_g \qquad (25)$$

$$C_2 = f_g/\alpha_f \qquad (26)$$

where f_g is the fractional free volume at T_g, α_f is the thermal expansion of the free volume relative to total volume, and the constant B is close to unity. It has been found that C_1 and C_2 are almost independent of materials and the average values for them are 17.44 and 51.6, respectively.[26] Then, assuming that $B = 1$, one can find that f_g is almost independent of material and is approximately equal to 0.025. This result strongly supports the concept that the glass transition can be recognized as an iso-free-volume state.

Recently Adam and Gibbs[27] have developed a quantitative theory relating molecular configurational relaxation time to the variation of size of the cooperatively rearranging region. The size of this region is determined by configurational restrictions in the glass-forming liquids and is expressed in terms of their configurational entropy:

$$\ln \tau = A' + C/S_c T \qquad (27)$$

where A' and C are constants and S_c is the configurational entropy. According to the Gibbs–DiMarzio theory,[25] S_c is equal to zero at the second-order transition temperature T_2. Thus S_c at temperatures above T_2 is given by

$$S_c(T) \simeq \Delta C_p \ln(T/T_2) \qquad (28)$$

where ΔC_p is the difference in specific heat between the supercooled liquid and the glass at T_g. From this theory the WLF constants are given by

$$C_1 = \frac{C}{2.303(\Delta C_p) T_g \ln(T_g/T_2)} \qquad (29)$$

$$C_2 \simeq \frac{T_g \ln(T_g/T_2)}{1 + \ln(T_g/T_2)} \qquad (30)$$

By the application of this theory to many experimental results concerning the molecular relaxation time, an interesting conclusion that $T_g/T_2 \simeq 1.30$ has been derived.[27]

4.3. Glass Transition from the Point of View of Free Volume in Amorphous Phase

In the preceding section it was stated that the fractional free volume in many amorphous polymers at T_g is nearly equal to 0.025. Some authors have reported quite different values for the fractional free volume at T_g. For instance, Simha and Boyer[28] showed that the product of T_g (°K) and $\Delta \alpha$ (where $\Delta \alpha$ is the difference in the cubic expansion coefficient above and below T_g) is a constant for amorphous polymers, having a value of 0.113, with all materials studied falling in the range 0.08–0.14. The consistency of this product is another manifestation of T_g as an iso-free-volume state. The difference in the values for the free volume at T_g is largely a matter of definition. Figure 17 shows a schematic specific volume–temperature plot to illustrate the different definitions of the free volume at T_g.

The free volume concept of the glass transition or the rate of micro-Brownian motion of amorphous chains is convenient to use in

Chapter 11

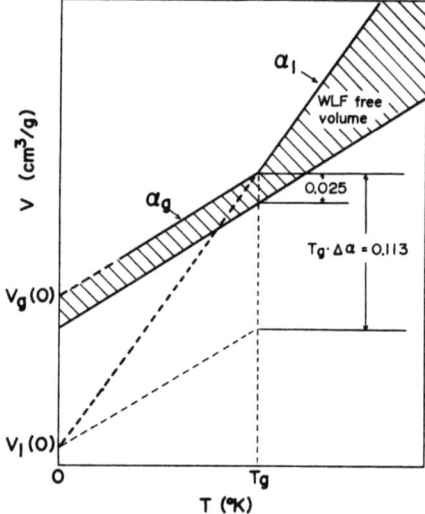

Fig. 17. Schematic representation of the different definitions of free volume in a polymer at the glass transition temperature.

interpreting the effect of pressure on these phenomena. When a hydrostatic pressure is applied to a supercooled liquid (at a temperature higher than T_g) a decrease of the specific volume and an increase of τ result mainly from a decrease of the free volume. If the applied pressure is sufficiently high, τ becomes extremely large and the supercooled liquid is vitrified. Thus the glass transition temperature increases with pressure.

Recently some authors have expressed doubt as to whether the free volume is the determining factor for the glass transition in polymers. The doubt arises from the following evidence:

1. The fractional free volume at T_g is not always constant but in some cases changes systematically with the degree of polymerization.[29]

2. For many polymers the densified glassy state can be formed by cooling the supercooled liquid state under high pressure[30,31]; if the glass transition is recognized as an iso-free-volume state also under high pressures, the densification is not expected to occur.

3. The temperature and pressure dependence of the relaxation time concerning micro-Brownian motion can be interpreted in terms of the variation of configurational entropy rather than that of the

free volume.[32] Nose[31] has proposed that the glass transition should be recognized as an isoconfigurational entropy state. This notion is not completely successful in interpreting all phenomena relating to the glass transition.

4.4. Dependence of the Glass Transition Temperature on Cooling Rate

The general method for the determination of T_g is by measurement of the specific volume–temperature relation. The effect of the cooling rate on T_g observed experimentally will be discussed in this section.

From the point of view of free volume, the glass transition of polymers is caused by the free volume dependence of the volume retardation process or the molecular relaxation time, as described in the preceding section. Assuming a linear dependence of the fractional free volume f on temperature T, f is given by

$$f = \alpha_f(T - T_s) = f_0 + \alpha_f(T - T_0) \tag{31}$$

where T_s is the temperature at which the free volume disappears, T_0 is a reference temperature, f_0 is f at T_0, and α_f is the expansion coefficient of free volume relative to the total volume. When the temperature is suddenly lowered from T_0 to T_1, f is considered to vary gradually from f_0 to f_1 (equilibrium value of f at T_1) with a relaxation time τ. Then the rate of the variation of f is expressed by

$$-df/dt = \{f - [f_0 - \alpha_f(T_0 - T)]\}/\tau \tag{32}$$

For cooling with a constant rate σ, $-dT/dt = \sigma$. Since $-df/dt = \sigma\, df/dT$, Eq. (32) can be written as

$$\sigma\, df/dt = [(f - f_0) - \alpha_f(T - T_0)]/\tau \tag{33}$$

Following Doolittle's analysis[33] on the melt viscosity, τ is described by the following equation:

$$\tau = \tau_0 e^{1/f} \tag{34}$$

Therefore Eq. (33) is the differential equation which describes the relationship between f and T. Examples of numerical solutions of Eq. (33) assuming that $\alpha_f = 4.8 \times 10^{-4}$ and $\tau_0 = 10^{-13}$ sec are shown in Figure 18.[34] Note that the values of f deviate from the line representing the equilibrium values of f at temperatures much higher than T_s. Even at an extremely slow rate of cooling, say 1° per several years, the deviation of f from equilibrium occurs at a temperature

Chapter 11

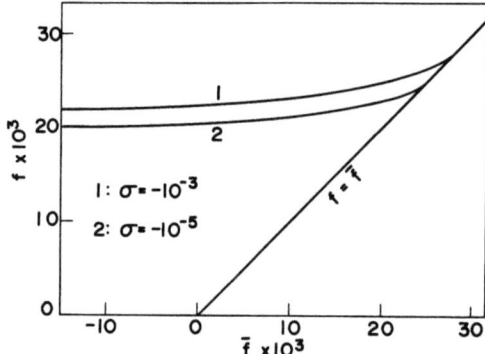

Fig. 18. Freezing-in of free volume (after Saito et al.[34]).

higher than T_s. The lowering of T_g by such an extremely slow experiment may be no more than 10°.

4.5. Dependence of the Glass Transition Temperature on Chemical Structure

The values of T_g in polymers are widely distributed: For polydimethyl siloxane $T_g = -123°C$, while $T_g = 208°C$ for polyvinyl carbazole. Careful observation of the value of T_g leads to a qualitative relationship between T_g and the chemical structure of polymers.

Beaman[35] has pointed out that there is a general empirical correlation between T_g and T_m (for temperatures in °K):

$$T_g = (2/3)T_m$$

for asymmetric polymers such as polychlorotrifluoroethylene or isotactic polypropylene, and

$$T_g = (1/2)T_m$$

for symmetric polymers such as polyethylene or polyvinylidene chloride. In general, T_g and T_m seem to depend on the same structural factors. Since T_m increases with intermolecular forces and chain stiffness, presumably T_g will do likewise.

Polymers of high cohesive energy density (CED) tend to have a high value of T_g. For instance, polydimethyl siloxane and polybutadiene have low values of T_g and CED. On the other hand, polyacrylonitrile and polyacrylic acid have high values of T_g and CED. Figure 19[36] shows the relationship between T_g and CED. For the

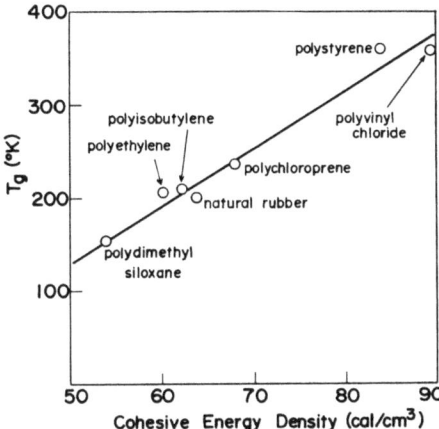

Fig. 19. Relation between glass transition temperature and cohesive energy density in polymers (after Boyer[36]).

polymers cited in Figure 19 there exists a linear relation between T_g and CED. However, there are many polymers whose T_g deviates from this linear relation. This means that other factors, such as the energy barrier impeding chain rotation and chain stiffness, should be taken into account for T_g. Polymers having a low energy barrier have a low value of T_g. Amorphous sulfur, with a very high value of CED, nevertheless has a T_g of about 0°C because of the relatively free rotation around sulfur–sulfur bonds. Polymers which consist of stiff molecular chains have higher values of T_g than more flexible polymers of similar chemical constitution and similar CED. For example, polyvinyl acetals have a much higher T_g than polyvinyl acetate. The difference in T_g for poly-α-methyl styrene and polystyrene could be attributed primarily to the difference in stiffness of the backbone chains.

Variation of T_g in polyalkyl methacrylates with length of the side chain provides some insight about the relation between T_g and the chemical structure. Depression of T_g with increase of the side-chain length is interpreted in terms of lowering of the interactions between backbone chains: long, flexible side chains assume the role of plasticizer. Nose[31] has shown theoretically that T_g in these polymers is proportional to CED.

The importance of ionic interaction as an intermolecular force has been pointed out by Eisenberg.[37] He investigated the effect of the counter ion on T_g of completely ionizable homopolymers, such as polyphosphates, polysilicates, and polyacrylates. The strength of the anion–cation interaction is considered to determine the glass transition in these systems. At T_g, kT is just large enough to allow the anion to leave the coordination sphere of the cation (or vice versa). Thus kT_g must be proportional to the electrostatic work of removing an anion from the coordination sphere of a cation,

$$T_g \propto \int F_{el}\, dr = \int (q_a q_c/r^2)\, dr \qquad (35)$$

where F_{el} is the electrostatic interaction, r is the internuclear distance between anion and cation, q is the charge, the subscripts a and c refer to anion and cation, and the limits of integration are the distance of closest approach δ and infinity. Since the anion charge is constant for any series of materials, the simple relation

$$T_g \propto q/\delta \qquad (36)$$

where q is the cation charge, is expected. In spite of the drastic differences in the backbone structures of these polymers, the experimental results are surprisingly similar, i.e.,

$$T_g = 625(q/\delta) - 12 \quad \text{for polyphosphates}$$
$$T_g = 635(q/\delta) + 132 \quad \text{for polysilicates}$$
$$T_g = 730(q/\delta) - 67 \quad \text{for polyacrylates}$$

where T_g is in °C, q is in units of electron charge, and δ is in Å. A similar result was obtained for aliphatic ionenes, i.e.,

$$T_g = 695(q/\delta) - 23$$

where q is the charge of the anion.

The value of T_g for a polymer can be changed by variation of the molecular weight. This phenomenon is interpreted in terms of the action of chain ends. Since each chain will necessarily possess more free volume than if it were chemically bound within a continuous chain, the presence of chain ends will increase the amount of free volume in the supercooled liquid state of the polymer at any given temperature. Remembering that the glass transition occurs at an iso-free-volume state, then polymers of very low degree of polymerization are expected to have lower values for T_g than the chemically identical polymer of infinite degree of polymerization. Variation of T_g for polystyrene as a function of molecular weight M is shown in

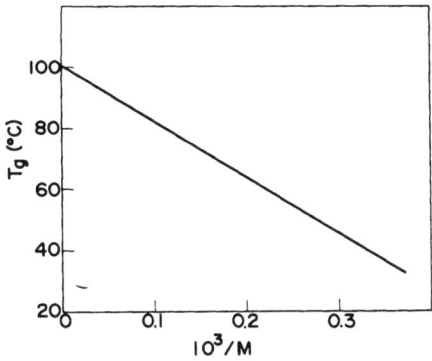

Fig. 20. Glass transition temperature for polystyrenes as a function of the molecular weight (after Fox and Flory[38]).

Figure 20.[38] This result can be described by

$$T_g = T_{g\infty} - K/M \qquad (37)$$

where $T_{g\infty}$ is the glass transition temperature of the infinite-molecular-weight polymer and K is a constant. Following Bueche,[39] K is given by $2\rho N_0 \theta/\alpha$, where ρ is the polymer density, N_0 is Avogadro's number, θ is the free volume contributed by a chain end, and α is the cubic expansion coefficient of the polymer at temperatures above T_g.

The lowering of T_g of a polymer on addition of a miscible liquid can be explained by the free volume concept. Low-molecular-weight liquids will contain more free volume than the pure polymer. If one assumes additivity of free volume, the polymer–diluent or the polymer–plasticizer system will contain more free volume in the supercooled liquid state at any given temperature than would the polymer alone. Consequently, the plasticized polymer must be cooled to a lower temperature in order to reduce its free volume to the critical amount at the glass transition.

For some partly crystalline polymers the effect of crystallinity on T_g has been investigated. However, a general trend of the crystallinity effect has not been obtained. In general, it would seem that the crystallites restrict micro-Brownian motion of the amorphous chains and then raise T_g. Kolb and Izard[40] have shown that T_g of polyethylene terephthalate increases with increase of crystallinity. On the other hand, Boyer and Spencer[41] have shown that T_g of a copolymer of vinylidene chloride and vinyl chloride is practically unaffected by crystallinity. These results suggest that the size of the motional unit

Chapter 11

in amorphous chains and of polymer crystallites should be taken into account. The distribution of the crystallites is also a very important factor for T_g: Even at an identical degree of crystallinity, many fine crystallites give stronger influences on the segmental mobility of amorphous chains than a small number of larger crystallites do.

4.6. Glass Transition in Copolymers

Variation of T_g by copolymerization has attracted considerable attention as an important method for control of T_g in polymers. For the copolymers in which the phase separation does not occur, the free volume concept is useful in considering the variation of T_g. If it is assumed that the amount of free volume contributed by a monomer unit is identical whether one is concerned with the copolymer or homopolymer, the value of T_g for the copolymer can be derived in the following manner.[39] On the assumption that the free volumes of each constituent are additive, the fractional free volume in the copolymer is

$$f = [f_g + \alpha_{f1}(T - T_{g1})]v_1 + [f_g + \alpha_{f2}(T - T_{g2})]v_2 \qquad (38)$$

where f_g is the critical value of the fractional free volume at T_g as an iso-free-volume state, v is the volume fraction, and subscripts 1 and 2 refer to two polymer constituents. At T_g of the copolymer, it is found that $T = T_g$ and $f = f_g$. Substitution of these values in Eq. (38) yields the following expression for the glass transition temperature of the

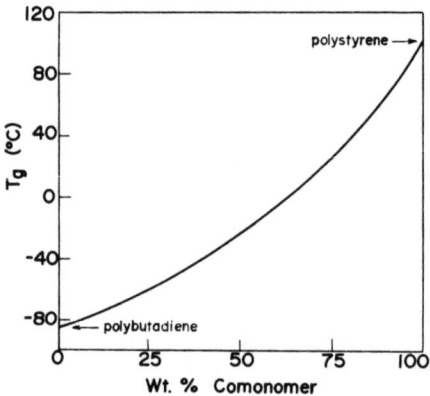

Fig. 21. Variation of glass transition temperature with copolymer composition of styrene and butadiene.

copolymer:

$$T_g = [T_{g1} + (KT_{g2} - T_{g1})v_2]/[1 + (K - 1)v_2] \qquad (39)$$

where the constant K is equal to $(\alpha_{f2}/\alpha_{f1})$. This equation has a form similar to the well-known Gordon–Taylor equation[42,44] and Mandelkern–Martin–Quinn equation[43,44] for T_g in copolymers. An example of the variation of T_g by copolymerization is shown in Figure 21.

Graft copolymers and block copolymers fall into a different category from random copolymers: Very often phase separation occurs in these two types of copolymer. In these heterogeneous polymers the amorphous region usually consists of chain sequences of one chain constituent precipitated as droplets in a matrix of the other constituent. Consequently, these two-phase systems will show physical properties analogous to a mixture of two different polymers. Each phase will show its glass transition temperature.

The glass transition of the polymers possessing intermolecular cross-links will be considered here. Since the cross-links restrict micro-Brownian motion of amorphous chains, T_g of such polymers will increase with increase of cross-link density. In addition, there will be a decrease in the specific volume, since introducing cross-links involves the exchange of van der Waals bonds for shorter, more compact primary bonds. Fox and Loshaek[45] discussed the effect of cross-links in a manner similar to the effect of an increase in molecular weight of the primary chains. The derived equation is

$$T_g(\chi) = T_g(0) + K_1 \chi \qquad (40)$$

where χ is number of cross-links per gram and K_1 is a constant.

DiMarzio[46] has discussed the change of configurational entropy of a polymer due to adding cross-links. The elevation of the transition temperature at which the configurational entropy becomes zero is found to be

$$\frac{T_2(\chi) - T_2(0)}{T_2(0)} = \frac{K_2 M \chi/\gamma}{1 - K_2 M \chi/\gamma} \qquad (41)$$

where T_2 is the Gibbs–Dimarzio second-order transition temperature,[25] χ is the number of cross-links per gram, M is the molecular weight of a repeating unit, γ is the number of flexible bonds per repeating unit, and K_2 is a constant independent of polymer and about

Chapter 11

equal to 1.3×10^{-23}. Although T_2 cannot be determined experimentally, the change of T_g with cross-link density can be found by Eq. (41) using $T_g(\chi)$ instead of $T_2(\chi)$.

5. Subsidiary Transitions in Solid Polymers

A number of instances are now available in which well-defined apparent transitions other than the melting and the glass transition occur in solid polymers. These subsidiary transitions are mainly due to localized molecular motions or limited motions within the backbone chains as compared with the micro-Brownian motion of backbone chains. Hence these apparent transitions are more readily observed by dynamic mechanical or dielectric measurements than by thermodynamic experiment. It should be remembered that these transitions are manifestations of relaxation phenomena and depend upon the time scale or the frequency of the measurement.

5.1. Side-Chain Motion

When dynamic mechanical or dielectric measurement is carried out for a polymer having long, flexible side chains, one can find an indication of the side-chain rotation at low temperatures at which the micro-Brownian motion of backbone chains is practically frozen in. Polymers in which rotational motion of side chains occurs are polyalkyl methacrylate, polyvinyl acetate, polybutene-1, cellulose derivatives, and so on.

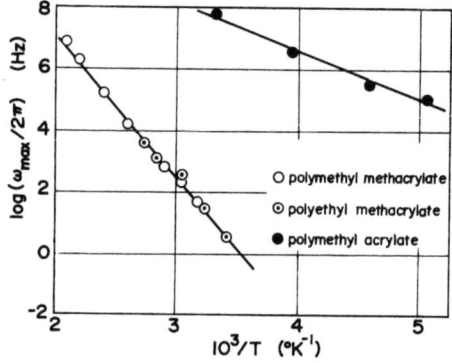

Fig. 22. Transition map for side-chain motions in methacrylate and acrylate polymers.

The side-chain motion has the following characteristics.

1. If the apparent transition temperature is defined as the temperature at which mechanical or dielectric loss becomes maximum, the transition temperature lies below T_g.

2. The transition temperature depends on backbone structure: The transition temperature in polymethyl acrylate is much lower than that in polymethyl methacrylate. However, the transition temperature remains nearly constant with minor variation of the side-chain structure; as an example, the transition map for the side-chain motions is shown in Figure 22.

3. For a random copolymer in which each constituent has a long side chain, the side-chain transition for each constituent can occur, while T_g of the copolymer varies continuously with the composition.[47] Therefore the motional unit for the side-chain motion is considered to be the individual side chain.

4. The apparent activation energy is much lower than that for the micro-Brownian motion of backbone chains. Consequently the transition temperature depends more strongly upon the frequency of the measurement than does T_g.

5. The rate of the side-chain motion decreases with increasing pressure. Consequently the transition temperature is raised with pressure: $(\partial T/\partial P)_\tau \simeq 0.01°/\text{atm}$, where $(\partial T/\partial P)_\tau \simeq T\Delta V/\Delta H$, with ΔH and ΔV the activation energy and the activation volume for the side-chain motion, respectively.[48]

The rotational motion of methyl groups is important as representative of motions of short side chains. The rotation is manifested most strikingly by nuclear magnetic resonance measurements because of the large magnetic dipolar interactions. The apparent transition temperature which can be evaluated by high-frequency (say 50 MHz) NMR absorption strongly depends upon the position of the methyl group in a polymer molecule. For instance, the transition temperature for the α-methyl rotation in polymethyl methacrylate is 270°K, while that for the methyl group at the end of the long side chain in polymethyl methacrylate is 90°K.

5.2. Torsional Oscillation of Frozen Backbone Chains

When dynamic mechanical or dielectric relaxation measurement is carried out for a polymer having no side chains, such as polyvinyl chloride and polyethylene terephthalate, one can find an indication of molecular motion even at temperatures below T_g. Some authors have postulated that this molecular motion in the glassy state can be attributed to motion of the chain ends. However, this opinion has been rejected because of the excessively high value of the observed relaxation strength. This molecular motion is now believed to be the torsional oscillation of frozen backbone chains.

In the glassy polymer and even in crystals of long-chain molecules the torsional oscillation around the chain axis is excited into large amplitude due to the nonlinear character of interatomic forces. The oscillation becomes rather incoherent to neighboring chains at higher temperatures. Since the oscillation meets viscous interchain resistance as in the liquid, this type of molecular motion is regarded as a damped oscillation of backbone chains about their equilibrium positions. This damped oscillation often occurs in the glassy amorphous region in polymers and has been termed the "local relaxation mode" process.[34]

The torsional oscillation of frozen backbone chains in the amorphous region has the following characteristics.

1. If the apparent transition temperature is defined as the temperature at which the mechanical or the dielectric loss peak appears, the transition temperature lies below T_g.

2. The apparent activation energy, activation volume, and the pressure dependence of the transition temperature are comparable with or less than those for the side-chain motion.[49]

3. This molecular motion occurs in many polymers without long side chains, especially in partly crystallizable polymers.

4. This molecular motion can occur for a polymer having a side chain if the side chain is rigid, such as polystyrene.

5. In disubstituted chains, in which both hydrogen atoms bounded to every other carbon atom of the polyethylene chain are substituted, the transition due to the torsional oscillation occurs at a higher temperature than in monosubstituted chains. However, the torsional oscillation cannot be observed in such disubstituted chains as polyvinylidene chloride.

The torsional oscillation of backbone chains has been discussed for $-(CH_2-CHR)_n-$ chains from a different point of view: Instead of the torsional oscillation, a rotational mechanism has been postulated and termed the "crankshaft" process. The mechanisms proposed by two authors[14,50] are illustrated in Figure 23. According to Schatski's model,[50] the two bonds 1 and 7 rotate simultaneously and the intervening carbon atoms move as a crankshaft. An essential criterion for the mechanism is that the bonds 1 and 7 are collinear. The chain

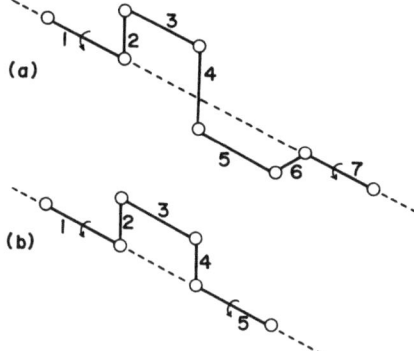

Fig. 23. Illustration of the crankshaft mechanism. (a) Schatzki's model,[50] (b) Boyer's model.[14]

bonds on either side may remain frozen and only a relatively small volume is required for the movement. According to Boyer's model,[14] the bonds 1 and 5 rotate simultaneously and there are only two intervening carbon atoms.

From a number of dynamic mechanical experiments Willbourn[51] postulated that the low-temperature transition for $-(CH_2)_n-$ chains arose from motion of 4–6 carbon atoms about the backbone chain axis. This correponds to the crankshaft mechanism for $-(CH_2-CH_2)_n-$ chains.

5.3. Molecular Motion in the Crystalline Phase

Thermal expansion of the crystalline lattice in a polymer becomes considerable at high temperatures due to nonlinearity of the interatomic forces. At these temperatures the torsional oscillation around the chain axis can be excited into large amplitude in the crystalline phase. Polymer crystals may be regarded as a plastic crystal from the point of view of molecular motion. Even in a single crystal of a polymer one can observe the presence of the torsional oscillation of the chains.

The dynamic mechanical loss of polyethylene is shown in Figure 24.[52] The peak height of the high-temperature loss for the single crystal is larger than that for bulk sample. This peak is attributable to the torsional oscillation of chains in the crystalline region. This type of molecular motion in the crystalline region has the

Chapter 11

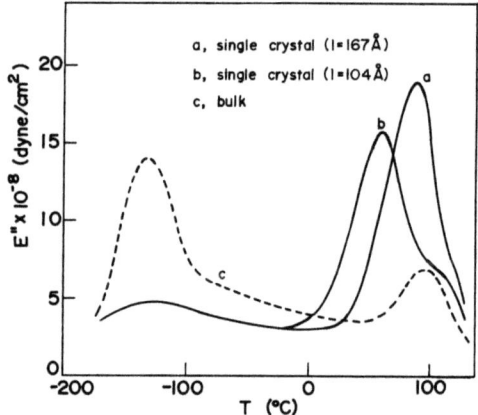

Fig. 24. Dynamic mechanical loss of various polyethylene samples, showing molecular motion in the crystalline phase (after Takayanagi and Matsuo[52]).

following characteristics:

1. The apparent transition temperature at which the loss peak appears lies above T_g and far above the transition temperature concerning the torsional oscillation in the glassy amorphous region.

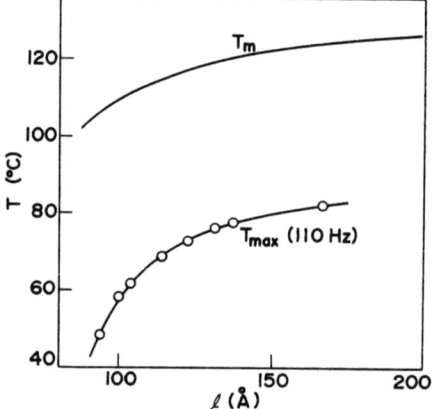

Fig. 25. Apparent transition temperature due to molecular motion in the crystalline phase and melting temperature for polyethylene as a function of thickness of crystalline lamella (after Takayanagi and Matsuo[52]).

2. The transition temperature strongly depends upon the thickness of crystalline lamella: As shown in Figure 25,[52] it increases with increase of the thickness. The existence of the thickness dependence in a range above 100 Å suggests that the molecular motion may involve longitudinal translation in addition to the torsional oscillation.
3. The apparent activation energy for this process is larger than that for the torsional oscillation in the glassy amorphous region.

This type of molecular motion in the crystalline region exists in several highly crystalline polymers, such as polytetrafluoroethylene, polychlorotrifluoroethylene, polyvinylidene fluoride, and so on.

As shown in Figure 24, a bulk sample of polyethylene exhibits a loss peak at very low temperatures. Since its height decreases gradually with increase of crystallinity, one may regard it as a molecular motion (torsional oscillation of frozen backbone chains) in the amorphous region. The peak is still observable for the single crystal, as shown in Figure 24. Consequently, it may be reasonable to consider that at least a part of the low-temperature relaxation is attributable to defects in the crystals. A number of experimental results indicate that the low-temperature relaxation involves simultaneous contributions from two kinds of molecular motion: One belongs to the glassy amorphous region and the other to defects in the crystalline region. Several authors have proposed possible mechanisms for the molecular motion connected with defects in the crystals: chain-end-induced row vacancies,[53] screw dislocations,[54] and motion of kinked chains.[55]

5.4. Molecular Motion in the Intermediate Phase

In a polymer consisting of stiff molecular chains such as polytetrafluoroethylene or polyacrylonitrile, an intermediate phase between crystalline and amorphous phases is formed when the polymer is solidified from the melt. In the intermediate phase the molecular chains may form in line but the spacing between chains may be irregular. The X-ray diffraction pattern for the intermediate phase is sharper than that for the amorphous phase but gives halos.

Molecular motion in the intermediate phase usually has a different value of the relaxation time from that in the amorphous phase. On the viscoelastic or the dielectric relaxation spectrum, accordingly, one can observe a discrete loss peak belonging to the

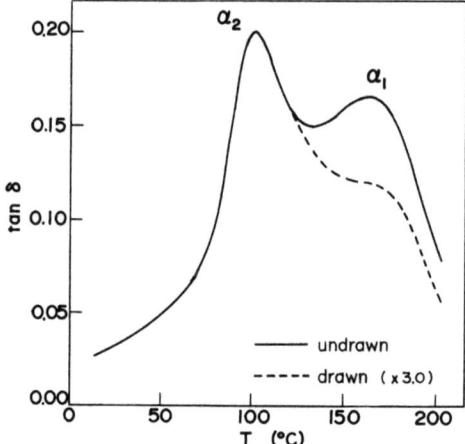

Fig. 26. Dynamic mechanical loss for polyacrylonitrile, showing molecular motions in intermediate phase and in amorphous phase (after Minami et al.[56]).

intermediate phase in addition to that due to the micro-Brownian motion of the molecular chains in the amorphous region. Polyacrylonitrile is a good example illustrating the apparent transition in the intermediate phase. Figure 26 shows the dynamic mechanical properties of polyacrylonitrile at 110 Hz. The loss tangent curve shows two peaks α_1 and α_2. Minami et al.[56] studied variations of the peak temperatures (apparent transition temperatures) by copolymerization with vinyl acetate.[56] The peak temperature of the α_1 process varies with the composition in accord with the Eq. (39) for T_g in copolymers. Therefore the α_1 process can be attributed to the micro-Brownian motion in the amorphous phase. On the other hand, the peak temperature of the α_2 process remains almost constant for vinyl acetate concentrations up to 36.9%. The peak height of the α_1 process was found to decrease with drawing of polyacrylonitrile in a similar manner to that for the amorphous peak, whereas the peak height of the α_2 process remains unchanged by drawing up to five times. Judging from these results and the insensitivity of the α_2 process to residual diluents, Minami et al. have concluded that the α_2 process is due to the molecular motion in the intermediate phase. Other examples for this type of molecular motion have been reported for polytetrafluoroethylene[57] and isotactic polypropylene.[58]

References

1. D. J. H. Sandiford, *J. Appl. Chem.* **8**, 188 (1958).
2. H. Fröhlich, *Theory of Dielectrics*, Clarendon, Oxford (1960).
3. J. D. Ferry, *Viscoelastic Properties of Polymers*, Wiley, New York (1961).
4. P. J. Flory, *J. Chem. Phys.* **17**, 223 (1949).
5. T. Arakawa, F. Nagatoshi, and N. Arai, *Polymer Lett.* **6**, 513 (1968).
6. S. Matsuoka, *J. Polymer Sci.* **42**, 511 (1960).
7. T. Davidson and B. Wunderlich, *J. Polymer Sci.* **A2**, **7**, 377 (1969).
8. J. I. Lauritzen, Jr. and J. D. Hoffman, *J. Res. NBS* **64A**, 73 (1960).
9. J. D. Hoffman and J. Weeks, *J. Res. NBS* **66A**, 13 (1962).
10. P. J. Flory, *J. Chem. Phys.* **17**, 223 (1949).
11. R. D. Evans, H. R. Mighton, and P. J. Flory, *J. Am. Chem. Soc.* **72**, 2018 (1950).
12. M. G. Broadhurst, *J. Chem. Phys.* **36**, 2578 (1962).
13. M. Dole, *Fortschr. Hochpolym. Forsch.* **2**, 221 (1960).
14. R. F. Boyer, *Rubber Chem. Tech.* **34**, 1303 (1963).
15. C. W. Bunn, *J. Polymer Sci.* **16**, 323 (1955).
16. R. Hill and E. E. Walker, *J. Polymer Sci.* **3**, 609 (1948).
17. L. Mandelkern, *Crystallization of Polymers*, McGraw-Hill, New York (1964).
18. P. J. Flory, *Trans. Faraday Soc.* **51**, 848 (1955).
19. D. H. Coffey and T. J. Meyrick, in *Proc. Rubber Tech. Conf. 2nd*, p. 170 (1954).
20. G. Natta, *Makromol. Chem.* **35**, 94 (1960).
21. J. P. Colson and R. K. Eby, *J. Appl. Phys.* **37**, 3511 (1966).
22. R. K. Eby, *J. Appl. Phys.* **34**, 219 (1948).
23. W. Kauzmann, *Chem. Revs.* **43**, 219 (1948).
24. F. E. Karasz, H. E. Bair, and J. M. O'Reilly, *J. Phys. Chem.* **69**, 2657 (1965).
25. J. H. Gibbs and E. A. DiMarzio, *J. Chem. Phys.* **28**, 373 (1958).
26. M. L. Williams, R. F. Landel, and J. D. Ferry, *J. Am. Chem. Soc.* **77**, 3701 (1955).
27. G. Adam and J. H. Gibbs, *J. Chem. Phys.* **43**, 139 (1965).
28. R. Simha and R. F. Boyer, *J. Chem. Phys.* **37**, 1003 (1962).
29. A. A. Miller, *J. Polymer Sci.* **A2**, **6**, 249 (1968).
30. S. Saito, Reps. *Progress Polymer Phys. Japan* **11**, 375 (1968).
31. T. Nose, *Polymer J.* **2**, 445, 530 (1971).
32. H. Sasabe and S. Saito, *Polymer J.* **3**, 631 (1972).
33. A. K. Doolittle, *J. Appl. Phys.* **22**, 1471 (1951); **23**, 236 (1952).
34. N. Saito, K. Okano, S. Iwayanagi, and T. Hideshima, in *Solid State Physics* (F. Seitz and D. Turnbull, eds.), Vol. 14, p. 468 (1963).
35. R. G. Beaman, *J. Polymer Sci.* **9**, 472 (1953).
36. R. F. Boyer, *J. Appl. Phys.* **25**, 825 (1954).
37. A. Eisenberg, *Macromolecules* **4**, 125 (1971).
38. T. G. Fox and P. J. Flory, *J. Appl. Phys.* **21**, 581 (1950).
39. F. Bueche, *Physical Properties of Polymers*, Interscience, New York (1962).
40. H. J. Kolb and E. F. Izard, *J. Appl. Phys.* **20**, 564 (1949).
41. R. F. Boyerr and R. S. Spencer, *J. Appl. Phys.* **15**, 398 (1944).
42. M. Gordon and J. S. Taylor, *J. Appl. Chem.* **2**, 493 (1952).
43. L. Mandelkern, G. M. Martin, and F. A. Quinn, *J. Res. NBS* **58**, 137 (1957).
44. L. A. Wood, *J. Polymer Sci.* **28**, 319 (1958).
45. T. G. Fox and S. Loshaek, *J. Polymer Sci.* **15**, 371 (1955).

Chapter 11

46. E. A. DiMarzio, *J. Res. NBS* **68A**, 611 (1964).
47. Y. Kawamura, S. Nagai, and Y. Wada, *J. Polymer Sci. A2*, **7**, 1559 (1969).
48. H. Sasabe and S. Saito, *J. Polymer Sci. A2*, **6**, 1401 (1968).
49. S. Saito, H. Sasabe, K. Yada, and T. Nakajima, *J. Polymer Sci. A2*, **6**, 1297 (1968).
50. T. F. Schatzki, *J. Polymer Sci.* **57**, 496 (1962), *Polymer Preprints* **6**, 646 (1965).
51. A. H. Willbourn, *Trans. Faraday Soc.* **54**, 717 (1958).
52. M. Takayanagi and T. Matsuo, *J. Macromol. Sci.* **B1**, 407 (1967).
53. J. D. Hoffman, G. Williams, and E. Passaglia, *J. Polymer Sci.* **C14**, 173 (1966).
54. R. Masui and Y. Wada, *Rep. Progress Polymer Phys. Japan* **13**, 271 (1970).
55. W. Pechhold, *Kolloid Z.* **228**, 1 (1968).
56. S. Minami, H. Sato, and N. Yamada, *Rep. Progr. Polymer Phys. Japan* **10**, 317 (1967).
57. Y. Ohzawa and Y. Wada, *Japan. J. Appl. Phys.* **3**, 436 (1964).
58. S. Minami, S. Uemura, and M. Takayanagi, *Rep. Progr. Polymer Phys. Japan* **7**, 253 (1964).

Index

Al–Ga–Ti system, 202-205
Alloys (*see also* Solid solutions, and specific system compositions)
  decomposition, 155-160
  order–disorder phenomena, 3-5, 24, 130
  martensitic transformations, 109-114
  massive transformations, 114-116
  morphology, 168-173
  pair energies in, 131-133
  phase diagrams, 179-229
  phase equilibria, 139-143
  thermochemical data, 179-229
Al–Ti–Sn system, 202-205
Al–Zn system, 142, 155-159
Amorphous films, 542
Antiferroelectricity, 13-15
Antiferromagnetism, 10-13, 21-22, 25
Aqueous solution growth of crystals (*see* Solution growth of crystals)
As_2S_3, 538-542
As_2Se_3, 535, 541-542
As_2Te_3, 535
As–Te–Ge glass, 537-538, 544
Au–Cd system, 109
Au–Ni system, 142
Avrami equation, 98, 100, 118, 167

Bain distortion, 191
Banding, in crystal growth, 250-251
$Ba_2NaNb_5O_{15}$, phase relations, 388-389
$BaTiO_3$, 14
Benzil, 261, 263

Bi, 261
Binary alloys (*see* Alloys)
Boron nitride, 499
Bragg–Williams model, 133-134, 272
Bridgman method, 347

Catalysis, in crystallization of viscous liquids, 527
Cb–Ni–Cr system, 205-211
Cellular growth, 237-239, 242, 334
Charge compensation, 412
Chemical potential, 9, 123, 139, 147
Chemical vapor deposition, 285-321
  epitaxial growth, 302-321
    GaAs, 293, 310-321
    kinetics, 291-302
    Si, 285-286, 293, 303-310
    thermodynamics, 286-291
Clustering in solid solutions, 129-174
Coarsening, 101-105, 168-170
Co–Fe system, 183-185
Co, free energy of, 185-186
Coherent interfaces, 77-78, 87-89, 139-143
Coherent phases, 77-78, 87-89, 139-143
Coherent solvus, 141
Common tangent construction, 139, 150, 186
Common tangent rule, 139, 150, 186
Compositional fluctuations in stable solid solutions, 152-153
Compressibility, 70, 472-474
Configurational free energy, 133-135

Index

Constitutional supercooling, 239-244, 333-335, 416-417
Convection, crystal growth, 236-237, 242, 329, 341-350, 366-372
Co_2O_3, 480-481
Coordination number, 4, 472-474
Copolymers, 568-572
Correlation function, 7, 19-20, 31-34, 46-48
Cr–Cb–Ni system, 205-211
Cr–Fe–Ni system, 190-200
Critical exponents, 6-7, 17-26, 29-35, 37, 40-44, 49-59
  definitions, 17-20
  experimental values, 20-25
  theory, 25-26, 29-35, 37, 40-44, 51-59
  thermodynamic inequalities for, 25-26
Critical nucleus, 83-84, 147-150
Critical phenomena, 1-59
  antiferromagnetism, 10-13, 21-22, 25
  critical exponents (*see* Critical exponents)
  critical points, 3, 5-6, 9-11, 13, 17, 20-23, 30-33
  ferroelectrics and antiferroelectrics, 13-15, 37-43
  ferromagnetism, 5-9, 24, 28, 43
  fluids, 9-10, 23-24
  general theories, 54-59
  helium, 15-16, 23-24, 56
  liquid crystals, 16-17
  order–disorder, 3-5, 24, 130
  superconductors, 16
  superfluid helium, 15-16, 23-24, 56
  theoretical models for, 16-59
Critical points, 3, 5-6, 9-11, 13, 17, 20-23, 30-33
Cr–Ni–Cb system, 205-211
Cr–Ni–Mo system, 200-202
Cr–Ni–Nb system, 205-211
Cr–Ni system, 183-185, 190-200
Crystal growth, 233-461, 497-503, 561-564
  banding, 250-251
  Bridgman, 347
  cellular, 237-239, 242, 334
  chemical vapor deposition (*see also* Chemical vapor deposition), 285-321
  convection effects, 236-237, 242, 329, 341-350, 366-372
  CVD, 285-321
  Czochralski, 237, 340-346, 361-364, 369, 371-372, 374, 376-381, 391-392, 395-397, 399
  dendritic, 237-240, 252-254
  diffusion effects, 234-251, 291-295, 307-308, 319-321, 327-345, 413-417

Crystal growth *(cont'd)*
  diffusion-controlled, 251-254
  epitaxial (*see also* Epitaxial growth), 302-321, 452-454
  facets, 252, 261-263
  floating zone, 348-351
  flux growth (*see also* Flux growth), 410, 437-450
  from liquids, 233-281, 325-400, 407-458
  from solutions (*see also* Solution growth of crystals), 233-281, 407-458, 497-503
  from the melt (*see also* Melt growth of crystals), 233-281, 325-400
  from the vapor (*see also* Vapor growth of crystals), 233-281, 283-321
  growth rate, 248-281, 288-302, 329-340, 413-417
  high pressure, 497-503
  horizontal melt growth, 346-347
  hydrothermal, 243-244, 410, 425-438, 464-468, 501-503
  heat transfer, 234-237, 251
  incorporation of impurities, 239-244, 250-251
  interface stability, 244-251
  kinetics, 248-281, 288-302, 329-340, 413-417, 432-435
  Kyropoulos method, 345
  liquid encapsulation, 345-346
  liquid–liquid–solid growth, 352-353
  liquids, 233-281, 325-400, 407-458
  mass transfer, 234-237, 284-286, 291-295, 307-308, 326-340, 413-417
  melt, 233-281, 325-400
  molten salt (*see also* Flux growth), 410, 437-450
  morphology, 237-251
  nonconservative methods, 351-354
  planar, 237-239
  polymers, 561-564
  solid–liquid–solid growth, 353-354
  solutions (*see also* Solution growth of crystals), 233-281, 407-458, 497-503
  spherulites, 261, 264
  Stockbarger method, 347
  theory of, 233-281
  vapor, 233-281, 283-321
  vapor–liquid–solid growth, 352, 456-458
  Verneuil method, 351
  zone melting, 332-334, 348-351
Crystallization in glasses, 544-545
Crystallization of polymers, 561-564
Crystallization of viscous systems, 523-529
Crystal morphology, 237-281

Index

Crystal pulling (*see* Czochralski growth)
Crystal structure, high pressures, 472-474
Crystals, liquid, 16-17
Cu–Ni–Fe system, 168-170
Cu–Zn system, 114-116
CVD processes, 285-321
Czochralski growth, 237, 340-346, 361-364, 369, 371-372, 374, 376-381, 391-392, 395-397, 399

Decorated Ising models, 35
Defects, melt growth of crystals, 391-393
Dendritic growth, 237-240, 252-254
Diamonds, high-pressure growth of, 497-499
Dielectric loss, 559
Dielectric relaxation, 559
Diffuse interface, 146-147
Diffusion, in crystal growth, 234-251, 291-295, 307-308, 319-321, 327-345, 413-417
Diffusion coefficients, liquids, 234-235
Diffusion-controlled crystal growth, 291-295
Diffusion-controlled growth in phase transformations, 70, 94-96, 98-99, 106
Diffusion-controlled transformations, 76
Diffusionless transformations, 108-114
Dimer model for critical phenomena, 35-37
Dislocation density, reduction of, 340, 391-392
Dislocation-free crystals, 340, 391-392
Dislocations, nucleation at, 91
Distribution coefficient, 241, 328-330, 411-412
Driving force, 68-69, 73, 82-84, 88, 90-94, 102, 108-109, 120, 123

Elastic energy, 78-80, 135-138, 141-143
Electrical memories in glasses, 533-538
Entropy of fusion, 237, 255-281
Epitaxial growth, 302-321, 452-454
  GaAs, 310-321
  heteroepitaxy, 302
  homoepitaxy, 302
  liquid-phase epitaxy (LPE), 452-454
  Si, 303-310
Epitaxy (*see* Epitaxial growth)
Exchange interactions, 5

Facets, crystal growth, 252, 261-263
Fe–Co system, 183-185
Fe–Cr–Ni system, 190-200
Fe–Cu–Ni system, 168-170
Fe, free energy of, 185-186
Fe–Mo–Ni system, 200-202

Fe–Ni–Mn system, 109
Fe–Ni system, 109, 190-193
Ferroelectricity, 13-15, 37-43
Ferromagnetism, 5-9, 24, 28, 43, 74-75
Fe–Ti–C system, 218-228
Films, 302-321, 452-454, 542
  amorphous, 542
  epitaxial, 302-321, 452-454
First-order phase line, 7, 9
First-order surface, 12
First-order transitions, 6, 9, 12, 15, 17, 30-31, 69-74
Fluids, critical phenomena, 9-10, 23-24
Flux growth, 410, 437-450
  electrolysis, 449
  equipment, 439
  materials applications, 444
  phase equilibria, 439-444
  solvents for, 439-440
Frank-Read source, 413
Free energy change, phase transformations, 68-73, 82-84, 88, 90-94, 107-108, 112, 121-123, 150-152, 183-190, 259-271, 287-291, 411
Free energy, configurational, 71-73
Free energy of mixing, 133-135, 189
Free volume, 575-577

Ga–Al–Ti system, 202-205
GaAs, 310-321, 452-453
  chemical vapor deposition, 310-321
  crystal growth, 310-321, 452-453
  liquid-phase epitaxy, 452-453
Gadolinium gallium garnet, 380
GaP, 384-385, 390-391, 451-453
  crystal growth, 452-453
  liquid-phase epitaxy, 452-453
  stoichiometry, 383-385, 390-391
Garnet, 380, 429-430, 440-444, 453-454
Gas–solid reactions, 283-321
  epitaxial crystal growth, 302-321
  kinetics, 291-302
  thermodynamics, 286-291
Gd–Ga garnet, 380
Ge, 352-353, 374, 376, 395
Glasses, 513-546, 572-584
  liquid–glass transition, 513-521
  memories, 533-546
  polymers, 572-584
  switching in, 529-533
  threshold switches, 529-533
  transitions in, 171-173, 513-546
Glass formation, 513-529

595

Index

Glass transition, 514, 558, 572-584
   inorganic, 514, 558
   polymers, 572-584
Glass transition in polymers, 572-584
   chemical structure, 578-582
   copolymers, 582-584
   cooling rate, 577-579
   free volume, 575-577
   molecular relaxation time, 574-575
Grain boundaries, nucleation at, 89-92, 100
Grain-boundary nucleation, 89-92, 100
Grain growth, 117-120
Growth, phase transformations, 92-99, 110, 119-120, 167-168
   crystal growth (*see* Crystal growth)
   diffusion-controlled, 94-96, 98-99
   interface-controlled, 92-94
   kinetics, 92-97, 115, 119-120, 167-168
Guinier–Preston zones, 141, 156-157, 173

Heisenberg model, 8, 18, 50-51, 57
Helium, 15-16, 23-24, 56
Heteroepitaxy, 302
Hf–Ta system, 186-189
High-pressure crystal growth, 497-503
   diamonds, 497-499
High-pressure transformations, 463-503
   experimental methods, 464-472
   phase diagrams, 474-501
   synthesis (*see also* Synthesis at high pressures), 477-497
Homoepitaxy, 302
Hydrogen-bonded model for critical phenomena, 37-43
Hydrogen bonds, 14
Hydrogen, metallic, 497
Hydrothermal growth, 243-244, 410, 425-438, 464-468, 501-503
   defects, 435-437
   equipment, 427-428, 464-468
   kinetics, 432-435
   materials applications, 438, 501-503
   phase equilibria, 429-432
   quartz, 431-437

Impurity segregation during crystal growth, 327-340, 372-381
InSb, 372-377
Interaction energies, 4
Interface-controlled growth, 92-94, 115-116
Interface-controlled transformations, 92-94, 115-116
Interface geometry, 77-78, 87-89, 139-143, 146-147, 237-254, 258-281

Interface geometry *(cont'd)*
   crystal growth, 237-254, 258-281
   diffuse, 146-147
   phase transformations, 77-78, 87-89, 139-143
Interfacial energy, 77-78, 103, 107-108, 119-120, 146
Interfacial structure, 77-78, 87-89, 139-143, 146-147, 237-254, 258-281
Interface morphology, crystal growth, 237-254
Interface stability, crystal growth, 244-251
Interfaces, coherent, 77-78, 87-89, 139-143
Intermetallic compounds, synthesis at high pressures, 496-497
Internal friction, 559
IPS transformation, 80-81
Invariant-plane strain transformation, 80-81
Ising model, 2, 4-5, 7-10, 12-13, 18-19, 30-32, 35, 46, 49-50, 56-58, 133, 259

Johnson–Mehl model, 97-98, 100, 118, 167

KH_2PO_4, 14-15
Kinetics, 76, 81-100, 106, 110-113, 115-120, 165-168, 248-281, 288-302, 329-340
   crystal growth, 248-281, 288-302, 329-340, 413-417, 432-435
   phase transformations in solids, 76, 81-100, 106, 110-113, 115-120, 165-168
K_2NiF_4, 34
Kyropoulos method, 345

Lamellar reactions, 105-108
Landau theory, 29
Lattice stability, 183, 185
$LiNbO_3$, 386-388
   phase relations, 387-388
   stoichiometry of, 386-388
Linear chains, 37, 46
Liquid crystals, 16-17
Liquid encapsulation, 345-346
Liquid–glass transformations, 513-521
Liquid–liquid–solid crystal growth, 352-353
Liquid-phase epitaxy (LPE), 452-454
Liquids, viscous (*see also* Viscous liquids), 513-546
Localized moments, 5
Loss tangent, 557, 559

Magnetism, 5-13, 21-25, 28, 43, 74-75
   antiferromagnetism, 10-13, 21-22, 25
   ferromagnetism, 5-9, 24, 28, 43, 74-75

Index

Martensitic transformations, 79, 108-114
  kinetics, 109-111
Massive transformations, 114-116
Mechanical loss tangent, 559
Melt growth of crystals, 233-281, 325-400
  boundary effects, 327-340, 342-345
  Bridgman, 347
  cellular, 334
  constitutional supercooling, 239-244, 333-335
  control of, 392-400
  convection, 236-237, 242, 329, 341-350, 366-372
  Czochralski method, 237, 340-346, 361-364, 369, 371-372, 374, 376-381, 391-392, 395-397, 399
  defects, 391-393
  floating zone, 348-351
  GaP, 384-385, 390
  growth rate fluctuations, 335-338
  horizontal growth, 346-347
  impurities, 327-340
  Kyropoulos method, 345
  liquid encapsulation, 345-346
  liquid–liquid–solid, 352-353
  materials applications, 354-360
  melt statics and dynamics, 361-372
  morphology, 360-361
  nonconservative methods, 351-354
  nonstoichiometry, 381-391
  normal freezing techniques, 340-347
  principles, 326-340
  solid–liquid–solid transport, 353-354
  steady-state segregation phenomena, 327-335
  Stockbarger method, 347
  striations, 335-340, 372-381
  theory of, 233-281, 326-340
  transient segregation phenomena, 335-340
  vapor–liquid–solid (VLS), 352, 456-458
  Verneuil method, 351
  vertical growth, 347
  zone melting, 332-334, 348-351
Melting of polymers, 561-572
  copolymers, 568-572
  effect of chemical structure, 565-568
  effect of molecular weight, 564-565
  melting temperature, 561-564
Memories in glasses, 533-546
Metallic hydrogen, 497
Metastable states, 139, 141, 143, 145-147
Mg–Ga–Al system, 179-181
Miscibility gap, 73-74, 121, 138-140
Mixing, free energy of, 133-135, 189

Mo–Fe–Ni system, 200-202
Molecular motion in polymers, 555-590
  in crystalline phase, 587-589
  in intermediate phase, 589-590
  melting, 561-572
  relaxation phenomena, 558-561, 572-590
  side-chain motion, 584-585
  torsional oscillation of main chains, 585-587
Molten salt growth, 410, 437-450
  electrolysis, 449
  equipment, 439
  materials applications, 444
  phase equilibria, 439-444
  solvents for, 439-440
Mo–Ni–Cr system, 200-202
Mo–Ni system, 200-202
Monte Carlo methods, 278-281
Morphology, 168-173, 237-251, 360-361
  interface (*see also* Crystal growth), 237-251, 360-361
  solid-solution decomposition, 168-173

Nb–Ni–Cr system, 205-211
Ni–Cr–Cb system, 205-211
Ni–Cr–Nb system, 205-211
Ni–Cr system, 183-185, 190-200
Ni–Fe–Cr system, 190-200
Ni–Fe–Cu system, 168-170
Ni, free energy of, 185
Ni, magnetic transition in, 186
Ni–Mo–Cr system, 200-202
Ni–Mo–Fe system, 200-202
Ni–Mo system, 200-202
Nonconservative crystal growth, 351-354
Nonlinear optical materials, 386-389
Nonstoichiometry, 381-391
Normal freezing techniques, 340-347
Nucleation, 81-92, 97-98, 100, 109-113, 115-118, 147-150, 158-159, 165-168, 257, 272-273, 291-292, 295-302, 413
  dislocations, 91
  grain boundaries, 89-92, 100
  kinetics, 81-92, 97, 100, 110-113, 115-118, 165-168
  Martensitic transformations, 110-113
  nonrandom, 89-92
  random, 84-85
  solid solutions, 147-150
  strain energy effects, 85-90
  surface, 257, 272-273, 291-292, 295-302, 413
  theory, 81-92
  viscous systems, 524

597

Index

Nucleation and growth (*see also* Nucleation, Growth), 81-98, 106-107, 109-116, 118-119, 153-154, 165-168
  kinetics, 81-98, 109-116, 165-168
  mechanisms, 106-107
  recrystallization, 118-119
Nucleation theory, 130, 147-152
Nucleus, critical, 83-84, 147-150

Onsager's solution, 33-34
Optical materials, nonlinear, 386-389
Optical memories in glasses, 538-546
Order–disorder, 3-5, 24, 130

Pair correlation function, 7, 19-20, 31-34, 46-48
Pair energies, 131-133
Pair interactions, 132-133
Particle coarsening, 101-105, 168-170
PbTe, stoichiometry of, 385-386
Phase diagrams, 179-229, 242, 327, 381-384, 409-412, 429-432, 439-444, 474-501, 521-523
  computation of, 179-229
  crystal growth, 242, 327, 381-384, 409-412, 429-432, 439-444
  high pressure, 474-501
  thermochemical data for, 179-229
  viscous liquids, 521-523
Phase equilibria (*see also* Phase diagrams), coherent and incoherent, 139-143
Phases, high pressure, 474-496
Phase separation, glasses, 521-523, 538
  viscous liquids, 521-523
Phase stability, 129-130
Phase transformations (*see also* Crystal growth)
  Buerger's classification of, 116-117
  chemical vapor deposition, 285-321
  clustering in solid solutions, 129-174
  coarsening, 101-105, 168-170
  critical phenomena, 1-59
  crystal growth (*see also* Crystal growth), 233-461
  crystal growth from liquids, 233-281, 325-400, 407-458
  crystal growth from the melt, 233-281, 325-400
  crystal growth from solutions, 233-281, 407-458, 497-503
  crystal growth from the vapor, 233-281, 283-321
  crystallography of, 75-81
  diffusion-controlled, 76, 94-96, 98-99, 106, 291-295
  diffusionless, 108-114

Phase transformations *(cont'd)*
  driving force, 68-69, 73, 82-84, 88, 90-94, 102, 108-109, 120, 123
  first-order, 69-74
  free energy change in, 68-73, 82-84, 88, 90-94, 107-108, 112, 121-123, 150-152, 183-190, 259-271, 287-291, 411
  glasses, 513-546, 572-584
  grain growth, 117-120
  growth, 92-99, 115, 119-120, 167-168
  heterogeneous, 68
  high pressures, 463-503
  homogeneous, 67-68
  interface-controlled, 92-94, 115-116
  interfacial structure, 77-78, 87-89, 139-143, 146-147, 237-254, 258-281
  invariant-plane strain, 80-81
  IPS, 80-81
  kinetics, 76, 81-100, 106, 110-113, 115-120, 165-168, 248-281, 288-302, 329-340
  liquid glass, 513-521
  martensitic, 79, 108-114
  massive, 114-116
  melting of polymers, 561-572
  nucleation, 81-92, 97-98, 100, 109-113, 115-118, 147-150, 153-154, 158-159, 165-168, 257, 272-273, 291-292, 295-302, 413
  nucleation and growth, 81-98, 106-107, 109-116, 118-119, 153-154, 165-168
  order–disorder, 3-5, 24, 130
  order of, 69-71
  phase separation, 521-523, 538
  polymers (*see also* Transitions in polymers), 555-590
  polymorphic, 116-117
  precipitation, 147-174
  recrystallization, 117-119
  second order, 74-75
  shear, 76, 80-81
  solidification (*see* Crystal growth)
  spinodal decomposition, 74, 121-124, 153-165, 168-173
  strain energy effects, 85-89
  thermodynamics of, 68-75
  transformational force, 69
  viscous liquids, 521-523
Phase transitions (*see* Critical phenomena, Phase transformations)
Planar crystal growth, 237-239
Polarization, spontaneous, 13
Polyethylene, 261, 264, 566-567, 587-589
Polymers, 555-590
  amorphous, 555-556

Polymers *(cont'd)*
  dielectric relaxation, 559
  glass transition, 572-584
  glasses, 572-584
  mechanical loss, 559, 587-590
  melting of, 561-572
  molecular motion in crystalline phase, 587-589
  molecular motion in intermediate phase, 589-590
  relaxation phenomena, 558-561, 572-590
  semicrystalline, 555-556
  side-chain motion, 584-585
  torsional oscillation of chains, 585-587
  transitions in (*see also* Transitions in polymers), 555-590
Polymorphic transformations, 116-117
Polymorphs, high pressure, 484-492
Polystyrene, 572-573, 580-581
Prandtl number, 268
Precipitation, 81-101, 105-108, 125, 147-174
  cellular, 105-108
  continuous, 83-101
  discontinuous, 105-108
  growth, 92-96
  kinetics, 81-101, 123-125, 153-168
  morphology, 168-173
  nucleation, 81-92, 147-152
  spinodal decomposition, 121-125, 153-154

Quartz crystal growth, 431-437
Quenchable phases, 474-476

Rayleigh number, 367
Recrystallization, 117-119
Relaxation phenomena in polymers, 558-561, 572-590

Scaling theory, 54-59
Screw dislocations, crystal growth, 257-258
Second-order transitions, 12, 15, 70, 74-75
Segregation coefficient (*see* Distribution coefficient)
Semiconductors, high-pressure synthesis of, 495-496
Series expansion calculations of critical phenomena, 45-48
Shear transformations, 76, 80-81
Si, 285-286, 303-310, 352, 371, 379, 457
  Czochralski growth, 371, 379
  epitaxial growth, 303-310
  vapor growth, 285-286, 303-310
  whiskers, 352, 457

Side-chain motion in polymers, 584-585
Slater model, 14-15
Sn–Al–Ti system, 202-205
Solidification (*see* Melt growth of crystals, Crystal growth)
Solid–liquid–solid crystal growth, 353-354
Solid solutions, 72-74, 81-92, 97-98, 100, 121-124, 139-174
  Bragg–Williams model, 133-134
  clustering in, 129-174
  compositional fluctuations, 152-153
  concentration profiles, 143-146
  configurational free energy, 133-135
  elastic energy of, 135-138, 141-143
  first-order transitions in, 72-74
  nonuniform, 134-135
  nucleation in, 81-92, 97-98, 100, 147-150, 165-168
  pair energies in, 131-133
  phase diagrams, 179-229
  spinodal decomposition of, 74, 121-124, 153-165, 168-173
  stationary states, 139-152
  strain energy of, 135-138, 141-143
Solids under pressure, 472-497
Solution growth of crystals, 407-458
  aqueous solutions, 417-425
  boundary effects, 413-417
  classification of, 407-409
  flux growth, 410, 437-450
  hydrothermal growth, 243-244, 410, 425-438, 464-468, 501-503
  incorporation of impurities, 416-417
  kinetics, 413-417
  liquid-phase epitaxy (LPE), 452-454
  materials applications, 420-421, 438, 444
  molten salt, 410, 437-450
  principles, 409-417
  temperature gradient zone melting, 454-456
  tipping, 453
  vapor–liquid–solid (VLS), 456-458
Specific heat, 19, 22-23, 70-71
Spherical model for critical phenomena, 43-45
Spherulites, 261, 264
Spinodal, 139, 148-149
Spinodal decomposition, 74, 121-124, 153-165, 168-173
  experimental results, 124-125
  kinetics, 123-127, 153-155
  morphology, 168-173
  thermodynamics, 121-123
Stockbarger method, 347

Index

Strain energy, 78, 80, 85-89, 109, 123, 135-138, 141-143
Striations, 335-340, 372-381
Superconductors, 16
Supercooling, constitutional, 239-244, 333-335, 416-417
Superfluid helium, 15-16, 23-24, 56
Surface nucleation, 257, 272-273, 291-292, 295-302, 413
Surface reactions, 295-302
Surface roughness, crystal growth, 258-281
Synthesis at high pressures, 477-497
   high-pressure polymorphs, 484-492
   intermetallic compounds, 496-497
   metallic hydrogen, 497
   normally unstable compounds, 492-494
   semiconductors, 495-496
   transition metal compounds, 478-484

Ta–Hf system, 186-189
Tan δ, 559
Ti–Al–Ga system, 202-205
Temperature-gradient zone melting, 454-456
Thermal expansion coefficient, 70
Thermochemical data, 179-229
Threshold switching in glasses, 529-533
Ti–Al–Sn system, 202-205
Ti–Fe–C system, 218-228
Tipping, 453
Torsional oscillation of polymer chains, 585-587
Transformational force, 69
Transformational stress, 113-114
Transitions (*see also* Phase transformations), 1-59, 69-75
   first-order, 6, 9, 12, 15, 17, 30-31, 69-75
   in polymers (*see* Transitions in polymers)
   in viscous liquids and glasses (*see* Transitions in viscous liquids and glasses)
   second-order, 12, 15, 70, 74-75
Transitions in polymers, 555-590
   classification of, 556-558
   glass transition (*see also* Glass transition in polymers), 572-584
   melting, 561-572
   molecular motion in crystalline phase, 587-589
   molecular motion in intermediate phase, 589-590
   relaxation phenomena, 558-561, 572-590
   side-chain motion, 584-585
   torsion in backbone chain, 585-587
Transitions in viscous liquids and glasses, 513-546
   crystallization of viscous systems, 523-529

Transitions in viscous liquids *(cont'd)*
   liquid–glass, 513-521
   phase separation of viscous melts, 521-523
   polymers, 572-584
Trichloroacetic acid, 261-262
Tricritical point, 11, 22, 30-33
Triple point, 9

Unquenchable phases, 474-476

Vapor deposition (*see also* Vapor growth of crystals)
   chemical, 285-321
   principles, 284-285
Vapor growth of crystals, 283-321
   diffusion-controlled, 292-295
   epitaxial, 302-321
   GaAs, 310-321
   kinetics, 288-302
   mass-input-limited, 292-295
   mass transport models, 307-308
   mechanisms, 284-286, 291-292, 307, 310, 313-321
   principles, 284-285
   Si, 285-286, 303-310
   surface-limited, 295-302
   surface reaction models, 308-310
   thermodynamics, 286-291
Vapor–liquid–solid (VLS) crystal growth, 352, 456-458
Verneuil method, 351
Viscous liquids, 513-546
   crystallization, 523-529
   heat capacity, 516
   liquid–glass transition, 513-521
   phase separation, 521-523
   self-diffusion, 518-520
   transport behavior, 516-521
   viscosity, 515-516

Whiskers, 352, 457
Wilson–Frenkel theory, 255
Wilson method for critical phenomena, 52-53

Yttrium iron garnet (YIG), crystal growth, 429-430, 440-444, 448

Zone melting, 332-334, 348-351
   floating zone, 348-351
   horizontal, 348
   temperature gradient, 454-456
   vertical, 348-351

MIX
Papier aus verantwortungsvollen Quellen
Paper from responsible sources
FSC® C105338

If you have any concerns about our products,
you can contact us on
ProductSafety@springernature.com

In case Publisher is established outside the EU,
the EU authorized representative is:
**Springer Nature Customer Service Center GmbH
Europaplatz 3, 69115 Heidelberg, Germany**

Printed by Libri Plureos GmbH
in Hamburg, Germany